Contemporary Mathematicians

Gian-Carlo Rota
Editor

Springer Science+Business Media, LLC

An Homage
to
Leonhard Euler

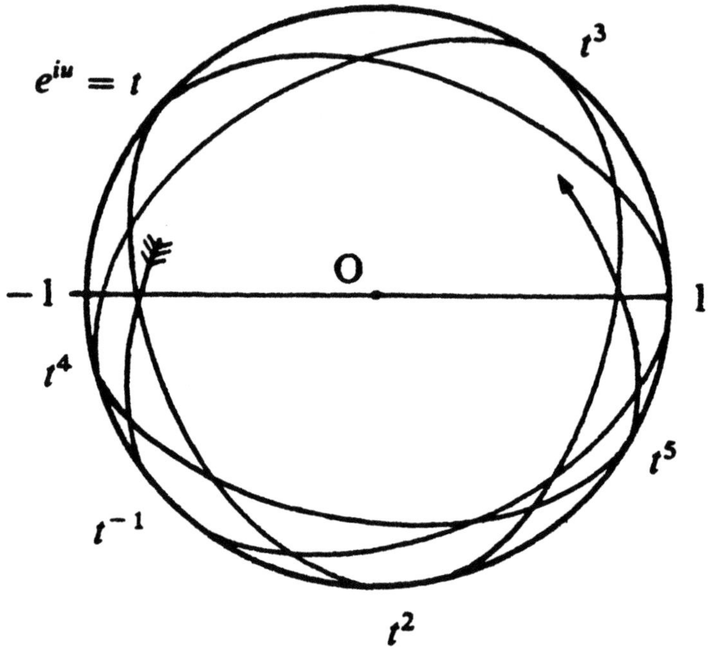

A bounded exponential Euler spline
(page 29).

Frontispiece from [114] *Cardinal Spline Interpolation* CBMS Regional Conference Monograph No. 12, SIAM, 1973. Reprinted by permission of SIAM.

I.J. Schoenberg
Selected Papers
Volume 2

Edited by
Carl de Boor

Springer Science+Business Media, LLC 1988

I.J. Schoenberg
Carl de Boor (*Editor*)
University of Wisconsin-Madison
Madison, WI 53705
U.S.A.

Library of Congress Cataloging-in-Publication Data
Schoenberg, I.J.
[Selections. 1988]
Selected papers / I.J. Schoenberg ; edited by Carl de Boor.
p. cm.—(Contemporary mathematicians)
Includes bibliographies.
ISBN 978-1-4899-0435-5

1. Mathematics—Collected works. I. De Boor, Carl. II. Title.
III. Title: I.J. Schoenberg, selected papers. IV. Series.
QA3.S365A25 1988
510—dc19 88-2860

CIP-Titelaufnahme der Deutschen Bibliothek
Schoenberg, Isaac J.:
Selected papers / I. J. Schoenberg. Ed. by Carl de Boor. —
Boston ; Basel : Birkhäuser.
(Contemporary mathematicians)
ISBN 978-1-4899-0435-5

NE: Schoenberg, Isaac J.: [Sammlung]
WG: 27 DBN 55.138120.5
6954/01 rr

WG: 27 DBN 88.021072.9 88.01.20
6954/02* rr

Originally published by Birkhäuser Boston Inc. in 1988
Softcover reprint of the hardcover 1st edition 1988

ISBN 978-1-4899-0435-5 ISBN 978-1-4899-0433-1 (eBook)
DOI 10.1007/978-1-4899-0433-1

Table of Contents

Volume 2

Volume 1

Publications of I. J. Schoenberg

1* Über die asymptotische Verteilung reeller Zahlen mod 1, *Math. Z.* **28** (1928), 171–199.

2. Sur un théorème de Steiner relatif à la quadrature des courbes roulettes, *Ann. Sci. Univ. Jassy* **16** (1930), 6–13.

3. Über total monotone Folgen mit stetiger Belegungsfunktion, *Math. Z.* **30** (1929), 761–767.

4* Über variationsvermindernde lineare Transformationen, *Math. Z.* **32** (1930), 321–328.

5. (With G. A. Bliss), On separation, comparison and oscillation theorems for self-adjoint systems of linear second order differential equations, *Amer. J. Math.* **LIII** (1931), 781–800.

6. The minimizing properties of geodesic arcs with conjugate end points, *Ann. of Math. (2)* **32** (1931), 763–776.

7* On finite and infinite completely monotonic sequences, *Bull. Amer. Math. Soc.* (Feb. 1932), 72–76.

8. On finite-rowed systems of linear inequalities in infinitely many variables. I, *Trans. Amer. Math. Soc.* **34** (1932), 594–619.

9. On finite-rowed systems of linear inequalities in infinitely many variables. II, *Trans. Amer. Math. Soc.* **35** (1932), 452–478.

10. (With G. A. Bliss), On the derivation of necessary conditions for the problem of Bolza, *Bull. Amer. Math. Soc.* (Dec. 1932), 858–864.

11. Some applications of the calculus of variations to Riemannian geometry, *Ann. of Math.* **33** (1932), 485–495.

12. Convex domains and linear combinations of continuous functions, *Bull. Amer. Math. Soc.* (Apr. 1933), 273–280.

13. (With T. H. Hildebrandt), On linear functional operations and the moment problem for a finite interval in one or several dimensions, *Ann. of Math.* **34** (1933), 317–328.

14. A remark on the preceding note by Bochner, *Bull. Amer. Math. Soc.* (Apr. 1934), 277–278.

15* Zur Abzählung der reellen Wurzeln algebraischer Gleichungen, *Math. Z.* **38** (1934), 546–564.

16* Remarks to Maurice Fréchet's article 'Sur la définition axiomatique d'une classe d'espace distanciés vectoriellement applicable sur l'espace de Hilbert,' *Ann. of Math.* **36** (1935), 724–732.

17. On the zeros of the successive derivatives of integral functions, *Proc. Nat. Acad. Sci.* **21** (1935), 674–676.

18* On asymptotic distributions of arithmetical functions, *Trans. Amer. Math. Soc.* **39** (1936), 315–330.

19. Extensions of theorems of Descartes and Laguerre to the complex domain, *Duke Math. J.* **2** (1936), 84–94.

20. On certain two-point expansions of integral functions of exponential type, *Bull. Amer. Math. Soc.* (1936), 284–288.

21. On the zeros of successive derivatives of integral functions, *Trans. Amer. Math. Soc.* **40** (1936), 12–23.

22* Regular simplices and quadratic forms, *J. London Math. Soc.* **12** (1937), 48–55.

23. On certain metric spaces arising from Euclidean spaces by a change of metric and their imbedding in Hilbert space, *Ann. of Math.* **38** (1937), 787–793.

24* On the Peano curve of Lebesgue, *Bull. Amer. Math. Soc.* **44** (1938), 519.

25* Metric spaces and positive definite functions, *Trans. Amer. Math. Soc.* **44** (1938), 522–536.

26* Metric spaces and completely monotone functions, *Ann. of Math.* **39** (1938), 811–841.

27. On metric arcs of vanishing Menger curvature, *Ann. of Math.* **41** (1940), 715–726.

28* (With J. von Neumann), Fourier integrals and metric geometry, *Trans. Amer. Math. Soc.* **50** (1941), 226–251.

29* Positive definite functions on spheres, *Duke Math. J.* **9** (1942), 96–108.

30. On local convexity in Hilbert space, *Bull. Amer. Math. Soc.* **48** (1942), 432–436.

31* Contributions to the problem of approximation of equidistant data by analytic functions. Part A: On the problem of smoothing or graduation. A first class of analytic approximation formulae, *Quart. Appl. Math.* **IV** (1946), 45–99.

32* Contributions to the problem of approximation of equidistant data by analytic functions. Part B: On the problem of osculatory interpolation. A second class of analytic approximation formulae, *Quart. Appl. Math.* **IV** (1946), 112–141.

33. (With H. Rademacher), An iteration method for calculation with Laurent series, *Quart. Appl. Math.* **IV** (1946), 142–159.

34. On totally positive functions, Laplace integrals and entire functions of the Laguerre-Pólya-Schur type, *Proc. Nat. Acad. Sci.* **33** (1947), 11–17.

35. Some analytic aspects of the problem of smoothing, *Courant Anniv. Vol.* (1948), 351–370.

36. On variation diminishing integral operators of the convolution type, *Proc. Nat. Acad. Sci.* **34** (1948), 164–169.

37. (With A. Whitney), Sur la positivité des déterminants de translations des fonctions de fréquence de Pólya..., *C. R. Acad. Sci. Paris* **228** (1949), 1996–1998.

38. (With H. Rademacher), Convex domains and Chebychev's approximation problem, *Canad. J. Math.* **2** (1950), 245–256.

39* On Pólya frequency functions. II. Variation-diminishing integral operators of the convolution type, *Acta Sci. Math. (Szeged)* **12** (1950), 97–106.

40. The finite Fourier series and elementary geometry, *Amer. Math. Monthly* **57** (1950), 390–404.

41. (With M. Aissen, A. Edrei, and A. M. Whitney), On the generating functions of totally positive sequences, *Proc. Nat. Acad. Sci.* **37** (1951), 303–307.

42. (With A. M. Whitney), A theorem on polygons in n dimensions with applications to variation-diminishing and cyclic variation-diminishing linear transformations, *Compositio Math.* **9** (1951), 141–160.

43* On Pólya frequency functions. I. The totally positive functions and their Laplace transforms, *J. Analyse Math.* **1** (1951), 331–374.

44* (With T. S. Motzkin), On lineal entire functions of n complex variables, *Proc. Amer. Math. Soc.* **3** (1952), 517–526.

45* (With M. Aissen and A. M. Whitney), On the generating functions of totally positive sequences. I, *J. Analyse Math.* **2** (1952), 93–103.

46. A remark on M. M. Day's characterization of inner-product spaces and a conjecture of L. M. Blumenthal, *Proc. Amer. Math. Soc.* **3** (1952), 961–964.

47* (With Anne Whitney), On Pólya frequency functions. III. The positivity of translation determinants with an application to the interpolation problem by spline curves, *Trans. Amer. Math. Soc.* **74** (1953), 246–259.

48* On smoothing operations and their generating functions, *Bull. Amer. Math. Soc.* **59** (1953), 199–230.

49* On a theorem of Kirzbraun and Valentine, *Amer. Math. Monthly* **60** (1953), 620–622.

50* (With T. S. Motzkin), The relaxation method for linear inequalities, *Canad. J. Math.* **6** (1954), 393–404.

51* An isoperimetric inequality for closed curves convex in even-dimensional Euclidean spaces, *Acta Math.* **91** (1954), 143–164.

52. A note on multiply positive sequences and the Descartes rule of signs, *Rend. Circ. Mat. Palermo (2)* **4** (1955), 123–131.

53* On the zeros of the generating functions of multiply positive sequences and functions, *Ann. of Math.* **62** (1955), 447–471.

54. Some extremal problems for positive definite sequences and related extremal convex conformal maps of the circle, *Kon. Ned. Akad. Amsterdam* **61** (1958), 28–37.

55* (With G. Pólya), Remarks on de la Vallée Poussin means and convex conformal maps of the circle, *Pacific J. Math.* **8** (1958), 295–334.

56. Spline functions, convex curves and mechanical quadratures, *Bull. Amer. Math. Soc.* **64** (1958), 352–357.

57. On variation diminishing approximation methods, in *On Numerical Approximation*, pp. 249–274, Univ. of Wisconsin Press, Madison, Wisconsin, 1959.

58. The integrability of certain functions and related summability methods, *Amer. Math. Monthly* **66** (1959), 361–375 and 562–563.

59. On the maxima of certain Hankel determinants and the zeros of the classical orthogonal polynomials, *Kon. Ned. Akad. Amsterdam* **21** (1959), 282–290.

60* (With B. Epstein), On a conjecture concerning schlicht functions, *Bull. Amer. Math. Soc.* **65** (1959), 273–275.

61. (With A. S. Besicovitch), Sur les arcs ascendants à pente partout nulle et des problémes qui s'y rattachent, *C. R. Acad. Sci. Paris* **249** (1959), 1079–1080.

62. On the question of unicity in the theory of best approximation, *Ann. New York Acad. Sci.* **86** (1960), 682–692.

63. (With J. C. Mairhuber and R. E. Williamson), On variation diminishing transformations on the circle, *Rend. Circ. Mat. Palermo (2)* **8** (1959), 241–270.

64. (With G. Szegö), An extremum problem for polynomials, *Compositio Math.* **14** (1960), 260–268.

65. (With C. T. Yang), On the unicity of solutions of problems of best approximation, *Ann. Mat. Pura Appl. (4)* **54** (1961), 1–12.

66. (With A. S. Besicovitch), On Jordan arcs and Lipschitz classes of functions defined on them, *Acta Math.* **106** (1961), 113–136.

67. On two theorems of P. Erdös and A. Renyi, *Illinois J. Math.* **6** (1962), 53–58.

68. On two theorems of Archimedes and F. J. van den Berg, *Simon Stevin* **5** (1962), 133–138.

69. Extrema for gap power series of positve real part, *Proc. Nat. Acad. Sci.* **48** (1962), 1151–1154.

70* On the Besicovitch-Perron solution of the Kakeya problem, in *Studies in Mathematical Analysis and Related Topics*, Pólya volume, pp. 359–363, Stanford Univ. Press,

Stanford, CA, 1962.

71. On certain minima related to the Besicovitch-Kakeya problem, *Mathematica (Cluj)* **4** (1962), 145–148.

72* Spline interpolation and best quadrature formulae, *Bull. Amer. Math. Soc.* **70** (1964), 143–148.

73. Spline interpolation and the higher derivatives, *Proc. Nat. Acad. Sci.* **51** (1964), 24–28.

74. (With Ch. Pisot), Arithmetic problems concerning Cauchy's functional equation, *Illinois J. Math.* **8** (1964), 40–56.

75* On best approximations of linear operators, *Indag. Math.* **26** (1964), 155–163.

76* Spline functions and the problem of graduation, *Proc. Nat. Acad. Sci.* **52** (1964), 947–950.

77. On trigonometric spline interpolation, *J. Math. Mech.* **13** (1964), 795–826.

78. A note on the cyclotomic polynomial, *Mathématika* **11** (1964), 131–136.

79. Arithmetic problems concerning Cauchy's functional equation, *Compositio Math.* **16** (1964), 169–175.

80* On interpolation by spline functions and its minimal properties, *Internat. Schriftenreihe Numer. Math. (Birkhäuser)* **5** (1964), 109–129.

81. (With Ch. Pisot), Arithmetic problems concerning Cauchy's functional equation II, *Illinois J. Math.* **9** (1965), 129–136.

82. Extrema for gap power series of positive real part, *J. Analyse Math.* **14** (1965), 379–391.

83. On monosplines of least deviation and best quadrature formulas, *SIAM J. Numer. Anal., Ser. B.* **2** (1965), 144–170.

84. (With F. Cunningham), On the Kakeya constant, *Canad. J. Math.* **17** (1965), 946–956.

85. Extremum problems for mass-distributions in a finite interval, *Rend. Circ. Mat. Palermo (2)* **13** (1964), 1–12.

86* (With H. B. Curry), On Pólya frequency functions. IV. The fundamental spline functions and their limits, *J. Analyse Math.* **17** (1966), 71–107.

87. On monosplines of least square deviation and best quadrature formulae. II, *SIAM J. Numer. Anal.* **3** (1966), 321–328.

88. (With S. J. Einhorn), On Euclidean sets having only two distances between points, *Indag. Math.* **28** (1966), 479–488 and 489–504.

89* On Hermite-Birkhoff interpolation, *J. Math. Anal. Appl.* **16** (1966), 538–543.

90* (With M. Marsden), On variation diminishing spline approximation methods, *Mathematica (Cluj)* **8** (1966), 61–82.

91. (With S. K. Zaremba), On Cauchy's lemma concerning convex polygons, *Canad. J. Math.* **19** (1967), 1062–1071.

92. On spline functions (with a supplement by T. N. E. Greville), in *Inequalities*, O. Shisha, ed., pp. 255–291, Academic Press, NY, 1967.

93. On the Ahlberg-Nilson extension of spline interpolation: The g-splines and their optimal properties, *J. Math. Anal. Appl.* **21** (1968), 207–231.

94. On spline interpolation at all integer points of the real axis, *Mathematica (Cluj)* **10** (1968), 151–170. Also in *Seminaire Delange-Pisot-Poitou: 1967/68, Theorie des Nombres*, Fasc.1, Exp.1, 18pp. Secrétariat mathematique, Paris, 1969.

95. Monosplines and quadrature formulae, in *Theory and Applications of Spline Functions*, T. N. E. Greville, ed., 157–207, Academic Press, NY, 1969.

96. Spline interpolation and the higher derivatives, in *Abhandlungen zur Zahlentheorie and Analysis, zur Erinnerung an Edmund Landau (1877–1938)*, P. Turan, ed., 281–295, VEB Deutscher Verlag der Wiss., Berlin, 1968.
97. Linkages and distance geometry. I and II, *Indag. Math.* **31** (1969), 43–63.
98. Cardinal interpolation and spline functions, *J. Approximation Theory* **2** (1969), 167–206.
99. (With Zvi Ziegler), On cardinal monosplines of least L_∞-norm on the real axis, *J. Analyse Math.* **23** (1970), 409–436.
100* (With A. Cavaretta), Solution of Landau's problem concerning higher derivatives on the halfline, MRC Technical Summary Report #1050; also in *Proc. Internat. Conference on Constructive Function Theory, Varna, May 1972*, Sofia, 297–308.
101. A second look at approximate quadrature formulae and spline interpolation, *Adv. in Math.* **4** (1970), 277–300.
102. On equidistant cubic spline interpolation, *Bull. Amer. Math. Soc.* **77** (1971), 1039–1044.
103. (With A. Sharma), The interpolatory background of the Euler-MacLaurin quadrature formula, *Bull. Amer. Math. Soc.* **77** (1971), 1034–1038.
104. On polynomial spline functions on the circle. I and II, Proc. of the Conference on Constructive Function Theory, August 1969, Akademiai Kiado, Budapest, 1971, pp. 403–433.
105* The perfect B-splines and a time-optimal control problem, *Israel J. Math.* **10** (1971), 261–274.
106* Norm inequalities for a certain class of C^∞ functions, *Israel J. Math.* **10** (1971), 364–372.
107. Notes on spline functions I. The limits of the interpolating periodic spline functions as their degree tends to infinity, *Indag. Math.* **34** (1972), 412–422.
108. Cardinal interpolation and spline functions II. Interpolation of data of power growth, *J. Approximation Theory* **6** (1972), 404–420.
109. (With P. R. Lipow), Cardinal interpolation and spline functions III. Cardinal Hermite interpolation, *Linear Algebra Appl.* **6** (1973), 273–304.
110* Cardinal interpolation and spline functions IV. The exponential Euler splines, ISNM (Birkhäuser Verlag) **20** (1972), 382–404.
111. (With A. Sharma), Cardinal interpolation and spline functions V. The B-splines for cardinal Hermite interpolation, *Linear Algebra Appl.* **7** (1973), 1–42.
112. Cardinal interpolation and spline functions VI. Semicardinal interpolation and quadrature formulas, *J. Analyse Math.* **XXVII** (1974), 159–204.
113* Cardinal interpolation and spline functions VII. The behavior of cardinal spline interpolants as their degree tends to infinity, *J. Analyse Math.* **XXVII** (1974), 205–229.
114. *Cardinal Spline Interpolation*, CBMS Regional Conference Monograph No. 12, vi + 125 pages, SIAM, Philadelphia, 1973.
115. Notes on spline functions III. On the convergence of the interpolating cardinal splines as their degree tends to infinity, *Israel J. Math.* **16** (1973), 87–93.
116. (With F. B. Richards) Notes on spline functions IV. A cardinal spline analogue of the theorem of the brothers Markov, *Israel J. Math.* **16** (1973), 94–102.
117. (With S. D. Silliman) On semicardinal quadrature formulas, *Math. Comp.* **28** (1974), 483–497. See also *Approximation Theory*, G. G. Lorentz ed., Academic Press, 1973, 461–467.

118. Splines and histograms, with an Appendix by C. de Boor. In *Spline Functions and Approximation Theory*, A. Meir and A. Sharma, eds., ISNM Vol. 21, Birkhäuser Verlag, (1973), 277– 327, 329–358.
119. Spline functions and differential equations – First order equations, in *Studies in Numerical Analysis*, B.K.P. Scaife, ed., Academic Press, London (1974), 311–324.
120. Remarks concerning the numerical inversion of the Laplace transform due to Bellman, Kalaba and Lockett, *J. Math. Anal. Appl.* **43** (1973), 823–828.
121. (With C. de Boor) Unique prime factorization and lattice points, *Math. Mag.* **46** (1973), 198–203.
122. The elementary cases of Landau's problem of inequalities between derivatives, *Amer. Math. Monthly* **80** (1973), 121– 158.
123. Notes on spline functions V. Orthogonal or Legendre splines, *J. Approximation Theory* **13** (1975), 84–104.
124. Notes on spline functions VI. Extremum problems of the Landau-type for the differential operator $D^2 \pm 1$, in *Studies in Spline Functions and Approximation Theory*, S. Karlin, C.A. Micchelli, A. Pinkus and I.J. Schoenberg, eds., Academic Press (1976), 353–369.
125. Remarks on two geometric conjectures of L. Fejes Tóth, Analele ştiinţifice ale Universităţii "Al.I. Cuza" din Iaşi **XXI** (1975), 9–13.
126. (With D. J. Newman) Splines and the logarithmic function, *Pacific J. Math.* **61** (1975), 241–258.
127. On the motion of a billiard ball in two dimensions, *Delta* **5** (1975), 1–17.
128. On Micchelli's theory of cardinal L-splines, in *Studies in Spline Functions and Approximation Theory*, S. Karlin et al., eds., Academic Press (1976), 251–276.
129* On the remainders and the convergence of cardinal spline interpolation for almost periodic functions, in *Studies in Spline Functions and Approximation Theory*, S. Karlin et al., eds., Academic Press (1976), 277–303.
130. On the location of the frets on the guitar, *Amer. Math. Monthly* **83** (1976), 550–552.
131. Extremum problems for the motions of a billiard ball I. The L_p norm, $1 \leq p < \infty$, *Indag. Math.* **38** (1976), 66–75.
132. Extremum problems for the motions of a billiard ball II. The L_∞ norm, *Indag. Math.* **38** (1976), 263–279.
133. On Chebyshev and Markov-type problems for polynomials in a circular ring, in *Colloquia Mathematica Societatis Janos Bolyai 19, Fourier Analysis and Approximation Theory*, Budapest (1976), 697–712.
134. (With C. de Boor) Cardinal interpolation and spline functions VIII. The Budan-Fourier theorem for splines and applications, in *Spline Functions, Karlsruhe 1975*, K. Böhmer, G. Meinardus and W. Schempp, eds., *Lecture Notes in Mathematics* **501**, Springer-Verlag (1976), 1–79.
135. The Landau problem for motions in a ring and in bounded continua, *Amer. Math. Monthly* **84** (1977), 1–12. A somewhat expanded version appeared as MRC Technical Summary Report #1563, Oct. 1975.
136. The Landau problem for the differential operator $D^2 - a^2$ in a circular ring, in *Colloquia Mathematica Societatis Janos Bolyai, 19, Fourier Analysis and Approximation Theory*, Budapest (1976), 713–723.
137. On the zeros of successive derivatives of integral functions II, in *Complex Analysis, Kentucky 1976*, J.D. Buchholz and T.J. Suffridge eds., *Lecture Notes in Mathematics*

599, Springer-Verlag (1977), 109–116.
138. Approximating lengths, areas and volumes by polygons and polyhedra, *Delta* **7** (1977), 32–46.
139. (With T.N.E. Greville and A. Sharma) The spline interpolation of sequences satisfying a linear recurrence relation, *J. Approximation Theory* **17** (1976), 200–221. See also *Approximation Theory*, G. G. Lorentz ed., Academic Press, 1973, 365–367.
140. (With J. M. Pollin) On the matrix approach to Fibonacci numbers and Fibonacci pseudoprimes, *Fibonacci Quart.* **18** (1980), 261–268.
141* The Landau problem I. The case of motions on sets, *Indag. Math.* **40** (1978), 276–286.
142. On the arithmetic-geometric mean, *Delta* **7** (1977), 49–65.
143. On cardinal spline smoothing, in Proc. of the International Symposium on Approximation Theory, Campinas, Brazil, 1977, J. B. Prolla, ed., North-Holland Publ. Co. (1979), 383–407.
144. On a problem of Steinhaus on lattice points, *Amer. Math. Monthly* **86** (1979), 765–766.
145. On Jacobi-Bertrand's proof of a theorem of Poncelet, in *Studies in Pure Mathematics to the Memory of Paul Turán*, publ. by the Hungarian Academy of Sciences (1980), 623–627.
146. Extremum problems for the multi-dimensional case of König and Szücs of billiard ball motions, *Math. Rep. Acad. Sci. Canada* **1** (1978), 37–40.
147. Extremum problems for the motions of a billiard ball III. The multi-dimensional case of König and Szücs, *Studia Sci. Math. Hungar.* **13** (1978), 53–78.
148. Extremum problems for the motions of a billiard ball IV. A higher-dimensional analogue of Kepler's Stella octangula, *Studia Sci. Math. Hungar.* **14** (1979), 273–292.
149* The Landau problem II. The case of motions on curves, *Indag. Math.* **43** (1981), 325–335.
150* The Landau problem III. Motions on special curves and time-optimal control problems, *Indag. Math.* **43** (1981), 337–351.
151. The harmonic analysis of skew polygons as a source of outdoor sculptures, in *The Geometric Vein*, The Coxeter Festschrift, C. Davis, B. Grünbaum, F. A. Sherk eds., Springer-Verlag, New York (1982), 165–176.
152. The Landau problem for motions on curves and time-optimal control problems, in *Approximation Theory*, E.W. Cheney, ed., Academic Press (1980), 811–821.
153. (With T. S. Motzkin) On Fejér sets in linear and spherical spaces, NBS Report of August 25, 1952, 19 pages. *Colloq. Math. Soc. János Bolyai* **35** (1980), 861–875.
154. (With R. Askey and A. Sharma) Hausdorff's moment problem and expansions in Legendre polynomials, *J. Math. Anal. Appl.* **86** (1982), 237–245.
155. A direct derivation of a Jacobian identity from elliptic functions, *Amer. Math. Monthly* **88** (1981), 616–618.
156. Two applications of approximate differentiation formulae: An extremum problem for multiply monotone functions and the differentiation of asymptotic expansions, *J. Math. Anal. Appl.* **89** (1982), 251–261.
157* On polynomial interpolation at the points of a geometric progression, *Proc. Roy. Soc. Edinburgh Sect. A* **90A** (1981), 195–207.
158. Interpolating splines as limits of polynomials, *Linear Algebra Appl.* **52/53** (1983), 617–628.
159. A new approach to Euler splines, *J. Approximation Theory* **39** (1983), 324–337.

160. (With T.N.T. Goodman and A. Sharma), Piecewise smooth solutions of some difference-differential equations, *J. Approximation Theory* **48** (1986), 262–271.

161. (With T.N.E. Greville and A. Sharma) The behavior of the exponential Euler spline $S_n(x;t)$ as $n \to \infty$ for negative values of the base t, in *Approximation Theory*, CMS Conference Proceedings, Vol. 3, Amer. Math. Soc., Providence RI (1982), 185–198.

162. Euler's contributions to cardinal spline interpolation, in *Leonardt Euler 1707-1783, Beiträge zu Leben und Werk*, Birkhäuser Verlag, Basel (1984), 199–213.

163. A direct approach to the Villarceau circles of a torus, *Simon Stevin* **59** (1985), 365–372.

164. Self-reflecting skew polygons and polytopes in the 4-dimensional hypercube, *Geom. Dedicata* **14** (1983), 355–373.

165. (With A. W. Goodman) On a theorem of Szegö on univalent convex maps of the unit circle, *J. Analyse Math.* **44** (1984/5), 200–204.

166. (With D. W. Crowe) On the equidecomposability of a regular triangle and a square of equal area, *Giessener Mitteilungen* **164** (1984), 59–64.

167. On hypocycloids and their diameters, *J. College Math.* **16** (1985), 262–267.

168. A conjectured analogue of Rolle's theorem for polynomials with real and complex coefficients, *Amer. Math. Monthly* **93** (1986), 8–13.

169. On the anti-cylinder, MRC Technical Summary Report #2842, (1985). There is related material in: On the turning lozange of constant side whose vertices alternate between two fixed circles of a ring, MRC Technical Summary Report #2813, (1985).

170. (With A. W. Goodman) A proof of Grace's theorem by induction, Honam Math. J. **xx** (1987), xxx-xxx.

171. Outdoor sculptures, MRC Technical Summary Report #2900 (1986); submitted.

172. The Chinese remainder problem and polynomial interpolation, *College Math. J.* **xx** (1987), xxx-xxx.

173. On vector indices mod m, a circular slide rule mod 100, *Mathematics Magazine* **xx** (1987), xxx-xxx.

Books

Mathematical Time Exposures, 270 pages, The Mathematical Association of America, 1982.

CARDINAL SPLINES

CONTRIBUTIONS TO THE PROBLEM OF APPROXIMATION OF EQUIDISTANT DATA BY ANALYTIC FUNCTIONS*

PART A.—ON THE PROBLEM OF SMOOTHING OR GRADUATION. A FIRST CLASS OF ANALYTIC APPROXIMATION FORMULAE

BY

I. J. SCHOENBERG

University of Pennsylvania and Ballistic Research Laboratories, Aberdeen Proving Ground

Introduction. Let there be given a sequence of ordinates

$$\{y_n\} \qquad (n = 0, \pm 1, \pm 2, \cdots),$$

corresponding to all integral values of the variable $x=n$. If these ordinates are the values of a known analytic function $F(x)$, then the problem of interpolation between these ordinates has an obvious and precise meaning: we are required to compute intermediate values $F(x)$ to the same accuracy to which the ordinates are known. Undoubtedly, the most convenient tool for the solution of this problem is the polynomial central interpolation method. It uses the polynomial of degree $k-1$, interpolating k successive ordinates, as an approximation to $F(x)$ only within a unit interval in x, centrally located with respect to its k defining ordinates. Assuming k fixed, successive approximating arcs for $F(x)$ are thus obtained which present discontinuities on passing from one arc to the next if k is odd, or discontinuities in their first derivatives if k is even (see section 2.121). Actually these discontinuities are irrelevant in our present case of an analytic function $F(x)$. Indeed, if the interpolated values obtained are sufficiently accurate, these discontinuities will be apparent only if we force the computation beyond the intrinsic accuracy of the y_n.

The situation is quite different if y_n are empirical data. In this case we are to determine an approximation $F(x)$ which, for $x=n$, may disagree with y_n by amounts depending on the accuracy of the data, provided we thereby improve the smoothness of the resulting approximation $F(x)$. In various applied fields such as Ballistics and Actuarial mathematics it is at times desirable to compute very smooth approximations $F(x)$ to an accuracy surpassing by far the accuracy to which the physical or statistical function involved may be determined. This physically unjustified accuracy becomes desirable whenever the approximation $F(x)$ enters into numerical processes of some complexity, such as the numerical solution of differential equations. Modern electronic computing machines, especially, require a good amount of forced mathematical accuracy in such auxiliary tables in order to avoid the excessive accumulation of rounding errors in the computation of the solution. These remarks justify the desirability of approximation methods to empirical data furnishing easily computed approximations $F(x)$ which are very smooth functions of x. Approximations meeting these requirements are of two kinds: 1. *Polynomial approximation*, where $F(x)$ is composed of a succession of polynomial arcs meeting with a certain number of continuous derivatives. 2. *Analytic approximations*, where $F(x)$ is an analytic and regular function of x for all real values of x.

* Received Oct. 18, 1945.

Important work concerning *polynomial approximations* is to be found in the actuarial literature under the subject of *osculatory interpolation*. Of the extensive literature we mention especially the fundamental work of W. A. Jenkins and the valuable systematization of the subject by T. N. E. Greville.[1] Especially important are those formulae derived by these authors which do not strictly interpolate the given ordinates, but rather combine the operation of smoothing the data and the operation of interpolation in one formula. Mr. Jenkins discusses interpolation formulae written in the convenient Everett (or Steffensen) form. Mr. Greville's starting point is his elegant expression of each polynomial arc in terms of the end point values of those derivatives which are to be continuous on passing from one arc to the next. Each of these two modes of attack has its peculiar advantages and one or the other seem indispensable for an algebraic treatment of the subject. The present writer has found the Lagrange form (explicitly in terms of the ordinates y_n) of such formulae preferable for two reasons: 1. The Lagrange form seems better adapted to computation with modern desk computing machines and undoubtedly superior for computation with punch-card machines. 2. The Lagrange form suggests a treatment of the subject by means of elementary concepts of Fourier analysis which, firstly, affords a more exhaustive treatment of the problem of *polynomial approximations*, secondly, shows how to extend these methods so as to furnish *analytic approximations*.

The explicit Lagrange form of the k-point central interpolation method, as well as of all the interpolation formulae of osculatory interpolation, is extremely simple in its formal appearance. Indeed, to every such formula corresponds an *even* function $L(x)$, defined for all real values of x, in terms of which the corresponding formula may be written as follows

$$F(x) = \sum_{n=-\infty}^{\infty} y_n L(x-n). \tag{1}$$

The simplicity of this formula springs from the fact that it depends on the single function $L(x)$ which describes the formula completely. Incidentally $F(x)=L(x)$ if

$$y_0 = 1, \qquad y_n = 0 \quad (n \neq 0). \tag{2}$$

Thus every interpolation method of this kind exhibits its corresponding $L(x)$ if we apply the method to the ordinates (2) (for an example see section 2.121).

The polynomial interpolation formulae arise from (1) if $L(x)$ is a composite polynomial function of arcs defined by various polynomials in successive unit intervals, such that $L(x)=0$ for sufficiently large values of $|x|$ (for an important example see chapter II, formula (11)). The number of continuous derivatives of $F(x)$ is, of course, equal to the number of continuous derivatives of $L(x)$ for all real x.

We obtain the formally simplest interpolation formula (1) if we choose

$$L(x) = \frac{\sin \pi x}{\pi x} \tag{3}$$

[1] W. A. Jenkins, *Osculatory interpolation: New derivation and formulae*, Record of the American Institute of Actuaries, **15**, 87 (1926).

Thomas N. E. Greville, *The general theory of osculatory interpolation*, Transactions of the Actuarial Society of America, **45**, 202–265 (1944).

W. A. Jenkins wrote four papers on this subject of which the above paper is the first. References to the other three papers are found in the excellent bibliography in Greville's paper.

in which case (1) becomes

$$F(x) = \sum_{n=-\infty}^{\infty} y_n \frac{\sin \pi(x-n)}{\pi(x-n)}. \tag{4}$$

This expression which interpolates the ordinates y_n, is known to mathematicians under the name of the *cardinal series*.[2] For this reason we wish to call the general formula (1) a formula of the *cardinal type*, referring to $L(x)$ as the *basic function* of the formula.

The aim of the paper, of which the present Part A is the first, is twofold. Firstly, we propose to carry through to a certain stage of completion the important actuarial work concerning polynomial approximations. Incidentally, our work will answer Mr. Greville's conjecture (loc. cit. pp. 212–213) concerning the existence of an "ordinary" interpolation formula furnishing an approximation $F(x)$ composed of polynomial arcs of degree $m+2$, having m continuous derivatives and such that if the data y_n are the values of a polynomial of degree $m-1$ then $F(x)$ reduces identically to that polynomial. In Part B it will be shown how to obtain such formulae for every value of m. (The case of $m=2$ reduces to Jenkins' formula mentioned in section 2. 122.) Secondly, we shall derive formulae of the cardinal type (1) with basic functions $L(x)$ which are analytic and regular for all real or complex values of x. The classical basic function (3) is of course analytic; however, its excessively slow rate of damping, for increasing x, makes the classical cardinal series (4) inadequate for numerical purposes. Our analytic $L(x)$, derived in chapter IV, dampen out exponentially. In Part B we will derive similar $L(x)$ which will dampen out even faster: like $\exp(-C^2x^2)$.

The paper is divided into five chapters. In chapter I we discuss the general problem of smoothing by means of a linear compound formula. This discussion, by no means exhaustive, is to serve as a guide to what is likely to be useful among formulae of the cardinal type (1) which smooth and interpolate at the same time. It serves to restrict somewhat the arbitrariness of the problem. The rather obvious idea of using cosine polynomials (or series) in this connection affords the possibility of a brief exposition of this subject in the more scientific manner of E. de Forest, W. F. Sheppard, E. T. Whittaker, and others, and may be followed up elsewhere.

Chapters II and III form the common foundation of both parts A and B. In chapter II we describe the interpolatory properties of the formula (1) in terms of extremely simple properties of the Fourier-transform

$$g(u) = \int_{-\infty}^{\infty} L(x) \cos uxdx \tag{5}$$

of the basic function $L(x)$ (Theorem 4). Thus we are assured that our formula (1) will be exact for (i.e., reproduce) polynomials of degree $k-1$, provided $g(u)-1$ has a zero of order k at $u=0$ and $g(u)$ has zeros of order k at all points $u=2\pi n$ $(n=\pm 1, \pm 2, \cdots)$. This elementary fact is reminiscent of N. Wiener's fundamental

[2] See J. M. Whittaker, *Interpolatory function theory*, Cambridge Tracts in Mathematics, 1935, pp. 62–64, for a discussion of the relation between the cardinal series and Stirling's interpolation series. The cardinal series was probably first investigated in an important mémoire by Ch. J. de la Vallée Poussin, *Sur la convergence des formules d'interpolation entre ordonnées équidistantes*, Bull. Acad. Roy. Belgique, 1908, 319–410.

description of the closure properties of the family of translation functions $\{L(x-\lambda)\}$ in terms of the zeros of $g(u)$. Chapter III contains a somewhat general discussion of polygonal lines, the individual arcs of which are polynomials of degree $k-1$, joined together with $k-2$ continuous derivatives. A general parametric representation of such curves is obtained (Theorem 5) which greatly facilitates their use for the purpose of approximation of data. For $k=4$ they represent approximately the curves drawn by means of a spline and for this reason we propose to call them *spline curves of order k*. These polynomial spline curves are finally smoothed out, by means of one-dimensional heat flow during the time interval t, into *analytic spline curves of order k*. An analytic spline curve of order k is represented by a series of the cardinal type

$$F(x) = \sum_{n=-\infty}^{\infty} f_n M_k(x - n, t), \tag{6}$$

where the basic function $M_k(x, t)$ is defined as

$$M_k(x, t) = \frac{1}{2\pi}\int_{-\infty}^{\infty} e^{-t(u/2)^2}\left(\frac{2 \sin u/2}{u}\right)^k \cos uxdx, \tag{7}$$

while the coefficients f_n may be thought of as arbitrary parameters.

The family of functions (6) forms the basis of our work. Its principal advantages for purposes of numerical approximation spring from two sources: 1) *The basic function* $M_k(x, t)$ *dampens out like* $\exp(-x^2t^{-1})$ (see III, formula (39)). As seen from our Table I, for $k=4$ and $t=0.5$, we have $M_4(x, 1/2)=0$ to something like 10 decimal places for $|x| \geqq 5$. This causes the great flexibility of the graph of $F(x)$ on varying the parameters f_n and the ease in computing $F(x)$. 2) *The family* (6) *contains* (or represents) *all polynomials of degree* $k-1$. The simplest analytic family of this type is obtained for $k=0$ and $t>0$ when (6) becomes

$$F(x) = \sum_{n=-\infty}^{\infty} f_n \frac{1}{\sqrt{\pi t}} e^{-(x-n)^2/t}. \tag{8}$$

This family obviously still enjoys the first property. However, (8) fails badly in its ability of representing even the simplest types of curves because of the low value of $k=0$. Indeed $F(x)\equiv 0$, for all $f_n=0$, is the only constant value (8) is capable of representing.

Chapter IV contains the chief results of the present Part A. We show how the family of curves (6) can be used to approximate given data. First we derive an analytic interpolation formula of the cardinal type (1) which leaves the given ordinates unchanged (Theorem 8). Secondly we extend the result to a family of formulae depending on a positive smoothing parameter ϵ such as to combine a certain variable amount of smoothing (depending on ϵ) with the operation of interpolation (Theorem 9).

In collaboration with Lt. J. H. Levin, the author has had the opportunity of applying on a large scale this analytic approximation method at the Ballistic Research Laboratory, Aberdeen Proving Ground, Maryland. The computations were performed on punch-card machines. The given equidistant data y_n were the values of the drag coefficient of a projectile as a function of its velocity. Since very accurately computed values of the derivatives $F'(x)$, $F''(x)$ of the approximation $F(x)$ were also desired, it seems doubtful if any of the existing osculatory interpolation formulae would have furnished satisfactory results in view of the complicated trend of the data to be approximated.

In the last chapter we discuss procedures for the accurate computation of the functions and constants tabulated at the end of the paper. The most noteworthy problem encountered in this connection is the following: Let

$$F(z) = \sum_{-\infty}^{\infty} a_n z^n \tag{9}$$

be a Laurent series which converges in a ring $\alpha < |z| < \beta$. We assume furthermore that $f(z)$ does not vanish in this ring:

$$F(z) \neq 0, \qquad (\alpha < |z| < \beta). \tag{10}$$

Under these circumstances we have an expansion of the reciprocal

$$\frac{1}{F(z)} = \sum_{-\infty}^{\infty} \omega_n z^n. \tag{11}$$

If the coefficients a_n of the expansion (9) are given numerically the problem consists in finding very accurate numerical values of the ω_n.[3] A very efficient iteration method solving this problem has been developed by H. A. Rademacher and the author. It solves the similar problem of finding the expansion of the nth root of $f(z)$ and generally of any algebraic function of Laurent series. This subject will be discussed elsewhere in a joint publication with Professor Rademacher.

In a sequel to these papers we expect to discuss the fitting of curves of the form (6) to data, in the sense of least squares. This will be accomplished by constructing series of the cardinal type (1) which also enjoy the orthogonality property

$$\int_{-\infty}^{\infty} L(x)L(x-n)dx = \begin{cases} 1 & \text{if} \quad n = 0 \\ 0 & \text{if} \quad n \neq 0. \end{cases}$$

This construction reduces to the problem of computing the Laurent expansion of the square root $\sqrt{f(z)}$ of an expansion (9).

The author wishes to express his appreciation for the encouraging interest shown in his work by Dr. A. N. Lowan of the Mathematical Tables Project. He has benefited much by the helpful advice of Dr. L. S. Dederick, Major A. A. Bennett, Lt. J. H. Levin and others. Especially valuable were the author's frequent discussions with Dr. C. B. Morrey. The tables were computed by Mrs. Mildred Young. The author takes this opportunity of expressing his thanks to the officials of the Ballistic Research Laboratory for their permission to publish these tables.

The reader who is mainly interested in the numerical applications, may pass directly from this point to the Appendix, where the use of the tables is fully explained and one example is worked out.

[3] It should be remarked here that also A. C. Aitken's computation in 1925 of the coefficients of E. T. Whittaker's smoothing method amounted to the expansion in a Laurent series of a certain simple rational function. See E. T. Whittaker and G. Robinson, *Calculus of observations*, London and Glasgow, 1940, pp. 308–312.

I. DEFINITIONS OF SMOOTHING AND SMOOTHING FORMULAE

1.1. A definition of smoothing formulae. Let $\{y_n\}$ $(n = \cdots -2, -1, 0, 1, 2, \cdots)$ be a given sequence or "table" which we wish to smooth. This smoothing operation is ordinarily performed by means of a formula of the following type

$$F_n = y_{n-p}L_p + \cdots + y_{n-1}L_1 + y_nL_0 + y_{n+1}L_{-1} + \cdots + y_{n+p}L_{-p}, \tag{1}$$

where the numerical coefficients L_ν are symmetric about the middle term L_0, i.e., $L_\nu = L_{-\nu}$. The linear transformation (1) if applied to the original sequence $\{y_n\}$ will transform it into the smoothed sequence $\{F_n\}$. By extending the definition of $L_\nu = 0$ for $|\nu| > p$ we may rewrite (1) as

$$F_n = \sum_{\nu=-\infty}^{\infty} y_\nu L_{n-\nu}. \tag{2}$$

If $y_\nu = \text{const.} = c$, we also wish that $F_n = c$; therefore

$$\sum_\nu L_\nu = 1 \tag{3}$$

is a natural requirement.

When does the formula (1) actually smooth? As an example let $p = 1$ and let the coefficients L_ν be $(-1, 3, -1)$. If we now apply the formula (1) to the periodic sequence

$$\{y_n\} = \{\cdots, 0, 1, 0, 1, 0, 1, \cdots\}$$

we obtain

$$\{F_n\} = \{\cdots, -2, 3, -2, 3, -2, 3, \cdots\}$$

which is a good deal rougher than the original sequence. Obviously this situation deserves some clarification.

There seems no doubt that the "smoothness" of a sequence $\{y_n\}$ depends in some way on its differences of higher order, especially on the sums of their squares. We also notice that the formula (2) agrees with the rule of multiplication of Fourier series. This suggests the use of such series.

Let us assume for the moment that

$$\sum_{n=-\infty}^{\infty} |y_n| < \infty. \tag{4}$$

We now define a function $T(u)$ by

$$T(u) = \sum_{n=-\infty}^{\infty} y_n e^{inu} \tag{5}$$

and call it the *characteristic function* of our sequence $\{y_n\}$; it is a complex-valued continuous function of u of period 2π.

Now (5) implies

$$e^{-iu}T(u) = \sum y_n e^{i(n-1)u} = \sum y_{n+1}e^{inu}$$

and by subtracting (5) we get

$$(e^{-iu} - 1)T(u) = \sum_{n=-\infty}^{\infty} \Delta y_n e^{inu}. \tag{6}$$

This shows that we obtain the characteristic function of the sequence $\{\Delta y_n\}$ of first differences of $\{y_n\}$ by multiplying the characteristic function $T(u)$ of $\{y_n\}$ by the factor $e^{-iu}-1$. Generally

$$(e^{-iu} - 1)^m T(u) = \sum_{n=-\infty}^{\infty} \Delta^m y_n e^{inu} \qquad (m = 0, 1, 2, \cdots). \tag{7}$$

Since $|e^{-iu}-1| = 2\ |\sin(u/2)|$, the Parseval relation furnishes the equation

$$\sum_{n=-\infty}^{\infty} (\Delta^m y_n)^2 = \frac{1}{2\pi}\int_0^{2\pi} [2 \sin (u/2)]^{2m} |\ T(u)\ |^2 du \qquad (m \geqq 0). \tag{8}$$

These formulae furnish an expression for the sums of the squares of the differences of any order in terms of the characteristic function $T(u)$ of the sequence.

Let us now turn to the "smoothed" sequence $\{F_n\}$. Let

$$\phi(u) = \sum_{n=-\infty}^{\infty} L_n e^{inu} = L_0 + 2L_1 \cos u + 2L_2 \cos 2u + \cdots \tag{9}$$

be the characterstic function of the sequence $\{L_n\}$. We shall also refer to $\phi(u)$ as the *characteristic function of the smoothing formula* (2). Notice that $\phi(u)$ is always real and even. By multiplication of the two Fourier series (5) and (9) we obtain, in view of (2),

$$T(u)\phi(u) = \sum_{n=-\infty}^{\infty} F_n e^{inu}. \tag{10}$$

Hence the characteristic function of the "smoothed" sequence $\{F_n\}$ is obtained by multiplying the characteristic function $T(u)$ of $\{y_n\}$ by the characteristic function $\phi(u)$ of the smoothing formula (2). By now applying (8) to the sequence $\{F_n\}$ we obtain

$$\sum_{n=-\infty}^{\infty} (\Delta^m F_n)^2 = \frac{1}{2\pi}\int_0^{2\pi} (2 \sin (u/2))^{2m} |\ T(u)\ |^2 (\phi(u))^2 du, \qquad (m \geqq 0). \tag{11}$$

A comparison of the relations (8) and (11) will readily furnish an answer to the question: what is a smoothing formula? Indeed, we notice that the integrands in (8) and (11) differ only, for each fixed value of m, by the factor $\phi(u)^2$ in (11). This justifies the following definition.

DEFINITION 1. *Let L_n be a symmetric sequence of coefficients, i.e., $L_{-n}=L_n$. The formation of the weighted means*

$$F_n = \sum_{\nu=-\infty}^{\infty} y_\nu L_{n-\nu} \qquad (n = 0, \pm 1, \pm 2, \cdots) \tag{12}$$

is said to be a smoothing formula if

$$\sum_n L_n = 1, \tag{13}$$

$$\sum_n |\ L_n\ | < \infty, \tag{14}$$

while the characteristic function

$$\phi(u) = \sum_{-\infty}^{\infty} L_n e^{inu} = L_0 + 2L_1 \cos u + 2L_2 \cos 2u + \cdots \tag{15}$$

satisfies the condition

$$-1 \leqq \phi(u) \leqq 1, \qquad (0 \leqq u \leqq 2\pi). \tag{16}$$

The necessity of the condition (16) is justified as follows: By a comparison of (8) and (11), in view of (16), we obtain the inequalities

$$\sum_{n=-\infty}^{\infty} (\Delta^m F_n)^2 \leqq \sum_{n=-\infty}^{\infty} (\Delta^m y_n)^2, \qquad (m = 0, 1, 2, \cdots).$$

Actually the equality sign in one of these relations will arise only under highly exceptional or else trivial conditions. This remark should make it clear why the smoothing quality of a formula violating (16) should be highly questionable.

So far we were concerned merely with the ability of a formula (2) to smooth the sequence. However, the discrepancies between the two sequences also deserve attention. By subtracting (10) from (5) we obtain

$$T(u)(1 - \phi(u)) = \sum (y_n - F_n)e^{inu}$$

and therefore

$$\sum_{n=-\infty}^{\infty} (y_n - F_n)^2 = \frac{1}{2\pi}\int_0^{2\pi} | T(u) |^2(1 - \phi(u))^2 du. \tag{17}$$

A comparison of the integrands of (17) and (11) reveals the obvious fact that strong smoothing may be achieved only if we allow relatively large discrepancies between F_n and y_n. Indeed, the integral of (17) will be small only if $\phi(u)$ differs but little from 1, while strong smoothing requires as small a $\phi(u)$ as possible.

1.11. *Examples of smoothing formulae.* (a) Our trivial example $L_0=3$, $L_1=L_{-1}=-1$, $L_n=0$ $(n>1)$ has the characteristic function $\phi(u)=3-2\cos u$. We find $\phi(u)\geqq 1$, with $\phi(\pi)=5$, which rules it out as a smoothing formula.

(b) If $L_n \geqq 0$ for all n, and $\sum L_r=1$, then (12) is always a smoothing formula. Indeed

$$| \phi(u) | = | \sum L_n e^{inu} | \leqq \sum | L_n | = 1.$$

Thus

$$F_n = (y_{n-1} + y_n + y_{n+1})/3 \tag{18}$$

is a smoothing formula with

$$\phi(u) = (1 + 2 \cos u)/3.$$

Let

$$\phi_1(u) = | \phi(u) | = | 1 + 2 \cos u | /3 = \sum L_n^{(1)} \cos nu.$$

Since $(\phi(u))^2=(\phi_1(u))^2$ it is clear from (11) that the formula (18) and the formula of characteristic function $\phi_1(u)$ have identical smoothing powers. However, since $0<1-\phi_1(u)<1-\phi(u)$ for $2\pi/3<u<4\pi/3$, we see by (17) that the formula

$$F_n^{(1)} = \sum y_\nu L_{n-\nu}^{(1)}$$

will alter the sequence $\{y_n\}$ much less than (18) will.

(c) Generally, our formula (17) shows that it is desirable for an efficient smoothing formula to have its characteristic function satisfy the more restrictive condition

$$0 \leqq \phi(u) \leqq 1. \tag{19}$$

1.12. *A comparison of smoothing formulae.* Again our relations (11) justify the following definition.

DEFINITION 2. *Let $\phi_1(u)$ and $\phi_2(u)$ be the characteristic functions of two smoothing formulae. We say that the first is stronger than the second if*

$$| \phi_1(u) | \leqq | \phi_2(u) |. \tag{20}$$

with the inequality sign holding for some value of u.

Later in this paper we shall set up a basic sequence of smoothing formulae of progressively greater strength according to this definition. Here we remark only that two smoothing formulae cannot in general be compared on the basis of this definition. However, the following remark seems obvious. Let

$$F_n = \sum_\nu y_\nu L_{n-\nu} \tag{21}$$

be a smoothing formula of characteristic function $\phi(u)$. The iteration, or repetition, of (21) may be thought of as another smoothing formula and its characteristic function is found to be $(\phi(u))^2$. Since $|\phi(u)| \leqq 1$ obviously

$$(\phi(u))^2 \leqq | \phi(u) |.$$

This shows that the formula (21) and the sequence of its successive iterates form a sequence of smoothing formula of progressively increasing strength.

1.13. *Smoothing formulae which are exact for polynomial values of a given degree.* The following definition is in common use.

DEFINITION 3. *A smoothing formula* (2) *is said to be exact for the degree m if it reproduces exactly the values $\{y_n\}$ of a polynomial of degree not exceeding m.*

If

$$F_n = \sum y_\nu L_{n-\nu} \tag{22}$$

is to be exact for the degree m, it is obviously sufficient that it be exact for the basic monomials $1, x, \cdots, x^m$. Thus the exactness for the degree m is equivalent to the relations

$$n^s = \sum_{\nu=-\infty}^{\infty} \nu^s L_{n-\nu} \qquad (s = 0, 1, \cdots, m). \tag{23}$$

Let us now assume for simplicity that the sequence of coefficients L_n tends to zero exponentially as $n \to \infty$, i.e., we assume the existence of two positive constants A and B such that

$$| L_n | \leqq A e^{-B|n|}$$

for all values of n. This implies that the function

$$\phi(u) = \sum_n L_n e^{inu}$$

is regular in the strip

$$|Iu| < B$$

of the complex u-plane. Now

$$\phi(u) = \sum L_\nu e^{i\nu u} = \sum L_{\nu-n} e^{i\nu u} \cdot e^{-inu}$$

and

$$e^{inu}\phi(u) = \sum_\nu e^{i\nu u} L_{n-\nu}.$$

We now expand both sides in ascending powers of u and compare like coefficients. Since

$$\phi(u) = 1 + \frac{u^2}{2!}\phi'' + \frac{u^4}{4!}\phi^{(4)} + \cdots,$$

we get the identities in n

$$n^s - \binom{s}{2}\phi'' n^{s-2} + \binom{s}{4}\phi^{(4)} n^{s-4} - \cdots = \sum_{\nu=-\infty}^{\infty} \nu^s L_{n-\nu} \qquad (s = 0, 1, 2, \cdots).$$

A comparison with (23) will show that a smoothing formula is always exact for a highest degree which is always odd. It also proves the following proposition which may evidently be established under conditions less stringent than the ones we used.

THEOREM 1. *A smoothing formula* (22) *is exact for a degree* $2\nu+1$ *if and only if* $\phi(u)-1$ *has at* $u=0$ *a zero of order* $2\nu+2$, *i.e.*,

$$\phi''(0) = \phi^{(4)}(0) = \cdots = \phi^{(2\nu)}(0) = 0. \tag{24}$$

As an illustration we mention the formula

$$F_n = \frac{1}{32}(-y_{n-3} + 9y_{n-1} + 16y_n + 9y_{n+1} - y_{n+3}) = y_n - \frac{3}{16}\delta^4 y_n - \frac{1}{32}\delta^6 y_n \tag{25}$$

of characteristic function

$$\phi(u) = (8 + 9\cos u - \cos 3u)/16. \tag{26}$$

We find that $\phi''(0)=0$, hence (25) is exact for cubics. The symmetry property $\phi(u)+\phi(\pi-u)=1$ shows that

$$\phi(\pi) = \phi'(\pi) = \phi''(\pi) = \phi'''(\pi) = 0.$$

This results in rather strong smoothing power. The formula (25) is part of a sequence of formulae, the next one of this kind being

$$F_n = \frac{1}{512}(3y_{n-5} - 25y_{n-3} + 150y_{n-1} + 256y_n + 150y_{n+1} - 25y_{n+3} + 3y_{n+5}) \tag{27}$$

or

$$F_n = y_n + \frac{5}{32}\delta^6 y_n + \frac{15}{256}\delta^8 y_n + \frac{3}{512}\delta^{10} y_n.$$

Its characteristic function

$$\phi(u) = (128 + 150\cos u - 25\cos 3u + 3\cos 5u)/256$$

again enjoys the symmetry property $\phi(u)+\phi(\pi-u)=1$. Also $\phi(u)-1$ has a zero of order 6 at $u=0$, hence (25) is exact for quintics, while $\phi(u)$ has a zero of order 6 at $u=\pi$ resulting in strong smoothing power.

1.2. Smoothing a finite table. In 1.1 we have discussed the smoothing of an infinite table $\{y_n\}$ which is such that the series of the absolute values of its entries converges. By (8), (11) and the inequalities (16) we have found that the sum of the squares of the differences of order m is diminished by smoothing. This is true for $m=0, 1, 2, \cdots$. Now we shall discuss briefly the practically most important case of a given finite table

$$\{y_n\} \qquad (n = 0, 1, \cdots, N). \tag{28}$$

To fix the ideas we assume the following simplest concrete situation: the third differences $\Delta^3 y_n$ are slowly varying and of slowly varying signs, while the $\Delta^4 y_n$ are of random signs. In this situation we naturally wish to minimize the 4th differences of the table. Now we form an average value of the $\Delta^3 y_n$ at each of the two ends of the table and we extend the column of $\Delta^3 y_n$ with the corresponding constant average value at each end.[4] Thus the $\Delta^3 y_n$ are defined for all n having one constant value for $n>N-3$ and another constant value for $n<0$. Now we extend the definition of y_n for all n from the values of the third differences. Also, we compute the $\Delta^4 y_n$ for the extended infinite table. Clearly $\Delta^4 y_n=0$ for $n<-1$ or $n>N-3$. Let

$$T_4(u) = \sum \Delta^4 y_n e^{inu}$$

be the characteristic function of the sequence of 4th differences, the series containing really a finite sum of terms only.

Let us now apply to the extended table y_n a smoothing formula

$$F_n = \sum_\nu y_\nu L_{n-\nu} \tag{29}$$

of characteristic function $\phi(u)$, *which is exact for cubics.* The result is the new sequence $\{F_n\}$ $(-\infty<n<\infty)$. Evidently y_n are the values of cubics for large $|n|$ and therefore $F_n=y_n$ for large $|n|$, hence also $\Delta^3 F_n=\Delta^3 y_n$ and $\Delta^4 F_n=0$ for large $|n|$. Notice also that we may think of the sequence $\{\Delta^4 F_n\}$ as arising from $\{\Delta^4 y_n\}$ by the smoothing operation (29). Therefore

$$\sum_{n=-\infty}^{\infty} (\Delta^4 y_n)^2 = \frac{1}{2\pi}\int_0^{2\pi} |T_4(u)|^2 du$$

and

$$\sum_{n=-\infty}^{\infty} (\Delta^4 F_n)^2 = \frac{1}{2\pi}\int_0^{2\pi} |T_4(u)|^2 \phi(u)^2 du.$$

Generally

$$\sum_{n=-\infty}^{\infty} (\Delta^{4+m} y_n)^2 = \frac{1}{2\pi}\int_0^{2\pi} (2\sin u/2)^{2m} |T_4(u)|^2 du, \qquad (m \geqq 0) \tag{30}$$

$$\sum_{n=-\infty}^{\infty} (\Delta^{4+m} F_n)^2 = \frac{1}{2\pi}\int_0^{2\pi} (2\sin u/2)^{2m} |T_4(u)|^2 \phi(u)^2 du, \qquad (m \geqq 0). \tag{31}$$

[4] Compare G. J. Lidstone, *Note on the computation of terminal values in graduation by Jenkins' modified osculatory formula*, Transactions of the Faculty of Actuaries (Scotland), **12**, 277 (1930).

A comparison of (30) and (31) shows that the sums of the squares of the fourths and subsequent differences have been decreased by the smoothing operation. No such statement can or should be inferred concerning the finite sums of relevant differences of orders 0, 1, 2, and 3.

II. INTERPOLATION FORMULAE

2.1. Interpolation formulae of the cardinal type. Let

$$F_n = \sum_\nu y_\nu L_{n-\nu} \tag{1}$$

be a smoothing formula. If we apply it to the "elementary" table

$$y_0 = 1, \qquad y_n = 0 \quad (n \neq 0), \tag{2}$$

then $F_n = L_n$. The even sequence $\{L_n\}$ may therefore be regarded as the smoothed version by (1) of the elementary table. Now suppose that we are given not only the even sequence of ordinates L_n but an even function $L(x)$ defined for all real x and such that $L(n) = L_n$. Then we may replace the integral variable n in (1) by the continuous variable x and we obtain the formula

$$F(x) = \sum_{\nu=-\infty}^{\infty} y_\nu L(x - \nu). \tag{3}$$

We call $L(x)$ the basic function of the formula (3). The chief aim of this paper is to point out that the subject of interpolation is truly dominated by the formulae of the type (3), the kind of approximation desired depending only on the choice of the basic function $L(x)$. The particular basic function

$$L(x) = \frac{\sin \pi x}{\pi x} \tag{4}$$

gives rise to the series

$$F(x) = \sum_{\nu=-\infty}^{\infty} y_\nu \frac{\sin \pi(x - \nu)}{\pi(x - \nu)} \tag{5}$$

which is well known to mathematicians and referred to as the *cardinal series*. For this reason we wish to call (3) a series, or formula of *cardinal type*.

We notice here for further reference that the basic function (4) may also be written as a Fourier integral as follows.

$$\frac{\sin \pi x}{\pi x} = \frac{1}{2\pi}\int_{-\pi}^{\pi} e^{ixu} du. \tag{6}$$

2.11. *The two kinds of interpolation formulae, ordinary or smoothing.* For integral values of $x = n$ our formula (3) becomes

$$F(n) = \sum_{\nu=-\infty}^{\infty} y_\nu L(n - \nu). \tag{7}$$

Equation (3) is an interpolation formula in the usual sense if $F(n) = y_n$, for all n, and this is the case if and only if $L(x)$ satisfies the conditions

$$L(0) = 1, \qquad L(n) = 0 \quad (n \neq 0). \tag{8}$$

Otherwise, (7) is a smoothing formula. We shall follow the accepted actuarial practice of referring to (3) as an *ordinary interpolation formula* if (3) reproduces exactly the given ordinates $\{y_n\}$. Otherwise we call (3) a *smoothing interpolation formula.*

2.12. *Examples of interpolation formulae of the cardinal type.* Later in this paper we shall discuss various classes of such interpolation formulas all arising from a common general theory. For purposes of orientation and illustration we mention here a few concrete examples.

2.121. *The k-point central interpolation formula.* Let k be a fixed integer ($=1, 2, 3, \cdots$). By k-point central interpolation we mean the interpolation method whereby the polynomial of degree at most $k-1$, defined by k consecutive ordinates y_n, is utilized within an interval of unit length centrally located with respect to the set of defining ordinates y_n. This set of k defining ordinates y_n is shifted up by one unit in the subscript for interpolation in the next unit interval. It seems obvious that this kind of interpolation is performed for any real value of x by a formula of the cardinal type

$$F(x) = \sum_{n=-\infty}^{\infty} y_n C_k(x - n) \tag{9}$$

with a suitable function $C_k(x)$. To obtain this function, it is sufficient to interpolate the elementary table (2) by means of this method of k-point central interpolation. The graphs indicate the resulting $C_k(x)$ for $k=1, 2, 3,$ and 4.[5] It is found that $C_k(x)$

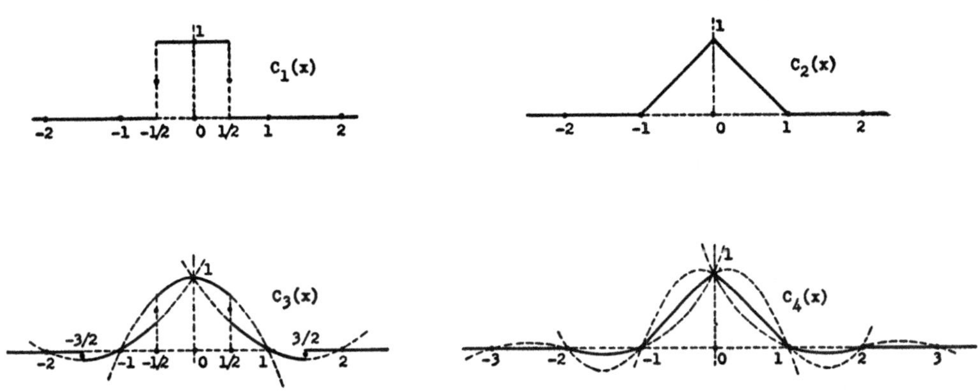

[5] These graphs indicate geometrically the construction of the successive arcs of these curves. Thus $C_3(x)$ is defined in the interval $1/2 < x < 3/2$ by the parabola passing through the points (0, 1), (1, 0), (2, 0). Similarly $C_4(x)$ is defined in $-1 < x < 0$ by the cubic which takes the values 0, 0, 1, 0 at $x = -2, -1, 0, 1$ respectively. We mention incidentally the following general analytic expression of the basic function $C_k(x)$ of k-point central interpolation. In terms of the "central" factorial

$$x^{[k]-1} = \begin{cases} x(x^2 - 1^2)(x^2 - 2^2) \cdots (x^2 - (\nu - 1)^2) & \text{if } k = 2\nu, \\ \left(x^2 - \dfrac{1}{4}\right)\left(x^2 - \dfrac{9}{4}\right) \cdots \left(x^2 - \dfrac{(2\nu - 1)^2}{4}\right) & \text{if } k = 2\nu + 1 \end{cases}$$

we define the corresponding truncated function

$$x_+^{[k]-1} = \begin{cases} x^{[k]-1} & \text{if } x > 0 \\ 0 & \text{if } x < 0 \end{cases} \qquad (k = 1, 2, 3, \cdots).$$

is a polygonal line composed of arcs of degree $k-1$. Also $C_k(x)$ itself is continuous with a discontinuous first derivative if k is even. For an odd k, $C_k(x)$ itself is discontinuous, the value assigned at a point of discontinuity being the arithmetic mean of the two local limits. Evidently for $k=2$ our formula (9) is identical with the method of linear interpolation and the graph of $F(x)$, as given by (9), is identical with the polygonal line of vertices (n, y_n).

For further reference we mention the following formulae which are valid for all real values of x

$$\begin{aligned} C_1(x) &= \frac{1}{2\pi}\int_{-\infty}^{\infty} \frac{\sin u/2}{u/2}\, e^{iux}du, \\ C_2(x) &= \frac{1}{2\pi}\int_{-\infty}^{\infty} \left(\frac{\sin u/2}{u/2}\right)^2 e^{iux}du, \\ C_3(x) &= \frac{1}{2\pi}\int_{-\infty}^{\infty} \left(\frac{\sin u/2}{u/2}\right)^3 \left(1+\frac{1}{8}u^2\right) e^{iux}du, \\ C_4(x) &= \frac{1}{2\pi}\int_{-\infty}^{\infty} \left(\frac{\sin u/2}{u/2}\right)^4 \left(1+\frac{1}{6}u^2\right) e^{iux}du. \end{aligned} \tag{10}$$

The k-point central interpolation method is the most important method for the interpolation and the construction of tables of analytic and regular functions. However, for the construction of tables of empirical functions, the low order of continuity of $C_k(x)$ is at times a serious limitation of this method. It seems indeed evident that the continuity properties of the linear compound

$$F(x) = \sum_n y_n L(x-n)$$

are directly dependent on the continuity properties of the basic function $L(x)$. We turn now to an interesting example of an "osculatory" interpolation formula having a basic $L(x)$ enjoying stronger continuity properties.

2.122. *An osculatory interpolation formula of W. A. Jenkins.* We define a basic function $L(x)$ for $x \leqq 0$ by

$$L(x) = \begin{cases} 0 & \text{if } \quad x \leqq -3 \\ -\dfrac{1}{12}(x+3)^3(x+2) & \text{if } \quad -3 \leqq x \leqq -2 \\ \dfrac{1}{12}(x+1)(x+2)(x+3)(3x+7) & \text{if } \quad -2 \leqq x \leqq -1 \\ \dfrac{1}{6}(x+1)(6-6x-9x^2-x^3) & \text{if } \quad -1 \leqq x \leqq 0 \end{cases} \tag{11}$$

The definition of this function is to be completed for $x=0$ by continuity, if k is given, and by the arithmetic mean of the two limits, if k is odd. Then

$$C_k(x) = \frac{1}{(k-1)!}\,\delta^k x_+^{[k]-1} \qquad (-\infty < x < \infty),$$

where δ^k is the symbol of the kth central difference of step unity (compare Theorem 3 of section 3.11). We will return to this subject in Part B where the extremely simple law of formation of the integrals (10) will also be given.

and extend its definition by the requirement $L(-x)=L(x)$ to all real values of x. Since the conditions (8) are visibly verified we see that

$$F(x) = \sum y_n L(x-n) \tag{11'}$$

is an ordinary interpolation formula. A closer inspection of the composite polynomial function (11) will show that $L(x)$, $L'(x)$, and $L''(x)$ are all continuous for all real values of x. Using a customary mathematical terminology we may say that $L(x)$ is of class C''. Moreover, in various ways it may be shown that the formula (11′) is exact if the y_n are the ordinates of a polynomial of degree 3 or less, i.e., $F(x)$ becomes identical with that cubic polynomial.

It is of interest to compare Jenkins' formula (11′) with the 4-point central interpolation formula (9) (for $k=4$). Both are exact for cubics. $C_4(x)$ is of class C, while the present $L(x)$ of class C''. This was achieved by increasing the complexity of the basic function in two ways: 1) The interval where $L(x)$ is non-vanishing was increased from $|x|\leqq 2$ to $|x|\leqq 3$. 2) The degree of the polynomial arcs has increased from 3 to 4. Later Jenkins' formula (11′) will appear as a member of a sequence of interpolation formulae of similar characteristics. Here we mention that the basic function (11) may be expressed in the form

$$L(x) = \frac{1}{2\pi}\int_{-\infty}^{\infty}\left(\frac{\sin u/2}{u/2}\right)^4\left(4+\cos u - 4\frac{\sin u}{u}\right)e^{iux}du \tag{11''}$$

for all real values of x.

2.123. *A smoothing interpolation formula of W. A. Jenkins.* We define a basic function $L(x)$ by

$$L(x) = \begin{cases} 0 & \text{if } x \leqq -3, \\ -\frac{1}{36}(x+3)^3 & \text{if } -3 \leqq x \leqq -2, \\ \frac{1}{36}(69+117x+63x^2+11x^3) & \text{if } -2 \leqq x \leqq -1, \\ \frac{1}{18}(15-27x^2-14x^3) & \text{if } -1 \leqq x \leqq 0, \end{cases} \tag{12}$$

$$L(-x) = L(x).$$

This particular $L(x)$, composed of cubic arcs, is of class C''. The formula[6]

[6] W. A. Jenkins writes his interpolation formula (11′) in the following Everett form

$$F(n+x) = y_n\xi + \delta^2 y_n\frac{\xi(\xi^2-1)}{6} - \delta^4 y_n\frac{\xi^3(\xi-1)}{12}$$
$$+ y_{n+1}x + \delta^2 y_{n+1}\frac{x(x^2-1)}{2} - \delta^4 y_{n+1}\frac{x^3(x-1)}{12}, \qquad (0\leqq x\leqq 1,\ x+\xi=1).$$

Likewise his formula (12′) takes the form

$$F(n+x) = y_n\xi + \delta^2 y_n\frac{\xi(\xi^2-1)}{6} - \delta^4 y_n\frac{\xi^3}{36}$$
$$+ y_{n+1}x + \delta^2 y_{n+1}\frac{x(x^2-1)}{6} - \delta^4 y_{n+1}\frac{x^3}{36}.$$

$$F(x) = \sum_\nu y_\nu L(x - \nu) \tag{12'}$$

corresponding to (12) is exact for polynomials of degree 3 or less. However, while (11′) was an ordinary interpolation foımula, the formula (12′) is a smoothing interpolation formula. Since

$$L(0) = 15/18, \qquad L(1) = 2/18, \qquad L(2) = -1/36,$$

while $L(n)=0$ for $n \geqq 3$, we see that for $x=n$ (12′) reduces to a smoothing formula of characteristic function

$$\phi(u) = \frac{1}{18}(15 + 4\cos u - \cos 2u).$$

We readily verify that $\phi''(0)=0$ and $\phi(\pi)=5/9 \leqq \phi(u) \leqq 1$. Hence (12′) reduces for integral $x=n$ to a smoothing formula, according to our Definition 1. On comparing Jenkins' two formulae (11′) and (12′) we notice that they are both exact for cubics, giving rise to curves of class C''. Since (12′) is only a *smoothing* interpolation formula while (11′) is an *ordinary* interpolation formula, it has been possible to lower the degree of $L(x)$ from 4 to 3. We finally mention that the function (12) may be expressed as

$$L(x) = \frac{1}{2\pi}\int_{-\infty}^{\infty}\left(\frac{\sin u/2}{u/2}\right)^4\left(\frac{4}{3} - \frac{1}{3}\cos u\right)e^{iux}du. \tag{12''}$$

2.2. A general theory of interpolation formulae of the cardinal type. In this section we shall discuss various characteristic properties of interpolation formulae of the cardinal type in terms of the Fourier transform of the basic function $L(x)$. This discussion will provide a sufficiently broad foundation for the subsequent development of specific formulae in the latter part of this paper.

2.21. *Characteristic properties of interpolation formulae.* Some of the following definitions have already occurred in the previous sections. For convenient reference we include them in our present enumeration of properties of an interpolation formula

$$F(x) = \sum_{\nu=-\infty}^{\infty} y_\nu L(x - \nu). \tag{13}$$

a. We say that (13) is an *ordinary interpolation formula* if $F(x)$ interpolates exactly the given ordinates y_n, i.e., if

$$L(0) = 1, \qquad L(\nu) = 0 \quad (\nu \neq 0). \tag{14}$$

b. We say that (13) is a *smoothing interpolation formula* if for $x=n$ (13) turns into a smoothing formula

$$F(n) = \sum_\nu y_\nu L(n - \nu). \tag{15}$$

The term "smoothing formula" is meant, of course, in the sense of our Definition 1, section 1.1.

c. We say that (13) is *exact for the degree* $k-1$ if the relation

$$P(x) = \sum_{n=-\infty}^{\infty} P(n)L(x - n)$$

is an identity for any polynomial $P(x)$ of degree at most $k-1$. The last condition is in turn equivalent with the k identities

$$x^\nu = \sum_{n=-\infty}^{\infty} n^\nu L(x-n) \qquad (\nu = 0, 1, \cdots, k-1) \tag{16}$$

out of which it can always be recovered by means of suitable linear combinations.

d. We say that (13) *preserves the degree* $k-1$,[7] if for any polynomial $P(x)$ of degree $\nu \leqq k-1$ we have an identity

$$\sum_{n=-\infty}^{\infty} P(n)L(x-n) = P(x) + (\text{a polynomial of degree} < \nu). \tag{17}$$

Notice that the leading term of $P(x)$ is not altered by (13). Again in terms of the monomials x^ν we may say: (13) preserves the degree $k-1$ whenever the k functions

$$Q_\nu(x) = \sum_{n=-\infty}^{\infty} n^\nu L(x-n), \qquad (\nu = 0, 1, \cdots, k-1), \tag{18}$$

are polynomials of the form

$$Q_\nu(x) = x^\nu + a_{\nu 1}x^{\nu-1} + \cdots + a_{\nu\nu}, \qquad (\nu = 0, 1, \cdots, k-1). \tag{19}$$

e. We say that (13) is of *degree* m *and of class* C^μ, if the basic function $L(x)$ is a polygonal line of polynomial arcs of degree at most m joining in such a way as to result in a function $L(x)$ having μ continuous derivatives. In the sequel, the junction points will always be either for integral values $x=n$ or else for $x=n+1/2$. Consequently no "condensation" of discontinuities will result by the formation of the linear compound (13). Hence the interpolation curve $F(x)$ will again be of degree m and of class C^μ. As examples we recall the formulae (11′) and (12′) of W. A. Jenkins, which are both of class C'' and of degree 4 and 3, respectively.

f. A formula (13) whose basic function $L(x)$ is composed of polynomial arcs will also be referred to as a *polynomial interpolation formula*. We shall say that it has the *span* s if the even function $L(x)$ vanishes identically for $x > s/2$, but not for $x > s'$ with $0 < s' < s/2$. Thus the k-point central interpolation formula (9) is of span k, while both formulae (11′), (12′) of Jenkins' have the span 6. For obvious practical reasons it is desirable to work with polynomial formulae having as small a span as possible.

g. We say that (13) *is an analytic interpolation formula* if the basic function $L(x)$ is analytic and regular for all real x. The original cardinal series (5) is an example of this type. Obviously no analytic formula can possibly have a finite span. The role of the span is taken over by the rate of damping of $L(x)$ as x increases. For obvious practical reasons it is desirable to work with analytic $L(x)$ damping out as fast as possible.

2.22. *The characteristic function of the basic function* $L(x)$. It was shown in chapter

[7] This property of interpolation formulae seems to have been neglected so far. It represents an important weaker form of the condition of exactness for the degree $k-1$. Compare T. N. E. Greville, loc. cit. pp. 210–211, for our slight departure from the standard terminology. Jenkins speaks of a *modified* interpolation formula in case the formula is not *ordinary*. The term "modified" seems natural in view of Jenkins' construction of such formulae by *modifying* certain terms of Everett's formula (the author is indebted for this last remark to Mr. Chalmers L. Weaver). It seems, however, less desirable if their construction is, as here, otherwise performed.

I that various properties of a smoothing formula

$$F_n = \sum_\nu y_\nu L_{n-\nu}$$

are readily expressible in terms of its characteristic function

$$\phi(u) = \sum_n L_n e^{inu}.$$

Likewise, the properties of the interpolation formula

$$F(x) = \sum_\nu y_\nu L(x - \nu) \tag{20}$$

will largely depend on the behaviour of the function

$$g(u) = \int_{-\infty}^{\infty} L(x)e^{iux}dx = \int_{-\infty}^{\infty} L(x) \cos uxdx. \tag{21}$$

This even function $g(u)$ is the Fourier transform of $L(x)$. Following a terminology used in probability theory we shall refer to $g(u)$ as the *characteristic function* of $L(x)$.

Under certain general assumptions which will always be verified in our application, the relation (21) may be inverted[8] to

$$L(x) = \frac{1}{2\pi}\int_{-\infty}^{\infty} g(u)e^{iux}du. \tag{22}$$

However, it should be remarked that at times our integrals are not absolutely convergent and that they then converge only as a principal value in the sense of Cauchy: $\lim_{A\to\infty}\int_{-A}^{A}$. An example of this kind is our first formula (10)

$$C_1(x) = \frac{1}{2\pi}\int_{-\infty}^{\infty} \frac{\sin u/2}{u/2} e^{iux}du.$$

Changing u and x to $2\pi u$ and $x/2\pi$ respectively we see

$$C_1\left(\frac{x}{2\pi}\right) = \int_{-\infty}^{\infty} \frac{\sin \pi u}{\pi u} e^{iux}du.$$

Inverting this relation we obtain

$$\frac{\sin \pi x}{\pi x} = \frac{1}{2\pi}\int_{-\infty}^{\infty} C_1\left(\frac{u}{2\pi}\right) e^{iux}du = \frac{1}{2\pi}\int_{-\pi}^{\pi} e^{iux}du \tag{23}$$

which is identical with (6) and shows that

$$g(u) = C_1\left(\frac{u}{2\pi}\right) = \begin{cases} 1 & \text{if} \quad |u| < \pi \\ \frac{1}{2} & \text{if} \quad |u| = \pi \\ 0 & \text{if} \quad |u| > \pi \end{cases}$$

is the characteristic function of the original cardinal series (5). It is precisely the discontinuity of its characteristic function which causes the extremely slow damping of the basic function (23). (See 2.21, g.)

[8] See e.g. S. Bochner, *Vorlesungen über Fouriersche Integrale*, Leipzig, 1932, Satz 11b on p. 42.

Similar reasons of slow damping will rule out the following rather obvious method of turning a given smoothing formula

$$F_n(x) = \sum_\nu y_\nu L_{n-\nu}$$

into a smoothing interpolation formula

$$F(x) = \sum_\nu y_\nu L(x - \nu).$$

From

$$\phi(u) = \sum_n L_n e^{inu}$$

we derive

$$L_n = \frac{1}{2\pi} \int_{-\pi}^{\pi} \phi(u) e^{inu} du.$$

Now we simply define a basic function $L(x)$ by

$$L(x) = \frac{1}{2\pi} \int_{-\pi}^{\pi} \phi(u) e^{ixu} du.$$

The corresponding characteristic function $g(u)$ is found to be

$$g(u) = \begin{cases} \phi(u) & \text{if} \quad |u| < \pi \\ 0 & \text{if} \quad |u| > \pi. \end{cases}$$

Again the discontinuities of $g(u)$, or of one of its higher derivatives, will imply that the damping of $L(x)$ is too slow for numerical purposes. Indeed, by partial integrations, $L(x)$ is found to tend to zero as a certain negative power of x only, as x tends to infinity. (Concerning the order of magnitude of Fourier integrals for large values of x, see the theorem on page 11 of Bochner's book quoted in our footnote 8.)

2.23. *Fundamental criteria in terms of characteristic functions.* We shall now restrict ourselves to basic functions $L(x)$ which are everywhere continuous with the exception of possible "discontinuities of the first kind" (such as were exhibited by the basic functions $L(x)$ of section 2.121). Moreover, we shall assume that $L(x)$ dampens out exponentially. This means that we assume the existence of two positive constants A and B such that the inequality

$$|L(x)| < Ae^{-B|x|} \tag{24}$$

holds for all real values of x. This clearly rules out the basic function (23) of the cardinal series. The assumption (24) implies that the characteristic function

$$g(u) = \int_{-\infty}^{\infty} L(x) e^{iux} dx \tag{25}$$

is analytic and regular not only on the real u-axis but also in the infinite strip

$$|Iu| < B \tag{26}$$

of the complex u-plane. It also implies that the expression

$$\lim_{p\to\infty} \sum_{n=-p}^{p} g(u + 2\pi n)e^{2\pi i n x} \tag{27}$$

converges uniformly in a circular neighborhood of $u=0$ and for every real value of x.

The following theorem will demonstrate the usefulness of the characteristic function of an interpolation formula.

THEOREM 2. *Let the basic function* $L(x)$ *satisfy the condition* (24). *Let the corresponding interpolation formula be*

$$F(x) = \sum_{\nu} y_\nu L(x - \nu). \tag{28}$$

For integral $x=n$ (28) *reduces to the smoothing formula*

$$F(n) = \sum y_\nu L(n - \nu). \tag{29}$$

A. *The characteristic function* $\phi(u)$, *of the smoothing formula* (29) *is given by the relation*

$$\phi(u) = \sum_{\nu=-\infty}^{\infty} g(u + 2\pi\nu). \tag{30}$$

In particular (28) *is an ordinary interpolation formula* (see 2.21, a) *if and only if*

$$\sum_{\nu=-\infty}^{\infty} g(u + 2\pi\nu) \equiv 1. \tag{31}$$

B. *The formula* (28) *is exact for the degree* $k-1$ (see 2.21, c) *if the following two conditions hold simultaneously:*

$g(u) - 1$ *has a zero of order* k *at* $u = 0$, (32)

$g(u)$ *has zeros of order* k *for all non-vanishing integral multiples of* 2π; $u=2\pi n$ $(n\neq 0)$. (33)

C. *The formula* (28) *preserves the degree* $k-1$ (see 2.21, d) *if the condition* (33) *holds, together with the additional condition*

$$g(0) = 1. \tag{34}$$

Remark. For some applications it is important to notice that an *ordinary* interpolation formula which *preserves* the degree $k-1$ is automatically *exact* for the degree $k-1$. This seems evident a priori. It is also evident in terms of our criteria, for (31) implies

$$g(u) - 1 = -\sum_{\nu\neq 0} g(u + 2\pi\nu)$$

and the right-hand side has a zero of order k at $u=0$ by (33).

Proof of A. Our formula (25) implies

$$L(x) = \frac{1}{2\pi}\int_{-\infty}^{\infty} g(u)e^{ixu}du = \lim_{p\to\infty}\frac{1}{2\pi}\int_{-(2p+1)\pi}^{(2p+1)\pi} g(u)e^{ixu}du$$

$$= \lim_{p\to\infty}\frac{1}{2\pi}\sum_{\nu=-p}^{p}\int_{-\pi}^{\pi} g(u + 2\pi\nu)e^{iux}e^{2\pi i\nu x}du$$

$$= \lim_{p\to\infty} \frac{1}{2\pi} \int_{-\pi}^{\pi} \left\{ \sum_{\nu=-p}^{p} g(u + 2\pi\nu) e^{2\pi i x\nu} \right\} e^{iux} du,$$

$$= \frac{1}{2\pi} \int_{-\pi}^{\pi} \left\{ \sum_{\nu=-\infty}^{\infty} g(u + 2\pi\nu) e^{2\pi i x\nu} \right\} e^{iux} du.$$

In particular, if $x = n$ is an integer, we find

$$L(n) = \frac{1}{2\pi} \int_{-\pi}^{\pi} \left\{ \sum_{-\infty}^{\infty} g(u + 2\pi\nu) \right\} e^{inu} du. \tag{35}$$

Since the characteristic function $\phi(u)$ of (29) is by definition the function of Fourier coefficients $L(n)$ (see 1.1, (9)), the relation (30) is established.

Proof of B. We wish to apply Poisson's summation formula[9]

$$\sum_{n=-\infty}^{\infty} f(x - n) = \sum_{n=-\infty}^{\infty} e^{2\pi i n x} \int_{-\infty}^{\infty} f(v) e^{-2\pi i v n} dv \tag{36}$$

to the function

$$f(x) = e^{-ixu} L(x). \tag{37}$$

By (37) and (25) we find

$$\int_{-\infty}^{\infty} f(v) e^{-2\pi i v n} dv = \int_{-\infty}^{\infty} L(v) e^{-i(u+2\pi n)v} dv = g(u + 2\pi n)$$

hence by (36)

$$e^{-ixu} \sum_{n=-\infty}^{\infty} e^{iun} L(x - n) = \sum_{n=-\infty}^{\infty} g(u + 2\pi n) e^{2\pi i n x}$$

and finally

$$\sum_{n=-\infty}^{\infty} e^{iun} L(x - n) = e^{iux} \sum_{n=-\infty}^{\infty} g(u + 2\pi n) e^{2\pi i n x}. \tag{38}$$

This identity actually holds for all real x and all real or complex values of u within the strip (26). It contains implicitly all the statements of Theorem 1. Thus for $x = 0$ it reduces to (30). To prove our statement B we assume x fixed and regard both members of (38) as functions of u, which we expand in series of powers of u, then equating the respective coefficients on both sides. On the left-hand side we have the expansion

$$\sum_{\nu=0}^{\infty} \frac{i^\nu u^\nu}{\nu!} \sum_{n=-\infty}^{\infty} n^\nu L(x - n).$$

On the right-hand side, our assumption (33) implies that the terms $g(u + 2\pi n) e^{2\pi i n x}$ $(n \neq 0)$ do not contribute any terms in u of order less than k. Thus our identity (38) becomes

$$\sum_{\nu=0}^{\infty} \frac{i^\nu u^\nu}{\nu!} \sum_{n=-\infty}^{\infty} n^\nu L(x - n) = e^{ixu} g(u) + u^k \text{ (regular function of } u\text{)}. \tag{39}$$

On the other hand our assumption (32) amounts to

[9] See S. Bochner, loc. cit., theorem 10 on page 35.

$$g(u) = 1 + u^k \text{ (regular function).}$$

This and (39) imply

$$\sum_{\nu=0}^{\infty} \frac{i^\nu u^\nu}{\nu!} \sum_{n=-\infty}^{\infty} n^\nu L(x-n) = \sum_{\nu=0}^{\infty} \frac{i^\nu u^\nu}{\nu!} x^\nu + u^k \text{ (regular function).} \tag{40}$$

A comparison of the coefficients of the first k terms on each side of (40) furnishes the identities (16). This concludes a proof of B.

Proof of C. Since $g(u)$ is regular at $u=0$, and even, it has in view of (34) an expansion of the form

$$g(u) = 1 - \frac{a_2}{2!} u^2 + \frac{a_4}{4!} u^4 - \frac{a_\mu}{6!} u_6 + \cdots .$$

We now define a sequence of polynomials by means of the generating function

$$e^{ixu} g(u) = \sum_{\nu=0}^{\infty} Q_\nu(x) \frac{(iu)^\nu}{\nu!} \tag{41}$$

or

$$e^{xu} g(u/i) = \sum_{\nu=0}^{\infty} Q_\nu(x) \frac{u^\nu}{\nu!} \tag{42}$$

where

$$g(u/i) = 1 + \frac{a_2}{2!} u^2 + \frac{a_4}{4!} u^4 + \cdots . \tag{43}$$

A comparison of terms on both sides of (42), using (43), shows that

$$Q_\nu(x) = x^\nu + \binom{\nu}{2} a_2 x^{\nu-2} + \binom{\nu}{4} a_4 x^{\nu-4} + \cdots . \tag{44}$$

On substituting the expansion (41) into the right-hand side of (39) and by comparison of the first k terms on both sides we find that the identities (18) and (19) are established. This completes the proof of our theorem.

As a brief illustration of our criteria let us consider again Jenkins' smoothing interpolation formula (12′) of 2.123. Its basic function is

$$L(x) = \frac{1}{2\pi} \int_{-\infty}^{\infty} \left(\frac{2 \sin u/2}{u}\right)^4 \left(\frac{4}{3} - \frac{1}{3} \cos u\right) e^{iux} du. \tag{45}$$

A simple method of evaluating explicitly such integrals in order to find the polynomial expressions (12) will be discussed later. An inspection of the characteristic function

$$g(u) = \left(\frac{2 \sin u/2}{u}\right)^4 (4 - \cos u)/3$$

immediately reveals that our condition (33) is verified for $k=4$. Direct expansion shows that $g(u) = 1 - (7/240)u^4 + \cdots$ and (32) is also verified for $k=4$. The interpolation formula (12′) is therefore exact for cubics. Also the fact that $L(x)$ is of class C'' is revealed by an inspection of the integral (45). Indeed we notice that $g(u)$ vanishes for $u=\infty$ like u^{-4}. This implies that we may differentiate (45) twice under the integral sign and that the integral

$$L''(x) = \frac{1}{2\pi}\int_{-\infty}^{\infty} g(u)(iu)^2 e^{iux}du$$

is also continuous since it converges absolutely.

III. THE THEORY OF SPLINE CURVES

The previous chapter provides a formal theory of interpolation formulae in terms of a basic function $L(x)$ which, as yet, is largely arbitrary. The present chapter will furnish the foundation for the derivation of special basic functions which are readily computed with great accuracy and lead to interpolation formulae enjoying the properties described in the previous chapter.

3.1. Polynomial spline curves of order k. A spline is a simple mechanical device for drawing smooth curves. It is a slender flexible bar made of wood or some other elastic material. The spline is place on the sheet of graph paper and held in place at various points by means of certain heavy objects (called "dogs" or "rats") such as to take the shape of the curve we wish to draw. Let us assume that the spline is so placed and supported as to take the shape of a curve which is nearly parallel to the x-axis. If we denote by $y=y(x)$ the equation of this curve then we may neglect its small slope y', whereby its curvature becomes

$$1/R = y''/(1+y'^2)^{3/2} \approx y''.$$

The elementary theory of the beam will then show that the curve $y=y(x)$ is a polygonal line composed of cubic arcs which join continuously, with a continuous first and second derivative.[10] These junction points are precisely the points where the heavy supporting objects are placed.

3.11. *Description of spline curves of order k.* Our last remark suggests the following definition.

DEFINITION 4. *A real function $F(x)$ defined for all real x is called a spline curve of order k and denoted by $\Pi_k(x)$ if it enjoys the following properties:*

1) *It is composed of polynomial arcs of degree at most $k-1$.*
2) *It is of class C^{k-2}, i.e., $F(x)$ has $k-2$ continuous derivatives.*
3) *The only possible junction points of the various polynomial arcs are the integer points $x=n$ if k is even, or else the points $x=n+1/2$ if k is odd.*

Thus a $\Pi_1(x)$ is a step function with possible discontinuities at the points $x=n+1/2$. A $\Pi_2(x)$ has an ordinary polygonal graph with vertices only at the integer points $x=n$. A $\Pi_4(x)$ corresponds to the elementary mathematical description of an ordinary (infinite) spline with the "dogs" placed at all or only some of the points with $x=n$.

It should be noticed that if a $\Pi_k(x)$ is of class C^{k-1}, then $\Pi^{(k-1)}(x)$ must necessarily be constant for all x. Thus such a $\Pi_k(x)$ reduces to a polynomial of degree $k-1$. It is just this relaxation of the requirement of the continuity of the $(k-1)$-order derivative of $\Pi_k(x)$ which turns the spline curve into a flexible and versatile instrument of approximation. Likewise, only the "dogs" (or "rats") enable the ordinary spline to trace curves differing from the graph of a cubic polynomial.

The special importance of spline curves will be due to the fact that by the addi-

[10] The author is indebted for this suggestion to Professor L. H. Thomas of Ohio State University.

tion of several spline curves of successive orders we may get any desired polygonal line of given degree m and class C^μ.

3.12. *The evaluation of certain Fourier integrals.* Our further work is based on the consideration of the functions

$$M_k(x) = \frac{1}{2\pi}\int_{-\infty}^{\infty}\left(\frac{2\sin u/2}{u}\right)^k e^{iux}du \qquad (k = 1, 2, \cdots; -\infty < x < \infty). \tag{1}$$

They have been evaluated explicitly for low values of k by various authors.[11] The following general explicit representation is essentially due to Laplace (see J. V. Uspensky, *Introduction to mathematical probability*, 1937, Example 3, pp. 277–278).

THEOREM 3. *Let k be a positive integer. Define the function x_+^{k-1} by*

$$x_+^{k-1} = \begin{cases} x^{k-1} & \text{if} \quad x \geqq 0 \\ 0 & \text{if} \quad x < 0. \end{cases} \tag{2}$$

For $k=1$ and $x=0$ this definition is modified to $0_+^{k-1}=1/2$.

The following identity holds for all real values of x

$$\frac{1}{2\pi}\int_{-\infty}^{\infty}\left(\frac{2\sin u/2}{u}\right)^k e^{iux}du = \frac{1}{(k-1)!}\delta^k x_+^{k-1}, \tag{3}$$

where δ^k stands for the usual symbol of the kth order central difference of step equal to unity.

The identity (3) is correct for $k=1$. Indeed it is well known that

$$\frac{1}{2\pi}\int_{-\infty}^{\infty}\frac{\sin u}{u}du = \tfrac{1}{2}.$$

On replacing u by ux we get that

$$\frac{1}{2\pi}\int_{-\infty}^{\infty}\frac{\sin xu}{u}du = \begin{cases} \tfrac{1}{2} & \text{if} \quad x > 0 \\ 0 & \text{if} \quad x = 0 \\ -\tfrac{1}{2} & \text{if} \quad x < 0, \end{cases}$$

or

$$\frac{1}{2\pi}\int_{-\infty}^{\infty}\frac{\sin xu}{u}du = x_+^0 - \tfrac{1}{2}.$$

Therefore

$$\frac{1}{2\pi}\int_{-\infty}^{\infty}\frac{2\sin u/2}{u}\cos uxdu = \frac{1}{2\pi}\int_{-\infty}^{\infty}\frac{\sin(x+\frac{1}{2})u}{u}du - \frac{1}{2\pi}\int_{-\infty}^{\infty}\frac{\sin(x-\frac{1}{2})u}{u}du$$

$$= \frac{1}{2\pi}\delta\int_{-\infty}^{\infty}\frac{\sin xu}{u}du = \delta(x_+^0 - \tfrac{1}{2}) = \delta x_+^0,$$

[11] See S. Bochner, *Fourier analysis*, Princeton University lectures, 1936–1937, where our $M_k(x)$ are worked out for $k=1, 2, 3$, and where the increasing smoothness qualities of these functions are clearly noted. Bochner also considers the integrals

$$M_k^*(x) = \frac{1}{2\pi}\int_{-\infty}^{\infty}\left(\frac{2\sin u/2}{u}\right)^k \frac{\sin ux}{u}du.$$

Using Theorem 5 below, we readily obtain the identity $M_k^*(x) = -\frac{1}{2}+(1/k!)\delta^k x_+^k$.

and (3) is established for $k=1$.

Let us consider for the moment the sequence of functions

$$N_k(x) = \frac{1}{(k-1)!}\delta^k x_+^{k-1}. \tag{4}$$

We have already shown that

$$M_k(x) = N_k(x) \tag{5}$$

holds for $k=1$. Assume now that (5) holds. We wish to show that the similar identity for $k+1$, rather than k, arises from (5) by performing the operation

$$\int_{x-1/2}^{x+1/2}$$

on both sides of (5). This will be accomplished if we prove that

$$M_{k+1}(x) = \int_{x-1/2}^{x+1/2} M_k(x)dx \tag{6}$$

and

$$N_{k+1}(x) = \int_{x-1/2}^{x+1/2} N_k(x)dx. \tag{7}$$

In view of

$$\int_{x-1/2}^{x+1/2} e^{iux}dx = \frac{2\sin u/2}{u}e^{iux}$$

we obtain (6) by an integration under the integral sign as follows

$$\int_{x-1/2}^{x+1/2} M_k(x)dx = \frac{1}{2\pi}\int_{-\infty}^{\infty}\left(\frac{2\sin u/2}{u}\right)^k\left(\int_{x-1/2}^{x+1/2} e^{iux}dx\right)du = M_{k+1}(x).$$

To prove (7) we notice that

$$\begin{aligned}\int_{x-1/2}^{x+1/2} N_k(x)dx &= \delta\int_{-\infty}^{x} N_k(x)dx = \delta\int_{-\infty}^{x}\delta^k\frac{x_+^{k-1}}{(k-1)!}dx \\ &= \delta\delta^k\int_{-\infty}^{x}\frac{x_+^{k-1}}{(k-1)!}dx = \delta^{k+1}\frac{x_+^k}{k!} = N_{k+1}(x).\end{aligned}$$

This concludes the proof of Theorem 3.

3.13. *Explicit polynomial expressions for $M_k(x)$*. The formula

$$M_k(x) = \frac{1}{(k-1)!}\delta^k x_+^{k-1} \tag{8}$$

will readily show that $y=M_k(x)$ *represents a spline curve of order k*. Indeed, if k is even, then

$$(x+n)_+^{k-1}$$

is a Π_k and therefore also their linear combination (8). If k is odd the same conclusion holds because

$$(x+n-\tfrac{1}{2})_+^{k-1}$$

is a Π_k.

It also follows from (8) that the explicit polynomial expressions for

$$(k-1)!M_k(x),$$

in successive unit intervals, are identical with the successive partial sums of the expansion

$$\delta^k x^{k-1} = \left(x+\frac{k}{2}\right)^{k-1} - \binom{k}{1}\left(x+\frac{k}{2}-1\right)^{k-1} + \cdots + (-1)^k\left(x-\frac{k}{2}\right)^{k-1},$$

an expression which incidentally vanishes identically, being the kth order difference of a polynomial of degree $k-1$. We thus get

$$(k-1)!M_k(x) = \begin{cases} 0 & \text{if} \quad x \leqq -\frac{k}{2} \\ \left(x+\frac{k}{2}\right)^{k-1} & \text{if} \quad -\frac{k}{2} \leqq x \leqq -\frac{k}{2}+1 \\ \left(x+\frac{k}{2}\right)^{k-1} - \binom{k}{1}\left(x+\frac{k}{2}-1\right)^{k-1} & \text{if} \quad -\frac{k}{2}+1 \leqq x \leqq -\frac{k}{2}+2 \\ \cdots\cdots\cdots\cdots & \\ \left(x+\frac{k}{2}\right)^{k-1} - \binom{k}{1}\left(x+\frac{k}{2}-1\right)^{k-1} + \cdots & \\ \quad + (-1)^{k-1}\binom{k}{1}\left(x-\frac{k}{2}+1\right)^{k-1} & \text{if} \quad \frac{k}{2}-1 \leqq x \leqq \frac{k}{2} \\ \delta^k x^{k-1} = 0 & \text{if} \quad \frac{k}{2} \leqq x. \end{cases} \tag{9}$$

For future reference we work out explicitly the cases $k=1, 2, 3$, and 4. The expansions

$$\begin{aligned} \delta x^0 &= 1 - 1 \\ \delta^2 x &= (x+1) - 2x + (x-1) \\ \delta^3 x^2 &= (x+\tfrac{3}{2})^2 - 3(x+\tfrac{1}{2})^2 + 3(x-\tfrac{1}{2})^2 - (x-\tfrac{3}{2})^2 \\ \delta^4 x^3 &= (x+2)^3 - 4(x+1)^3 + 6x^3 - 4(x-1)^3 + (x-2)^3 \end{aligned} \tag{10}$$

now furnish the following expressions

$$M_1(x) = \begin{cases} 0 & \text{if} \quad x < -\frac{1}{2} \\ 1 & \text{if} \quad -\frac{1}{2} < x < \frac{1}{2} \\ 0 & \text{if} \quad \frac{1}{2} < x, \end{cases} \tag{11}$$

to which must be added $M_1(\pm 1/2) = 1/2$ as required by (1) for $k=1$.

$$M_2(x) = \begin{cases} 0 & \text{if} \quad x \leqq -1 \\ x+1 & \text{if} \quad -1 \leqq x \leqq 0 \\ -x+1 & \text{if} \quad 0 \leqq x \leqq 1 \\ 0 & \text{if} \quad 1 \leqq x, \end{cases} \tag{12}$$

$$M_3(x) = \begin{cases} 0 & \text{if} \quad x \leqq -\frac{3}{2} \\ (1/2)(x+3/2)^2 & \text{if} \quad -\frac{3}{2} \leqq x \leqq -\frac{1}{2} \\ (1/2)(x+3/2)^2 - (3/2)(x+1/2)^2 & \text{if} \quad -\frac{1}{2} \leqq x \leqq \frac{1}{2} \\ (1/2)(-x+3/2)^2 & \text{if} \quad \frac{1}{2} \leqq x \leqq \frac{3}{2} \\ 0 & \text{if} \quad \frac{3}{2} \leqq x, \end{cases} \tag{13}$$

$$M_4(x) = \begin{cases} 0 & \text{if} \quad x \leqq -2 \\ (1/6)(x+2)^3 & \text{if} \quad -2 \leqq x \leqq -1 \\ (1/6)(x+2)^3 - (4/6)(x+1)^3 & \text{if} \quad -1 \leqq x \leqq 0 \\ (1/6)(-x+2)^3 - (4/6)(-x+1)^3 & \text{if} \quad 0 \leqq x \leqq 1 \\ (1/6)(-x+2)^3 & \text{if} \quad 1 \leqq x \leqq 2 \\ 0 & \text{if} \quad 2 \leqq x. \end{cases} \tag{14}$$

In deriving these expressions the expansions (10) were used up to the point from where the evenness of the functions $M_k(x)$ allowed us to complete their definition for all x by symmetry.

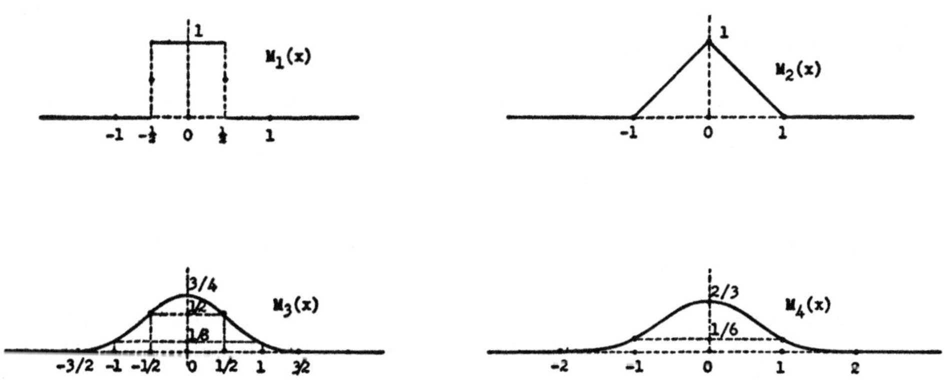

3.14. *Interpolation formula with $M_k(x)$ as basic function.* The interpolation formula

$$F(x) = \sum_{\nu=-\infty}^{\infty} y_\nu M_k(x-\nu) \tag{15}$$

will play an important role in our subsequent work. We mention it at this place because it contributes to our investigation of k-order spline curves.

To the basic function

$$L(x) = M_k(x)$$

corresponds the characteristic function

$$g(u) = \left(\frac{2 \sin u/2}{u}\right)^k, \tag{16}$$

as seen by comparing III (1) with II (22). The characteristic function of the formula (15) for integral $x=n$ is

$$\phi_k(u) = M_k(0) + 2M_k(1)\cos u + 2M_k(2)\cos 2u + \cdots. \tag{17}$$

The $\{\phi_k(u)\}$ represent an interesting sequence of cosine polynomials which we will investigate more closely later in this paper. Here we mention without proof that

$$1 = \phi_1(u) = \phi_2(u) > \cdots > \phi_k(u) > \cdots > 0 \qquad (0 < u < 2\pi) \tag{18}$$

while, of course, $\phi_k(0) = 1$. Hence (15) is a smoothing interpolation formula of progressively increasing strength as k increases. We assemble the various properties of (15) in the form of a theorem.

THEOREM 4.

$$F(x) = \sum_{\nu=-\infty}^{\infty} y_\nu M_k(x - \nu) \tag{19}$$

is a polynomial smoothing interpolation formula of degree $k-1$, class C^{k-2} and span $s = k$ (see sections 2.21 and 3.13). It is exact for the degree 1 and preserves the degree $k-1$. The smoothing power of (19) increases progressively for increasing values of k.

The exactness of (19) for the degree 1 and the preservation of the degree $k-1$ follow by Theorem 2 (B and C). Indeed, by (16), $g(u)-1$ has a double zero for $u=0$ while $g(u)$ has zeros of order k for $u = 2\pi n$ $(n \neq 0)$. Since the preservation of the degree $k-1$ implies the identities II(18) and (19), the following corollary results.

COROLLARY. *Any given polynomial $P_{k-1}(x)$ of degree at most $k-1$ may be represented in the form*

$$P_{k-1}(x) = \sum_{n=-\infty}^{\infty} y_n M_k(x - n) \tag{20}$$

where $\{y_n\}$ are the ordinates of some other suitably chosen polynomial of the same degree as P_{k-1}. This representation is unique.

3.15. *The analytic representation of spline curves of order k.* We know that if $\{y_n\}$ is an arbitrary sequence of ordinates, then our interpolation formula

$$F(x) = \sum_n y_n M_k(x - n) \tag{21}$$

represents a spline curve of order k. This is true because all $M_k(x-n)$ are such curves. The following question arises: *Let $F(x)$ be a given Π_k; can we always represent it in the form (21) for an appropriate sequence $\{y_n\}$?*

This question is answered affirmatively by the following theorem.

THEOREM 5. *Any spline curve $\Pi_k(x)$ may be represented in one and only one way in the form*

$$\Pi_k(x) = \sum_{n=-\infty}^{\infty} y_n M_k(x - n) \tag{22}$$

for appropriate values of the coefficients y_n. There are no convergence difficulties since $M_k(x)$ vanishes for $|x| > k/2$. Thus (22) represents a Π_k for arbitrary $\{y_n\}$ and represents the most general one.

In order to prove this theorem we return to the interpolation formula (21) and differentiate it repeatedly. By (8) we have

$$M_k'(x) = \delta^k \frac{d}{dx} \frac{1}{(k-1)!} x_+^{k-1} = \delta^k \frac{1}{(k-2)!} x_+^{k-2} = \delta M_{k-1}(x)$$

and repeating we get

$$M_k^{(\nu)}(x) = \delta^\nu M_{k-\nu}(x) \qquad (0 \leqq \nu \leqq k-1). \tag{23}$$

From (21) and (23) we obtain by partial summation

$$F'(x) = \sum_n y_n \delta M_{k-1}(x-n) = \sum_n \delta y_{n+1/2} M_{k-1}(x-n-\tfrac{1}{2})$$

or

$$F'(x+\tfrac{1}{2}) = \sum_n \delta y_{n+1/2} M_{k-1}(x-n). \tag{24}$$

If $k>2$, this formal rule of differentiation of a spline curve may now again be applied to (24) with the result

$$F''(x+1) = \sum \delta^2 y_{n+1} M_{k-2}(x-n)$$

or

$$F''(x) = \sum \delta^2 y_n M_{k-2}(x-n).$$

Generally for $0 \leqq \nu \leqq k-1$

$$F^{(\nu)}(x) = \begin{cases} \sum_n \delta^\nu y_n M_{k-\nu}(x-n) & \text{if } \nu \text{ is even,} \\ \sum_n \delta^\nu y_{n+1/2} M_{k-\nu}(x-n-\tfrac{1}{2}) & \text{if } \nu \text{ is odd.} \end{cases} \tag{25}$$

This result may be stated as follows: *The νth derivative of the spline curve* (21) *may be obtained directly by applying the same interpolation formula* (21) *with $k-\nu$, rather than k, to the sequence of the νth central differences $\delta^\nu y$ properly centered according to the parity of ν. In particularly: $F^{(k-2)}(x)$ is obtained by interpolating linearly among the $\delta^{k-2}y$. $F^{(k-1)}(x)$ is a step function whose successive values agree with those of the corresponding $\delta^{k-1}y$.*

Now let $F(x)$ be a given Π_k. We are to show the existence of a sequence $\{y_n\}$ such that (21) holds identically. Suppose for the moment that such y_n have been found which do make (21) hold. Then by (25) for $\nu = k-1$ we have

$$F^{(k-1)}(x) = \begin{cases} \delta^{k-1} y_n & \text{for } n-\tfrac{1}{2} < x < n+\tfrac{1}{2} \text{ if } k \text{ is odd.} \\ \delta^{k-1} y_{n+1/2} & \text{for } n < x < n+1 \quad \text{if } k \text{ is even.} \end{cases} \tag{26}$$

In either case the successive constant values of the step function $F^{(k-1)}(x)$ determine uniquely the values of the differences of order $k-1$ of the sequence of the as yet unknown coefficients y_n. These differences in turn determine the coefficients y_n uniquely up to an additive sequence of vanishing differences of order $k-1$. Let $\bar{y}_n$ be one sequence such that

$$\delta^{k-1}\bar{y}_{n+1/2} = F^{(k-1)}(n+\tfrac{1}{2}) \quad \text{if} \quad k \text{ is even}$$

or else

$$\delta^{k-1}\bar{y}_n = F^{(k-1)}(n) \quad \text{if} \quad k \text{ is odd.}$$

Consider the k-order spline curve

$$\bar{F}(x) = \sum_n \bar{y}_n M_k(x-n)$$

and let

$$R(x) = F(x) - \bar{F}(x).$$

From the way $\bar{F}(x)$ was defined it is clear that $\bar{F}^{(k-1)}(x)$ and $F^{(k-1)}(x)$ agree in their successive unit intervals of constancy. Hence $R(x)$ is a Π_k whose various polynomial arcs are of degree $k-2$ or lower. Therefore $R(x)$ is *identical* with a polynomial of degree $k-2$. As such it allows of a representation of the form (21) in view of our Corollary of section 3.14. Therefore also

$$F(x) = \bar{F}(x) + R(x)$$

may be represented by our formula (21).

The unicity of the representation (21) is readily established. Indeed two different such representations would imply a representation of zero

$$0 = \sum_n y_n M_k(x-n)$$

without all y_n vanishing. However the $\delta^{k-1}y$ all vanish, and our conclusion would contradict the uniqueness of the representation (20) of polynomials.

A simple example might illustrate our proof of Theorem 5. Let us find the representation of the spline curve of order 4

$$F(x) = \frac{1}{3!} x_+^3.$$

By (26) we have

$$\delta^3 y_{n+1/2} = F'''(n+\tfrac{1}{2}) = (n+\tfrac{1}{2})_+^0 = \begin{cases} 1 & \text{if } n \geqq 0 \\ 0 & \text{if } n < 0. \end{cases}$$

A sequence having these third differences is

$$y_n = \begin{cases} 0 & \text{if } n < 0 \\ \binom{n+1}{3} = \dfrac{n(n^2-1)}{6} & \text{if } n \geqq 0. \end{cases}$$

Hence

$$x_+^3 = \sum_{n=0}^{\infty} n(n^2-1) M_4(x-n), \tag{27}$$

with the possibility still open that both member might differ by a third degree polynomial. However this possibility is excluded by the remark that both sides vanish identically for $x \leqq 0$. Incidentally, (27) implies

$$-(-x)_+^3 = \sum_{n=-\infty}^{0} n(n^2-1) M_4(x-n)$$

and by addition and subtraction of the two relations we get the identities

$$x^3 = \sum_{n=-\infty}^{\infty} n(n^2 - 1)M_4(x - n), \qquad |x|^3 = \sum_{n=-\infty}^{\infty} |n(n^2 - 1)| M_4(x - n).$$

In later applications we shall frequently operate with polygonal lines $F(x)$ of degree $k-1$ not having continuity properties as strong as a Π_k. Thus Jenkins' $L(x)$ defined by II(11) is of degree 4 and class C''. Let $F(x)$ be a polygonal line of degree $k-1$, having vertices at integral point $x=n$, and being for all real x a function of class C^μ $(-1 \leqq \mu \leqq k-2)$. We certainly obtain a curve of degree $k-1$ and class C^μ by addition of spline curves.

$$F(x) = \Pi_{\mu+2} + \Pi_{\mu+3} + \cdots + \Pi_k, \tag{28}$$

where Π_ν stands for $\Pi_\nu(x)$ or $\Pi_\nu(x+1/2)$, according to whether ν is even or odd.

THEOREM 6. *Any given polygonal line $F(x)$ of degree $k-1$ and class C^μ may be represented as a sum* (28) *of $k-\mu-1$ appropriate spline curves of orders $\mu+2, \mu+3, \cdots, k$.*

This theorem is a corollary to Theorem 5. Indeed, $F^{(\mu+1)}(x)$ may have certain discontinuities. We determine a $\Pi_{\mu+2}$ having the same discontinuities in its $(\mu+1)$st derivative. Then

$$F(x) - \Pi_{\mu+2}$$

is of degree $k-1$ and class $C^{\mu+1}$. Proceeding in this way the theorem is readily established.

Substituting for the Π_ν in (28) their expressions in terms of the M_ν we obtain an explicit (parametric) representation of such polygonal lines. Thus Jenkins' function II(11) may be represented as

$$L(x) = 4M_4(x) + \tfrac{1}{2}M_4(x + 1) + \tfrac{1}{2}M_4(x - 1) - 2M_5(x + \tfrac{1}{2}) - 2M_5(x - \tfrac{1}{2}).$$

At a glance we recognize a curve of degree 4, class C'', span $s=6$.

3.16. *A summation property of spline curves.* The degree of a polynomial is decreased or increased by one unit if we difference or else sum the polynomial. Not so with our spline curves Π_k. Indeed let

$$\Pi_k(x) = \sum_n y_n M_k(x - n) \tag{29}$$

be a given spline curve. Then

$$\Pi_k(x + 1) = \sum y_n M_k(x + 1 - n) = \sum y_{n+1} M_k(x - n)$$

and subtracting (29) we have

$$\Delta\Pi_k(x) = \Pi_k(x + 1) - \Pi_k(x) = \sum \Delta y_n M_k(x - n). \tag{30}$$

Hence $\Delta\Pi_k(x)$ will in general be also a Π_k. Now let the spline curve $\Pi_k^*(x)$ be given and we wish to find $\Pi_k(x)$ such that

$$\Delta\Pi_k(x) = \Pi_k^*(x). \tag{31}$$

By using Theorem 5 the solution is immediately found. Indeed let

$$\Pi_k^*(x) = \sum y_n^* M_k(x - n).$$

Now (29) will be a solution of (31) provided $\Delta y_n = y_n^*$, for all n. This relation defines the y_n up to an additive constant which appears as an arbitrary additive constant in the solution $\Pi_k(x)$. It is thus seen that the operation of differencing or summing spline curves (29) reduces to performing the same operations on the sequence of the coefficients $\{y_n\}$.

3.17. *An interpretation of $M_k(x)$ in probability theory*. In concluding our discussion of polynomial spline curves we mention briefly the following interpretation of $M_k(x)$. As seen from its values as given by III(11) it is clear that $M_1(x)$ may be interpreted as the probability density function of the error committed on a random real variable x, if that variable is rounded off to its nearest integral value. Now III(1) shows that the characteristic function of $M_k(x)$ is the kth power of the characteristic function of $M_1(x)$. From a known proposition in probability theory we may conclude that $M_k(x)$ *is the density distribution function of the error committed on the sum*

$$x_1 + x_2 + \cdots + x_k$$

of k statistically independent real random variables $x_1, \cdots, x_k$, if each variable is replaced by its nearest integral value.

This interpretation, otherwise entirely irrelevant for our purpose, does make a few of the properties of $M_k(x)$ intuitively obvious, such as

$$M_k(x) \begin{cases} > 0 & \text{for} \quad |x| > k/2 \\ = 0 & \text{for} \quad |x| < k/2, \end{cases}$$

$$\int_{-\infty}^{\infty} M_k(x)dx = 1.$$

In concluding we note the identity

$$\int_{x-1/2}^{x+1/2} dx_1 \int_{x_1-1/2}^{x_1+1/2} dx_2 \cdots \int_{x_{k-1}-1/2}^{x_{k-1}+1/2} f(x_k)dx_k = \int_{-\infty}^{\infty} M_k(u - x)f(u)du.$$

3.2. Analytic spline curves of order k. The polynomial spline curves $\Pi_k(x)$ described in section 3.1 will be shown to be sufficient for the derivation of polynomial approximations to equidistant data enjoying various desirable properties. These polynomial approximations will have any a priori assigned number of continuous derivatives. However, in order to obtain analytic approximations we shall now proceed to derive from our spline curves $\Pi_k(x)$ an analoguous family of analytic functions.

To achieve this end we shall smooth out our $\Pi_k(x)$ by means of one-dimensional heat flow. Consider an infinite homogeneous bar (the x-axis) in which the temperature at the point x at the time t is denoted by $F(x, t)$. We assume the flow of heat to be governed by the equation

$$\frac{\partial F}{\partial t} = \frac{1}{4} \frac{\partial^2 F}{\partial x^2}. \tag{32}$$

If $F(x) = F(x, 0)$ is given, i.e., the temperature distribution at the time $t = 0$ is known, then $F(x, t)$ is determined by the following integral[12]

$$F(x, t) = \frac{1}{\sqrt{\pi t}} \int_{-\infty}^{\infty} e^{-(u-x)^2 t^{-1}} F(u, 0) du. \tag{33}$$

This result is easy to verify: by partial differentiations we find that $F(x, t)$ as defined by (33) indeed satisfies the differential equation (32) while familiar arguments originated by Weierstrass will show that (33) implies

$$\lim_{t \to +0} F(x, t) = F(x, 0),$$

provided $F(x, 0)$ is continuous and, e.g., bounded.

The solution of the problem of finding $F(x, t)$, if $F(x, 0)$ is given, is especially simple in the case when $F(x, 0)$ is defined by a Fourier integral

$$F(x, 0) = \frac{1}{2\pi} \int_{-\infty}^{\infty} \psi(u) e^{iux} du. \tag{34}$$

Indeed, in this case we find

$$F(x, t) = \frac{1}{2\pi} \int_{-\infty}^{\infty} e^{-t(u/2)^2} \psi(u) e^{iux} du. \tag{35}$$

Notice that the temperature t enters only in the additional exponential factor. This can be proved in two ways, either by substituting (34) into (33) or else by verifying directly that (35) satisfies the differential equation (32). Obviously (35) reduces to (34) for $t=0$, as it should.

We may (and wish to) think of $F(x, t)$, for a fixed $t>0$, as a smoothed version of $F(x) = F(x; 0)$. In fact $F(x, t)$ is analytic and regular for all real or complex values of x if $\psi(u)$ is, e.g., bounded.

If we now apply this heat-flow transformation to our basic k-order spline curve

$$M_k(x) = \frac{1}{2\pi} \int_{-\infty}^{\infty} \left(\frac{2 \sin u/2}{u} \right)^k e^{iux} du \tag{36}$$

we obtain by (35) its smoothed version

$$M_k(x, t) = \frac{1}{2\pi} \int_{-\infty}^{\infty} e^{-t(u/2)^2} \left(\frac{2 \sin u/2}{u} \right)^k e^{iux} du. \tag{37}$$

Obviously

$$M_k(x, 0) = M_k(x). \tag{38}$$

[12] See H. S. Carslaw, *Mathematical theory of the conduction of heat*, Dover Publications, New York, 1945, Chapter III, Section 16. Certain smoothing properties of heat flow were already noticed by Ch. Sturm in 1886. See in this connection G. Pólya, *Qualitatives über Wärmeausgleich*, Z. angew. Math. u. Mech. **13**, 125–128 (1933). It should be mentioned here that Weierstrass derived his famous approximation theorem by means of the integral (33). Finally see E. Czuber, *Wahrscheinlichkeitsrechnung*, vol. I, Leipzig-Berlin, 1924, pp. 417–418, for a brief sketch of a method of using (33) to derive analytic approximations to given data.

The graph of $y=M_k(x, t)$ $(t>0)$ is a bell-shaped curve which dampens out very fast. Later we shall learn how to compute its values very accurately. Here we mention that

$$0 < M_k(x, t) < \frac{1}{\sqrt{\pi t}} e^{-(x-k/2)^2 \cdot t^{-1}} \quad \text{if} \quad x \geqq k/2. \tag{39}$$

Also, the recurrence relation III(6) generalizes so that $M_{k+1}(x, t)$ is obtained by the averaging operation

$$M_{k+1}(x, t) = \int_{x-1/2}^{x+1/2} M_k(x, t)dx. \tag{40}$$

If now

$$\Pi_k(x) = \sum_{n=-\infty}^{\infty} y_n M_k(x - n) \tag{41}$$

is a spline curve of order k, then its heat-flow transform is

$$\Pi_k(x, t) = \sum_{n=-\infty}^{\infty} y_n M_k(x - n, t). \tag{42}$$

The graph of this function may be called an *analytic spline curve of order k*. We notice that the series (42) fails to converge only if y_n increases very fast with $|n|$.

Summarizing we see that the curve (42) arises from the step function

$$\sum_n y_n M_1(x - n)$$

by $k-1$ successive applications of the averaging operation

$$\int_{x-1/2}^{x+1/2} (\quad)dx$$

followed by the heat-flow transformation during a time interval t. The order of application of these k operations is of course irrelevant.

The remaining parts of the paper are devoted to the problem of utilizing the two families of curves (41) and (42) for the purpose of approximating given equidistant data.

IV. A FIRST CLASS OF ANALYTIC INTERPOLATION FORMULAE

We shall now use the analytic spline curves III(42) to obtain

1. A smoothing interpolation formula which is exact for the degree 1 only.
2. An ordinary interpolation formula exact for the degree $k-1$.
3. A smoothing interpolation formula depending on a positive parameter ϵ, of strictly increasing smoothing power as ϵ increases. For $\epsilon=0$ this formula reduces to 2, while for $\epsilon=\infty$ it is identical with 1.

4.1. A smoothing interpolation formula exact for the degree 1. A comparison of the formulae III(15) and III(42) immediately suggests the following analytic interpolation formula

$$F(x) = \sum_{n=-\infty}^{\infty} y_n M_k(x - n, t) \tag{1}$$

where $M_k(x, t)$ is defined by III(37). For the sequel we shall use the following notation

$$\psi_k(u, t) = e^{-t(u/2)^2}\left(\frac{2 \sin u/2}{u}\right)^k, \tag{2}$$

$$\psi_k(u) = \psi_k(u, 0) = \left(\frac{2 \sin u/2}{u}\right)^k, \tag{3}$$

in terms of which III(37) becomes

$$M_k(x, t) = \frac{1}{2\pi}\int_{-\infty}^{\infty} \psi_k(u, t)e^{iux}du. \tag{4}$$

The characteristic function of the smoothing formula (1) for integral x is

$$\phi_k(u, t) = \sum_{n=-\infty}^{\infty} M_k(n, t) \cos nu. \tag{5}$$

The general relation II(30) furnishes the following equivalent expression

$$\phi_k(u, t) = \sum_{\nu=-\infty}^{\infty} \psi_k(u + 2\pi\nu, t). \tag{6}$$

The properties III(18) for the case $t=0$ generalize for $t>0$ as follows

$$1 > \phi_1(u, t) > \phi_2(u, t) > \cdots > \phi_k(u, t) > \cdots > 0, \qquad (0 < u < 2\pi, t > 0). \tag{7}$$

Moreover, for each fixed u, $0<u<2\pi$, $\phi_k(u, t)$ is strictly decreasing as t increases. These last two properties we state without proving them here.

The arguments which lead to Theorem 4 now allow us to state the following theorem.

THEOREM 7. *For $t>0$, $k=1, 2, \cdots$,*

$$F(x) = \sum_{n=-\infty}^{\infty} y_n M_k(x - n, t) \tag{8}$$

is an analytic smoothing interpolation formula which is exact for the degree 1 *and preserves the degree $k-1$. The smoothing power of* (8) *increases whenever either k or t is increased.*

4.2. An ordinary interpolation formula exact for the degree $k-1$. We shall now use the important property (7) to the effect that the periodic function $\phi_k(u, t)$ is positive for real u. This allows us to define the basic function

$$L_k(x, t) = \frac{1}{2\pi}\int_{-\infty}^{\infty} \frac{\psi_k(u, t)}{\phi_k(u, t)} e^{iux}du \tag{9}$$

whose characteristic function is

$$g(u) = \frac{\psi_k(u, t)}{\phi_k(u, t)}. \tag{10}$$

Our Theorem 2 of Chapter II will readily yield the following result.

THEOREM 8. *For $t\geqq 0$, $k=1, 2, 3, \cdots$*

$$F(x) = \sum_{n=-\infty}^{\infty} y_n L_k(x-n, t) \tag{11}$$

is an ordinary interpolation formula which is exact for the degree $k-1$.

Firstly we realize by (10) and (2) that the conditions II(33), (34) of Theorem 2 are verified. Therefore (11) preserves the degree $k-1$. Secondly we have by (6), (10), and II(30)

$$\phi(u) = \sum_\nu g(u+2\pi\nu) = \sum_\nu \frac{\psi_k(u+2\pi\nu, t)}{\phi_k(u+2\pi\nu, t)} = \frac{1}{\phi_k(u, t)} \sum_\nu \psi_k(u+2\pi\nu, t) = 1.$$

Thus II(31) holds and (11) is therefore an *ordinary* interpolation formula. This concludes the proof of our theorem. Indeed, by the remark following Theorem 2 our formula (11) must also be exact for the degree $k-1$.

We also mention without further details the following two limiting relations

$$\lim_{t\to\infty} \frac{\psi_k(u, t)}{\phi_k(u, t)} = \lim_{k\to\infty} \frac{\psi_k(u, t)}{\phi_k(u, t)} = \begin{cases} 1 & \text{if } |u| < \pi \\ 0 & \text{if } |u| > \pi. \end{cases} \tag{12}$$

They show, in view of the integral representation (9), that our present basic function $L_k(x, t)$ converges towards the basic function II(6) of the original cardinal series whenever either k or else t tends to infinity.

4.3. A family of smoothing interpolation formulae depending on a smoothing parameter ϵ. In section 4.1 we have derived the smoothing interpolation formula (8) in the derivation of which no attempt was made to compromise between smoothness of results and goodness of fit. Such a compromise is afforded by the following basic function

$$L_k(x, t, \epsilon) = \frac{1}{2\pi}\int_{-\infty}^{\infty} \frac{\epsilon + \phi_k(u, t)}{\epsilon + \phi_k(u, t)^2} \psi_k(u, t) e^{iux} du \qquad (0 \leqq \epsilon \leqq \infty) \tag{13}$$

which depends also on the smoothing parameter ϵ. The corresponding interpolation formula

$$F(x) = \sum_{n=-\infty}^{\infty} y_n L_k(x-n, t, \epsilon) \tag{14}$$

includes our previous formula (8) and (11) as special cases. Indeed by (13) ,(4) and (9) we find

$$L_k(x, t, 0) = L_k(x, t), \tag{15}$$

$$L_k(x, t, \infty) = M_k(x, t). \tag{16}$$

Let us now investigate the characteristic function of the smoothing formula (14) for integral x. By (13) and II(30) this characteristic function is ($\phi_k(u, t)$ is periodic!):

$$\phi_k(u, t, \epsilon) = \frac{\epsilon + \phi_k(u, t)}{\epsilon + \phi_k(u, t)^2} \sum_\nu \psi_k(u + 2/\pi\nu, t).$$

This and (6) give

$$\phi_k(u, t, \epsilon) = (\epsilon\phi_k(u, t) + \phi_k(u, t)^2)/(\epsilon + \phi_k(u, t)^2). \tag{17}$$

On the other hand we have by (7) the inequalities

$$0 < \phi_k(u, t) < 1, \qquad (0 < u < 2\pi, t > 0). \tag{18}$$

Now (17), (18) imply

$$0 < \phi_k(u, t, \epsilon) < 1, \qquad (0 < u < 2\pi, t > 0, \epsilon > 0), \tag{19}$$

and therefore (14) is a smoothing interpolation formula in the sense of section 2.21 b. Moreover, we see from (17) that for fixed t and u $(0<u<2\pi)$ $\phi_k(u, t, \epsilon)$ decreases monotonically from

$$\phi_k(u, t, 0) = 1 \quad \text{to} \quad \phi_k(u, t, \infty) = \phi_k(u, t)$$

as ϵ varies from $\epsilon=0$ to $\epsilon=\infty$. Finally (14) is exact for the degree 1 and preserves the degree $k-1$ for the same reason as mentioned in the case of (8).

We summarize these properties in the following theorem.[13]

THEOREM 9. *For* $t\geqq 0$, $k=1, 2, \cdots$,

$$F(x) = \sum_{n=-\infty}^{\infty} y_n L_k(x - n, t, \epsilon), \tag{20}$$

of basic function (13), *is a smoothing interpolation formula which is exact for the degree* 1 *and preserves the degree* $k-1$. *For* $\epsilon=0$ (20) *is identical with the ordinary interpolation formula* (11). *For increasing values of* ϵ *it increases in smoothing power until for* $\epsilon=\infty$ (20) *is identical with our smoothing formula* (8).

4.31. *A property of the derivatives of the approximation* $F(x)$. Let the given data $\{y_n\}$ satisfy the additional condition

$$\sum_{n=-\infty}^{\infty} |y_n| < \infty. \tag{21}$$

Assume also $t>0$, $\epsilon>0$. We know that the sequence $\{F(n)\}$ obtained by (20) is smoother than $\{y_n\}$ in the sense of our discussion in I Section 1.1. However, it appears

[13] The method used in deriving Theorems 8 and 9 obviously generalizes as follows. Let

$$L(x) = \frac{1}{2\pi}\int_{-\infty}^{\infty} g(u)e^{iux}du$$

be a basic function. Let the periodic function

$$\phi(u) = \sum_{\nu} g(u + 2\pi\nu)$$

satisfy the inequality

$$0 < \phi(u) \leqq 1.$$

Then

$$L_1(x) = \frac{1}{2\pi}\int_{-\infty}^{\infty} \frac{g(u)}{\phi(u)} e^{iux}du$$

is the basic function of an *ordinary* interpolation formula. Moreover

$$L_1(x, \epsilon) = \frac{1}{2\pi}\int_{-\infty}^{\infty} \frac{\epsilon + \phi(u)}{\epsilon + \phi(u)^2} g(u)e^{iux}du$$

gives rise to a family of smoothing formulae of increasing strength, as ϵ increases.

to be of some interest to discuss here the smoothness of the function $F(x)$, rather than that of the sequence $\{F(n)\}$, as a function of the smoothing parameter ϵ. In this connection we prove the following

THEOREM 10. *Denote by $F(x, \epsilon)$ the approximation* (20) *so as to indicate its dependence on ϵ. The condition* (21) *insures the convergence of the integrals of the squares of the derivatives*

$$\int_{-\infty}^{\infty} (F^{(m)}(x, \epsilon))^2 dx, \qquad (m = 0, 1, 2, \cdots). \tag{22}$$

Also each of these integrals is a monotonically decreasing function of ϵ in the range $0<\epsilon<\infty$.

Indeed, let

$$T(u) = \sum_{n=-\infty}^{\infty} y_n e^{inu} \tag{23}$$

be the characteristic function of the sequence $\{y_n\}$. For convenience we define

$$\Omega_k(u, t, \epsilon) = \frac{\epsilon + \phi_k(u, t)}{\epsilon + \phi_k(u, t)^2}. \tag{24}$$

Then (13) becomes

$$L_k(x, t, \epsilon) = \frac{1}{2\pi} \int_{-\infty}^{\infty} \Omega_k(u, t, \epsilon) \psi_k(u, t) e^{-iux} du. \tag{25}$$

By substitution of (25) into (20) we obtain

$$F(x, \epsilon) = \frac{1}{2\pi} \int_{-\infty}^{\infty} \Omega_k(u, t, \epsilon) \psi_k(u, t) T(u) e^{-iux} du. \tag{26}$$

We may evidently differentiate under the integral sign obtaining

$$F^{(m)}(x, \epsilon) = \frac{1}{2\pi} \int_{-\infty}^{\infty} \Omega_k(u, t, \epsilon) \psi_k(u, t) T(u)(-iu)^m e^{-iux} du.$$

This formula exhibits the Fourier transform of $F^{(m)}(x, \epsilon)$. We now use the analogue of the Parseval relation for Fourier integrals finally obtaining

$$\int_{-\infty}^{\infty} (F^{(m)}(x, \epsilon)^2 dx = \frac{1}{2\pi} \int_{-\infty}^{\infty} (\Omega_k(u, t, \epsilon) \psi_k(u, t))^2 |\, T(u)\,|^2 u^{2m} du. \tag{27}$$

This relation establishes our theorem. Indeed, by (24) the behaviour of the function $\Omega_k(u, t, \epsilon)$ of period of 2π, is as follows: $\Omega_k(0, t, \epsilon) = 1$, while for each fixed u, $0<u<2\pi$, it decreases from

$$\Omega_k(u, t, 0) = 1/\phi_k(u, t) \quad \text{to} \quad \Omega_k(u, t, \infty) = 1,$$

as ϵ increases from $\epsilon = 0$ to $\epsilon = \infty$.

The discussion of I section 1.2 concerning the smoothing of a finite table also has an analogue concerning the derivatives of the approximations $F(x, \epsilon)$. We state the result without further details. We assume the concrete situation of I section 1.2 where a finite table was extended to an infinite table by constant third differences at each end. To this extended table we apply our formula (20) with such a value of k which will insure that the formula (20) preserves cubics, i.e., $k \geqq 4$. Then we can prove that the integrals (22) converge for $m=4, 5, 6, \cdots$ and represent decreasing functions of ϵ.

4.32. *Formula* (20) *as applied to subtabulation*. Our formula (20) is excellently suited for the systematic interpolation, or subtabulation of given ordinates y_n. It is less suited for interpolation. The reason is obvious: For subtabulation to tenths we need only a table of the basic function $L_k(x, t, \epsilon)$ for the step $h=0.1$ only, while interpolation would call for a much more elaborate table of this function.

The following transformation of the formula (20), in terms of the function $M_k(x, t)$, is of importance for numerical applications. First of all we expand the even periodic function (24) in a Fourier (cosine series)

$$\Omega_k(u, t, \epsilon) = \frac{\epsilon + \phi_k(u, t)}{\epsilon + \phi_k(u, t)^2} = \sum_{\nu=-\infty}^{\infty} \omega_\nu^{(k)}(t, \epsilon) e^{i\nu u} = \sum_{\nu=-\infty}^{\infty} \omega_\nu^{(k)}(t, \epsilon) \cos \nu u. \tag{28}$$

Substituting this expansion into (25) we get

$$L_k(x, t, \epsilon) = \sum_{\nu=-\infty}^{\infty} \omega_\nu^{(k)}(t, \epsilon) \frac{1}{2\pi} \int_{-\infty}^{\infty} \psi_k(u, t) e^{i(\nu-x)u} du,$$

which in view of (4) becomes

$$L_k(x, t, \epsilon) = \sum_{\nu=-\infty}^{\infty} \omega_\nu^{(k)}(t, \epsilon) M_k(x - \nu, t). \tag{29}$$

This formula expresses the basic function $L_k(x, t, \epsilon)$ in terms of the $M_k(x)$. If we substitute this expansion into our formula (20) we obtain

$$F(x) = \sum_n y_n L_k(x - n, t, \epsilon) = \sum_{n,\nu} y_n \omega_\nu^{(k)}(t, \epsilon) M_k(x - n - \nu, t)$$

and replacing ν by $\nu - n$

$$F(x) = \sum_{n,\nu} y_n \omega_{\nu-n}^{(k)}(t, \epsilon) M_k(x - \nu, t).$$

A first summation by n introduces the sums

$$f_\nu = \sum_{n=-\infty}^{\infty} y_n \omega_{\nu-n}^{(k)}(t, \epsilon) \tag{30}$$

in terms of which our last expression becomes

$$F(x) = \sum_{\nu=-\infty}^{\infty} f_\nu M_k(x - \nu, t). \tag{31}$$

The pair of relations (30) and (31) is equivalent to (20) and represents its practical form. The reason for this is that the basic function $M_k(x, t)$ dampens out like

$\exp(-x^2)$ (see III (39)) while $L_k(x, t, \epsilon)$, $\epsilon < \infty$, dampens out like $\exp(-x)$ only. We notice incidentally that (31) is identical with our formula (8), to be applied to the new computed ordinates $\{f_n\}$ given by (30).

Frequently we require also tables of the derivatives $F'(x)$ and $F''(x)$, of the approximation $F(x)$. These are then computed by the formulae

$$F'(x) = \sum_\nu f_\nu M_k'(x - \nu, t), \tag{31'}$$

$$F''(x) = \sum_\nu f_\nu M_k''(x - \nu, t), \tag{31''}$$

from corresponding tables of M_k' and M_k''.

4.33. *The least squares origin of formula* (20). We want to sketch briefly the genesis of our formula (20). Let the sequence $\{y_n\}$ be given and consider the spline curve $F(x)$ given by (31), where the coefficients $\{f_\nu\}$ are as yet unknown. If we try to determine these unknowns by the requirement that $F(x)$ should interpolate strictly the given ordinates y_n, i.e.,

$$F(n) = y_n \qquad (n = 0, \pm 1, \pm 2, \cdots), \tag{32}$$

we obtain an infinite system of linear equations in the unknown f_ν, the solution of which was found to be given by

$$f_\nu = \sum_{n=-\infty}^{\infty} y_\nu \overset{(k)}{\omega}_{\nu-n}(t, 0).$$

This leads to our ordinary interpolation formula (11).

In view of Whittaker's well known smoothing method it seemed natural to proceed now as follows: *Let ϵ be a given positive number. To determine the unknown coefficients f_ν of* (31) *as solutions of the following minimal problem*: ~~we set~~

$$\sum_n (F(n) - y_n)^2 + \epsilon \cdot \sum_n (f_n - y_n)^2 = \text{minimum}. \tag{33}$$

For $\epsilon = 0$ the solution is of course identical with the solution of the ordinary interpolation problem (32). For $\epsilon = \infty$ the solution is obviously $f_n = y_n$ in which case (31) reduces to (8). For $0 \leqq \epsilon \leqq \infty$ a system of "normal equations" arises whose solution is found to be given by (30). The explicit solution of these normal equations (matrix inversion) is performed by the numerical determination of the cosine coefficients ω_ν of the expansion (28) for each given set of values of k, t and ϵ.

V. THE COMPUTATION OF THE TABLES I, II, III

In this last chapter we shall discuss the methods used in the computation of our tables which allow us to use our formulae (30), (31) or

$$f_n = \sum_{\nu=-\infty}^{\infty} y_\nu \overset{(k)}{\omega}_{n-\nu}(t, \epsilon), \tag{1}$$

$$F(x) = \sum_{n=-\infty}^{\infty} f_n M_k(x - n, t), \tag{2}$$

for subtabulation to tenths.

5.1. The computation of the function $M_k(x, t)$ and its derivatives. This function is defined (see III(37)) by the integral

$$M_k(x, t) = \frac{1}{2\pi}\int_{-\infty}^{\infty} e^{-t(u/2)^2}\left(\frac{2 \sin u/2}{u}\right)^k e^{iux}du, \qquad (t > 0). \tag{3}$$

For the special case of $t=0$ and $k=1, 2, 3, \cdots$ we found previously the explicit polynomial expressions III(3). It now so happens that for our present case of $t>0$ (3) allows us to define our function also for $k=0$ as

$$M_0(x, t) = \frac{1}{2\pi}\int_{-\infty}^{\infty} e^{-t(u/2)^2}e^{iux}du.$$

This last integral is known to be identical to

$$M_0(x, t) = \frac{1}{\sqrt{\pi t}}\, e^{-x^2/t}. \tag{4}$$

The recurrence relation III(40) shows that (3) is obtained from (4) by repeating k times the averaging operation

$$\int_{x-1/2}^{x+1/2} \qquad \text{or} \qquad \delta\int_{-\infty}^{x}.$$

The result, however, is not changed if we perform all k integrations first to be followed by the operation δ^k of kth central differencing. This proves the following result:

If we define a sequence of functions $g_k(x, t)$ by

$$g_0(x, t) = \frac{1}{\sqrt{\pi t}}\, e^{-x^2/t} \tag{5}$$

and the recurrence relation

$$g_k(x, t) = \int_{-\infty}^{x} g_{k-1}(x, t)dx \qquad (k = 1, 2, 3, \cdots), \tag{6}$$

then

$$M_k(x, t) = \delta^k g_k(x, t). \tag{7}$$

This relation reduces the problem to the problem of computing the repeated integral $g_k(x, t)$ of the error function (5). This we do as follows. It is easy to prove by induction or otherwise that (5) and (6) imply

$$g_k(x, t) = \frac{1}{(k-1)!}\,\frac{1}{\sqrt{\pi t}}\int_{-\infty}^{x}(x-u)^{k-1}e^{-u^2/t}du. \tag{8}$$

With $x-u=v$ this becomes

$$g_k(x, t) = \frac{1}{(k-1)!}\,\frac{1}{\sqrt{\pi t}}\int_{0}^{\infty} e^{-(x-v)^2/t}v^{k-1}dv. \tag{9}$$

By differentiating this with respect to x we get

$$g_k'(x, t) = \frac{1}{(k-1)!}\frac{1}{\sqrt{\pi t}}\int_0^\infty e^{-(x-v)^2/t}v^{k-1}(-2xt^{-1}+2vt^{-1})dv$$

$$= -\frac{2x}{t}g_k(x, t) + \frac{2k}{t}g_{k+1}(x, t)$$

and therefore the recurrence relation

$$g_{k+1}(x, t) = \frac{t}{2k}g_k'(x, t) + \frac{x}{k}g_k(x, t), \qquad (k = 1, 2, \cdots) \tag{10}$$

which allows us to compute the successive values $g_k(x, t)$ by the operation of differentiation rather than integration. Indeed from (5) and (6) for $k=1$ we get

$$g_1(x, t) = \frac{1}{\sqrt{\pi t}}\int_{-\infty}^{x} e^{-x^2/t}dx. \tag{11}$$

Now (11) and (10) for $k=1$ will give

$$g_2(x, t) = \frac{t}{2}\frac{1}{\sqrt{\pi t}}e^{-x^2/t} + x\frac{1}{\sqrt{\pi t}}\int_{-\infty}^{x} e^{-x^2/t}dx$$

from which $g_3(x, t)$ is readily determined.

This progressive computation is greatly simplified if we realize that $g_k(x, t)$ will appear as an expression of the form

$$g_k(x, t) = P_k(x, t)g_0(x, t) + Q_k(x, t)g_1(x, t), \tag{12}$$

where P_k, Q_k are polynomials in x and t, while $g_0(x, t)$ is the error function (5) and $g_1(x, t)$ is the error integral (11). Substituting (12) into (10) we find

$$g_{k+1}(x, t) = \frac{t}{2k}(P_k' + Q_k)g_0(x, t) + \left(\frac{t}{2k}Q_k' + \frac{x}{k}Q_k\right)g_1(x, t).$$

On comparing with (12) for $k+1$, rather than k, we obtain the recurrence relations

$$P_{k+1} = \frac{t}{2k}(P_k' + Q_k)$$

$$Q_{k+1} = \frac{t}{2k}Q_k' + \frac{x}{k}Q_k \qquad (k = 1, 2, \cdots). \tag{13}$$

Since $P_1=0$, $Q_1=1$ we readily obtain the following explicit expressions

$$\begin{aligned} P_2 &= t/2, & Q_2 &= x, \\ P_3 &= tx/4, & Q_3 &= t/4 + x^2/2, \\ P_4 &= t(t + x^2)/12, & Q_4 &= x(3t + 2x^2)/12, \\ P_5 &= tx(5t + 2x^2)/96, & Q_5 &= (3t^2 + 12tx^2 + 4x^4)/96, \\ P_6 &= t(4t^2 + 9tx^2 + 2x^4)/480, & Q_6 &= x(15t^2 + 20tx^2 + 4x^4)/480. \end{aligned} \tag{14}$$

Excellent tables of the probability function (5) and its integral (11) are now available.

By means of these tables the formulae (12) and (14) allow us to compute readily the function $g_k(x, t)$.[14] It seems worth while to point out that the relation (7) goes over into III(8) if $t \to +0$. Indeed by an obvious change of variable we see that (11) becomes

$$g_1(x, t) = \frac{1}{\sqrt{\pi}} \int_{-\infty}^{x/\sqrt{t}} e^{-u^2} du$$

and therefore

$$\lim_{t \to +0} g_1(x, t) = \overset{0}{x_+}.$$

Now by induction we prove by (6), on letting $t \to +0$, that

$$\lim_{t \to +0} g_k(x, t) = \frac{1}{(k-1)!} \overset{k-1}{x_+}$$

which proves our last statement by continuity.

The computation of the derivatives of $M_k(x, t)$ is immediately settled by the relation

$$M_k^{(\nu)}(x, t) = \delta^{\nu} g_{k-\nu}(x, t), \tag{15}$$

which is implied by (6).

5.2. The computation of the cosine coefficients $\omega_n^{(k)}(t, \epsilon)$. By IV(5) and IV(28) we can see that the problem consists in computing the values of the coefficients of the cosine expansion of the function

$$\Omega_k(u, t, \epsilon) = \frac{\epsilon + \phi_k(u, t)}{\epsilon + \phi_k(u, t)^2} = \sum_{n=-\infty}^{\infty} \omega_n^{(k)}(t, \epsilon) \cos nu, \tag{16}$$

where the even periodic function $\phi_k(u, t)$ is defined by its cosine expansion

$$\phi_k(u, t) = \sum_{n=-\infty}^{\infty} M_k(n, t) \cos nu. \tag{17}$$

---

[14] I owe to D. H. Lehmer the reference to the functions $Hh_n(x)$ defined by

$$Hh_0(x) = \int_x^{\infty} e^{-x^2/2} dx, \qquad Hh_n(x) = \int_x^{\infty} Hh_{n-1}(x) dx.$$

Tables of these functions were published by J. R. Airey as Tables XV, Group IV, of the Mathematical Tables of the British Association for the Advancement of Science. The relation between our $g_k(x, t)$ and these new functions is

$$g_k(x, t) = \frac{1}{\sqrt{2\pi}} (t/2)^{(k-1)/2} Hh_{k-1}(-x\sqrt{2/t}).$$

This relation, for $k=4$, $t=\frac{1}{2}$, would readily allow us to compute our Table I by means of Airey's tables of Hh_3, Hh_2 and Hh_1. However, for other sets of values of k and t, such as $k=8$, $t=\frac{1}{2}$, which are needed for other purposes, the range of x in Airey's table becomes insufficient. In this case tables of $M_k(x, t)$ and its derivatives are computed by our formula (12) and the excellent *Tables of Probability Functions*, vol. I (1941), vol. II (1942), prepared by the Mathematical Tables Project under the direction of A. N. Lowan.

The coefficients $M_k(n, t)$ of this extremely fast convergent series are readily computed to 10 decimal places. The obvious procedure would be to compute from (17) a table of $\phi_k(u, t)$, then compute a similar table of $\Omega_k(u, t, \epsilon)$ which is then to be used in computing ω_n by some method of numerical harmonic analysis. It would be hard to achieve accuracy by this method and for this reason we proceeded differently. It should be born in mind that the cosine expansion of the denominator of (16) is readily obtained by the simple operation of multiplication of Fourier series. The only troublesome part is the computation of the expansion

$$\frac{1}{\epsilon + \phi_k(u, t)^2} = \sum_{-\infty}^{\infty} c_n \cos nu, \tag{18}$$

i.e., the reciprocation of a given cosine series. This was done as follows. The above-mentioned method of a 24-ordinate harmonic analysis scheme was used for obtaining values of the c_n's accurate to 4–5 decimal places. These values were then improved to values accurate to 8 or 9 places by an iteration method developed by H. A. Rademacher and the author. This method is closely related to the method recommended by H. Hotelling for the reciprocation of ordinary matrices and will be described elsewhere.

In concluding this paper we want to point out two special cases of our ordinary interpolation formula (11), or (20) for $\epsilon = 0$, which are of mathematical interest. We mention first the case of $k = 0$, $t > 0$. This corresponds to interpolating our ordinates y_n by means of a function $F(x)$ as described by the formula (8) of the Introduction. Although, as remarked there, the resulting interpolation formula is useless for practical purposes, it has the remarkable feature that the expansion coefficients $\omega_n^{(0)}(t, 0)$ of (16) may be obtained explicitly. Indeed the function $\phi_0(u, t)$ reduces to a Theta function which is a regular and uniform function of

$$z = e^{iu}$$

with singularities only at $z = 0$ and $z = \infty$. The simple zeros of this function are real, negative and form a geometric progression. As a result we are able to find explicitly the decomposition in partial fractions of the reciprocal

$$1/\phi_0(u, t).$$

The expansion of these partial fractions into geometric power series furnishes explicitly the Laurent expansion in powers of z and therefore also the cosine expansion (16).

The second special case of interest is $k > 0$, $t = 0$. In this case our formula (11) reduces to an *ordinary polynomial interpolation formula of degree* $k-1$ *and class* $k-2$. This does not contradict Mr. Greville's statement (loc. cit. page 212) to the effect that such formulae do not exist. Indeed Mr. Greville considers only basic polynomial functions $L(x)$ of finite span s only, while our basic $L_k(x, 0)$ are of infinite span. This case, which is of considerable interest, requires a more detailed investigation of the cosine polynomials $\phi_k(u, 0)$. We postpone this discussion to the second part B of this paper.

APPENDIX

Description of the tables and their use for the analytic approximation of equidistant data.

Tables I and II. In Table I we find the 8-place values of the even function

$$M(x) = \frac{1}{2\pi}\int_{-\infty}^{\infty} e^{-u^2/8}\left(\frac{2\sin u/2}{u}\right)^4 \cos uxdu \tag{1}$$

and its derivatives $M'(x)$, $M''(x)$ for the step of $\Delta x = 0.1$. The graph of $M(x)$ is a bell-shaped curve and $M(x)$ vanishes to 8 places for $x \leqq -4.3$ and $x \geqq 4.3$. We now define a function of period 2π by the cosine series

$$\phi(u) = M(0) + 2M(1)\cos u + 2M(2)\cos 2u + \cdots \tag{2}$$

and expand in cosine series the following functions

$$\frac{\epsilon + \phi(u)}{\epsilon + \phi(u)^2} = \omega_0(\epsilon) + 2\omega_1(\epsilon)\cos u + 2\omega_2(\epsilon)\cos 2u + \cdots, \tag{3}$$

where ϵ is a non-negative parameter. Our Table II gives the 8-plane values of these coefficients for $\epsilon = 0, 0.1, 0.2, \cdots, 1.0$.

These tables may be used as follows to obtain an analytic approximation $F(x)$ to our ordinates y_n. We discuss first the case when $F(x)$ is to interpolate the ordinates, in the usual sense, i.e.,

$$F(n) = y_n. \tag{4}$$

For this end we compute first from the sequence $\{y_n\}$ a new sequence of coefficients $\{f_n\}$ by means of the formula

$$f_r = \cdots + y_{n-2}\omega_2(0) + y_{n-1}\omega_1(0) + y_n\omega_0(0) + y_{n+1}\omega_1(0) + y_{n+2}\omega_2(0) + \cdots \tag{5}$$

or

$$f_n = \sum_\nu y_\nu \omega_{n-\nu}(0), \tag{5'}$$

where $\omega_m = \omega_{-m}$. The analytic approximation of the ordinates y_n is then given by

$$F(x) = \sum_n f_n M(x - n). \tag{6}$$

The values of $F(x)$, x to even tenths, are now readily computed. Thus

$$F(2.3) = f_{-1}M(3.3) + f_0M(2.3) + f_1M(1.3) + f_2M(0.3)$$
$$+ f_3M(-1 + .3) + f_4M(-2 + .3) + f_5M(-3 + .3) + f_6M(-4 + .3).$$

The tabular values of $M(x)$ are so arranged that all 8 values needed in this computation are found in the fourth column headed $x + .3$. Generally, if the values of f_n are written in a vertical column, we compute the values of $F(m + \nu \cdot 10^{-1})$ $(\nu = 0, 1, \cdots, 9)$ by matching the column of values of f_n with the νth column of the table of $M(x)$ in such a way that f_m corresponds to the row for $x = 0$. The products

$$f_n M(m - n + \nu \cdot 10^{-1})$$

are then accumulated in the products counter of a desk computing machine. Also the values f_n are best computed by (5′) in a similar way if the column of values of $\omega_n(0)$ is extended upwards by symmetry for negative values of n.

From the tables of $M'(x)$ and $M''(x)$ we may likewise compute tables of the derivatives of $F(x)$ by

$$F^{(\nu)}(x) = \sum_n f_n M^{(\nu)}(x - n). \tag{7}$$

A check of the computation of the coefficients f_n is afforded by (4). Indeed the values $F(n)$ computed by (6) should agree with the y_n to about eight significant figures.

The formula (6) is exact for cubics, i.e., if the y_n are the ordinates of a polynomial of degree at most 3, then $F(x)$ is identical with that polynomial.

If the conditions (4) of strict interpolation are not required, then we have the possibility of obtaining an approximation $F(x)$ which is such that the sequence $\{F(n)\}$ is smoother than the given $\{y_n\}$. The approximation $F(x)$ is then given by the pair of formulae

$$f_n = \sum_\nu y_\nu \omega_{n-\nu}(\epsilon), \tag{8}$$

$$F(x) = \sum_n f_n M(x - n), \tag{9}$$

which are applied as above. The choice of the value of the smoothing parameter ϵ depends on the amount of smoothing desired. The strongest smoothing afforded by our table is obtained for $\epsilon = +\infty$. Then (3) shows that $\omega_0(\infty) = 1$, $\omega_1(\infty) = \omega_2(\infty) = \cdots = 0$. Thus (8) becomes $f_n = y_n$ and (9) reduces to

$$F(x) = \sum_n y_n M(x - n). \tag{10}$$

This formula is especially simple to apply. It should be remarked however that, if $\epsilon > 0$, our formula (9) is exact only for linear functions and the same is true of (10).

Table III. We may eliminate the coefficients f_n between (8) and (9). In terms of the new even function

$$L(x, \epsilon) = \sum_{n=-\infty}^{\infty} \omega_n(\epsilon) M(x - n), \tag{11}$$

our formulae (8), (9), then reduce to

$$F(x) = \sum_n y_n L(x - n, \epsilon). \tag{12}$$

Table III gives the values of $L(x, \epsilon)$ and $L''(x, \epsilon)$ for $\epsilon = 0, 0.1, \cdots, 1.0$ for the step $\Delta x = 0.5$. These may be used for subtabulation to halves in preference to (5), (6) or (8), (9). For subtabulation to fifths or tens, the use of formulae (8), (9) is preferable because of the slower damping of the function $L(x, \epsilon)$. Even so, formula (12) and Table III allow us to estimate quickly how well $F(x)$ approximates the y_n. By (12) we have

$$F''(x) = \sum_n y_n L''(x - n, \epsilon). \tag{13}$$

The table of $L''(x, \epsilon)$ then allows us to compute quickly a table of $F''(x)$ for the step $\Delta x = 0.5$ or else only isolated values if such are needed.

Example of subtabulation to tenths. We consider the following fairly smooth sequence of 64 ordinates y_n:

n	y_n	n	y_n	n	y_n	n	y_n	n	y_n	n	y_n
1	24614	12	25370	23	29290	34	73820	45	82450	56	79962
2	24644	13	25504	24	30160	35	77830	46	82290	57	79698
3	24680	14	25660	25	31320	36	80240	47	82110	58	79431
4	24723	15	25850	26	32840	37	81660	48	81911	59	79161
5	24772	16	26080	27	34790	38	82330	49	81699	60	78889
6	24828	17	26350	28	37260	39	82680	50	81472	61	78614
7	24892	18	26660	29	40440	40	82840	51	81234	62	78338
8	24966	19	27040	30	44750	41	82830	52	80987	63	78060
9	25048	20	27490	31	51120	42	82780	53	80736	64	77780
10	25143	21	28010	32	59390	43	82700	54	80481		
11	25250	22	28600	33	67550	44	82590	55	80223		

The differences of the section of this table with which we will be concerned are as follows:

n	y_n	Δ	Δ^2	Δ^3	Δ^4	Δ^5
27	34790					
28	37260	2470				
29	40440	3180	710			
30	44750	4310	1130	420		
31	51120	6370	2060	930	510	
32	59390	8270	1900	−160	−1090	−1600
33	67550	8160	−110	−2010	−1850	−760
34	73820	6270	−1890	−1780	230	2080
35	77830	4010	−2260	−370	1410	1180
36	80240	2410	−1600	660	1030	−380
37	81660	1420	−990	610	−50	−1080
38	82330	670	−750	240	−370	−320

We illustrate the case of strict interpolation, i.e., we use our Tables II for $\epsilon=0$. From our formula (5) and the values of ω_n as given in the column of Table II, with the heading $\epsilon=0$, we obtain the following coefficients.

n	f_n
27	34662.222
28	37031.355
29	40215.195
30	44060.182
31	50349.304
32	59490.524
33	68212.510
34	74566.216
35	78283.074
36	80460.234
37	81953.811
38	82356.888

From these values and our Table I of $M(x)$ and $M''(x)$, we obtain by the formulae (6) and (7) the following tables of $F(x)$ and $F''(x)$ with their differences.

Table of the function $F(x)$ and of its second derivative $F''(x)$.

x	$F(x)$	Δ	Δ^2	Δ^3	Δ^4	$F''(x)$	Δ	Δ^2	Δ^3	Δ^4
31.0	51120.00					2117.97				
31.1	51884.17	76417				1966.48	−15149			
31.2	52667.97	78380	1963			1787.44	−17904	−2755		
31.3	53469.63	80166	1786	−177		1583.71	−20373	−2469	286	
31.4	54287.11	81748	1582	−204	−27	1359.15	−22456	−2083	386	100
31.5	55118.17	83106	1358	−224	−20	1118.30	−24085	−1629	454	68
31.6	55960.40	84223	1117	−241	−17	866.08	−25222	−1137	492	38
31.7	56811.29	85089	866	−251	−10	607.04	−25868	− 646	491	− 1
31.8	57668.25	85696	607	−259	− 8	346.89	−26051	− 183	463	−28
31.9	58528.68	86043	347	−260	− 1	88.63	−25826	225	408	−55
32.0	59390.00	86132	89	−258	2	− 163.98	−25261	565	340	−68
32.1	60249.69	85969	− 163	−252	6	− 408.22	−24424	837	272	−68
32.2	61105.30	85561	− 408	−245	7	− 642.14	−23392	1032	195	−77
32.3	61954.51	84921	− 640	−232	13	− 864.26	−22212	1180	148	−47
32.4	62795.08	84057	− 864	−224	8	−1073.51	−20925	1287	107	−41
32.5	63624.93	82985	−1072	−208	16	−1269.11	−19560	1365	78	−29
32.6	64442.10	81717	−1268	−196	12	−1450.39	−18128	1432	67	−11
32.7	65244.77	80267	−1450	−182	14	−1616.76	−16637	1491	59	− 8
32.8	66031.30	78653	−1614	−164	18	−1767.70	−15094	1543	52	− 7
32.9	66800.16	76886	−1767	−153	11	−1902.77	−13507	1587	44	− 8
33.0	67550.00	74984	−1902	−135	18	−2021.68	−11891	1616	29	−15
33.1	68279.64	72964	−2020	−118	17	−2124.30	−10262	1629	13	−16
33.2	68988.05	70841	−2123	−103	15	−2210.71	− 8641	1621	− 8	−21
33.3	69674.37	68632	−2209	− 86	17	−2281.13	− 7042	1599	−22	−14
33.4	70337.91	65354	−2278	− 69	17	−2335.91	− 5478	1564	−35	−13
33.5	70978.07	64016	−2338	− 60	9	−2375.46	− 3955	1523	−41	− 6
33.6	71594.50	61643	−2373	− 35	25	−2400.17	− 2471	1484	−39	2
33.7	72186.94	59244	−2399	− 26	9	−2410.41	− 1024	1447	−37	2
33.8	72755.29	56835	−2409	− 10	16	−2406.55	386	1410	−37	0
33.9	73299.58	54429	−2406	3	13	−2389.01	1754	1368	−42	− 5
34.0	73820.00	52042	−2387	19	16	−2358.32	3069	1315	−53	−11

An inspection of these tables shows that they are very smooth and that they define $F(x)$ and $F''(x)$ to 7 significant figures by 4-point central interpolation. We have chosen on purpose an example for which it would be hard to obtain similar results by standard methods, if we are to maintain the forced accuracy requirement, and the same high degree of consistency between the function $F(x)$ and its second derivative $F''(x)$. For purposes of comparison we show also the interpolated values $F_c(x)$ for the range $x=31.6-32.5$ obtained by the 10-point central interpolation method.
On comparing with our table of $F(x)$ we notice that

$$F_c(x) < F(x)$$

throughout this range, with the exception of the point $x=32.0$ where, of course, both values agree. The curve $F_c(x)$ has a corner at $x=32$. This is the typical discontinuity in the first derivative due to central interpolation methods (see the first paragraph of our Introduction).

x	$F_c(x)$	Δ	Δ^2	Δ^3	Δ^4
31.6	55959.90				
31.7	56810.60	85070			
31.8	57667.55	85695	625		
31.9	58528.22	86067	372	−253	
32.0	59390.00	86178	111	−261	− 8
32.1	60248.72	85872	− 306	−417	−156
32.2	61103.61	85489	− 383	− 77	340
32.3	61952.37	84876	− 613	−230	−153
32.4	62792.77	84040	− 836	−223	7
32.5	63622.68	82991	−1049	−213	10

Notice that we needed 12 coefficients f_n for the subtabulation of three panels. Each additional coefficient f_n ($n=39, 40, \cdots$) allows the subtabulation of an additional panel.

It should be remarked that 53 ordinates y_n enter into the computation of each coefficient f_n. This is due to the slow rate of damping of the $\omega_n(\epsilon)$ for $\epsilon=0$. Thus for $\epsilon=.1$ (very moderate smoothing) only 35 ordinates y_n are needed, for $\epsilon=1.0$ only 23, for $\epsilon=\infty$ only 1. Concerning the important matter of dealing with the ends of a table see section 1.2 and the last paragraph of section 4.31.

TABLE I: $M_k(x, t)$, $M_k'(x, t)$, $M_k''(x, t)$ for $k=4$, $t=0.5$, $\Delta x=0.1$.
$M_4(x, 1/2)$

x	$x+0$	$x+.1$	$x+.2$	$x+.3$	$x+.4$
4	.00000004	.00000002	.00000001		
3	.00011325	.00005910	.00002991	.00001467	.00000697
2	.01616917	.01105340	.00737858	.00480621	.00305258
1	.22597004	.18940616	.15590118	.12596479	.09986387
0	.51549499	.51132566	.49901141	.47911917	.45254731
−1	.22597004	.26483185	.30499058	.34523755	.38420963
−2	.01616917	.02311310	.03230776	.04418973	.05917998
−3	.00011325	.00021062	.00038032	.00066726	.00113822
−4	.00000004	.00000010	.00000026	.00000062	.00000143

x	$x+.5$	$x+.6$	$x+.7$	$x+.8$	$x+.9$
3	.00000321	.00000143	.00000062	.00000026	.00000010
2	.00188907	.00113822	.00066726	.00038032	.00021062
1	.07764689	.05917998	.04418973	.03230776	.02311310
0	.42046084	.38420963	.34523755	.30499058	.26483185
−1	.42046084	.45254731	.47911917	.49901141	.51132566
−2	.07764689	.09986387	.12596479	.15590118	.18940616
−3	.00188907	.00305258	.00480621	.00737858	.01105340
−4	.00000321	.00000697	.00001467	.00002991	.00005910
−5				.00000001	.00000002

$M_4'(x, 1/2)$

x	$x+0$	$x+.1$	$x+.2$	$x+.3$	$x+.4$
4	−.00000039	−.00000015	−.00000006	−.00000002	−.00000001
3	−.00071955	−.00039340	−.00020833	−.00010680	−.00005298
2	−.05961795	−.04334506	−.03071644	−.02120376	−.01424920
1	−.37860391	−.35140346	−.31784825	−.28043885	−.24150489
0	.00000000	−.08306134	−.16227165	−.23406492	−.29541674
−1	.37860391	.39695855	.40419276	.39846265	.37855467
−2	.05961795	.07996844	.10465732	.13368990	.16673619
−3	.00071955	.00127546	.00219236	.00365652	.00592117
−4	.00000039	.00000096	.00000229	.00000529	.00001179

x	$x+.5$	$x+.6$	$x+.7$	$x+.8$	$x+.9$
3	−.00002542	−.00001179	−.00000529	−.00000229	−.00000096
2	−.00931577	−.00592117	−.00365652	−.00219236	−.00217546
1	−.20306520	−.16673619	−.13368990	−.10465732	−.07996844
0	−.34404758	−.37855467	−.39846265	−.40419276	−.39695855
−1	.34404758	.29541674	.23406492	.16227165	.08306134
−2	.20306520	.24150489	.28043885	.31784825	.35140346
−3	.00931577	.01424920	.02120376	.03071644	.04334506
−4	.00002542	.00005298	.00010680	.00020833	.00039340
−5		.00000001	.00000002	.00000006	.00000015

$M_4''(x, 1/2)$

x	$x+.0$	$x+.1$	$x+.2$	$x+.3$	$x+.4$
4	.00000357	.00000145	.00000056	.00000021	.00000008
3	.00423106	.00243772	.00135797	.00073109	.00038024
2	.18251117	.14368197	.10978191	.08140988	.05858190
1	.23181861	.30800376	.35890239	.38537940	.38991971
0	−.83712882	−.81763132	−.76058819	−.67020231	−.55301267
−1	.23181861	.13144694	.01013023	−.12678241	−.27209706
−2	.18251117	.22494722	.26885281	.31126005	.34845442
−3	.00423106	.00710375	.01154277	.01816077	.02768079
−4	.00000357	.00000851	.00001955	.00004332	.00009259
−5					.00000001

x	$x+.5$	$x+.6$	$x+.7$	$x+.8$	$x+.9$
4	.00000003	.00000001			
3	.00019097	.00009259	.00004332	.00001955	.00000851
2	.04089359	.02768079	.01816077	.01154277	.00710375
1	.37617315	.34845442	.31126005	.26885281	.22494722
0	−.41725773	−.27209706	−.12678241	.01013023	.13144694
−1	−.41725773	−.55301267	−.67020231	−.76058819	−.81763132
−2	.37617315	.38991971	.38537940	.35890239	.30800376
−3	.04089359	.05858190	.08140988	.10978191	.14368197
−4	.00019097	.00038024	.00073109	.00135797	.00243772
−5	.00000003	.00000008	.00000021	.00000056	.00000145

TABLE II: $\omega_n^{(k)}(t, \epsilon)$ for $k=4$, $t=0.5$, $\epsilon=0$ (0.1) 1.0.

n	$\epsilon=.0$	$\epsilon=.1$	$\epsilon=.2$	$\epsilon=.3$	$\epsilon=.4$
0	3.50637741	1.61378653	1.39953009	1.30308904	1.24631521
1	−1.84900618	−.26890929	−.14132793	−.09293505	−.06784110
2	.87238793	−.08981772	−.09332063	−.08242675	−.07223261
3	−.40443570	.07027891	.04023694	.02480160	.01624479
4	.18693997	−.02078617	−.00397484	.00050538	.00184671
5	−.08636451	.00133949	−.00223908	−.00188468	−.00132999
6	.03989615	.00169234	.00099749	.00037463	.00010684
7	−.01842978	−.00088114	−.00010447	.00005298	.00006541
8	.00851350	.00019734	−.00005372	−.00003828	−.00001761
9	−.00393275	.00001073	.00002469	.00000455	−.00000116
10	.00181670	−.00002625	−.00000273	.00000174	.00000130
11	−.00083921	.00001022	−.00000129	−.00000070	−.00000014
12	.00038767	−.00000156	.00000061	.00000003	−.00000006
13	−.00017908	−.00000044	−.00000007	.00000004	.00000002
14	.00008272	.00000036	−.00000003	−.00000001	
15	−.00003821	−.00000011	.00000002		
16	.00001765	.00000001			
17	−.00000815	.00000001			
18	.00000377				
19	−.00000174				
20	.00000080				
21	−.00000037				
22	.00000017				
23	−.00000008				
24	.00000004				
25	−.00000002				
26	.00000001				

n	$\epsilon=.5$	$\epsilon=.6$	$\epsilon=.7$	$\epsilon=.8$	$\epsilon=.9$	$\epsilon=1.0$
0	1.20834767	1.18095463	1.16016154	1.14379093	1.13054096	1.11958158
1	−.05268720	−.04264921	−.03556878	−.03034057	−.02634263	−.02319971
2	−.06387537	−.05710538	−.05156939	−.04698035	−.04312407	−.03984269
3	.01107108	.00773445	.00547641	.00389078	.00274451	.00189634
4	.00219079	.00218246	.00204868	.00187682	.00170183	.00153738
5	−.00091344	−.00062674	−.00043120	−.00029659	−.00020265	−.00013620
6	−.00000351	−.00004704	−.00006163	−.00006358	−.00006018	−.00005476
7	.00005180	.00003700	.00002545	.00001716	.00001138	.00000739
8	−.00000656	−.00000140	.00000085	.00000171	.00000193	.00000187
9	−.00000206	−.00000175	−.00000127	−.00000087	−.00000057	−.00000036
10	.00000064	.00000025	.00000006	−.00000002	−.00000005	−.00000006
11	.00000003	.00000006	.00000006	.00000004	.00000002	.00000001
12	−.00000004	−.00000002	−.00000001			

TABLE III: $L_k(x, t, \epsilon)$, $L_k''(x, t, \epsilon)$ for $k=4$, $t=0.5$, $\epsilon=0$ (0.1) 1.0, $\Delta x=0.5$.
$L_4(x, 1/2, \epsilon)$

x	$\epsilon=.0$	$\epsilon=.1$	$\epsilon=.2$	$\epsilon=.3$	$\epsilon=.4$
0.0	1.00000000	.70747935	.65457028	.62707488	.60947694
0.5	.62191163	.53757743	.51070485	.49509729	.48452022
1.0	.00000000	.20252568	.22066478	.22681461	.22949350
1.5	−.17291085	−.02061576	.01285810	.02919893	.03901344
2.0	.00000000	−.06545791	−.04840114	−.03681939	−.02872086
2.5	.07415615	−.02765903	−.03096280	−.02894823	−.02631330
3.0	.00000000	.00709183	−.00340667	−.00711221	−.00850828
3.5	−.03382251	.01344008	.00756627	.00392340	.00177090
4.0	.00000000	.00401304	.00502862	.00410188	.00314843
4.5	.01558996	−.00276042	.00041203	.00121901	.00135036
5.0	.00000000	−.00251219	−.00118870	−.00038015	.00001138
5.5	−.00719897	−.00027479	−.00076318	−.00054505	−.00033539
6.0	.00000000	.00065102	−.00007596	−.00021043	−.00019927
6.5	.00332530	.00042139	.00019012	.00003152	−.00002867
7.0	.00000000	−.00000773	.00012316	.00007297	.00003257
7.5	−.00153608	−.00015286	.00000861	.00003207	.00002580
8.0	.00000000	−.00006787	−.00002989	−.00000086	.00000704
8.5	.00070958	.00002025	−.00001865	−.00000923	−.00000251
9.0	.00000000	.00003031	−.00000162	−.00000502	−.00000321
9.5	−.00032779	.00000793	.00000477	−.00000028	−.00000120
10.0	.00000000	−.00000573	.00000301	.00000115	.00000010
10.5	.00015142	−.00000566	.00000017	.00000072	.00000036
11.0	.00000000	−.00000083	−.00000075	.00000011	.00000018
11.5	−.00006995	.00000159	−.00000045	−.00000014	.00000001
12.0	.00000000	.00000099	−.00000003	−.00000011	−.00000004
12.5	.00003231	−.00000007	.00000012	−.00000002	−.00000002
13.0	.00000000	−.00000034	.00000007	.00000001	−.00000001
13.5	−.00001492	−.00000014	.00000000	.00000001	
14.0	.00000000	.00000004	−.00000002		
14.5	.00000690	.00000007	−.00000001		
15.0	.00000000	.00000002			
15.5	−.00000318	−.00000001			
16.0	.00000000	−.00000001			
16.5	.00000147				
17.0	.00000000				
17.5	−.00000068				
18.0	.00000000				
18.5	.00000031				
19.0	.00000000				
19.5	−.00000014				
20.0	.00000000				
20.5	.00000007				
21.0	.00000000				
21.5	−.00000003				
22.0	.00000000				
22.5	.00000001				
23.0	.00000000				
23.5	−.00000001				

$L_4(x, 1/2, \epsilon)$

x	$\epsilon=.5$	$\epsilon=.6$	$\epsilon=.7$	$\epsilon=.8$	$\epsilon=.9$	$\epsilon=1.0$
0.0	.59702260	.58765637	.58031608	.57438799	.56948898	.56536580
0.5	.47675954	.47077398	.46599416	.46207717	.45880202	.45601892
1.0	.23077657	.23140004	.23168090	.23177314	.23175789	.23168050
1.5	.04557847	.05027848	.05380648	.05654957	.05874137	.06053136
2.0	−.02276409	−.01820178	−.01459584	−.01167397	−.00925829	−.00722771
2.5	−.02384227	−.02167108	−.01979233	−.01816756	−.01675611	−.01552233
3.0	−.00896161	−.00898986	−.00881784	−.00855221	−.00824659	−.00792883
3.5	.00044979	−.00038991	−.00093726	−.00129962	−.00154092	−.00170083
4.0	.00238588	.00180229	.00135734	.00101562	.00075048	.00054256
4.5	.00127570	.00114308	.00100305	.00087278	.00075725	.00065681
5.0	.00019600	.00027865	.00030985	.00031471	.00030617	.00029109
5.5	−.00019071	−.00009654	−.00003598	.00000283	.00002756	.00004311
6.0	−.00015995	−.00012137	−.00008970	−.00006510	−.00004639	−.00003223
6.5	−.00004696	−.00004885	−.00004474	−.00003886	−.00003288	−.00002744
7.0	.00000987	−.00000179	−.00000738	−.00000973	−.00001039	−.00001018
7.5	.00001687	.00001000	.00000536	.00000238	.00000050	−.00000064
8.0	.00000771	.00000641	.00000486	.00000350	.00000244	.00000165
8.5	.00000044	.00000148	.00000168	.00000156	.00000134	.00000110
9.0	−.00000156	−.00000057	−.00000004	.00000021	.00000031	.00000034
9.5	−.00000101	−.00000067	−.00000039	−.00000020	−.00000009	−.00000002
10.0	−.00000023	−.00000027	−.00000022	−.00000017	−.00000011	−.00000008
10.5	.00000011	−.00000001	−.00000005	−.00000006	−.00000005	−.00000004
11.0	.00000012	.00000006	.00000002	.00000000	−.00000001	−.00000001
11.5	.00000004	.00000004	.00000002	.00000001	.00000001	
12.0	.00000000	.00000001	.00000001	.00000001		
12.5	−.00000001					
13.0	−.00000001					

$L_4''(x, 1/2, \epsilon)$

x	$\epsilon=.0$	$\epsilon=.1$	$\epsilon=.2$	$\epsilon=.3$	$\epsilon=.4$
0.0	−3.47753764	−1.50781449	−1.27083552	−1.16381926	−1.10100908
0.5	−1.03983382	−.69689346	−.61542693	−.57326418	−.54670549
1.0	2.15613767	.54167607	.40225158	.34798919	.31925128
1.5	1.50655095	.77131824	.63355043	.56889199	.53067521
2.0	−.58680458	.31875073	.30878327	.29072626	.27601758
2.5	−.68097024	−.03482567	.02460402	.04246822	.04943501
3.0	.24590776	−.12647279	−.07651577	−.05154399	−.03726679
3.5	.31311773	−.06455381	−.05654901	−.04581150	−.03775297
4.0	−.11165611	.01678358	−.00530735	−.01047426	−.01153447
4.5	−.14452585	.03142765	.01425664	.00665986	.00297134
5.0	.05142591	.00673809	.00811324	.00596871	.00423710
5.5	.06675341	−.00655085	.00060896	.00183704	.00187944
6.0	−.02374375	−.00477107	−.00194355	−.00054832	.00001970
6.5	−.03083551	−.00059653	−.00138896	−.00087620	−.00050023
7.0	.01096729	.00133431	−.00011460	−.00030729	−.00026914
7.5	.01424417	.00097571	.00035809	.00005960	−.00003417
8.0	−.00506618	−.00005726	.00019854	.00010609	.00004375
8.5	−.00657998	−.00035828	.00001209	.00004959	.00003703
9.0	.00234028	−.00012112	−.00004884	−.00000116	.00000954
9.5	.00303957	.00004878	−.00003391	−.00001505	−.00000404
10.0	−.00108107	.00005884	−.00000242	−.00000732	−.00000433
10.5	−.00140411	.00001802	.00000898	−.00000026	−.00000164
11.0	.00049939	−.00001229	.00000485	.00000168	.00000013
11.5	.00064862	−.00001316	.00000023	.00000113	.00000053
12.0	−.00023069	−.00000110	−.00000123	.00000016	.00000025
12.5	−.00029962	.00000374	−.00000083	−.00000023	.00000001
13.0	.00010657	.00000182	−.00000005	−.00000015	−.00000005
13.5	.00013841	−.00000018	.00000022	−.00000003	−.00000004
14.0	−.00004923	−.00000067	.00000012	.00000002	−.00000001
14.5	−.00006394	−.00000033	.00000000	.00000002	.00000001
15.0	.00002274	.00000009	−.00000003	.00000001	
15.5	.00002954	.00000016	−.00000002		
16.0	−.00001050	.00000003			
16.5	−.00001364	−.00000003			
17.0	.00000485	−.00000002			
17.5	.00000630	.00000000			
18.0	−.00000224	.00000001			
18.5	−.00000291				
19.0	.00000104				
19.5	.00000134				
20.0	−.00000048				
20.5	−.00000062				
21.0	.00000022				
21.5	.00000029				
22.0	−.00000010				
22.5	−.00000013				
23.0	.00000005				
23.5	.00000006				
24.0	−.00000002				
24.5	−.00000003				
25.0	.00000001				
25.5	.00000001				
26.0	−.00000001				

$L_4''(x, 1/2, \epsilon)$

x	$\epsilon=.5$	$\epsilon=.6$	$\epsilon=.7$	$\epsilon=.8$	$\epsilon=.9$	$\epsilon=1.0$
0.0	−1.05919265	−1.02916420	−1.00647330	−.98868331	−.97433986	−.96251767
0.5	−.52821280	−.51450880	−.50390754	−.49544281	−.48851722	−.48273978
1.0	.30155959	.28962736	.28106624	.27464186	.26965348	.26567454
1.5	.50527188	.48711044	.47346009	.46281678	.45428116	.44728133
2.0	.26453429	.25546310	.24815810	.24216442	.23716440	.23293281
2.5	.05240385	.05363758	.05404202	.05402794	.05379802	.05345832
3.0	−.02823788	−.02210889	−.01772635	−.01446547	−.01196172	−.00998973
3.5	−.03182937	−.02737565	−.02393662	−.02121579	−.01901742	−.01720888
4.0	−.01135624	−.01078633	−.01011492	−.00944820	−.00882438	−.00825481
4.5	.00101034	−.00009918	−.00075259	−.00114593	−.00138383	−.00152531
5.0	.00303129	.00219629	.00160744	.00118299	.00087064	.00063651
5.5	.00166647	.00142136	.00120000	.00101295	.00085821	.00073079
6.0	.00024844	.00033324	.00035457	.00034759	.00032845	.00030479
6.5	−.00027461	−.00014145	−.00006202	−.00001410	.00001495	.00003245
7.0	−.00020347	−.00014764	−.00010563	−.00007505	−.00005291	−.00003686
7.5	−.00005788	−.00005841	−.00005187	−.00004389	−.00003637	−.00002987
8.0	.00001265	−.00000184	−.00000816	−.00001051	−.00001095	−.00001051
8.5	.00002301	.00001329	.00000717	.00000344	.00000121	−.00000014
9.0	.00000979	.00000777	.00000567	.00000399	.00000274	.00000184
9.5	.00000025	.00000163	.00000187	.00000172	.00000145	.00000117
10.0	−.00000199	−.00000071	−.00000008	.00000021	.00000032	.00000034
10.5	−.00000133	−.00000085	−.00000049	−.00000026	−.00000012	−.00000004
11.0	−.00000029	−.00000033	−.00000026	−.00000019	−.00000013	−.00000008
11.5	.00000016	.00000000	−.00000005	−.00000006	−.00000005	−.00000004
12.0	.00000015	.00000007	.00000002	.00000000	−.00000001	−.00000001
12.5	.00000005	.00000004	.00000003	.00000001	.00000001	
13.0	.00000000	.00000001	.00000001	.00000001		
13.5	−.00000002					
14.0	−.00000001					

Reprinted from
Quart. Appl. Math. **IV** (1946), pp. 45–99

Reprinted from *Quart. Appl. Math.* **IV** (1946), pp. 112–141

CONTRIBUTIONS TO THE PROBLEM OF APPROXIMATION OF EQUIDISTANT DATA BY ANALYTIC FUNCTIONS*

PART B—ON THE PROBLEM OF OSCULATORY INTERPOLATION. A SECOND CLASS OF ANALYTIC APPROXIMATION FORMULAE

BY

I. J. SCHOENBERG

University of Pennsylvania and Ballistic Research Laboratories, Aberdeen Proving Ground

Introduction. The present second part of the paper has two objectives. Firstly, we wish to carry further the important actuarial work on the subject of osculatory interpolation (Chapters I and II). Secondly, we construct even analytic functions $L(x)$, of extremely fast damping rate, such that the interpolation formula of cardinal type

$$F(x) = \sum_{\nu=-\infty}^{\infty} y_\nu L(x - \nu) \tag{1}$$

reproduces polynomials of a certain degree and reduces to a smoothing formula for integral values of the variable x (Chapter III). This second problem is found to be intimately connected with the subject of osculatory interpolation.

A preliminary remark concerning our notation is necessary. In Part A, Section 2.21 we described various characteristic properties (or "type characteristics") of a polynomial interpolation formula of the form (1), such as: (i) The degree m of the composite polynomial function $L(x)$; (ii) its class C^μ, i.e., order of contact is μ; (iii) the highest degree k of polynomials for which the formula (1) is exact; (iv) the span s of the even polynomial function $L(x)$. For convenience we propose to summarize all these statements by saying that (1) is a formula of type[1]

* Received Oct. 18, 1945. Part A of this paper appeared in this Quarterly **4**, 45–99 (1946).

[1] The connection of these types characteristics with the notation as used by Greville in his paper *The general theory of osculatory interpolation*, Trans. Actuar. Soc. Amer., **45**, pp. 202–265 (1944), especially pp. 210–211, is as follows: The first three symbols D^m, C^μ, E^k, require no further comment since they are identical respectively with the characteristics 4, 1, and 6 of Greville's classification, pp. 210–211. There remain three further characteristics to be discussed: (i) Whether the formula (1) is an "end-point" or "mid-point" formula. This point is of importance if (1) is written in terms of certral differences, since it, then reduces to either the Everett or else the Steffensen form. The following statement is obvious: *The formula* (1) *is an "end point" or "mid-point" formula depending on whether the span s is even or odd.*

(ii) Greville's adjectives "ordinary" and "modified" agree respectively with our "ordinary" and "smoothing."

(iii) The highest order d of differences involved (explicitly or implicitly). We start with the following question: *Let x be given. How many ordinates y_n enter into the computation of $F(x)$ by* (1)? Assuming that $L(x)$ is continuous, hence $=0$, at the end point $x=s/2$ of its span, we have $L(x-n)\neq 0$, as long as n in such that

$$|x-n| < s/2.$$

This inequality is found to be equivalent to

$$-\frac{s}{2} + x < n < \frac{s}{2} + x. \tag{*}$$

Let $s=2\sigma$ be even (end-point formula) and let now x be anywhere within $0 \leqq x \leqq 1$. By (*) $F(x)$ then re-

$$D^m, \quad C^\mu, \quad E^k, \quad s. \tag{2}$$

As an instance we may describe the k-point central interpolation formula of Part A, Section 2.121 as a formula of type[2]

$$D^{k-1}, \quad \begin{Bmatrix} C^0 \text{ if } k \text{ is even} \\ C^{-1} \text{ if } k \text{ is odd} \end{Bmatrix}, \quad E^{k-1}, \quad s = k.$$

In Chapter II we construct a class of ordinary interpolation formulae and two classes of smoothing interpolation formulae. These classes by no means describe all possible osculatory interpolation formulae. Furthermore, a number of interesting problems concerning remainder terms and orders of approximations await solution. No attempt has been made to see which of the numerous formulae tabulated by Greville, loc. cit., are contained in the three classes of formulae developed in Chapter II. The essential progress made in this direction may perhaps be briefly described as follows. The construction of an interpolation formula usually requires the solution of a more or less complicated system of linear equations, unless, as in Lagrange's formula, the basic interpolating functions are obvious from the start. These systems of equations are especially troublesome if one wishes to construct an osculatory interpolation formula of any *general* class. As Greville correctly points out, loc. cit., pp. 255–256, the mere agreement between the number of unknowns with the number of equations which they should satisfy, will, by itself, never prove the existence of a solution. Basically, our parametric representation of spline curves of order k (Part A, Section 3.15, Theorem 5) circumvents this difficulty.

An example which illustrates the operation of this representation is as follows. Let $F(x)$ be defined as equal to 0 for $x \leqq 0$, as well as for $x \geqq 4$. We propose to complete the definition of $F(x)$ in the range $0 \leqq x \leqq 4$ by four cubic arcs joining at $x=1, 2, 3$, in such a way that $F(x)$ be of class C'' for all real x. Of course, we are not interested in the obvious but trivial solution $F(x) \equiv 0$. Let us now count the available parameters and the number of conditions. The 4 cubic arcs furnish $4 \cdot 4 = 16$ parameters. The second order contact requirements at $x=0, 1, 2, 3, 4$ lead to a system of $3 \cdot 5 = 15$ homogeneous equations. The solution of a homogeneous system of 15 equations in 16 unknowns depends on anything from 1 to 16 arbitrary parameters, depending on the rank of the system. Our Theorem 5, for $k=4$, furnishes immediately the one-parameter solution

$$F(x) = c \cdot M_4(x - 2) \tag{3}$$

the graph of which is given in Part A, Section 3.13. Again Theorem 5 will easily show that this is the most general solution of the problem. We see how this complicated system of 15 equations in 16 unknowns is explicitly solved by (3). As a variation of the problem, let us now define $F(x)$ to be equal to 0 for $x \leqq 0$, as well as for $x \geqq 3$, and let us propose now to bridge this gap by 3 cubic arcs giving a $F(x)$ of class C''. Now we find that the problem amounts to a system of 12 homogeneous equations in 12 unknowns. This tells us precisely nothing. Again by Theorem 5 we can readily show that

quires all y_n such that $-\sigma < n < \sigma + 1$ that is $s = 2\sigma$ consecutive ordinates. Let $s = 2\sigma + 1$ be odd (mid-point formula) and let x be anywhere within $-\frac{1}{2} \leqq x \leqq \frac{1}{2}$. Again by (*) $F(x)$ now requires all y_n such that $-\sigma - 1 < n < \sigma + 1$, hence again $s = 2\sigma + 1$ consecutive ordinates. We have therefore proved the following: *The highest order d of differences involved is always related with the span s by the relation $s = d + 1$.*

[2] The symbol C^{-1} is to indicate the class of piecewise continuous functions.

the trivial solution $F(x)\equiv 0$ is the only solution. These considerations generalize and allow to characterize our basic functions

$$M_k(x) = \frac{1}{(k-1)!}\,\delta^k x_+^{k-1},$$

up to a multiplicative constant and a shift along the x-axis, as follows:[3] *Let $F(x)$ be $=0$ for $x\leqq 0$, as well as for $x\geqq n$, where n is a positive integer. We wish to complete the definition of $F(x)$ by a succession of n arcs, of degree $k-1$, joining at $x=1, 2, \cdots, n-1$ such as to furnish a $F(x)$ of C^{k-2}. Then $n=k$ is the smallest value of n for which this can be done in a non-trivial way and for this minimal value $n=k$ the gap is bridged by*

$$F(x) = c\cdot M_k(x-k/2)$$

and in no other way.

The reader who is mainly interested in the numerical applications may pass directly from here to the Appendix where the use of the tables is fully explained and one example is worked out.

I. THE COSINE POLYNOMIALS $\phi_k(u)$ AND CERTAIN RELATED SETS OF POLYNOMIALS

In the present chapter we propose to study further properties of the cosine polynomials

$$\phi_k(u) = \sum_{n=-\infty}^{\infty} M_k(n)\cos nu \tag{1}$$

which were mentioned in Part A, sections 3.14 and 4.1 (for $t=0$). By Part A, section 4.1, formula (6) (for $t=0$) we may also write

$$\phi_k(u) = \sum_{\nu=-\infty}^{\infty} \psi_k(u+2\pi\nu)$$

and therefore

$$\phi_k(u) = (2\sin u/2)^k\cdot \sum_{\nu=-\infty}^{\infty} \frac{(-1)^{\nu k}}{(u+2\pi\nu)^k}. \tag{2}$$

1.1. **Expression of $\phi_k(u)$ in terms of rational polynomials.** We introduce two new sets of periodic functions defined by

$$\rho_k(u) = (2\sin u/2)^k\cdot \sum_{\nu=-\infty}^{\infty} \frac{1}{(u+2\pi\nu)^k}, \tag{3}$$

$$\sigma_k(u) = (2\sin u/2)^k\cdot \sum_{\nu=-\infty}^{\infty} \frac{(-1)^{\nu}}{(u+2\pi\nu)^k}. \tag{3'}$$

A comparison with (2) shows that

$$\phi_k(u) = \begin{cases}\rho_k(u) & \text{if } k \text{ is even,}\\ \sigma_k(u) & \text{if } k \text{ is odd.}\end{cases} \tag{4}$$

[3] Problems of this kind concerning polygonal lines of a certain degree and class are of importance for the theory of formulae of mechanical quadratures. The author expects to discuss this connection elsewhere.

By differentiation of (3) and also (3′) we readily find the recurrence relations

$$\rho_{k+1}(u) = \cos\frac{u}{2}\,\rho_k(u) - \frac{2}{k}\sin\frac{u}{2}\,\rho_k'(u), \tag{5}$$

$$\sigma_{k+1}(u) = \cos\frac{u}{2}\,\sigma_k(u) - \frac{2}{k}\sin\frac{u}{2}\,\sigma_k'(u). \tag{5'}$$

These may be used in a progressive computation of ρ_k and σ_k if we start with

$$\rho_2(u) = 1, \qquad \sigma_1(u) = 1. \tag{6}$$

We prefer, however, to express $\rho_k(u)$ and $\sigma_k(u)$ as polynomials in the variable

$$x = \cos(u/2) \tag{7}$$

by means of

$$\rho_k(u) = U_{k-2}(\cos u/2), \qquad \sigma_k(u) = V_{k-1}(\cos u/2). \tag{8}$$

Substituting into (5), and (5′) respectively we find that the two sequences of polynomials $U_n(x)$, $V_n(x)$, both of exact degree n, satisfy the recurrence relations

$$U_{k+1}(x) = xU_k(x) + \frac{1}{k+2}(1-x^2)U_k'(x), \tag{9}$$

$$V_{k+1}(x) = xV_k(x) + \frac{1}{k+1}(1-x^2)V_k'(x), \tag{9'}$$

with initial values which by (6) and (8) are

$$U_0(x) = 1, \qquad V_0(x) = 1. \tag{10}$$

A simple calculation now gives

$$\left.\begin{aligned}
U_1(x) &= x, & V_1(x) &= x,\\
U_2(x) &= \frac{1}{3}(1+2x^2), & V_2(x) &= \frac{1}{2}(1+x^2),\\
U_3(x) &= \frac{1}{3}(2x+x^3), & V_3(x) &= \frac{1}{6}(5x+x^3)\\
U_4(x) &= \frac{1}{15}(2+11x^2+2x^4), & V_4(x) &= \frac{1}{24}(5+18x^2+x^4)\\
U_5(x) &= \frac{1}{45}(17x+26x^3+2x^5), & V_5(x) &= \frac{1}{120}(61x+58x^3+x^5).
\end{aligned}\right\} \tag{11}$$

We record as a lemma the following properties which are readily established by induction.

LEMMA 1. *$U_k(x)$ and $V_k(x)$ are polynomials of exact degree k which are even or odd according as k is even or odd. The coefficients of their highest terms are positive. Also*

$$U_k(1) = V_k(1) = 1, \qquad U_k(-1) = V_k(-1) = (-1)^k. \tag{12}$$

In view of (4) and (8) we find the following expression of $\phi_k(u)$ in terms of our new polynomials:

$$\phi_k(u) = \begin{cases} U_{k-2}(x) & \text{if } k \text{ is even,} \\ V_{k-1}(x) & \text{if } k \text{ is odd.} \end{cases} \tag{13}$$

This shows that the even polynomials $U_{2\nu}$, $V_{2\nu}$ are of special interest.

1.2. **The zeros of the polynomials** U_k **and** V_k. We propose to prove the following proposition:

LEMMA 2. *The zeros of the even polynomials* $U_{2\nu}(x)$ *and* $V_{2\nu}(x)$ *are all simple and purely imaginary.*

We carry through the proof for $U_{2\nu}(x)$ only since the proof for $V_{2\nu}(x)$ is entirely similar. In order to deal with real zeros, we define a new sequence of polynomials $u_k(x)$ by

$$u_k(x) = i^{-k}U_k(xi), \qquad (k = 0, 1, \cdots). \tag{14}$$

These new polynomials are also real and satisfy a recurrence relation which in view of (9) is readily found to be

$$u_{k+1}(x) = xu_k(x) - \frac{1}{k+2}(1 + x^2)u_k'(x). \tag{15}$$

From (11) we find

$$u_0(x) = 1, \qquad u_1(x) = x, \qquad u_2(x) = \tfrac{1}{3}(2x^2 - 1), \qquad u_3(x) = \tfrac{1}{3}(x^3 - 2x), \cdots .$$

In view of (14) it obviously suffices to show that the zeros of $u_k(x)$ ar real and simple, while those of $u_{2\nu}(x)$ are also different from zero. This is readily done by induction as follows. Let $k = 2\nu$ be even and let us assume that the k zeros of $u_k(x)$ are

$$-\xi_\nu, -\xi_{\nu-1}, \cdots, -\xi_1, \xi_1, \xi_2, \cdots, \xi_\nu \qquad (0 < \xi_1 < \cdots < \xi_\nu), \tag{16}$$

and therefore simple. This, and the fact that $u_k(x)$ has a highest term of positive coefficient (Lemma 1 and (14)), imply that

$$u_k'(\xi_\nu) > 0,$$

and that the sequence of values of $u_k'(x)$, at the k roots (16), alternate in sign. By (15) we therefore find

$$u_{k+1}(\xi_\nu) < 0$$

and that the values of $u_{k+1}(x)$,at the k roots (16), alternate in sign. Since $u_{k+1}(0) = 0$, we conclude that $u_{k+1}(x)$ has ν positive and ν negative zeros which must therefore be simple.

Let now $k = 2\nu + 1$ be odd and let u_k have the simple zeros

$$-\xi_\nu, \cdots, -\xi_1, 0, \xi_1, \cdots, \xi_\nu \qquad (0 < \xi_1 < \cdots < \xi_\nu). \tag{17}$$

Now we conclude as before that $u_{k+1}(\xi_\nu) < 0$ and that the values of $u_{k+1}(x)$, at the k roots (17), alternate in sign. Again the conclusion is that $u_{k+1}(x)$ has simple real roots none of which vanishes. This proves the theorem by complete induction.

1.3. **A few corollaries.** In this last section of the present chapter we prove several auxiliary propositions which will be used in the next chapter in the derivation of interpolation formulae of various kinds. These propositions represent the solutions of the algebraic problems arising by the Fourier integral transformation of the problems of the construction of those interpolation formulae.

LEMMA 3. *Let $k=2\nu$ be even. We can determine uniquely an even polynomial $P_k(x)$, of degree k, and an odd polynomial $P_{k-1}(x)$, of degree $k-1$, satisfying the identity*

$$U_k(x)P_k(x) + U_{k+1}(x)P_{k-1}(x) \equiv 1. \tag{18}$$

Likewise polynomials $Q_k(x)$ and $Q_{k-1}(x)$, even and odd respectively, exist uniquely such as to satisfy

$$V_k(x)Q_k(x) + V_{k+1}(x)Q_{k-1}(x) \equiv 1. \tag{19}$$

We wish to show first that U_k and U_{k+1} have no common zeros. Indeed a common zero x of U_k and U_{k+1} would, by (9), be a zero of

$$(1 - x^2)U_k'(x).$$

Since by (12) $x \neq \pm 1$, x must be a zero of $U_k'(x)$. But this contradicts our Lemma 2 to the effect that $U_k(x)$ has only simple zeros. The polynomials $U_k(x)$, $U_{k+1}(x)$ having no common divisors, the identity (18) is assured by the elementary theory of the greatest common divisor of two polynomials. We now show that $P_k(x)$ is even and $P_{k-1}(x)$ is odd as follows. Replacing x by $-x$ in (18) we find

$$U_k(x)P_k(-x) - U_{k+1}(x)P_{k-1}(-x) \equiv 1.$$

Since our polynomials P_k, P_{k-1} are uniquely defined by (18) we find

$$P_k(x) = P_k(-x), \qquad P_{k-1}(x) = -P_{k-1}(-x),$$

which prove our statement. An identical reasoning proves the existence of the polynomials Q_k and Q_{k+1} satisfying (19).

The polynomials P_k and P_{k+1} are easily determined for low values of k by the method of indeterminate coefficients. Thus for $k=2$ by (11),

$$U_2(x) = (1 + 2x^2)/3, \qquad U_3(x) = (2x + x^3)/3,$$

from which we find

$$P_2(x) = 2x^2 + 3, \qquad P_1(x) = -4x,$$

satisfying the identity

$$U_2(x)P_2(x) + U_3(x)P_1(x) \equiv 1. \tag{20}$$

Likewise for $k=4$ we have by (11)

$$V_4(x) = (5 + 18x^2 + x^4)/24, \qquad V_5(x) = (61x + 58x^3 + x^5)/120.$$

The corresponding polynomials Q_4, Q_3 are found to be

$$Q_4(x) = (3648 + 4789x^2 + 83x^4)/760, \qquad Q_3(x) = -(1469x + 83x^3)/152.$$

They satisfy the identity

$$V_4(x)Q_4(x) + V_5(x)Q_3(x) \equiv 1. \tag{21}$$

The identities (18), (19) will later be used in the following form. Again for an even k, but replacing k by $k-2$, we get by (18) and (8)

$$\rho_k(u)P_{k-2}(x) + \rho_{k+1}(u)P_{k-3}(x) \equiv 1, \qquad (k \text{ even}, x = \cos u/2).$$

Likewise for an odd k, but replacing k by $k-1$, we obtain from (19) and (8) the identity

$$\sigma_k(u)Q_{k-1}(u) + \sigma_{k+1}(u)Q_{k-2}(k) \equiv 1, \qquad (k \text{ odd}, x = \cos u/2).$$

The even polynomials P_{k-2}, $x^{-1}P_{k-3}$, and Q_{k-1}, $x^{-1}Q_{k-2}$, may now be expressed in powers of

$$1 - x^2 = (\sin u/2)^2.$$

We have therefore proved the following:

LEMMA 4. *We can find constants a_ν, a_ν', b_ν, b_ν' such as to satisfy the following two identities:*

For an even k

$$\begin{aligned}\rho_k(u)\{a_0 - a_2(2 \sin u/2)^2 + a_4(2 \sin u/2)^4 - \cdots \pm a_{k-2}(2 \sin u/2)^{k-2}\} \\ + \rho_{k+1}(u)\{a_0' - a_2'(2 \sin u/2)^2 + a_4'(2 \sin u/2)^4 - \cdots \\ \mp a_{k-4}'(2 \sin u/2)^{k-4}\}(2 \cos u/2) \equiv 1,\end{aligned} \tag{22}$$

and for k odd

$$\begin{aligned}\sigma_k(u)\{b_0 - b_2(2 \sin u/2)^2 + b_4(2 \sin u/2)^4 - \cdots \pm b_{k-1}(2 \sin u/2)^{k-1}\} \\ + \sigma_{k+1}(u)\{b_0' - b_2'(2 \sin u/2)^2 + b_4'(2 \sin u/2)^4 - \cdots \\ \mp b_{k-3}'(2 \sin u/2)^{k-3}\}(2 \cos u/2) \equiv 1.\end{aligned} \tag{22'}$$

As examples we mention that the identities (20) and (21) become on passing to the variable u

$$\rho_4(u)\{5 - \tfrac{1}{2}(2 \sin u/2)^2\} + \rho_5(u)\{-2\}(2 \cos u/2) \equiv 1 \tag{23}$$

and

$$\begin{aligned}\sigma_5(u)\left\{\frac{213}{19} - \frac{991}{760}(2 \sin u/2)^2 + \frac{83}{12160}(2 \sin u/2)^4\right\} \\ + \sigma_6(u)\left\{-\frac{194}{38} + \frac{83}{1216}\left(2 \sin \frac{u}{2}\right)^2\right\}\left(2 \cos \frac{u}{2}\right) \equiv 1.\end{aligned} \tag{24}$$

The last proposition which we wish to derive here concerns the expansion of $1/\phi_k(u)$ in ascending powers of the variable

$$s = \sin^2 u/2 = 1 - \cos^2 u/2 = 1 - x^2. \tag{25}$$

Let us assume for the moment that k is even. Then by (13)

$$\phi_k(u) = U_{k-2}(x), \qquad (k \text{ even}). \tag{26}$$

Now $U_{k-2}(x)$ is an even polynomial which, by Lemma 2, has purely imaginary zeros. Being an even polynomial, $U_{k-2}(x)$ may be expressed as a polynomial $U^*(s)$ in the variable

$$s = 1 - x^2 \tag{25'}$$

of degree $\kappa=(k-2)/2$. This change of variable transforms the purely imaginary zeros of $U_{k-2}(x)$ into the zeros

$$\alpha_1, \alpha_2, \cdots, \alpha_\kappa \qquad (\kappa = (k-2)/2)$$

of $U^*(s)$ which, by (25′), must all be positive and greater than 1. Finally, since $U_{k-2}(1)=U^*(0)=1$, we have the identity

$$\phi_k(u) = U_{k-2}(x) = \left(1-\frac{s}{\alpha_1}\right)\left(1-\frac{s}{\alpha_2}\right)\cdots\left(1-\frac{s}{\alpha_\kappa}\right). \tag{27}$$

An entirely similar identity is derived for an *odd* k by repeating our arguments for

$$\phi_k(u) = V_{k-1}(x),$$

instead of (26).

This establishes the following

LEMMA 5. *The reciprocal of the cosine polynomial $\phi_k(u)$ admits of an expansion*

$$\frac{1}{\phi_k(u)} = \sum_{n=0}^{\infty} c_{2n}^{(k)}(2\sin u/2)^{2n} \tag{28}$$

which converges for all real values of u and where the coefficients are positive rational numbers

$$c_{2n}^{(k)} > 0, \qquad (n = 0, 1, 2, \cdots). \tag{29}$$

Indeed, in view of (27), the expansion (28) may be obtained as

$$\frac{1}{\phi_k(u)} = \prod_{\nu=1}^{\kappa}\left(1+\frac{s}{\alpha_\nu}+\frac{s^2}{\alpha_\nu^2}+\cdots\right) = \sum_{n=0}^{\infty} c_{2n}^{(k)}(4s)^n$$

which reduces to (28), in view of (25). In conclusion we notice the following consequences of the identity (28). Since

$$\phi_k(u) = \sum_{\nu=-\infty}^{\infty} \psi_k(u+2\pi\nu),$$

where

$$\psi_k(u) = \left(\frac{2\sin u/2}{u}\right)^k,$$

we have in the neighborhood of the origin $u=0$

$$\phi_k(u) = \left(\frac{2\sin u/2}{u}\right)^k + u^k\cdot(\text{regular function of } u). \tag{30}$$

On multiplying (28) by $\phi_k(u)$ we therefore have

$$\left(\frac{2\sin u/2}{u}\right)^k\cdot\sum_{n=0}^{\infty} c_{2n}^{(k)}(2\sin u/2)^{2n} = 1 + u^k\cdot(\text{regular function})$$

and also

$$\left(\frac{2\sin u/2}{u}\right)^k\cdot\sum_{0\leqq 2n<k} c_{2n}^{(k)}(2\sin u/2)^{2n} = 1 + u^k\cdot(\text{regular function}). \tag{31}$$

It is of special interest to point out that if

$$g_{k,m}(u) \equiv \left(\frac{2 \sin u/2}{u}\right)^k \cdot \sum_{n=0}^{m-1} c_{2n}^{(k)} (2 \sin u/2)^{2n}$$

then

$$g_{k,m}(u) = \begin{cases} 1 + u^{2m} \cdot (\text{regular function}) & \text{if} \quad 2m < k \\ 1 + u^k \cdot (\text{regular function}) & \text{if} \quad 2m - 2 < k \leq 2m. \end{cases} \tag{32}$$

As an illustration we find for $k=6$ by (13), and (11), and (25)

$$\frac{1}{\phi_6(u)} = \frac{1}{U_4(x)} = \frac{15}{2 + 11x^2 + 2x^4} = \frac{30}{30 - 30s + 4s^2} = \frac{1}{1 - s + \frac{2}{15}s^2}$$

whence

$$\frac{1}{\phi_6(u)} = 1 + s + \frac{13}{15}s^2 + \cdots$$

or

$$\frac{1}{\phi_6(u)} = 1 + \frac{1}{4}(2 \sin u/2)^2 + \frac{13}{240}(2 \sin u/2)^4 + \cdots . \tag{33}$$

The relations (32) now become (for $k=6$, $m=1, 2, 3$)

$$\left(\frac{2 \sin u/2}{u}\right)^6 = 1 + u^2 \cdot (\text{regular function}),$$

$$\left(\frac{2 \sin u/2}{u}\right)^6 \left\{1 + \frac{1}{4}(2 \sin u/2)^2\right\} = 1 + u^4 \cdot (\text{regular function}),$$

$$\left(\frac{2 \sin u/2}{u}\right)^6 \left\{1 + \frac{1}{4}(2 \sin u/2)^2 + \frac{13}{240}(2 \sin u/2)^4\right\} = 1 + u^6 \cdot (\text{regular function}).$$

In the next chapter we shall need the numerical values of the coefficients

$$c_{2n}^{(k)} \quad \text{for} \quad 2n < k. \tag{34}$$

It is of interest then to point out that these coefficients (34) may also be otherwise computed as coefficients of a simple generating function. Indeed, from (30) it is clear that the coefficients (34) will not change if on the left-hand side of (28) we replace $\phi_k(u)$ by the first term on the right-hand side of (30). That means, if

$$\left(\frac{u}{2 \sin u/2}\right)^k = \sum_{n=0}^{\infty} d_{2n}^{(k)} (2 \sin u/2)^{2n} \tag{35}$$

then

$$c_{2n}^{(k)} = d_{2n}^{(k)} \qquad (2n < k). \tag{36}$$

However, the coefficients of (35) are readily determined. Indeed, if we set

$$v = 2 \sin u/2 \quad \text{or} \quad u = 2 \arcsin v/2, \tag{37}$$

then (35) becomes

$$\left(\frac{2 \arcsin v/2}{v}\right)^k = \sum_{n=0}^{\infty} d_{2n}^{(k)} v^{2n}. \tag{38}$$

Since

$$\frac{2 \arcsin v/2}{v} = 1 + \frac{1}{2}\cdot\frac{1}{3}\cdot\frac{v^2}{4} + \frac{1\cdot 3}{2\cdot 4}\cdot\frac{1}{5}\cdot\frac{v^4}{16} + \frac{1\cdot 3\cdot 5}{2\cdot 4\cdot 6}\cdot\frac{1}{7}\cdot\frac{v^6}{64} + \cdots, \tag{39}$$

we find the expansion (38) by raising (39) to the k-th power. Thus

$$d_0^{(k)} = 1, \qquad d_2^{(k)} = \frac{k}{24}, \qquad d_4^{(k)} = \frac{5k^2 + 22k}{5760}, \cdots. \tag{40}$$

Since all coefficients of (39) are positive, it is clear that the coefficients of (38) are likewise positive. This however does not imply the positivity of the coefficients of (28) beyond the kth term. For $k=6$ the values (40) agree with the coefficients of (33).

II. POLYNOMIAL INTERPOLATION FORMULAE

In this chapter we wish to apply our general Theorem 2 of Part A, section 2.23 and our Lemmas 4 and 5 of the last section in deriving three distinct classes of polynomial interpolation formulae for each value of the positive integer k. The formulae of the first class (Theorem 1 below) are of the ordinary kind (see Part A, section 2.21 a and b), and of the type

$$D^k, \quad C^{k-2}, \quad E^{k-1}, \quad s = \begin{cases} 2k-2 & \text{if } k \text{ is even} \\ 2k-1 & \text{if } k \text{ is odd.} \end{cases}$$

The existence of ordinary interpolation formulae of degree k and class $k-2$ was previously conjectured by Mr. Greville who verified their existence up to and including $k=6$. (See Greville, loc. cit., pp. 212–213.) The formulae of the second class are smoothing interpolation formulae (Theorem 2 below). For a given integral k and each integral m, such that $0 \leqq 2m-2 < k$, a formula is derived which is of the type

$$D^{k-1}, \qquad C^{k-2}, \qquad E^{\min(2m-1,k-1)}, \qquad s = k + 2m - 2.$$

These formulae are derived from an *ordinary* interpolation formula of type

$$D^{k-1}, \qquad C^{k-2}, \qquad E^{k-1}, \qquad s = \infty,$$

discussed in Part A.

The formulae of the third and last class are again *smoothing* interpolation formulae (Theorem 3 below). While in the second class the degree D^{k-1} and the "order of contact" C^{k-2} were fixed, while the degree of exactness E^{2m-1} and the span $s=k+2m-2$ increased apace, in the present class the span $s=k$ is constant. More precisely, a formula is derived for each m such that $0 \leqq 2m \leqq k-1$ which is of type

$$D^{k-1}, \qquad C^{k-2m}, \qquad E^{2m-1}, \qquad s = k.$$

These formulae are derived from the formula of ordinary k-point central interpolation in a manner somewhat reminiscent of Mr. Jenkins' original procedure.

2.1. Ordinary polynomial interpolation formulae of the Jenkins-Greville type. We are returning to our basic functions

$$M_k(x) = \frac{1}{2\pi}\int_{-\infty}^{\infty}\left(\frac{2 \sin u/2}{u}\right)^k e^{iux}du \tag{1}$$

and wish to show that the osculatory interpolation formulae of the type investigated by Jenkins and Greville may be readily derived in terms of these functions. We shall use the operational symbol σ to mean

$$\sigma f(x) = f(x + \tfrac{1}{2}) + f(x - \tfrac{1}{2}).$$

THEOREM 1. *We define the basic polynomial function $L(x)$ by the following two formulae according to the parity of k:*

$$L(x) = a_0 M_k(x) + a_2\delta^2 M_k(x) + \cdots + a_{k-2}\delta^{k-2}M_k(x) + a_0'\sigma M_{k+1}(x) + a_2'\sigma\delta^2 M_{k+1}(x) + \cdots + a_{k-4}'\sigma\delta^{k-4}M_{k+1}(x) \quad (k \text{ even}) \tag{2}$$

$$L(x) = b_0 M_k(x) + b_2\delta^2 M_k(x) + \cdots + b_{k-1}\delta^{k-1}M_k(x) + b_0'\sigma M_{k+1}(x) + b_2'\sigma\delta^2 M_{k+1}(x) + \cdots + b_{k-3}'\sigma\delta^{k-3}M_{k+1}(x), \quad (k \text{ odd}), \tag{3}$$

where the numerical constants a_ν, a_ν', b_ν, b_ν' are those defined in Lemma 4. Then

$$F(x) = \sum_{n=-\infty}^{\infty} y_n L(x - n) \tag{4}$$

is an ordinary polynomial interpolation formula of type

$$D^k, \quad C^{k-2}, \quad E^{k-1}, \quad s = \begin{cases} 2k - 2 & \text{if } k \text{ is even,} \\ 2k - 1 & \text{if } k \text{ is odd.} \end{cases} \tag{5}$$

Indeed, we notice that

$$\begin{aligned} \delta e^{iux} &= (e^{iu/2} - e^{-iu/2})e^{iux} = 2i \sin u/2 e^{iux}, \\ \sigma e^{iux} &= (e^{iu/2} + e^{-iu/2})e^{iux} = 2 \cos u/2 e^{iux}. \end{aligned} \tag{6}$$

Let k now be even, hence $L(x)$ defined by (2), and let

$$L(x) = \frac{1}{2\pi}\int_{-\infty}^{\infty} g(u)e^{iux}du.$$

We evidently obtain this integral representation by performing the operation

$$a_0 + a_2\delta^2 + \cdots + a_{k-2}\delta^{k-2}$$

on the relation (1) and add to it the result of performing the operation

$$a_0'\sigma + a_2'\sigma\delta^2 + \cdots + a_{k-4}'\sigma\delta^{k-4}$$

on (1), with k replaced by $k+1$. In view of (6) we have

$$\begin{aligned} \delta^\nu e^{iux} &= (2i \sin u/2)^\nu e^{iux} \\ \sigma\delta^\nu e^{iux} &= (2i \sin u/2)^\nu (2 \cos u/2)e^{iux}, \end{aligned}$$

and therefore

$$g(u) = \left(\frac{2 \sin u/2}{u}\right)^k \{a_0 - a_2(2 \sin u/2)^2 + \cdots \pm a_{k-2}(2 \sin u/2)^{k-2}\} + \left(\frac{2 \sin u/2}{u}\right)^{k+1} \{a_0' - a_2'(2 \sin u/2)^2 + \cdots \mp a_{k-4}'(2 \sin u/2)^{k-4}\}(2 \cos u/2). \tag{7}$$

We now turn to Theorem 2 of Part A, section 2.23, which states that (4) is an *ordinary* interpolation formula if and only if the following identity holds:

$$\sum_{\nu=-\infty}^{\infty} g(u + 2\pi\nu) = 1. \tag{8}$$

It should be noticed now that both expressions in (7) contained within braces are periodic functions of period 2π. Since k is even we find that $\sum g(u+2\pi\nu)$ is identical with the left-hand side of our relation I (22). This proves (8) for even k. A precisely similar reasoning for odd k will show that $\sum g(u+2\pi\nu)$ is identical with the left-hand side of I (22′).

There remains the problem of showing that (4) is of the type as stated in the Theorem. Since $L(x)$ is by (2), (3), a linear combination of functions of the form

$$M_k(x + n), \qquad M_{k+1}(x + \tfrac{1}{2} + n),$$

it is clear that $L(x)$ is a polynomial line of degree k, of class C^{k-2}, with discontinuities at $x=n$, or $x=n+1/2$, according to whether k is even or odd. Finally (4) is exact for the degree $k-1$, again by Theorem 2 of Part A, section 2.23. There remains the discussion of the span s of $L(x)$. Now the span of $M_k(x)$ is $=k$ (see Part A, section 3.13) and therefore the span of $\delta^\nu M_k(x)$ is equal to $k+\nu$, while the span of $\sigma\delta^\nu M_k(x)$ is equal to $k+\nu+1$. Now it is immediately verified that the two terms of (2) involving δ^{k-2} and $\sigma\delta^{k-4}$ are both of span $2k-2$, while the similar two terms of (3) are both of span $2k-1$. This completes a proof of the Theorem.

As illustrations we mention that the identities I(23) and I(24) corresponding to the cases $k=4$ and $k=5$ give rise to the basic functions

$$L(x) = 5M_4(x) + \tfrac{1}{2}\delta^2 M_4(x) - 2\sigma M_5(x) \tag{9}$$

and

$$L(x) = \frac{213}{19} M_5(x) + \frac{991}{760}\delta^2 M_5(x) + \frac{83}{12160}\delta^4 M_5(x) - \frac{194}{38}\sigma M_6(x) - \frac{83}{1216}\sigma\delta^2 M_6(x). \tag{10}$$

These two basic functions give rise to ordinary interpolation formulae (4) which are of the types

$$D^4,\ C^2,\ E^3,\ s = 6 \quad \text{and} \quad D^5,\ C^3,\ E^4,\ s = 9, \tag{11}$$

respectively.

Incidentally, the characteristic function of (9) is, by I(23),

$$g(u) = \left(\frac{2 \sin u/2}{u}\right)^4 \left(5 - 2 \sin^2 \frac{u}{2}\right) - 4\left(\frac{2 \sin u/2}{u}\right)^5 \cos u/2$$

or

$$g(u) = \left(\frac{2 \sin u/2}{u}\right)^4 \left(4 + \cos u - 4\frac{\sin u}{u}\right).$$

This agrees with our formula (11″) of Part A, section 2.122, as in fact the basic function (9) is identical with Jenkins' function there described by formula (11).

In concluding we wish to mention the numerical results for $k=6$. In this case we need the identity

$$U_4(x)P_4(x) + U_5(x)P_3(x) \equiv 1. \tag{12}$$

By (11) we have

$$U_4(x) = \frac{1}{15}(2 + 11x^2 + 2x^4), \qquad U_5(x) = \frac{1}{45}(17x + 26x^3 + 2x^5).$$

By indeterminate coefficients we find

$$P_4(x) = \frac{15}{2} + \frac{115}{7}x^2 + \frac{27}{21}x^4, \qquad P_3(x) = -\frac{285}{14}x - \frac{27}{7}x^3,$$

from which, on passing to the variable I(25), the identity (12) becomes

$$\rho_6(u)\left\{\frac{353}{14} - \frac{133}{28}(2 \sin u/2)^2 + \frac{9}{448}(2 \sin u/2)^4\right\}$$
$$+ \rho_7(u)\left\{-\frac{339}{28} + \frac{27}{56}(2 \sin u/2)^2\right\}(2 \cos u/2) \equiv 1.$$

The basic function corresponding to $k=6$ is therefore

$$L(x) = \frac{353}{14}M_6(x) + \frac{133}{28}\delta^2 M_6(x) + \frac{9}{448}\delta^4 M_6(x)$$
$$- \frac{339}{28}\sigma M_7(x) - \frac{27}{56}\sigma\delta^2 M_7(x), \tag{13}$$

giving rise to an ordinary interpolation formula of type

$$D^6, \ C^4, \ E^5 \ \text{ and } \ s = 10.$$

2.2. **A first class of smoothing interpolation formulae derived from an ordinary interpolation formula of type D^{k-1}, C^{k-2}.** We start by recalling an ordinary polynomial interpolation formula derived in Part A, section 4.2. Indeed, the formula (9) of that section furnishes, for $t=0$, the following polynomial basic function

$$L_k(x) = \frac{1}{2\pi}\int_{-\infty}^{\infty} \frac{\psi_k(u)}{\phi_k(u)} e^{iux} du. \tag{14}$$

The corresponding formula

$$F(x) = \sum_{n=-\infty}^{\infty} y_n L_k(x - n) \tag{15}$$

is, as we know, an ordinary polynomial interpolation formula of type

$$D^{k-1}, \qquad C^{k-2}, \qquad E^{k-1}, \qquad s = \begin{cases} \infty & \text{if } k \geqq 3, \\ k & \text{if } k = 1, 2. \end{cases} \tag{16}$$

We turn now to the expansion I(28) of Lemma 5, substituting I(28) into the integral (14) and integrating term-wise we obtain the expansion

$$L_k(x) = M_k(x) - c_2^{(k)}\delta^2 M_k(x) + c_4^{(k)}\delta^4 M_k(x) - \cdots, \qquad (k \geqq 3). \tag{17}$$

On comparing the present interpolation formula (15) with the formula (4) of the Jenkins-Greville type, we notice, by (5) and (16), that they are both of class C^{k-2} and that they are both exact for the degree $k-1$. The degree of (15) is lower by 1 than the degree of (4). This reduction of the degree to the lowest possible value $k-1$, for a formula of class C^{k-2}, was achieved at the price of having an infinite span. The infinite span of (15) clearly disqualifies this interpolation formula as far as numerical purposes are concerned.

We now turn to the *partial sums* of the series (17). They will yield smoothing interpolation formulae of considerable practical importance. Indeed, let

$$L_{k,m}(x) = M_k(x) - c_2^{(k)}\delta^2 M_k(x) + \cdots + (-1)^{m-1} c_{2m-2}^{(k)}\delta^{2m-2} M_k(x), \quad (2m-2 < k). \tag{18}$$

The characteristic function of this basic function is identical with the left-hand side of the identity I(32). In view of our Theorem 2 of Part A, this identity I(32) proves that (18) is the basic function of a smoothing interpolation formula which is exact for the degree equal to $\min(2m-1, k-1)$. It is, moreover, visibly of degree $k-1$, of class C^{k-2}, and of span $s=k+2m-2$. One further important point is in need of proof, namely that the formulae based on (18) actually do smooth any given sequence (see Definition b of Part A, section 2.2). This will readily follow from Lemma 5. Indeed the characteristic function $\phi_{m,k}(u)$ of the formula

$$F(n) = \sum_\nu y_\nu L_{k,m}(n-\nu) \tag{19}$$

is, by Theorem 2, Part A, given by

$$\phi_{k,m}(u) = \sum_{\nu=-\infty}^{\infty} g_{k,m}(u+2\pi\nu).$$

By I(32) and I(28) we now have

$$\phi_{k,m}(u) = \sum_{\nu=-\infty}^{\infty} g_{k,m}(u+2\pi\nu) = \phi_k(u)\sum_{n=0}^{m-1} c_{2n}^{(k)}(2\sin u/2)^{2n}$$

$$< \phi_k(u)\sum_{n=0}^{\infty} c_{2n}^{(k)}(2\sin u/2)^{2n} = 1, \qquad (0 < u < 2\pi). \tag{20}$$

Since obviously $\phi_{m,k}(u) > 0$, for all u, we see that (19) is indeed a smoothing formula according to our definition. Recalling the relations I(36), I(38), we may therefore state the following Theorem:

THEOREM 2. *Let k be a positive integer and m an integer such that $0<2m<k+2$. Let the positive rational numbers $d_{2n}^{(k)}$ be defined by the expansion*

$$\left(\frac{2\arcsin v/2}{v}\right)^k = \sum_{n=0}^{\infty} d_{2n}^{(k)} v^{2n}. \tag{21}$$

Then

$$L_{k,m}(x) = M_k(x) - d_2^{(k)}\delta^2 M_k(x) + d_4^{(k)}\delta^4 M_k(x) - \cdots + (-1)^{m-1} d_{2m-2}^{(k)}\delta^{2m-2} M_k(x) \tag{22}$$

gives rise to a smoothing interpolation formula

$$F(x) = \sum_{n=-\infty}^{\infty} y_n L_{k,m}(x-n) \tag{23}$$

of type

$$D^{k-1}, \quad C^{k-2}, \quad E^{\min(2m-1,k-1)}, \quad s = k + 2m - 2. \tag{24}$$

Moreover, the formula (23) *preserves the degree* $k-1$. (*See Part* A, *section* 2.21, *Definition* d.)

If k is fixed and m increases then the smoothing power of our formula (23) *decreases according to our definition of Part* A, *section* 1.12, *Definition* 2.

The last statement concerning the decreasing smoothing power of (23) follows from (20), since $\phi_{k,m}(u)$ increases strictly as m increases while u is constant $(0<u<2\pi)$.

Notice, by (24), how on increasing m by one unit both the degree of exactness, as well as the span, increase by two units.

As illustration we find from I(40) that

$$L_{k,2}(x) = M_k(x) - \frac{k}{24}\delta^2 M_k(x), \qquad (k \geqq 4), \tag{25}$$

yields a smoothing formula of type

$$D^{k-1}, \quad C^{k-2}, \quad E^3, \quad s = k + 2. \tag{26}$$

The characteristic function of (25) is

$$g_{k,2}(u) = \left(\frac{2\sin u/2}{u}\right)^k \left(1 + \frac{k}{6}\sin^2\frac{u}{2}\right) \tag{27}$$

or

$$g_{k,2}(u) = \left(\frac{2\sin u/2}{u}\right)^k \left\{1 + \frac{k}{12} - \frac{k}{12}\cos u\right\}.$$

For $k=4$ this function $g_{4,2}(u)$ agrees with the integrand of our formula (12″) of Part A, section 2.123. Also Mr. Jenkins' basic function, as given by formula (12) of Part A, section 2.123, may be derived by working out the various polynomial expressions of

$$L_{4,2}(x) = M_4(x) - \tfrac{1}{6}\delta^2 M_4(x) \tag{28}$$

from the explicit expressions of $M_4(x)$ (see Part A, section 3.13, (14)).

Likewise, by I(40)

$$L_{k,3}(x) = M_k(x) - \frac{k}{24}\delta^2 M_k(x) + \frac{k(5k+22)}{5760}\delta^4 M_k(x), \qquad (k \geqq 6), \tag{29}$$

yields a smoothing formula of type

$$D^{k-1}, \quad C^{k-2}, \quad E^5, \quad s = k + 4. \tag{30}$$

2.3. **A second class of smoothing interpolation formulae derived from the ordinary k-point central interpolation formula.** Among the smoothing interpolation formulae (23) described by Theorem 2 the one of most interest is obtained by letting m

assume its largest value. If k is even, m is maximal if $2m-2=k-2$ or $m=k/2$. If k is odd, m is maximal if $2m-2=k-1$ or $m=(k+1)/2$. In either case

$$\max m = \mu = \left[\frac{k+1}{2}\right], \tag{31}$$

where $[x]$ represents the largest integer not exceeding x. The corresponding basic function (22) is

$$L_{k,\mu}(x) = \frac{1}{2\pi}\int_{-\infty}^{\infty}\left(\frac{2\sin u/2}{u}\right)^k \{1 + d_2^{(k)}(2\sin u/2)^2 + \cdots + d_{2\mu-2}^{(k)}(2\sin u/2)^{2\mu-2}\}e^{iux}du. \tag{32}$$

We recall that the smoothing interpolation formula based on this function is by (24) of the type

$$D^{k-1}, \qquad C^{k-2}, \qquad E^{k-1}, \qquad s = k + 2\mu - 2. \tag{33}$$

Indeed, the formula is exact for the degree $k-1$ because of

$$\left(\frac{2\sin u/2}{u}\right)^k\{1 + d_2^{(k)}(2\sin u/2)^2 + \cdots + d_{2\mu-2}^{(k)}(2\sin u/2)^{2\mu-2}\} = 1 + u^k\cdot(\text{regular function}). \tag{34}$$

An interesting counterpart to (32) is obtained as follows. An identity of the type (34) may also be obtained if in the expression within braces we replace $2\sin u/2$ by u. Indeed, rational constants $\gamma_{2\nu}^{(k)}$ may be determined such that

$$\left(\frac{2\sin u/2}{u}\right)^k\{1 + \gamma_2^{(k)}u^2 + \gamma_4^{(k)}u^4 + \cdots + \gamma_{2\mu-2}^{(k)}u^{2\mu-2}\} = 1 + u^k\cdot(\text{regular function}). \tag{35}$$

LEMMA 6. *The basic function*

$$\Gamma_{k,\mu}(x) = \frac{1}{2\pi}\int_{-\infty}^{\infty}\left(\frac{2\sin u/2}{u}\right)^k\{1 + \gamma_2^{(k)}u^2 + \cdots + \gamma_{2\mu-2}^{(k)}u^{2\mu-2}\}e^{iux}du \tag{36}$$

is identical, for all real values of x, with the basic function $C_k(x)$ of the k-point central interpolation method (*see Part* A, *section* 2.121).

Notice first that by differentiation of

$$M_k(x) = \frac{1}{2\pi}\int_{-\infty}^{\infty}\left(\frac{2\sin u/2}{u}\right)^k e^{iux}du$$

we obtain

$$M_k^{(2\nu)}(x) = (-1)^\nu \frac{1}{2\pi}\int_{-\infty}^{\infty}\left(\frac{2\sin u/2}{u}\right)^k u^{2\nu}e^{iux}du, \qquad (2\nu < k). \tag{37}$$

Therefore the integral (36) may also be written as[4]

[4] Assuming (41) already established, we see by (38) and the relations

$$M_k^{(\nu)}(x) = \delta^\nu M_{k-\nu}(x), \qquad (0 \leq \nu \leq k-1), \tag{*}$$

(see Part A, section 3.15, formula (23)) that we may express $C_k(x)$ as follows

$$\Gamma_{k,\mu}(x) = M_k(x) - \gamma_2^{(k)} M_k''(x) + \gamma_4^{(k)} M_k^{(4)}(x) - \cdots + (-1)^{\mu-1} \gamma_{2\mu-2}^{(k)} M_k^{(2\mu-2)}(x). \quad (38)$$

Now $M_k(x)$ and all its derivatives are functions of span $s=k$. Therefore also $\Gamma_{k,\mu}(x)$ has the span $s=k$. Furthermore by (35), and Theorem 2 of Part A, we conclude that

$$F(x) = \sum_{n=-\infty}^{\infty} y_n \Gamma_{k,\mu}(x-n) \quad (39)$$

is an interpolation formula of the following type characteristics:

$$E^{k-1}, \qquad s = k. \quad (40)$$

These last two properties (40) allow us to show readily that

$$\Gamma_{k,\mu}(x) = C_k(x) \quad \text{for all real } x. \quad (41)$$

Indeed, let k be even, $k=2\kappa$. Let $P_0(x)$ be the polynomial of degree $k-1$ defined by the following k conditions

$$P_0(-\kappa+1) = P_0(-\kappa+2) = \cdots = P_0(-1) = 0, \qquad P_0(0) = 1,$$
$$P_0(1) = P_0(2) = \cdots = P_0(\kappa) = 0. \quad (42)$$

Since (39) is exact for the degree $k-1$ we have the identity

$$P_0(x) = \sum_{n=-\infty}^{\infty} P_0(n)\Gamma_{k,\mu}(x-n), \quad \text{for all real } x. \quad (43)$$

We now restrict x to the range

$$0 \leqq x \leqq 1. \quad (44)$$

Then we may write (43) as

$$P_0(x) = \sum_{n=-\kappa+1}^{n=\kappa} P_0(n)\Gamma_{k,\mu}(x-n)$$

since $\Gamma_{k,\mu}(x-n)=0$ if $|x-n| \geqq \kappa$. In view of (42) this identity reduces to the single term, for $n=0$:

$$P_0(x) = \Gamma_{k,\mu}(x), \qquad (0 \leqq x \leqq 1),$$

and therefore (41) holds for the range (44). Likewise, applying the formula (39) to the polynomial $P_1(x)$, of degree k, defined by

$$P_1(-\kappa+2) = \cdots = P_1(-1) = 0, \qquad P_1(0) = 1, \qquad P_1(1) = \cdots = P_1(\kappa+1) = 0,$$

we find that (41) holds in the range $1 \leqq x \leqq 2$, and so forth. Similar arguments obviously apply, with obvious modifications, to the case of an odd k.

The coefficients $\gamma_{2\nu}^{(k)}$ are the expansion coefficients of

$$\left(\frac{u}{2\sin u/2}\right)^k = \sum_{\nu=0}^{\infty} \gamma_{2\nu}^{(k)} u^{2\nu}. \quad (45)$$

$$C_k(x) = M_k(x) - \gamma_2^{(k)}\delta^2 M_{k-2}(x) + \gamma_4^{(k)}\delta^4 M_{k-4}(x) + \cdots + (-1)^{\mu-1}\gamma_{2\mu-2}^{(k)}\delta^{2\mu-2} M_{k-2\mu+2}(x). \quad (**)$$

This formula reveals at a glance the following fact: *If k is even, then $C_k^{(2\nu)}(x)$ ($\nu=0, 1, 2, \cdots$) are continuous. If k is odd then $C_k^{(2\nu+1)}(x)$ ($\nu=0, 1, 2, \cdots$) are continuous.* The author learned this property from Kingsland Camp, *Notes on Interpolation*, Trans. Actuar. Soc. Amer., **38**, p. 22 (1937).

In N. E. Nörlund's *Differenzenrechnung*, page 143, we find the expansion

$$\left(\frac{t}{\sin t}\right)^k = \sum_{\nu=0}^{\infty} (-1)^\nu \frac{t^{2\nu}}{(2\nu)!} D_{2\nu}^{(k)}.$$

The coefficient $D_{2\nu}^{(k)}$ is a polynomial in k of degree ν. Nörlund's Table 6 on page 460, loc. cit., lists these polynomials for $\nu=0, 1, \cdots, 6$. We therefore have

$$\left(\frac{u}{2\sin u/2}\right)^k = \sum_{\nu=0}^{\infty} (-1)^\nu \frac{u^{2\nu}}{(2\nu)!}\cdot\frac{D_{2\nu}^{(k)}}{2^{2\nu}} \tag{46}$$

whence

$$\gamma_{2\nu}^{(k)} = (-1)^\nu \frac{D_{2\nu}^{(k)}}{(2\nu)!2^{2\nu}}\cdot \tag{47}$$

The first few values are

$$\gamma_0^{(k)} = 1, \qquad \gamma_2^{(k)} = \frac{k}{24}, \qquad \gamma_4^{(k)} = \frac{k(5k+2)}{5760}\cdot \tag{48}$$

The expansion coefficients of $t/\sin t$ are positive (see Nörlund, loc. cit., Chapter II, sections 2, 3). Therefore *the coefficients* $\gamma_{2\nu}^{(k)}$ *are all positive*. We shall use this fact later.

In view of the results of our last section it seems natural to consider the partial sums

$$\Gamma_{k,m}(x) = M_k(x) - \gamma_2^{(k)} M_k''(x) + \cdots + (-1)^{m-1}\gamma_{2m-2}^{(k)} M_k^{(2m-2)}(x),$$
$$\left(1 \leqq m < \mu = \left[\frac{k+1}{2}\right]\right), \tag{49}$$

of the sum (38). The properties of the interpolation formulae based on these functions are described by the following theorem:

THEOREM 3. *The formula*

$$F(x) = \sum y_\nu \Gamma_{k,m}(x-\nu), \qquad \left(1 \leqq m < \mu = \left[\frac{k+1}{2}\right]\right), \tag{50}$$

is a smoothing interpolation formula of the type

$$D^{k-1}, \qquad C^{k-2m}, \qquad E^{2m-1}, \qquad s = k. \tag{51}$$

The smoothing power of (50) *decreases, as m increases from $m=1$ to $m=\mu-1$, until for $m=\mu$* (50) *reduces to the* (*ordinary*) *k-point central interpolation formula.*

Indeed, the characteristic function of (49) is

$$g(u) = \left(\frac{2\sin u/2}{u}\right)^k \left\{1 + \gamma_2^{(k)} u^2 + \cdots + \gamma_{2m-2}^{(k)} u^{2m-2}\right\}. \tag{52}$$

From (45) we conclude that

$$g(u) = 1 + u^{2m}\cdot(\text{regular function})$$

and therefore (50) is exact for the degree $2m-1$, by Theorem 2 of Part A. The remaining three characteristics

$$D^{k-1}, \qquad C^{k-2m}, \qquad s = k,$$

are evident on inspection of (49). There remains the investigation of the characteristic function of the corresponding smoothing formula

$$F(n) = \sum_{\nu} y_\nu \Gamma_{k,m}(n-\nu). \tag{53}$$

By Theorem 2, Part A, this characteristic function is

$$\chi_m(u) = \sum_{\nu=-\infty}^{\infty} g(u+2\pi\nu).$$

From (52) and I(3), we obtain

$$\chi_m(u) = \phi_k(u) + \gamma_2^{(k)}(2 \sin u/2)^2 \phi_{k-2}(u) + \cdots + \gamma_{2m-2}^{(k)}(2 \sin u/2)^{2m-2} \phi_{k-2m+2}(u). \tag{54}$$

This expression is obviously positive for all values of u. For $m=\mu$, however, we obtain the identity

$$\chi_\mu(u) \equiv 1 \tag{55}$$

since, by Lemma 6, we have before us an *ordinary* interpolation formula. Since (54) is a partial sum of the left-hand side of (55) we have therefore proved the inequalities

$$0 < \chi_m(u) < 1 \qquad (1 \leqq m < \mu,\ 0 < u < 2\pi). \tag{56}$$

Then (53) is indeed a smoothing formula. The final statement of the Theorem is evident from (54), since $\chi_m(u)$ increases with m.

As illustrations we mention the following four special cases, two from each end of the range of values of m.

(i) $m=2$. The formula based on

$$\Gamma_{k,2}(x) = M_k(x) - \frac{k}{24} M_k''(x) \qquad (k \geqq 4) \tag{57}$$

has the type

$$D^{k-1}, \qquad C^{k-4}, \qquad E^3, \qquad s = k. \tag{57'}$$

(ii) $m=3$. The formula based on

$$\Gamma_{k,3}(x) = M_k(x) - \frac{k}{24} M_k''(x) + \frac{k(5k+2)}{5760} M_k^{(4)}(x), \qquad (k \geqq 6), \tag{58}$$

has the type

$$D^{k-1}, \qquad C^{k-6}, \qquad E^5, \qquad s = k. \tag{58'}$$

The values (48) were used.

(iii) $m=\mu-1$. The formula based on[5]

[5] The formula (59) is especially instructive because we can observe very clearly how the addition to $C_k(x)$ of the extra term removes the crudest discontinuities of $C_k(x)$. Indeed, by the formula (*) of our preceding footnote we may write (59) as

$$\Gamma_{k,\mu-1}(x) = C_k(x) + (-1)^{\mu} \gamma^{(k)}_{2\mu-2} M_k^{(2\mu-2)}(x) \qquad (k \geqq 4) \tag{59}$$

has the type

$$D^{k-1}, \qquad \begin{Bmatrix} C^2 & \text{if} & k \text{ is even} \\ C^1 & \text{if} & k \text{ is odd} \end{Bmatrix}, \qquad E^{k-3}, \qquad s = k. \tag{59'}$$

(iv) $m = \mu - 2$. The formula based on

$$\Gamma_{k,\mu-2}(x) = C_k(x) + (-1)^{\mu} \gamma^{(k)}_{2\mu-2} M_k^{(2\mu-2}(x) + (-1)^{\mu-1} \gamma^{(k)}_{2\mu-4} M_k^{(2\mu-4)}(x) \quad (k \geqq 6) \qquad (60$$

has the type

$$D^{k-1}, \qquad \begin{Bmatrix} C^4 & \text{if} & k \text{ is even} \\ C^3 & \text{if} & k \text{ is odd} \end{Bmatrix}, \qquad E^{k-5}, \qquad s = k. \tag{60'}$$

These formulae show clearly how an increase in "order of contact" is compensated by a corresponding loss in "reproductive power" and vice versa.

III. A SECOND CLASS OF ANALYTIC INTERPOLATION FORMULAE

In Part A, section 4.2, we described a class of *ordinary* analytic interpolation formulae of basic function

$$L_k(x, t) = \frac{1}{2\pi} \int_{-\infty}^{\infty} \frac{\psi_k(u, t)}{\phi_k(u, t)} e^{iux} du \qquad (k = 1, 2, \cdots; t > 0). \tag{1}$$

These interpolation formulae are exact for the degree $k-1$. The basic function (1) as well as those of the smoothing interpolation formulae derived from it in Part A, section 4.3, dampen out like a descending exponential function. In the present last chapter we wish to construct smoothing analytic interpolation formulae of basic functions dampening out like

$$\exp(-c^2x^2),$$

hence much more rapidly. In view of the development of section 2.2 it would seem fairly obvious how such formulae may be derived. We clearly need an analogue of Lemma 5 which we state as a **conjecture**: *The reciprocal of $\phi_k(u, t)$ admits of an expansion*

$$\Gamma_{k,\mu-1}(x) = C_k(x) + (-1)^{\mu} \gamma^{(k)}_{2\mu-2} \delta^{2\mu-2} M_{k-2\mu+2}(x). \tag{59'}$$

Let k be *odd*, hence $2\mu-2=k-1$ and therefore

$$\Gamma_{k,\mu-1}(x) = C_k(x) + (-1)^{\mu} \gamma^{(k)}_{k-1} \delta^{k-1} M_1(x). \tag{59''}$$

As seen from the graph of $M_1(x)$, the corrective term is a step-function with discontinuities at $x=n+1/2$ whose values are proportional with the binomial coefficients of order $k-1$: $\binom{k-1}{\nu}$. Their addition to $C_k(x)$ offsets the discontinuities of $C_k(x)$ and turn it into a function (59'') of class C^1. If k is *even*, hence $2\mu-2=k-2$, we have

$$\Gamma_{k,\mu-1}(x) = C_k(x) + (-1)^{\mu} \gamma^{(k)}_{k-2} \delta^{k-2} M_2(x). \tag{59'''}$$

As seen from the graph of $M_2(x)$, the corrective term is now an ordinary polygonal line with vertices at $x=n$, whose ordinates (at these vertices) are proportional to the binomial coefficients of order $k-2$: $\binom{k-2}{\nu}$. Again, the superposition of this polygonal line on $C_k(x)$ offsets the corners of $C_k(x)$ and turns it into a function (59''') of class C^2. The formulae (59'''), (59'') are especially convenient for constructing tables of these functions from existing tables of $C_k(x)$, i.e., tables of Lagrange interpolation coefficients.

$$\frac{1}{\phi_k(u,t)} = \sum_{n=0}^{\infty} c_{2n}^{(k)}(t)(2\sin u/2)^{2n}, \tag{2}$$

which converges for all real values of u and where the coefficients are all positive

$$c_{2n}^{(k)}(t) > 0, \qquad (n = 0, 1, 2, \cdots). \tag{3}$$

A proof of this conjecture would require a closer function-theoretic study of the entire periodic function $\phi_k(u, t)$ which has not been carried through as yet.

Since

$$\phi_k(u,t) = \sum_{\nu=-\infty}^{\infty} \psi_k(u + 2\pi\nu, t)$$

(see Part A, section 4.1, formula (6)) we have

$$\phi_k(u,t) = \psi_k(u,t) + u^k\cdot(\text{regular function}).$$

Therefore the expansion (2) agrees in its terms of order less than k with the similar terms of the expansion

$$\frac{1}{\psi_k(u,t)} = e^{t(u/2)^2}\left(\frac{u}{2\sin u/2}\right)^k = \sum_{n=0}^{\infty} d_{2n}^{(k)}(t)(2\sin u/2)^{2n}, \qquad (-\pi \leqq u \leqq \pi). \tag{4}$$

Hence

$$c_{2n}^{(k)}(t) = d_{2n}^{(k)}(t), \qquad (0 \leqq 2n < k). \tag{5}$$

The expansion (4) is readily determined and its coefficients are found to be positive as follows. We turn back to section 1.3 where in terms of the variable

$$v = 2\sin u/2 \tag{6}$$

we have by I(35)

$$\left(\frac{u}{2\sin u/2}\right)^k = \sum_{n=0}^{\infty} d_{2n}^{(k)} v^{2n} \qquad (-\pi \leqq u \leqq \pi). \tag{7}$$

Also by I(39)

$$u/2 = \arcsin v/2 = \frac{v}{2} + \frac{1}{2}\cdot\frac{1}{3}\cdot\frac{v^3}{8} + \frac{1\cdot 3}{2\cdot 4}\cdot\frac{1}{5}\cdot\frac{v^5}{32} + \cdots \qquad (-2 \leqq v \leqq 2). \tag{8}$$

On substituting (8) into the exponential series we find the expansion

$$e^{t(u/2)^2} = \sum_{n=0}^{\infty} e_{2n}(t)v^{2n} \qquad (-2 \leqq v \leqq 2) \tag{9}$$

with positive coefficients, the first three of which are found to be

$$e_0(t) = 1, \qquad e_2(t) = \frac{t}{4}, \qquad e_4(t) = \frac{t^2}{32} + \frac{t}{48}. \tag{10}$$

On multiplying the series (7) and (9) we obtain the expansion (4). From the values I(40) and (10) we readily find

$$d_0^{(k)}(t) = 1, \qquad d_2^{(k)}(t) = \frac{k}{24} + \frac{t}{4}, \qquad d_4^{(k)}(t) = \frac{5k^2 + 22k}{5760} + \frac{tk}{96} + \frac{t}{48} + \frac{t^2}{32}. \tag{11}$$

Our arguments of section 2.2 may now be repeated leading to the following theorem:

THEOREM 4. *Let k be a positive integer and m an integer such that* $0<2m<k+2$. *Then*

$$L_{k,m}(x, t) = M_k(x, t) - d_2^{(k)}(t)\delta^2 M_k(x, t) + \cdots + (-1)^{m-1} d_{2m-2}^{(k)}(t)\delta^{2m-2} M_k(x) \tag{12}$$

gives rise to a "smoothing" analytic interpolation formula

$$F(x) = \sum_{n=-\infty}^{\infty} y_n L_{k,m}(x - n, t) \tag{13}$$

which is exact for the degree $\min(2m-1, k-1)$. *Moreover* (13) *always preserves the degree* $k-1$.

The adjective "smoothing" was purposely written in quotation marks in order to indicate that there is no general proof as yet that (13) always reduces, for integral values of the variable x, to a smoothing formula in the sense of our Definition 1 of Part A, section 1.1. For indeed, (2) and (3), which imply such a proof, were only conjectured. In the Appendix we give 8-place tables of the three basic functions

$$\left.\begin{aligned} L_1(x) &= L_{4,2}(x, 1/8), \\ L_2(x) &= L_{4,2}(x, 1/2), \\ L_3(x) &= L_{6,3}(x, 1/2), \end{aligned}\right\} \tag{14}$$

as well as 7-place tables of their first and second derivatives. For these three sets of values of the parameters k, t, and m, the interpolation formula (13) is indeed a smoothing formula. This point is verified by an inspection of the corresponding characteristic functions

$$\phi_i(u) = L_i(0) + 2L_i(1)\cos u + 2L_i(2)\cos 2u + \cdots, \qquad (i = 1, 2, 3,). \tag{15}$$

From the values of $L_i(n)$, as given by our tables, we computed the following table for these characteristic functions:

u	$\phi_1(u)$	$\phi_2(u)$	$\phi_3(u)$
0°	1.00000	1.00000	1.00000
30°	.99734	.99519	.99952
60°	.96332	.93655	.97760
90°	.85492	.76500	.84693
120°	.67727	.51297	.56702
150°	.50474	.29296	.27879
180°	.43283	.20728	.16123

Since $0<\phi_i(u)<1$ for $0<u\leq 180°$, all three formulae (13) are smoothing formulae according to Part A, section 1.1. Also $\phi_2(u)<\phi_1(u)$ implies that L_2 gives a stronger smoothing formula as compared to L_1.

Our set of tables is intended mainly for the purpose of illustrating the method. A more complete set of tables would be needed in order to furnish smoothing of a desired strength, as required by the needs of the numerical data at hand.

APPENDIX

Description of the tables and their use for the analytic approximation of equidistant data. In the Tables I, II, and III, we have tabulated the following three functions

$$L_1(x) = M_4(x, 1/8) - \frac{19}{96}\delta^2 M_4(x, 1/8) \tag{1}$$

$$L_2(x) = M_4(x, 1/2) - \frac{7}{24}\delta^2 M_4(x, 1/2) \tag{2}$$

$$L_3(x) = M_6(x, 1/2) - \frac{3}{8}\delta^2 M_6(x, 1/2) + \frac{199}{1920}\delta^4 M_6(x, 1/2) \tag{3}$$

and their first two derivatives. The function $M_k(x, t)$ occurring in these definitions may be defined by the integral

$$M_k(x, t) = \frac{1}{2\pi}\int_{-\infty}^{\infty} e^{-t(u/2)^2}\left(\frac{2 \sin u/2}{u}\right)^k \cos uxdu.$$

A given sequence of equidistant ordinates

$$\{y_n\} \tag{4}$$

is approximated by either one of the three analytic functions

$$F_i(x) = \sum_{\nu=-\infty}^{\infty} y_\nu L_i(x - \nu), \qquad (i = 1, 2, 3). \tag{5}$$

The choice among these approximations depends on the amount of smoothing desired. The formula (5), for $i=1$ and $i=2$, is exact for (i.e., reproduces) cubic polynomials. For $i=3$ the formula (5) is exact for quintic polynomials. For the same data (4), the sequence $\{F_2(n)\}$ is always smoother than the sequence $\{F_1(n)\}$. Generally, the sequence $\{F_3(n)\}$ should be smoother than the sequence $\{F_1(n)\}$.

The first and second derivatives of the approximation (5) may be computed by the similar formulae

$$F_i'(x) = \sum y_\nu L_i'(x - \nu), \tag{6}$$

$$F_i''(x) = \sum y_\nu L_i''(x - \nu). \tag{7}$$

The arrangement of our tables is such as to facilitate the computation of $F_i(x)$ by (5), as explained in the Appendix to Part A.

An example of smoothing with subtabulation to tenths. We propose to compute a table of the approximation $F_2(x)$, in the range $31 \leq x \leq 34$, for the same ordinates $\{y_n\}$ as were used in our example of Part A (Appendix). The ordinates which we now require are given by the following table:

n	y_n	Δ	Δ^2	Δ^3	Δ^4	Δ^5
26	32840					
27	34790	1950				
28	37260	2470	520			
29	40440	3180	710	190		
30	44750	4310	1130	420	230	
31	51120	6370	2060	930	510	280
32	59390	8270	1900	− 160	−1090	−1600
33	67550	8160	− 110	−2010	−1850	− 760
34	73820	6270	−1890	−1780	230	2080
35	77830	4010	−2260	− 370	1410	1180
36	80240	2410	−1600	660	1030	− 380
37	81660	1420	− 990	610	− 50	−1080
38	82330	670	− 758	240	− 370	− 320
39	82680	350	− 320	430	190	560

From these values and the Table II of $L_2(x)$ and $L_2''(x)$ we obtain the following tables of the approximation $F_2(x)$ and its second derivative $F_2''(x)$ shown with their differences.

x	$F_2(x)$	Δ	Δ^2	Δ^3	Δ^4	$F_2''(x)$	Δ	Δ^2	Δ^3	Δ^4
31.0	51232.76					1901.77				
31.1	51989.86	75710				1767.20	−13457			
31.2	52764.62	77476	1766			1611.60	−15560	−2103		
31.3	53555.48	79086	1610	−156		1436.84	−17476	−1916	187	
31.4	54360.69	80521	1435	−175	−19	1245.42	−19142	−1666	250	63
31.5	55178.34	81765	1244	−191	−16	1040.29	−20513	−1371	295	45
31.6	56006.39	82805	1040	−204	−13	824.64	−21565	−1052	319	24
31.7	56842.68	83629	824	−216	−12	601.66	−22298	− 733	319	0
31.8	57684.98	84230	601	−223	− 7	374.38	−22728	− 430	303	−16
31.9	58531.03	84605	375	−226	− 3	145.59	−22879	− 151	279	−24
32.0	59378.53	84750	145	−230	− 4	− 82.29	−22788	91	242	−37
32.1	60225.21	84668	− 82	−227	3	− 307.15	−22486	302	211	−31
32.2	61068.82	84361	− 307	−225	2	− 527.12	−21997	489	187	−24
32.3	61907.17	83835	− 526	−219	6	− 740.51	−21339	658	169	−18
32.4	62738.12	83095	− 740	−214	5	− 945.74	−20523	816	158	−11
32.5	63559.62	82150	− 945	−205	9	−1141.23	−19549	974	158	0
32.6	64369.71	81009	−1141	−196	9	−1325.45	−18422	1127	153	− 5
32.7	65166.57	79686	−1323	−182	14	−1496.90	−17145	1277	150	− 3
32.8	65948.46	78189	−1497	−174	8	−1654.18	−15728	1417	140	−10
32.9	66713.83	76537	−1652	−155	19	−1796.14	−14196	1532	115	−25
33.0	67461.25	74742	−1795	−143	12	−1921.91	−12577	1619	87	−28
33.1	68189.46	72821	−1921	−126	17	−2030.98	−10907	1670	51	−36
33.2	68897.38	70792	−2029	−108	18	−2123.19	− 9221	1686	16	−35
33.3	69584.08	68670	−2122	− 93	15	−2198.74	− 7555	1666	− 20	−36
33.4	70248.81	66473	−2197	− 75	18	−2258.04	− 5930	1625	− 41	−21
33.5	70890.96	64215	−2258	− 61	14	−2301.69	− 4365	1565	− 60	−19
33.6	71510.12	61916	−2299	− 41	20	−2330.31	− 2862	1503	− 62	− 2
33.7	72105.98	59586	−2330	− 31	10	−2344.57	− 1426	1436	− 67	− 5
33.8	72678.41	57243	−2343	− 13	18	−2345.07	− 50	1376	− 60	7
33.9	73227.40	54899	−2344	− 1	12	−2332.47	1260	1310	− 66	− 6
34.0	73753.07	52567	−2332	12	13	−2307.47	2500	1240	− 70	− 4

A comparison of the approximation $F_2(x)$ with the strictly interpolating function $F(x)$, obtained in the Appendix of Part A, is of interest. The function $F(x)$ was obtained by the formula

$$F(x) = \sum_{\nu=-\infty}^{\infty} y_\nu L_k(x-\nu, t), \qquad (k = 4, t = 1/2),$$

where, in view of III(1) and III(2), we may define $L_k(x, t)$ by the expansion

$$L_k(x, t) = M_k(x, t) - c_2^{(k)}(t)\delta^2 M_k(x, t) + c_4^{(k)}(t)\delta^4 M_k(x, t) - \cdots. \tag{8}$$

Our present approximation $F_2(x)$ was computed by the formula (5), for $i=2$, where $L_2(x)$, by (2), III(5) and III(11), happens to be identical with the sum

$$L_2(x) = M_k(x, t) - c_2^{(k)}(t)\delta^2 M_k(x, t)$$

of the first two terms of the series (8). A comparison of the tables of $F(x)$ and $F_2(x)$ shows that their difference in the range $31 \leqq x \leqq 34$ nowhere exceeds 0.23% of the value of $F(x)$.

TABLE I. $L_1(x) = L_{4,2}(x, 1/8)$, $L_1'(x)$, $L_1''(x)$

$L_1(x)$

x	$x+.0$	$x+.1$	$x+.2$	$x+.3$	$x+.4$
3	−.00041123	−.00018550	−.00007621	−.00002830	−.00000943
2	−.03462580	−.02756299	−.02099452	−.01533104	−.01073240
1	.14220425	.08277004	.03516336	−.00066601	−.02556532
0	.78566556	.77475976	.74281594	.69204028	.62577328
−1	.14220425	.21251290	.29166041	.37661030	.46353840
−2	−.03462580	−.04145730	−.04699226	−.04985257	−.04841746
−3	−.00041123	−.00083691	−.00157669	−.00277249	−.00458632
−4			−.00000003	−.00000017	−.00000074

x	$x+.5$	$x+.6$	$x+.7$	$x+.8$	$x+.9$
3	−.00000280	−.00000074	−.00000017	−.00000003	
2	−.00718751	−.00458632	−.00277249	−.00157669	−.00083691
1	−.04092087	−.04841746	−.04985257	−.04699226	−.04145730
0	.54811118	.46353840	.37661030	.29166041	.21251290
−1	.54811118	.62577328	.69204028	.74281594	.77475976
−2	−.04092087	−.02556532	−.00066601	.03516336	.08277004
−3	−.00718751	−.01073240	−.01533104	−.02099452	−.02756299
−4	−.00000280	−.00000943	−.00002830	−.00007621	−.00018550

$L_1'(x)$

x	$x+.0$	$x+.1$	$x+.2$	$x+.3$	$x+.4$
3	.0030922	.0015611	.0007155	.0002953	.0001089
2	.0709642	.0690870	.0616092	.0514079	.0405956
1	−.6512052	−.5359047	−.4163489	−.3016954	−.1986416
0	.0000000	−.2168063	−.4183406	−.5915396	−.7269119
−1	.6512052	.7515608	.8262912	.8662852	.8650057
−2	−.0709642	−.0638857	−.0445045	−.0099839	.0416488
−3	−.0030922	−.0056121	−.0094213	−.0147669	−.0217952
−4		−.0000001	−.0000006	−.0000026	−.0000103

$L_1'(x)$

x	$x+.5$	$x+.6$	$x+.7$	$x+.8$	$x+.9$
3	.0000357	.0000103	.0000026	.0000006	.0000001
2	.0304945	.0217952	.0147669	.0094213	.0056121
1	−.1113051	−.0416488	.0099839	.0445045	.0638857
0	−.8188067	−.8650057	−.8662852	−.8262912	−.7515608
−1	.8188067	.7269119	.5915396	.4183406	.2168063
−2	.1113051	.1986416	.3016954	.4163489	.5359047
−3	−.0304945	−.0405956	−.0514079	−.0616092	−.0690870
−4	−.0000357	−.0001089	−.0002953	−.0007155	−.0015611

$L_1''(x)$

x	$x+.0$	$x+.1$	$x+.2$	$x+.3$	$x+.4$
4	−.0000004	−.0000001			
3	−.0197354	−.0114013	−.0059477	−.0027759	−.0011500
2	.0202421	−.0521077	−.0926044	−.1078983	−.1061742
1	1.0966561	1.1912649	1.1844282	1.0974670	.9568292
0	−2.1943249	−2.1159568	−1.8927238	−1.5554043	−1.1427100
−1	1.0966561	.8923803	.5868711	.2020705	−.2337191
−2	.0202421	.1269947	.2654156	.4283134	.6058876
−3	−.0197354	−.0311722	−.0454298	−.0617347	−.0788289
−4	−.0000004	−.0000019	−.0000092	−.0000377	−.0001346

x	$x+.5$	$x+.6$	$x+.7$	$x+.8$	$x+.9$
3	−.0004201	−.0001346	−.0000377	−.0000092	−.0000019
2	−.0947351	−.0788289	−.0617347	−.0454298	−.0311722
1	.7867260	.6058876	.4283134	.2654156	.1269947
0	−.6915708	−.2337191	.2020705	.5868711	.8923803
−1	−.6915708	−1.1427100	−1.5554043	−1.8927238	−2.1159568
−2	.7867260	.9568292	1.0974670	1.1844282	1.1912649
−3	−.0947351	−.1061742	−.1078983	−.0926044	−.0521077
−4	−.0004201	−.0011500	−.0027759	−.0059477	−.0114013
−5					−.0000001

TABLE II. $L_2(x) = L_{4,2}(x, 1/2)$, $L_2'(x)$, $L_2''(x)$

$L_2(x)$

x	$x+.0$	$x+.1$	$x+.2$	$x+.3$	$x+.4$
5	−.00000001				
4	−.00003297	−.00001721	−.00000871	−.00000428	−.00000203
3	−.00453670	−.00313034	−.00210474	−.00137859	−.00087930
2	−.04033977	−.03775949	−.03379714	−.02913417	−.02429574
1	.20271717	.14753253	.09914645	.05829935	.02523450
0	.68438455	.67711287	.65567464	.62117134	.57534514
−1	.20271717	.26343912	.32793367	.39399436	.45907812
−2	−.04033977	−.04070831	−.03791256	−.03092183	−.01869149
−3	−.00453670	−.00640787	−.00882100	−.01183236	−.01545906
−4	−.00003297	−.00006127	−.00011052	−.00019364	−.00032972
−5	−.00000001	−.00000003	−.00000007	−.00000018	−.00000042

x	$x+.5$	$x+.6$	$x+.7$	$x+.8$	$x+.9$
4	−.00000094	−.00000042	−.00000018	−.00000007	−.00000003
3	−.00054590	−.00032972	−.00019364	−.00011052	−.00006127
2	−.01965692	−.01545906	−.01183236	−.00882100	−.00640787
1	−.00024449	−.01869149	−.03092183	−.03791256	−.04070831
0	.52044824	.45907812	.39399436	.32793367	.26343912
−1	.52044824	.57534514	.62117134	.65567464	.67711287
−2	−.00024449	.02523450	.05829935	.09914645	.14753253
−3	−.01965692	−.02429574	−.02913417	−.03379714	−.03775949
−4	−.00054590	−.00087930	−.00137859	−.00210474	−.00313034
−5	−.00000094	−.00000203	−.00000428	−.00000871	−.00001721

$L_2'(x)$

x	$x+.0$	$x+.1$	$x+.2$	$x+.3$	$x+.4$
5	.0000001				
4	.0002093	.0001145	.0000607	.0000311	.0000154
3	.0162494	.0120195	.0086291	.0060153	.0040721
2	.0162409	.0339777	.0441321	.0482532	.0478931
1	−.5820676	−.5195203	−.4469715	−.3695748	−.2920635
0	.0000000	−.1448007	−.2821139	−.4050264	−.5077160
−1	.5820676	.6294198	.6567760	.6601752	.6369100
−2	−.0162409	.0104651	.0471784	.0943909	.1518602
−3	−.0162494	−.0213049	−.0270545	−.0332049	−.0392596
−4	−.0002093	−.0003705	−.0006358	−.0010581	−.0017083
−5	−.0000001	−.0000002	−.0000006	−.0000015	−.0000034

$L_2'(x)$

x	$x+.5$	$x+.6$	$x+.7$	$x+.8$	$x+.9$
4	.0000074	.0000034	.0000015	.0000006	.0000002
3	.0026769	.0017083	.0010581	.0006358	.0003705
2	.0444848	.0392596	.0332049	.0270545	.0213049
1	−.1876748	−.1518602	−.0943909	−.0471784	−.0104651
0	−.5858619	−.6369100	−.6601752	−.6567760	−.6294198
−1	.5858619	.5077160	.4050264	.2821139	.1448007
−2	.1876748	.2920635	.3695748	.4469715	.5195203
−3	−.0444848	−.0478931	−.0482532	−.0441321	−.0339777
−4	−.0026769	−.0040721	−.0060153	−.0086291	−.0120195
−5	−.0000074	−.0000154	−.0000311	−.0000607	−.0001145

$L_2''(x)$

x	$x+.0$	$x+.1$	$x+.2$	$x+.3$	$x+.4$
5	−.0000010	−.0000004	−.0000001		
4	−.0012284	−.0007087	−.0003952	−.0002129	−.0001108
3	−.0465343	−.0380479	−.0298698	−.0225871	−.0164844
2	.2201282	.1369510	.0687454	.0162834	−.0210828
1	.5579563	.6842412	.7580806	.7819152	.7615818
0	−1.4606815	−1.4227560	−1.3118991	−1.1365778	− .9099683
−1	.5579763	.3809905	.1594624	−.0960473	.3711575
−2	.2201282	.3157558	.4193623	.5245097	.6230076
−3	−.0465343	−.0543645	−.0601451	−.0620423	−.0578316
−4	−.0012284	−.0020585	−.0033357	−.0052283	−.0079270
−5	−.0000010	−.0000025	−.0000057	−.0000126	−.0000270

x	$x+.5$	$x+.6$	$x+.7$	$x+.8$	$x+.9$
4	−.0000557	−.0000270	−.0000126	−.0000057	−.0000025
3	−.0116249	−.0079270	−.0052283	−.0033357	−.0020585
2	−.0450247	−.0578316	−.0620423	−.0601451	−.0543645
1	.7053804	.6230076	.5245097	.4193623	.3157558
0	−.6486751	−.3711575	−.0960473	.1594624	.3809905
−1	−.6486751	−.9099683	−1.1365778	−1.3118991	−1.4227560
−2	.7053804	.7615818	.7819152	.7580806	.6842412
−3	−.0450247	−.0210828	.0162834	.0687454	.1369510
−4	−.0116249	−.0164844	−.0225871	−.0298698	−.0380479
−5	−.0000557	−.0001108	−.0002129	−.0003952	−.0007087
−6				−.0000001	−.0000004

TABLE III. $L_3(x)=L_{6,3}(x, 1/2)$, $L_3'(x)$, $L_3''(x)$

$L_3(x)$

x	$x+.0$	$x+.1$	$x+.2$	$x+.3$	$x+.4$
6	.00000024	.00000012	.00000005	.00000002	.00000001
5	.00008286	.00005169	.00003150	.00001874	.00001088
4	.00268003	.00209760	.00160285	.00119677	.00087361
3	.00146844	.00384345	.00530620	.00602933	.00618622
2	−.06657958	−.06134625	−.05394260	−.04532872	−.03631488
1	.20814167	.14737265	.09315196	.04655844	.00818720
0	.70841267	.70114512	.67967926	.64500315	.59869152
−1	.20814167	.27390708	.34268907	.41215975	.47975381
−2	−.06657958	−.06856746	−.06617125	−.05825575	−.04376882
−3	.00146844	−.00197317	−.00659556	−.01244745	−.01948472
−4	.00268003	.00333890	.00404886	.00476644	.00542627
−5	.00008286	.00012978	.00019870	.00029745	.00043550
−6	.00000024	.00000049	.00000096	.00000183	.00000340

x	$x+.5$	$x+.6$	$x+.7$	$x+.8$	$x+.9$
5	.00000616	.00000340	.00000183	.00000096	.00000049
4	.00062367	.00043550	.00029745	.00019870	.00012978
3	.00593823	.00542627	.00476644	.00404886	.00333890
2	−.02754330	−.01948472	−.01244745	−.00659556	−.00197317
1	−.02182865	−.04376882	−.05825575	−.06617125	−.06856746
0	.54280388	.47975381	.41215975	.34268907	.27390708
−1	.54280388	.59869152	.64500315	.67967926	.70114512
−2	−.02182865	.00818720	.04655844	.09315196	.14737265
−3	−.02754330	−.03631488	−.04532872	−.05394260	−.06134625
−4	.00593823	.00618622	.00602933	.00530620	.00384345
−5	.00062367	.00087361	.00119677	.00160285	.00209760
−6	.00000616	.00001088	.00001874	.00003150	.00005169
−7		.00000001	.00000002	.00000005	.00000012

$L_3'(x)$

x	$x+.0$	$x+.1$	$x+.2$	$x+.3$	$x+.4$
6	−.0000017	−.0000009	−.0000004	−.0000002	−.0000001
5	−.0003813	−.0002500	−.0001598	−.0000996	−.0000605
4	−.0062356	−.0053949	−.0044992	−.0036324	−.0028472
3	.0288514	.0189098	.0106360	.0041176	−.0007095
2	.0379708	.0648953	.0815717	.0893524	.0898535
1	−.6356363	−.5771501	−.5054638	−.4254184	−.3417929
0	.0000000	−.1447858	−.2828738	−.4080067	−.5147714
−1	.6356363	.6763588	.6953825	.6897327	.6576776
−2	−.0379708	.0001221	.0497076	.1103706	.1808527
−3	−.0288514	−.0401754	−.0523583	−.0646105	−.0758656
−4	.0062356	.0069002	.0072273	.0070158	.0060307
−5	.0003813	.0005675	.0008239	.0011670	.0016121
−6	.0000017	.0000033	.0000063	.0000116	.0000206

$L_3'(x)$

x	$x+.5$	$x+.6$	$x+.7$	$x+.8$	$x+.9$
5	−.0000358	−.0000206	−.0000116	−.0000063	−.0000033
4	−.0021709	−.0016121	−.0011670	−.0008239	−.0005675
3	−.0040148	−.0060307	−.0070158	−.0072273	−.0069002
2	.0847953	.0758656	.0646105	.0523583	.0401754
1	−.2590022	−.1808527	−.1103706	−.0497076	−.0001221
0	−.5989336	−.6576776	−.6897327	−.6953825	−.6763588
−1	.5989336	.5147714	.4080067	.2828738	.1447858
−2	.2590022	.3417929	.4254184	.5054638	.5771501
−3	−.0847953	−.0898535	−.0893524	−.0815717	−.0648953
−4	.0040148	.0007095	−.0041176	−.0106360	−.0189098
−5	.0021709	.0028472	.0036324	.0044992	.0053949
−6	.0000358	.0000605	.0000996	.0001598	.0002500
−7		.0000001	.0000002	.0000004	.0000009

$L_3'(x)$

x	$x+.0$	$x+.1$	$x+.2$	$x+.3$	$x+.4$
6	.0000117	.0000061	.0000031	.0000015	.0000007
5	.0015631	.0010868	.0007350	.0004835	.0003094
4	.0077612	.0088518	.0089242	.0083240	.0073357
3	−.1069900	−.0913730	−.0739648	−.0565175	−.0403140
2	.3239975	.2160536	.1197566	.0385786	−.0256675
1	.5032414	.6588169	.7667843	.8261081	.8390021
0	−1.4591699	−1.4253081	−1.3259553	−1.1676245	−.9605716
−1	.5032414	.3045939	.0708748	−.1867923	−.4548011
−2	.3239975	.4384255	.5526106	.6585611	.7474910
−3	−.1069900	−.1186224	−.1237228	−.1195410	−.1033391
−4	.0077612	.0052612	.0009447	−.0055676	−.0145672
−5	.0015631	.0021860	.0029705	.0039174	.0050049
−6	.0000117	.0000217	.0000392	.0000685	.0001165
−7				.0000001	.0000001

x	$x+.5$	$x+.6$	$x+.7$	$x+.8$	$x+.9$
6	.0000003	.0000001	.0000001		
5	.0001926	.0001165	.0000685	.0000392	.0000217
4	.0061776	.0050049	.0039174	.0029705	.0021860
3	−.0261859	−.0145672	−.0055676	.0009447	.0052612
2	−.0726654	−.1033391	−.1195410	−.1237228	−.1186224
1	.8104398	.7474910	.6585611	.5526106	.4384255
0	−.7179590	−.4548011	−.1867923	−.0708748	.3045939
−1	−.7179590	−.9605716	−1.1676245	−1.3259553	−1.4253081
−2	.8104398	.8390021	.8261081	.7667843	.6588169
−3	−.0726654	−.0256675	.0385786	.1197566	.2160536
−4	−.0261859	−.0403140	−.0565175	−.0739648	−.0913730
−5	.0061776	.0073357	.0083240	.0089242	.0088518
−6	.0001926	.0003094	.0004835	.0007350	.0010868
−7	.0000003	.0000007	.0000015	.0000031	.0000061

Cardinal Interpolation and Spline Functions IV. The Exponential Euler Splines*)

By

I. J. SCHOENBERG

MATH. RESEARCH CENTER
UNIVERSITY OF WISCONSIN
MADISON

For George Pólya on his 85th birthday, December 13, 1972

Contents

Introduction

As background for our discussion we recall a result from [10]. First a few definitions. Let $\mathscr{S}_n$ denote the class of cardinal spline functions $S(x)$ of degree n $(n\geqq 1)$ having their knots at the integer points of the real axis. This means that $S(x)\in\mathscr{S}_n$, provided that the restriction of $S(x)$ to every unit interval $(\nu, \nu+1)$ is a polynomial of degree n at most, and that

$$(1)\qquad S(x)\in C^{n-1}(-\infty, \infty).$$

*) Sponsored by the U. S. Army under Contract No. DA—31—124—ARO—D—462.

Furthermore, let

(2) $$\mathcal{S}_n^* = \{S(x);\ S(x+\tfrac{1}{2})\in\mathcal{S}_n\}.$$

The elements of $\mathcal{S}_n^*$ are again cardinal spline functions of degree n, but having their knots at $\nu+\frac{1}{2}$ half way between consecutive integers.

Let $y=(y_\nu)$ $(-\infty<\nu<\infty)$ be an infinite sequence of real or complex numbers. We call cardinal spline interpolation the problem of finding $S(x)$ in $\mathcal{S}_n$, or perhaps in $\mathcal{S}_n^*$, such that

(3) $$S(\nu)=y_\nu \quad \text{for all integers } \nu.$$

This problem has a unique solution subject to certain conditions to be described presently.

Let

(4) $$\tilde{\mathcal{Y}} = \{y=(y_\nu);\ y_\nu=O(|\nu|^s) \text{ as } \nu\to\pm\infty, \text{ for some appropriate } s\geqq 0\},$$

(5) $$\tilde{\mathcal{S}}_n = \{S(x);\ S(x)\in\mathcal{S}_n,\ S(x)=O(|x|^s) \text{ as } x\to\pm\infty, \text{ for some appropriate } s\geqq 0\}.$$

We define similarly the class $\tilde{\mathcal{S}}_n^*$ if we replace in (5) the condition $S(x)\in\mathcal{S}_n$ by $S(x)\in\mathcal{S}_n^*$. We may briefly describe the sequences y and functions $S(x)$ of these classes as being of power growth.

The main result of [10] is the following

THEOREM 1. 1. *Let n be odd. If*

(6) $$y = (y_\nu)\in\tilde{\mathcal{Y}},$$

then the interpolation problem (3) *has a unique solution $S(x)$ such that*

(7) $$S(x)\in\tilde{\mathcal{S}}_n.$$

2. *Let n be even. If* (6) *holds, then the problem* (3) *has a unique solution $S(x)$ such that*

(8) $$S(x)\in\tilde{\mathcal{S}}_n^*.$$

Only if $n=1$ is this result trivial, the graph of the solution $S(x)$ being the continuous polygonal line obtained by linear interpolation between consecutive data.

The purpose of the present paper is to study the possibility of spline interpolation of the geometric progression

(9) $$y_\nu = t^\nu \qquad (-\infty<\nu<\infty;\ t \text{ fixed},\ t\neq 0).$$

If t is such that $|t|=1$, then $y=(t^\nu)$ is a bounded sequence so that Theorem 1 applies.

If $|t| \neq 1$, then (t^ν) grows exponentially and a different special approach is needed to solve this problem. We shall see that already Euler has provided the essential tools for its solution.

I acknowledge with pleasure the helpful advice of my colleagues T. N. E. GREVILLE and A. SHARMA.

I. The cardinal spline interpolation of a geometric progression

1. *The exponential splines.* If we seek to find within the class $\mathcal{S}_n$ a not identically vanishing element having smallest support, then we find the so-called *B*-splines, up to a constant factor. The forward *B*-spline is the function

$$(1.1)\quad Q(x) = Q_{n+1}(x) = \frac{1}{n!}\left\{x_+^n - \binom{n+1}{1}(x-1)_+^n + \cdots + (-1)^{n+1}(x-n-1)_+^n\right\},$$

where $x_+ = \max(0, x)$. This is an element of $\mathcal{S}_n$ having several properties, the most obvious being that

$$(1.2)\qquad Q(x) > 0 \quad \text{if} \quad 0 < x < n+1, \quad Q(x) = 0 \quad \text{elsewhere},$$

and its symmetry about the point $x = (n+1)/2$ expressed by the identity

$$(1.3)\qquad Q(x) = Q(n+1-x).$$

Its most important property we state as

LEMMA 1. *Every $S(x) \in \mathcal{S}_n$ allows a unique representation of the form*

$$(1.4)\qquad S(x) = \sum_{-\infty}^{\infty} c_j Q(x-j),$$

with constant coefficients c_j. Conversely, for every sequence (c_j) (1. 4) defines an element of $\mathcal{S}_n$ (see [3] for a more general Theorem 4 on page 80).

We turn now to the problem of interpolating the sequence (9) by an element of $\mathcal{S}_n$. Equivalently, we are to interpolate at the integers the exponential function $f(x) = t^x$ by an element of $\mathcal{S}_n$. Because this exponential satisfies the functional equation

$$(1.5)\qquad f(x+1) = tf(x) \quad \text{for all real } x,$$

it seems natural to try to interpolate the sequence (9) by an element of $\mathcal{S}_n$ satisfying the same functional equation (1. 5).

Such elements are described by

LEMMA 2. *The most general element* $S(x)$ *of* $\mathscr{S}_n$ *satisfying the functional equation*

$$S(x+1) = tS(x) \tag{1.6}$$

is of the form

$$S(x) = c_0 \sum_{-\infty}^{\infty} t^j Q(x-j) \qquad (c_0 \text{ constant}). \tag{1.7}$$

PROOF. Starting from the expansion (1. 4) we find that

$$S(x+1) = \sum_j c_j Q(x+1-j) = \sum c_{j+1} Q(x-j), \tag{1.8}$$

and

$$tS(x) = \sum_j tc_j Q(x-j). \tag{1.9}$$

Since the right sides of (1. 8) and (1. 9) are to be identical by the requirement (1. 6), the unicity in Lemma 1 shows that we must have $c_{j+1}=tc_j$ for all j, or $c_j=c_0 t^j$ for all j, and the representation (1. 4) reduces to (1. 7). Conversely, the same argument shows that the function (1. 7) satisfies (1. 6).

DEFINITION 1. *The elements of* $\mathscr{S}_n$ *of the form* (1. 7) *are called* exponential splines. For convenience we write

$$\Phi_n(x;t) = \sum_{-\infty}^{\infty} t^j Q_{n+1}(x-j). \tag{1.10}$$

Similar arguments show that

$$S^*(x) = c_0 \Phi_n(x+\tfrac{1}{2};t) \tag{1.11}$$

is the most general element of $\mathscr{S}_n^*$ satisfying (1. 6).

2. *A preliminary solution of the interpolation problem.* A solution of the problem of interpolating the geometric progression (9) becomes now obvious: It all depends on whether

$$\Phi_n(0;t)\neq 0, \tag{2.1}$$

or

$$\Phi_n(0;t)=0, \tag{2.2}$$

and we may state the following: *If* (2. 1) *holds, then*

$$S_n(x;t)=\Phi_n(x;t)/\Phi_n(0;t) \tag{2.3}$$

is a solution of the problem. If (2. 2) *holds, then the attempt at interpolation fails.*

PROOF. If (2. 1) holds, then $S_n(0;t)=1$ and the relation (1. 6) shows that $S_n(\nu;t)=t^\nu$ for all integers ν.

Let us express the critical quantity $\Phi_n(0; t)$ in terms of the B-spline. Using (1. 10) and (1. 3) we find that

$$\Phi_n(0; t) = \sum t^j Q(-j) = \sum t^j Q(n+1+j) = \sum t^{j-n} Q(j+1)$$

and in view of (1. 2), we obtain

$$\Phi_n(0; t) = t^{-n} \sum_0^{n-1} Q(j+1) t^j. \tag{2. 4}$$

LEMMA 3. *The reciprocal algebraic equation of degree $n-1$*

$$\sum_0^{n-1} Q(j+1) t^j = 0 \tag{2. 5}$$

has only simple and negative roots λ_ν, which we label so as to satisfy the inequalities

$$\lambda_{n-1} < \lambda_{n-2} < \cdots < \lambda_2 < \lambda_1 < 0. \tag{2. 6}$$

This was established in [9, Lemma 8 on page 182] for odd values of n only. In Lemma 7 below we derive this for all n and point out that it was known to FROBENIUS [5] in 1910.

Similarly, an attempt of interpolation of (9) by an exponential spline (1. 11), of $\mathcal{S}_n^*$, depends on the condition

$$\Phi_n(\tfrac{1}{2}; t) \neq 0, \tag{2. 7}$$

in which case

$$S_n(x; t) = \Phi_n(x+\tfrac{1}{2}; t)/\Phi_n(\tfrac{1}{2}; t) \tag{2. 8}$$

is a solution in $\mathcal{S}_n^*$. From

$$\Phi_n(\tfrac{1}{2}; t) = \sum t^j Q(\tfrac{1}{2}-j) = \sum t^j Q(n+\tfrac{1}{2}+j) = \sum t^{j-n} Q(j+\tfrac{1}{2})$$

we obtain

$$\Phi_n(\tfrac{1}{2}; t) = t^{-n} \sum_0^{n} Q(j+\tfrac{1}{2}) t^j. \tag{2. 9}$$

The relevant result here is

LEMMA 3*. *The reciprocal algebraic equation of degree n*

$$\sum_0^{n} Q(j+\tfrac{1}{2}) t^j = 0 \tag{2. 10}$$

has only simple and negative roots μ_ν satisfying

$$\mu_n < \mu_{n-1} < \cdots < \mu_2 < \mu_1 < 0. \tag{2. 11}$$

This was established in [9, Lemma 8, page 182] for even values of n only. For a general proof see Lemma 7* below (§ 5).

We may summarize our results as follows.

THEOREM 2. 1. *If*

(2. 12) $$t \neq \lambda_\nu \qquad (\nu=1, \ldots, n-1),$$

then

(2. 13) $$S_n(x; t) = \Phi_n(x; t)/\Phi_n(0; t)$$

is an element of $\mathcal{S}_n$ *interpolating the sequence* (9).

2. *If*

(2. 14) $$t \neq \mu_\nu \qquad (\nu=1, \ldots, n),$$

then

(2. 15) $$S_n^*(x; t) = \Phi_n(x+\tfrac{1}{2}; t)/\Phi_n(\tfrac{1}{2}; t)$$

is an element of $\mathcal{S}_n^*$ *interpolating the sequence* (9).

DEFINITION 2. *We call* (2. 13) *the* exponential Euler spline of degree *n*, *while* (2. 15) *is the* midpoint exponential Euler spline. *Note the conditions* (2. 12) *and* (2. 14), *respectively, under which these functions exist.*

Here are the reasons for these terms:

1. Assume that n is odd. Since $n-1$ is even and because the λ_ν are reciprocal in pairs, it follows from (2. 6) that $-1 \neq \lambda_\nu$ for $\nu=1, \ldots, n-1$. By Theorem 2 we may therefore consider the function $S_n(x; -1)$ and find it to be identical with the odd degree *Euler spline* interpolating the sequence $y_\nu=(-1)^\nu$ (see [10, § 1B]).

2. Let n be even. From (2. 11), and the fact that the μ_ν are reciprocal in pairs, we conclude that $-1 \neq \mu_\nu$ for $\nu=1, \ldots, n$. By Theorem 2 we may therefore consider the function $S_n^*(x; -1)$ which is identical with the even degree Euler spline interpolating the sequence $y_\nu=(-1)^\nu$ (see [10, § 1 B]).

Further reasons for the term "exponential Euler splines" will become obvious in Part II of the present paper.

We add that the exponential splines

(2. 16) $$\Phi_n(x; \lambda_\nu) \quad (\nu=1, \ldots, n-1) \quad \text{and} \quad \Phi_n(x+\tfrac{1}{2}; \mu_\nu) \quad (\nu=1, \ldots, n)$$

are the eigensplines of the classes $\mathcal{S}_n$ and $\mathcal{S}_n^*$, respectively, and that they play a fundamental role in the proof of Theorem 1 (see [10]).

Theorem 2 furnishes an explicit spline interpolant for the geometric progression (9) within $\mathcal{S}_n$, as well as $\mathcal{S}_n^*$. However, we are not yet satisfied with this solution, which we call "preliminary", and this for several reasons. Firstly, the evaluation of the exponential Euler splines (2. 13) and (2. 15), is laborious. Secondly, it seems difficult to decide from these expressions how the spline functions $S_n(x; t)$ and $S_n^*(x; t)$ behave as we let $n\to\infty$. Indeed, the feeling seems inescapable that the relations

$$\lim_{n\to\infty} S_n(x; t) = \lim_{n\to\infty} S_n^*(x; t) = t^x$$

should hold, at least if $t>0$, when the conditions (2. 12) and (2. 14) are automatically fulfilled. For this reason we present below a computationally sounder approach to the exponential Euler splines. It will also show clearly the role of Euler's work in this matter.

II. The exponential Euler polynomials

3. *The exponential Euler polynomial $A_n(x; t)$.* We assume throughout that $t\neq 1$. Since n and t are kept fixed we shall sometimes omit them and write $\Phi_n(x; t)=$ $=\Phi_n(x)=\Phi(x)$.

LEMMA 4. *The polynomial $P(x)$ representing the spline function $\Phi_n(x; t)$ in the interval* [0, 1], *is of exact degree n.*

PROOF. Let us assume that

$$P(x)\in\pi_{n-1}. \tag{3. 1}$$

From the functional equation

$$\Phi_n(x+1; t) = t\Phi_n(x; t), \tag{3. 2}$$

we obtain $\Phi_n(x+\nu) = t^\nu\Phi(x)$ and therefore also

$$\Phi(x) = t^\nu\Phi(x-\nu) = t^\nu P(x-\nu) \quad \text{if} \quad \nu<x<\nu+1.$$

From (3. 1) we conclude that all polynomial components of $\Phi(x)$ are of degree $n-1$. But then the basic requirement (1) implies that the entire function $\Phi_n(x; t)$ reduces to a single polynomial of degree $n-1$, which is absurd in view of (3. 2) and the assumption that $t\neq 1$.

DEFINITION 3. *If*

$$\Phi_n(x; t) = P(x) = C_0(t)x^n + \text{lower degree terms}, \quad \text{in} \quad 0\leqq x\leqq 1,$$

we define

$$A_n(x; t) = \frac{1}{C_0(t)}P(x) = x^n + \text{lower degree terms},$$

and call it the exponential Euler polynomial *of degree n.*

By differentiation of (3. 2) and setting $x=0$ we obtain

$$\Phi^{(\nu)}(1) = t\Phi^{(\nu)}(0) \qquad (\nu=0, 1, \ldots, n-1). \tag{3. 3}$$

From the Definition 3 we conclude that $A_n(x;t)$ has the following properties:

(3. 4) $$A_n(x;t) = x^n + \text{lower degree terms},$$

(3. 5) $$A_n^{(\nu)}(1;t) = tA_n^{(\nu)}(0;t) \qquad (\nu=0, 1, \ldots, n-1).$$

LEMMA 5. *The exponential Euler polynomial $A_n(x;t)$ is uniquely defined by the properties* (3. 4) *and* (3. 5).

PROOF. Indeed, suppose we had two distinct polynomials $\tilde{A}(x)$ and $\tilde{\tilde{A}}(x)$, both having the properties (3. 4) and (3. 5). By means of the definition

(3. 6) $$\tilde{S}(x) = t^\nu \tilde{A}(x-\nu) \quad \text{if} \quad \nu \leq x < \nu+1, \quad \text{and all integer} \quad \nu,$$

we obtain a piecewise polynomial function $\tilde{S}(x)$ defined for all real x. The properties $\tilde{A}^{(\nu)}(1)=t\tilde{A}^{(\nu)}(0)$, $(\nu=0, \ldots, n-1)$, and (3. 6), will easily show that $\tilde{S}(x)$ is an element of $\mathcal{S}_n$ satisfying the relation $\tilde{S}(x+1) = t\tilde{S}(x)$. Similarly we obtain $\tilde{\tilde{S}}(x) \in \mathcal{S}_n$, such that $\tilde{\tilde{S}}(x) = \tilde{\tilde{A}}(x)$ in $[0, 1]$ and satisfies $\tilde{\tilde{S}}(x+1) = t\tilde{\tilde{S}}(x)$. It follows that $s(x) = \tilde{S}(x) - \tilde{\tilde{S}}(x)$ is also an element of $\mathcal{S}_n$ satisfying $s(x+1) = ts(x)$. Moreover, $s(x) \not\equiv 0$. By Lemma 2 we conclude that

$$s(x) = c_0 \Phi_n(x;t) \qquad (c_0 \neq 0).$$

However, now we reach a contradiction with Lemma 4, because

$$S(x) = \tilde{A}(x) - \tilde{\tilde{A}}(x) \in \pi_{n-1} \quad \text{in} \quad [0, 1].$$

Let us determine the generating function of the $A_n(x;t)$. In order to use more effectively the relations (3. 5) for this purpose, we write

(3. 7) $$A_n(x;t) = x_n + \binom{n}{1} a_1 x^{n-1} + \binom{n}{2} a_2 x^{n-2} + \cdots + a_n,$$

where the coefficients $a_\nu = a_\nu(t)$ depend on t and presumably also on n. Observe, that the relations (3. 5) are equivalent to the relations

(3. 8) $$1 + \binom{\nu}{1} a_1 + \binom{\nu}{2} a_2 + \cdots + a_\nu = ta_\nu \qquad (\nu=1, 2, \ldots, n).$$

Indeed, the relation (3. 5) is equivalent to the relation (3. 8) with ν replaced by $n-\nu$.

The relations (3. 8) show a surprising fact: Since $t \neq 1$, the coefficients $a_\nu = a_\nu(t)$ do not depend on n at all, but rather the coefficients $a_0=1, a_1, \ldots, a_n$ form the section of an infinite sequence $a_0=1, a_1, a_2, \ldots$ defined by the infinite system of relations

(3. 9) $$1 + \binom{\nu}{1} a_1 + \cdots + a_\nu = ta_\nu \qquad (\nu=1, 2, 3, \ldots).$$

Let us determine this infinite sequence. The structure of the relations (3. 9) suggest the use of the generating function

$$g(z) = \sum_0^\infty \frac{a_\nu}{\nu!} z^\nu. \tag{3.10}$$

The relations (3. 9) are equivalent to the identity

$$t-1 = \sum_0^\infty \frac{ta_\nu}{\nu!} z^\nu - \sum_0^\infty \frac{1+\binom{\nu}{1}a_1+\cdots+a_\nu}{\nu!} z^\nu,$$

or

$$t-1 = tg(z)-e^z g(z),$$

whence

$$g(z) = \frac{t-1}{t-e^z} = \sum_0^\infty \frac{a_\nu(t)}{\nu!} z^\nu. \tag{3.11}$$

Combining (3. 7) and (3. 11) we obtain

LEMMA 6. *The exponential Euler polynomials*

$$A_n(x;t) = x_n + \binom{n}{1} a_1(t)x^{n-1} + \cdots + a_n(t) \tag{3.12}$$

form an Appell sequence of polynomials generated by the expansion

$$\frac{t-1}{t-e^z} e^{xz} = \sum_0^\infty \frac{A_n(x;t)}{n!} z^n. \tag{3.13}$$

L. CARLITZ [2] uses the notation $A_n(x;t)=H_n(x|t)$ and calls them Eulerian polynomials. If $t=-1$, then (3. 13) becomes

$$\frac{2}{e^z+1} e^{xz} = \sum_0^\infty \frac{A_n(x;-1)}{n!} z^n$$

showing that $A_n(x;-1)=E_n(x)$ are the standard Euler polynomials (see NÖRLUND [7, 35]).

We introduce now a new set of polynomials $\Pi_n(z)$ by their generating function

$$\frac{t-1}{t-e^z} = \sum_0^\infty \frac{\Pi_n(t)}{(t-1)^n} \frac{z^n}{n!} \qquad (\Pi_0(t)=1). \tag{3.14}$$

In view of (3. 11), we see that

$$a_n(t) = \frac{\Pi_n(t)}{(t-1)^n}. \tag{3.15}$$

We shall see that $\Pi_n(t)$ $(n \geqq 1)$ is a reciprocal monic polynomial of degree $n-1$ having integer coefficients. These are remarkable polynomials indeed. They were introduced by EULER [4, Chap. VII, § 178] and were studied by several mathematicians prior to FROBENIUS [5] who showed that $\Pi_n(t)$ has simple negative zeros. For this reason we call them *Euler—Frobenius polynomials.* They were rediscovered by a number of mathematicians, in particular in connection with spline interpolation (see [8], [9], [1]). Also by E. HILLE [6, 47]. We retained the notation $\Pi_n(t)$ used by QUADE and COLLATZ in their outstanding paper [8].

LEMMA 7. 1. *The Euler—Frobenius polynomial $\Pi_n(t)$ defined by the generating function* (3. 14) *is a monic reciprocal polynomial of degree $n-1$ having integer coefficients. It is related to the B-spline* (1. 1) *by the relation*

$$\Pi_n(t) = n! \sum_0^{n-1} Q_{n+1}(\nu+1)t^\nu. \tag{3.16}$$

2. *$\Pi_n(t)$ can be independently defined by the expansion*

$$\frac{\Pi_n(t)}{(1-t)^{n+1}} = \sum_0^\infty (\nu+1)^n t^\nu. \tag{3.17}$$

3. *It satisfies the recurrence relation*

$$\Pi_{n+1}(t) = (1+nt)\Pi_n(t) + t(1-t)\Pi_n'(t) \qquad (\Pi_1(t)=1). \tag{3.18}$$

4. *The zeros λ_ν of $\Pi_n(t)$ are simple and negative. We label them so that*

$$\lambda_{n-1} < \lambda_{n-2} < \cdots < \lambda_2 < \lambda_1 < 0. \tag{3.19}$$

PROOF OF 2: Assuming $|t|<1$ and $|z|$ sufficiently small, the following expansions and interchanges of summations are valid

$$\frac{t-1}{t-e^z} = e^{-z}(1-t)/(1-te^{-z}) = e^{-z}(1-t)\sum_0^\infty t^\nu e^{-\nu z} = (1-t)\sum_0^\infty t^\nu e^{-(\nu+1)z} =$$

$$= (1-t)\sum_0^\infty t^\nu \sum_{n=0}^\infty \frac{(-1)^n(\nu+1)^n}{n!} z^n,$$

$$\frac{t-1}{t-e^z} = (1-t)\sum_{n=0}^\infty (-1)^n \frac{z^n}{n!} \sum_{\nu=0}^\infty (\nu+1)^n t^\nu.$$

On comparing with (3. 14) we find that (3. 17) holds.

PROOF OF 1. In (3. 17) we may replace $(\nu+1)^n$ by $(\nu+1)_+^n$ and then sum over all integer values of ν. Thus (3. 17) yields

$$\Pi_n(t) = \sum_{i=0}^{n+1} (-1)^i \binom{n+1}{i} t^i \sum_{\nu=-\infty}^\infty (\nu+1)_+^n t^\nu.$$

Setting $i+\nu=j$, whence $\nu=j-i$, and summing first by i we obtain

$$\Pi_n(t)=\sum_{j=0}^{\infty} t^j \sum_{i=0}^{n+1}(-1)^i\binom{n+1}{i}(j+1-i)_+^n.$$

In terms of the B-spline

$$Q_{n+1}(t)=\frac{1}{n!}\sum_{i=0}^{n+1}(-1)^i\binom{n+1}{i}(t-i)_+^n$$

we obtain

$$\Pi_n(t)=n!\sum_{j=0}^{\infty} t^j Q_{n+1}(j+1)$$

which establishes (3. 16) in view of (1. 2).

PROOF OF 3. Multiplying (3. 17) by t and differentiating we easily obtain (3. 18).

PROOF OF 4. We establish this by induction in n, as first done by FROBENIUS in [5]. Assuming that the zeros of $\Pi_n(t)$ satisfy (3. 19), we find from (3. 18) that

$$\Pi_{n+1}(\lambda_\nu)=\lambda_\nu(1-\lambda_\nu)\Pi_n'(\lambda_\nu) \qquad (\nu=1,\ldots,n-1).$$

Here $(-1)^{\nu-1}\Pi_n'(\lambda_\nu)>0$ for all ν, hence

$$(-1)^\nu \Pi_{n+1}(\lambda_\nu)>0 \qquad (\nu=1,\ldots,n-1),$$

and it follows $\Pi_{n+1}(t)$ has a zero in each of the n intervals

$$(-\infty,\lambda_{n-1}),(\lambda_{n-1},\lambda_{n-2}),\ldots,(\lambda_1,0).$$

Using (3. 18) we easily find that

$$\begin{aligned}
&\Pi_0(t)=\Pi_1(t)=1, && \Pi_4(t)=t^3+11t^2+11t+1,\\
&\Pi_2(t)=t+1, && \Pi_5(t)=t^4+26t^3+66t^2+26t+1,\\
&\Pi_3(t)=t^2+4t+1, && \Pi_6(t)=t^5+57t^4+302t^3+302t^2+57t+1.
\end{aligned}$$

4. *Solution of the interpolation problem in $\mathscr{S}_n$.* From its Definition 3 we know that the exponential Euler polynomial

$$A_n(x;t)=x_n+\binom{n}{1}a_1(t)x^{n-1}+\cdots+a_n(t) \tag{4. 1}$$

is, up to a non-vanishing factor, identical with the polynomial component of $\Phi_n(x,t)$ in [0, 1]. Moreover, by

$$A_n(0;t)=a_n(t)=(t-1)^{-n}\Pi_n(t)$$

and the identity (3. 16), we know that the λ_ν in (2. 12) are identical with those in (3. 19). We may therefore state the

COROLLARY 1. *If* $t \neq 1$ *and*

(4. 2) $$t \neq \lambda_\nu \qquad (\nu = 1, \ldots, n-1),$$

then the exponential Euler spline (2. 13) *may also be defined by*

(4. 3) $$S_n(x; t) = A_n(x; t)/A_n(0; t) \quad \text{in the interval } [0, 1].$$

The extension of $S_n(x; t)$ *to a real* x *may be accomplished by means of the functional equation*

(4. 4) $$S_n(x+1; t) = tS_n(x; t).$$

In § 11 below we give numerical results for the particularly simple case when $t=2$.

5. *The case of the class* $\mathcal{S}_n^*$. The case of $\mathcal{S}_n^*$ does not require an essentially new discussion, in view of the definition (2) of $\mathcal{S}_n^*$. A shift of origin to $x=\frac{1}{2}$ is all that is required. Writing

(5. 1) $$B_n(x; t) = A_n(x+\tfrac{1}{2}; t) = x_n + \binom{n}{1} b_1(t)x^{n-1} + \cdots + b_n(t),$$

we obtain from (3. 13) the generating function

(5. 2) $$\frac{t-1}{t-e^z} e^{(x+\frac{1}{2})z} = \sum_0^\infty \frac{B_n(x; t)}{n!} z^n,$$

whence

(5. 3) $$\frac{t-1}{t-e^z} e^{z/2} = \sum_0^\infty \frac{b_n(t)}{n!} z^n.$$

We now define a new sequence of polynomials $P_n(t)$ by setting

(5. 4) $$b_n(t) = \frac{P_n(t)}{2^n(t-1)^n}.$$

As an analogue of Lemma 7 we have here

LEMMA 7*. 1. *The* $P_n(t)$ *defined by the generating function*

(5. 5) $$\frac{t-1}{t-e^z} e^{z/2} = \sum_0^\infty \frac{P_n(t)}{2^n(t-1)^n} \frac{z^n}{n!}$$

are monic reciprocal polynomials of degree n having integer coefficients. In terms of the B-spline (1. 1) *we may write*

(5. 6) $$P_n(t) = 2^n n! \sum_{\nu=0}^{n} Q_{n+1}(\nu+\tfrac{1}{2})t^\nu.$$

2. *$P_n(t)$ can be independently defined by the expansion*

(5. 7)
$$\frac{P_n(t)}{(1-t)^{n+1}} = \sum_{\nu=0}^{\infty} (2\nu+1)^n t^\nu.$$

3. *$P_n(t)$ satisfies the recurrence relation*

(5. 8)
$$P_{n+1}(t) = \big(1+(2n+1)t\big)P_n(t)+2t(1-t)P_n'(t) \qquad \big(P_0(t)=1\big).$$

4. *The zeros μ_ν of $P_n(t)$ are simple and negative and we may assume that*

(5. 9)
$$\mu_n < \mu_{n-1} < \cdots < \mu_2 < \mu_1 < 0.$$

As in the case of Lemma 7, these statements are established in the order 2., 1., 3., and 4. As the arguments are entirely analogous, we may omit details. (We begin the proof of 3. by replacing t by t^2 in (5. 7)).

The polynomials $P_n(t)$ may be called *midpoint Euler—Frobenius polynomials.* We find by means of (5. 8)

$$P_0(t) = 1, \qquad P_3(t) = t^3+23t^2+23t+1,$$
$$P_1(t) = t+1, \qquad P_4(t) = t^4+76t^3+230t^2+76t+1.$$
$$P_2(t) = t^2+6t+1,$$

In view of the identity (5. 6) it is clear that the μ_ν in (2. 14) are identical with those in (5. 9). We may therefore state

COROLLARY 1*. *If $t\neq 1$ and*

(5. 10)
$$t\neq\mu_\nu \qquad (\nu=1, \ldots, n),$$

then the exponential Euler spline (2. 15) *may be written as*

(5. 11)
$$S_n^*(x; t) = B_n(x; t)/B_n(0; t) \quad \textit{in the interval} \quad -\tfrac{1}{2}\leq x\leq \tfrac{1}{2}.$$

The extension of $S_n^*(x; t)$ to all real x is made by the functional equation

(5. 12)
$$S_n^*(x+1; t) = tS_n^*(x; t).$$

6. *A characteristic extremum property of the polynomial $A_n(x;t)/A_n(0; t)$ if $t>1$.* We consider the polynomial

(6. 1)
$$\varphi_n(x)=\varphi_n(x; t)=A_n(x; t)/A_n(0; t),$$

which is the polynomial representing the exponential Euler spline (4. 3) in the interval [0, 1]. Here we assume that

(6. 2)
$$t>1.$$

Similar results hold if $0<t<1$, but we omit their discussion.

From (3. 5) we see that $\varphi_n(x)$ has the following properties:

$$\varphi_n(0) = 1, \tag{6. 3}$$

$$\varphi_n^{(\nu)}(1) = t\varphi_n^{(\nu)}(0) \qquad (\nu=0, 1, \dots, n-1). \tag{6. 4}$$

We define the class F_n of functions $f(x)$ satisfying the following three conditions:

$$f(x) \in C^{n-1}[0, 1] \quad \textit{and} \quad f^{(n-1)}(x) \quad \textit{satisfies a Lipschitz condition,} \tag{6. 5}$$

$$f(0) \geqq 1, \tag{6. 6}$$

$$f^{(\nu)}(1) \geqq tf^{(\nu)}(0) \qquad (\nu=0, 1, \dots, n-1). \tag{6. 7}$$

THEOREM 3. *The polynomial $\varphi_n(x)$ is the unique element of the class F_n that minimizes the norm*

$$\|f^{(n)}\| = \operatorname*{essential\,sup}_{0\leqq x\leqq 1} |f^{(n)}(x)| \quad \text{if} \quad f(x) \in F_n, \tag{6. 8}$$

giving it its least value

$$\|\varphi_n^{(n)}\| = \frac{n!}{a_n(t)} = n!(t-1)^n/\Pi_n(t). \tag{6. 9}$$

PROOF. We define the function $K(x)$ as a polynomial of degree $n-1$ by setting

$$K(x) = \varphi_{n-1}(x; t^{-1}) = A_{n-1}(x; t^{-1})/A_{n-1}(0; t^{-1}). \tag{6. 10}$$

We know that the $A_n(x; t)$ form an Appell sequence. This means that $A_n'(x; t) = nA_{n-1}(x; t)$. Repeated differentiation of (6. 10) therefore gives

$$K^{(\nu)}(0) = n(n-1)\cdots(n-\nu+1)A_{n-1-\nu}(0; t^{-1})/a_{n-1}(t^{-1}) =$$

$$= \nu!\binom{n}{\nu} a_{n-1-\nu}(t^{-1})/a_{n-1}(t^{-1}) = \nu!\binom{n}{\nu}(t^{-1}-1)^\nu \Pi_{n-1-\nu}(t^{-1})/\Pi_{n-1}(t^{-1})$$

in view of (3. 15). Since $\Pi_k(x)>0$ if $x>0$, and $t^{-1}-1<0$, by (6. 2), we conclude that

$$(-1)^\nu K^{(\nu)}(0)>0 \qquad (\nu=0, 1, \dots, n-1). \tag{6. 11}$$

Moreover, $K(x)$ satisfies, by (6. 10), (6. 3), and (6. 4), the relations

$$K(0)=1, \tag{6. 12}$$

$$K^{(\nu)}(1) = t^{-1}K^{(\nu)}(0) \qquad (\nu=0, 1, \dots, n-2). \tag{6. 13}$$

Evidently, we are to establish the following: *If*

$$f(x) \in F_n \quad \textit{and} \quad \|f^{(n)}\| \leqq \|\varphi_n^{(n)}\|, \tag{6. 14}$$

then

$$f(x) = \varphi_n(x) \quad \textit{in} \quad [0, 1]. \tag{6. 15}$$

We first derive a certain identity by integrations by parts:

$$\int_0^1 Kf^{(n)}\,dx = \int_0^1 K\,df^{(n-1)} = K(1)f^{(n-1)}(1) - K(0)f^{(n-1)}(0) - \int_0^1 K'f^{(n-1)}\,dx =$$

$$= t^{-1}K(0)\left(f^{(n-1)}(1) - tf^{(n-1)}(0)\right) - \int_0^1 K'\,df^{(n-2)} =$$

$$= t^{-1}K(0)\left(f^{(n-1)}(1) - tf^{(n-1)}(0)\right) - t^{-1}K'(0)\left(f^{(n-2)}(1) - tf^{(n-2)}(0)\right) + \int_0^1 K''\,df^{(n-3)}$$

and we continue in like manner until we reach the relation

$$\int_0^1 K(x)f^{(n)}(x)\,dx =$$

$$= \sum_{\nu=0}^{n-2} t^{-1}(-1)^{\nu}K^{(\nu)}(0)\left(f^{(n-\nu-1)}(1) - tf^{(n-\nu-1)}(0)\right) + (-1)^{n-1}\int_0^1 K^{(n-1)}(x)f'(x)\,dx.$$

But $K^{(n-1)}(x) = K^{(n-1)}$ is a constant and therefore

$$\int_0^1 K(x)f^{(n)}(x)\,dx = \sum_{\nu=0}^{n-2} t^{-1}(-1)^{\nu}K^{(\nu)}(0)\left(f^{(n-\nu-1)}(1) - tf^{(n-\nu-1)}(0)\right) + (-1)^{n-1}K^{(n-1)}\left(f(1) - f(0)\right). \tag{6.16}$$

This is the identity that we need. We apply it first to $f(x) = \varphi_n(x)$. From (6. 3) and (6. 4) we obtain that

$$\int_0^1 K(x)\varphi_n^{(n)}(x)\,dx = (-1)^{n-1}K^{(n-1)}\left(\varphi_n(1) - \varphi_n(0)\right) = |K^{(n-1)}|\,(t-1), \tag{6.17}$$

by (6. 11). Observe next that by (6. 11) and (6. 7) all terms of the sum in (6. 16) are $\geqq 0$. Moreover, by (6. 6) and (6. 7)

$$f(1) - f(0) \geqq tf(0) - f(0) = (t-1)f(0) \geqq t-1 \tag{6.18}$$

and therefore

$$f(1) - f(0) \geqq t-1. \tag{6.19}$$

Using these facts as well as the assumptions (6. 14) we get the following string of inequalities and relations

$$|K^{(n-1)}|\,(t-1) \leqq \sum_{0}^{n-2} t^{-1}(-1)^{\nu}K^{(\nu)}(0)\left(f^{(n-\nu-1)}(1) - tf^{(n-\nu-1)}(0)\right) + (-1)^{n-1}K^{(n-1)}\left(f(1) - f(0)\right) = \tag{6.20}$$

$$= \int_0^1 K(x)f^{(n)}(x)\,dx \leqq \int_0^1 K(x)\varphi_n^{(n)}(x)\,dx = |K^{(n-1)}|\,(t-1)$$

by (6. 19), (6. 16), (6. 14) and (6. 17). Since the extreme terms of (6. 20) are equal, we must have everywhere equalities, also in (6. 19). Therefore, by $f(1)-f(0) = t-1$ and (6. 18), we get

(6. 21) $$f(0)=1,$$

(6. 22) $$f^{(\nu)}(1) = tf^{(\nu)}(0) \qquad (\nu=0, \ldots, n-1).$$

Moreover, the relation

$$\int_0^1 K(x)\,(\varphi_n^{(n)}(x)-f^{(n)}(x))\,dx = 0,$$

in view of $K(x)>0$ in [0, 1], shows that $f^{(n)}(x)=\varphi_n^{(n)}(x)$ almost everywhere. We conclude that $f(x)$ is a polynomial of degree n having the same highest degree term at $\varphi_n(x)$. By a reasoning already used in the proofs of Lemmas 4 and 5, we conclude that $f(x)=\varphi_n(x)$ in [0, 1]. Thus (6. 15) is established and therefore also Theorem 3.

III. The limit of the exponential Euler spline $S_n(x; t)$ as $n\to\infty$

7. *The Fourier expansion of* $S_n(x; t)t^{-x}$. *Let*

(7. 1) $$t=\tau e^{i\alpha},\ \tau>0,\ -\pi<\alpha\leqq\pi, \quad \text{hence} \quad \tau=|t|,$$

and let t^x be the exponential function defined by

(7. 2) $$t^x=\tau^x e^{i\alpha x} \quad \text{for all real} \quad x.$$

LEMMA 8. *The function $\Omega(x)$ defined by the relation*

(7. 3) $$S_n(x; t)=t^x\,\Omega(x) \quad \textit{for all real } x,$$

is a periodic function of period 1.

PROOF. This fact is evident, as both $S_n(x; t)$ and t^x satisfy the same functional equation (1. 5). Explicitly

$$\Omega(x+1) = S_n(x+1; t)t^{-x-1} = tS_n(x; t)t^{-1}t^{-x} = \Omega(x).$$

Let us find the Fourier series expansion of $\Omega(x)$.
For this purpose we assume that

(7. 4) $$0\leqq x\leqq 1,$$

so that

(7. 5) $$S_n(x; t)=A_n(x; t)/A_n(0; t).$$

In writing this relation we assume that

$$(7.6)\qquad A_n(0;t) = a_n(t) = \Pi_n(t)(t-1)^{-n} \neq 0,$$

hence

$$(7.7)\qquad t\neq 1, \quad \text{and} \quad t\neq\lambda_\nu \qquad (\nu=1, \dots, n-1).$$

where the λ_ν are the zeros of the Euler—Frobenius polynomial $\Pi_n(t)$.

Our main tool is Euler's generating function (3. 13). Dividing both sides by z^{n+1} we obtain

$$(7.8)\qquad F(z) = \frac{t-1}{t-e^z}\frac{e^{xz}}{z^{n+1}} = \sum_{\nu=0}^{\infty}\frac{A_\nu(x;t)}{\nu!}\frac{z^\nu}{z^{n+1}}.$$

The left side is a meromorphic function having a pole of order $n+1$ at the origin and simple poles at the zeros of e^z-t. By (7. 1), these poles are at the points

$$(7.9)\qquad t_k = \log\tau + i\alpha + 2k\pi i \qquad (k=0, \pm 1, \pm 2, \dots).$$

Notice that $t_k\neq 0$, for all k, for if $\log\tau=0$ and $\alpha=0$, then $t=1$, a case excluded by (7. 7).

Let us evaluate the residues of $F(z)$. From (7. 8) we find that

$$(7.10)\qquad \text{Residue of } F(z) \text{ at } z=0 = \frac{A_n(x;t)}{n!},$$

$$(7.11)\qquad \text{Residue at } t_k = -\frac{t-1}{e^{t_k}}\frac{e^{xt_k}}{t_k^{n+1}} = -\frac{t-1}{t}\frac{e^{xt_k}}{t_k^{n+1}}.$$

LEMMA 9. *The sum of all the residues of* $F(z)$ *in the entire finite plane vanishes*

PROOF. Let $z = u+iv$, u and v real. Let L and l be large and positive, l being an integer, and let $R_{L,l}$ denote a rectangle bounded by the lines $u=\pm L$, $v=\alpha\pm$ $\pm(2l+1)\pi$. By Cauchy's residue theorem we obtain

$$(7.12)\qquad A_n(x;t)/n! - \sum_{k=-l}^{l}(t-1)t^{-1}e^{xt_k}t_k^{-n-1} = \frac{1}{2\pi i}\int_{R_{L,l}} F(z)\,dz.$$

Our assumption (7. 4) implies that the integral on the right side $\to 0$ as L and $l\to+\infty$. We may omit the simple details. Moreover, the resulting series on the left side converges absolutely, if $n\geqq 1$, and we obtain the expansion

$$(7.13)\qquad A_n(x;t)/n! = \sum_{-\infty}^{\infty}(t-1)t^{-1}e^{xt_k}t_k^{-n-1}.$$

Here

$$e^{xt_k} = e^{x(\log\tau + i\alpha + 2k\pi i)} = t^x e^{2k\pi i x}$$

and (7. 13) becomes

$$(7.14)\qquad A_n(x;t)/n! = (t-1)t^{-1}t^x \sum_{-\infty}^{\infty} e^{2k\pi i x}/t_k^{n+1} \qquad (0 \leqq x \leqq 1).$$

Using (7. 14), (7. 5), and (7. 3) we obtain

THEOREM 4. *The periodic function $\Omega(x)$ of Lemma 8 admits the Fourier series expansion*

$$(7.15)\qquad \Omega(x) = \sum_{-\infty}^{\infty} \frac{1}{t_k^{n+1}} e^{2\pi i k x} \Big/ \sum_{-\infty}^{\infty} \frac{1}{t_k^{n+1}} \quad \textit{for all real} \quad x.$$

This is obtained by setting $x=0$ in (7. 14) and using (7. 5) and (7. 3).

8. *The case when* $t>0$, $t\neq 1$. In this case $t=\tau$, $\alpha=0$, and (7. 9) becomes

$$(8.1)\qquad t_k = \log t + 2\pi i k,$$

and from (7. 15) we obtain

$$(8.2)\qquad \Omega_n(x) = \frac{1+2\sum_{k=1}^{\infty} \operatorname{Re}\left\{\left(\dfrac{\log t}{\log t+2\pi i k}\right)^{n+1} e^{2\pi i k x}\right\}}{1+2\sum_{k=1}^{\infty} \operatorname{Re}\left(\dfrac{\log t}{\log t+2\pi i k}\right)^{n+1}}.$$

To obtain more explicit results we assume that

$$(8.3)\qquad t>1,$$

and write

$$(8.4)\qquad t_k = \log t + 2\pi i k = \varrho_k e^{i\gamma_k} \qquad (-\pi<\gamma_k<\pi).$$

Thus

$$(\log t + 2\pi i k)^{-n-1} e^{2\pi i k x} = \varrho_k^{-n-1} e^{(2\pi k x - (n+1)\gamma_k) i},$$

and (8. 2) may be written as

$$(8.5)\qquad \Omega_n(x) = \frac{1+2\sum_{k=1}^{\infty} (\varrho_0/\varrho_k)^{n+1} \cos(2\pi k x - (n+1)\gamma_k)}{1+2\sum_{k=1}^{\infty} (\varrho_0/\varrho_k)^{n+1} \cos(n+1)\gamma_k}.$$

Notice that the first ratio

$$(8.6)\qquad \varrho_0/\varrho_1 = \left(1+\frac{4\pi^2}{(\log t)^2}\right)^{-\frac{1}{2}} < 1$$

is the largest of the ratios ϱ_0/ϱ_k.

Observe that the relative error in the approximation of t^x by $S_n(x;t)$ is by (7. 3) equal to

$$\frac{t^x - S_n(x;t)}{t^x} = 1 - \Omega_n(x). \tag{8. 7}$$

From (8. 5) we immediately obtain the

COROLLARY 2. *The relative error* (8. 7) *is estimated by*

$$1 - \Omega_n(x) = O(\varrho_0/\varrho_1)^n \quad as \quad n \to \infty, \tag{8. 8}$$

uniformly for all real x.

EXAMPLE. For $t=2$ we find

$$\varrho_0/\varrho_1 = \left(1 + \frac{4\pi^2}{(\log 2)^2}\right)^{-\frac{1}{2}} = 0.1096. \tag{8. 9}$$

This shows that $S_n(x;2)$ will approximate 2^x to about n significant figures.
Again from (8. 5) we obtain the following more precise result.

COROLLARY 3.

$$\Omega_n(x) = \frac{1 + 2(\varrho_0/\varrho_1)^{n+1}\cos(2\pi x - (n+1)\gamma_1)}{1 + 2(\varrho_0/\varrho_1)^{n+1}\cos(n+1)\gamma_1} + O(\varrho_0/\varrho_2)^{n+1} \quad \text{as} \quad n \to \infty.$$

9. *The convergence theorem for the case when t is not negative.* We return to the case of a general t assuming, however, that t is not negative and (7. 1) becomes

$$t = |t| e^{i\alpha} \qquad (-\pi < \alpha < \pi). \tag{9. 1}$$

If we mark the poles (7. 9) on the vertical line $\text{Re } z = \log |t|$ we see that

$$|t_0| < |t_k| \quad \text{if} \quad k \neq 0. \tag{9. 2}$$

This shows that in the two sums of (7. 15) the terms corresponding to $k=0$ dominate all others. In fact

$$\Omega_n(x) = \{1 + \sum_{k\neq 0} (t_0/t_k)^{n+1} e^{2\pi i k x}\}/\{1 + \sum_{k \neq 0} (t_0/t_k)^{n+1}\} = 1 + O(\gamma^n) \tag{9. 3}$$

where

$$\gamma = \max(|t_0/t_1|, |t_0/t_{-1}|) < 1. \tag{9. 4}$$

This proves

THEOREM 5. *If t is not negative and* (9. 1) *holds, then*

$$\lim_{n\to\infty} S_n(x;t) = |t|^x e^{i\alpha x}. \tag{9. 5}$$

From (9. 3) and (9. 4) it is clear that the rate of convergence deteriorates as t approaches the negative side of the real axis.

10. *The case when t is real and negative.* Now it becomes especially important to assume that t is such that

$$\Pi_n(t) \neq 0, \tag{10. 1}$$

and Theorem 4 is valid. In (7. 9) we now have that $\alpha = \pi$, hence

$$t_k = \log|t| + (2k+1)\pi i \tag{10. 2}$$

and

$$\bar{t}_k = t_{-k-1}. \tag{10. 3}$$

We also write

$$t_k = \varrho_k e^{i\gamma_k}, \quad \varrho_n > 0 \qquad (-\pi < \gamma_k < \pi). \tag{10. 4}$$

Now (7. 13) becomes

$$\frac{|t|}{1+|t|} \frac{A_n(x;t)}{n!} = 2\,\mathrm{Re}\,\frac{e^{xt_0}}{t_0^{n+1}} + 2\,\mathrm{Re}\,\frac{e^{xt_1}}{t_1^{n+1}} + \cdots$$

and using (10. 4) we find that

$$S_n(x;t) = |t|^x \frac{\cos(\pi x - (n+1)\gamma_0) + (\varrho_0/\varrho_1)^{n+1}\cos(3\pi x - (n+1)\gamma_1) + \cdots}{\cos(n+1)\gamma_0 + (\varrho_0/\varrho_1)^{n+1}\cos(n+1)\gamma_1 + \cdots}. \tag{10. 5}$$

Notice that if $t = -1$, hence by (10. 1) that n is odd, or $n = 2m-1$, then $\log|t| = 0$ and (10. 2) shows that $\gamma_0 = \gamma_1 = \gamma_2 = \cdots = \frac{\pi}{2}$. Then (10. 5) becomes

$$S_n(x;-1) = \frac{\cos(\pi x \quad m\pi) + (\varrho_0/\varrho_1)^{2m}\cos(3\pi x - m\pi) + \cdots}{\cos m\pi + (\varrho_0/\varrho_1)^{2m}\cos m\pi + \cdots}.$$

This is seen to converge to $\cos \pi x$ as $n = 2m-1 \to \infty$. Thus

$$\lim S_n(x;-1) = \cos \pi x \quad \text{as} \quad n = 2m-1 \to \infty. \tag{10. 6}$$

If t is negative, $t \neq -1$, then the question as to when we have convergence becomes much more delicate. Let us at least discuss a single example.

THEOREM 6. *If*

$$t = -e, \tag{10. 7}$$

then the

$$\lim_{n\to\infty} S_n(x;-e) \textit{ does not exist.} \tag{10. 8}$$

PROOF. Here $\log|t| = \log e = 1$, while (10. 2) and (10. 4) show that

(10. 9) $$\gamma_0 = \arctan \pi.$$

To begin with we observe that $t = -e$ is a transcendental number and therefore

(10. 10) $$\Pi_n(-e) \neq 0 \quad \text{for all } n.$$

Moreover, we claim that

(10. 11) γ_0 *is not a rational multiple of* π.

Indeed, suppose that $\gamma_0 = r\pi$, where r is rational. It follows that $e^{\gamma_0 i} = e^{r\pi i}$ is a root of unity, hence that $e^{\gamma_0 i}$ is an algebraic number. By (10. 9) we conclude that

$$\pi = \tan \gamma_0 = \frac{1}{i} \frac{e^{\gamma_0 i} - e^{-\gamma_0 i}}{e^{\gamma_0 i} + e^{-\gamma_0 i}}$$

is an algebraic number, which it is not. This contradiction establishes (10. 11).

Now we return to (10. 5) and may argue as follows: If we choose the angle ψ such that

(10. 12) $$\psi \not\equiv \frac{\pi}{2} \pmod{\pi}$$

hence

(10. 13) $$\cos \psi \neq 0,$$

then we can let n grow to infinity through an infinite sequence N such that

(10. 14) $$\lim_{\substack{n\to\infty \\ n\in N}} (n+1)\gamma_0 \equiv \psi \pmod{2\pi}$$

or

(10. 15) $$\lim_{\substack{n\to\infty \\ n\in N}} \cos(n+1)\gamma_0 = \cos\psi \neq 0.$$

This is the case because the numbers $(n+1)\gamma_0$ are everywhere dense, mod 2π, in fact they are equidistributed, mod 2π. Now (10. 5) and (10. 15) show that

(10. 16) $$\lim_{\substack{n\to\infty \\ n\in N}} S_n(x; -e) = e^x \frac{\cos(\pi x - \psi)}{\cos\psi} = e^x(\cos \pi x + \sin \pi x \tan \psi).$$

Since ψ was arbitrary, subject to (10. 13), the theorem is established.

Notice, however, that the limits on the right side of (10. 16) satisfy the functional equation

$$f(x+1) = -ef(x).$$

11. *Approximating the exponential 2^x by exponential Euler splines.* In the present last section we wish to apply Corollary 1 of § 4 to the special case when $t=2$. The corresponding exponential 2^x is remarkable in several ways. It is the finite difference analogue of e^x because $\Delta 2^x = 2^x$. We also mention the beautiful theorem of Polya and Hardy: "2^x *is the transcendental entire function of least growth which assumes integer values at* $x=0, 1, 2, \ldots$." (See [11, 55].)

From (3. 15) we see that if $t=2$ then $a_n(2)=\Pi_n(2)$ are all integers. They are readily computed from the recurrence relations (3. 9) which become

$$a_n = 1+\binom{n}{1}a_1+\cdots+\binom{n}{n-1}a_{n-1} \qquad (n=1, 2, \ldots, a_1=1). \tag{11.1}$$

We find the values

n	0	1	2	3	4	5	6	7	8
a_n	1	1	3	13	75	541	4683	47293	545835

From (4. 1) we see that the corresponding exponential Euler polynomials $A_n(x; 2)$ are monic polynomials with integer coefficients. In the interval [0, 1] the relation (4. 3) is valid. Using the above values for the a_n, we find in [0, 1]

$$S_3(x; 2) = (x^3+3x^2+9x+13)/13,$$

$$S_5(x; 2) = (x^5+5x^4+30x^3+130x^2+375x+541)/541,$$

$$S_7(x; 2) = (x^7+7x^6+63x^5+455x^4+2625x^3+11{,}361x^2+32{,}781x+47{,}293)/47{,}293.$$

For the other intervals $(\nu, \nu+1)$ (ν integer$\neq 0$), the spline function $S_n(x; 2)$ is to be computed from the relation

$$S_n(x; 2) = 2^\nu S_n(x-\nu; 2). \tag{11.2}$$

Remarks 1. From (8. 8) and (8. 9) we see that $S_n(x; 2)$ approximates 2^x in [0, 1] to almost n accurate decimal places. Using (11. 2) we get an approximation to almost n significant figures for all real x.

2. A much better approximation is obtained if we replace in [0, 1] $S_n(x; 2)$ by other appropriate polynomials of degree n, for instance the polynomial of best approximation to 2^x in [0, 1] afterwards extending the approximation to all real x by means of (11. 2). However, the global approximating function so obtained is discontinuous for all integer values of x, while our exponential Euler spline satisfies $S_n(x; 2)\in C^{n-1}(-\infty, \infty)$.

3. Similar results are obtained for the case when $t=-1$. In this case we obtain the approximation of $\cos \pi x$ for all real x by the classical Euler splines.

REFERENCES

[1] J. H. Ahlberg and E. N. Nilson, *Polynomial splines on the real line*. J. Approximation Theory **3** (1970), 398—409.

[2] L. Carlitz, *Eulerian numbers and polynomials*. Math. Mag. **32** (1959), 247—260.

[3] H. B. Curry and I. J. Schoenberg, *On Pólya frequency functions IV. The fundamental spline functions and their limits*. J. Analyse Math. **17** (1966), 71—107.

[4] L. Euler, *Institutionis Calculi Differentialis*, vol. II. Petersburg 1755.

[5] G. Frobenius, *Über die Bernoullischen Zahlen und die Eulerschen Polynome*. Sitzungsberichte der Preussischen Akad. der Wiss. Phys.-Math. Kl. (1910), 809—847. Also in volume 3 of Collected Works, 440—478.

[6] E. Hille, *Analytic function theory*, vol. II. Ginn and Company, Boston 1962.

[7] N. E. Nörlund, *Vorlesungen über Differenzenrechnung*. Springer, Berlin 1924.

[8] W. Quade and L. Collatz, *Zur Interpolationstheorie der reellen periodischen Funktionen*. Sitzungsber. der Preuss. Akad. der Wiss., Phys.-Math. Kl, (1938), 383—429.

[9] I. J. Schoenberg, *Cardinal interpolation and spline functions*. J. Approximation Theory **2** (1969), 167—206.

[10] I. J. Schoenberg, *Cardinal interpolation and spline functions II. Interpolation of data of power growth*. MRC-Tech. Sum. Rep. **1104**, Sept. 1970. To appear in J. Approximation Theory.

[11] J. M. Whittaker, *Interpolatory functions theory*. Cambridge Tract **33**, University Press, Cambridge 1935.

Reprinted from
ISNM (Birkhauser Verlag) **20** (1972) pp. 382–404

CARDINAL INTERPOLATION AND SPLINE FUNCTIONS VII. THE BEHAVIOR OF CARDINAL SPLINE INTERPOLANTS AS THEIR DEGREE TENDS TO INFINITY*

By

I. J. SCHOENBERG

CONTENTS

Introduction

Let $F(x)$ be a function from $\mathbb{R}$ to $\mathbb{C}$ and let

$$S_m(x) = \sum_{-\infty}^{\infty} F(v) L_m(x-v) \tag{1}$$

be the spline function of degree $2m-1$, with knots at the integers, that interpolates $F(x)$ at all the integers. We know that $S_m(x)$, as defined by (1), exists

* Sponsored by U.S. Army under contract No. DA-31-124-ARO-D-462.

if $F(x)$ grows at most like a power $|x|^\gamma$ as $x \to \pm\infty$ (see [7]). Here we investigate conditions which will insure that $S_m(x)$ will converge to $F(x)$ as $m \to \infty$.

A noteworthy contribution to this problem was made by Quade and Collatz as early as 1938 in their remarkable paper [3]. They showed that if $n = 2k + 1 (> 1)$ is odd, and if $F(x)$ is a trigonometric polynomial of period n and of order k, hence

$$F(x) = \sum_{-k}^{k} a_\nu e^{2\pi i \nu x/n},$$

then $S_m(x) \to F(x)$, as $m \to \infty$, for all real x.

Another problem of this kind was studied in [8]. Let us consider the exponential function

$$F(x) = t^x, \quad (t \in \mathbb{C}, \ t \text{ not a negative real}, \ t \neq 1).$$

A cardinal spline function $S_n(x;t)$, of degree n, the so-called *exponential Euler spline* that interpolates t^x at all integers was constructed there. It was also shown that

$$\lim_{n\to\infty} S_n(x;t) = t^x, \text{ for all real } x.$$

The present paper describes a fairly general class of functions $F(x)$ exhibiting the same convergence phenomenon. Let $F(x)$ satisfy the conditions

1. $F(x)$ is entire of exponential type $\leqq \pi$.
2. For a certain integer $k \geqq 0$, the restriction of $F^{(k)}(x)$ to the real axis $\mathbb{R}$ is in $L_2(\mathbb{R})$.

It is shown that with $S_m(x)$ defined by (1), we have

$$\lim_{m\to\infty} S_m^{(\nu)}(x) = F^{(\nu)}(x), \qquad (\nu = 0, 1, 2, \cdots), \tag{2}$$

uniformly in x on $\mathbb{R}$. The relations (2) show the strong affinity of $S_m(x)$ to $F(x)$ and all its derivatives. The relations (2) are established in Theorem 2, while the nature of the limit function is determined by Theorem 1. The elegant proof of Theorem 1 presented here is due to my colleague Richard A. Askey. These results were first announced in [5].

The paper is based on a few results of [6] and [7], and precise references are given below. Our references list all previous six papers on cardinal spline interpolation ([2], [6]–[10]).

I. The interpolatory properties of certain entire functions

1. The spaces $l_2^{(k)}$ and $PW_\pi^{(k)}$. Let

$$y = (y_\nu), \quad (\nu = 0, \pm 1, \cdots), \tag{1.1}$$

be a sequence of numbers, real or complex. As usual we say that

$$y \in l_2, \text{ provided that } \sum_{-\infty}^{\infty} |y_\nu|^2 < \infty .$$

We are interested in the space $l_2^{(k)}$ of sequences obtained from the elements of l_2 by a succession of k summations. A direct definition of $l_2^{(k)}$ is

$$l_2^{(k)} = \{(y_\nu); (\Delta^k y_\nu) \in l_2\}. \tag{1.2}$$

From the inequalities

$$|\Delta^{k+p} y_\nu|^2 = \left| \sum_{j=0}^{p} (-1)^j \binom{p}{j} \Delta^k y_{\nu+j} \right|^2 \leqq \sum_{j=0}^{p} \binom{p}{j}^2 \cdot \sum_{j=0}^{p} |\Delta^k y_{\nu+j}|^2$$

we obtain that

$$|\Delta^{k+p} y_\nu|^2 \leqq \binom{2p}{p} \sum_{j=0}^{p} |\Delta^k y_{\nu+j}|^2 \tag{1.3}$$

Assuming that $(y_\nu) \in l_2^{(k)}$ and summing (1.3) over ν we find that

$$\| \Delta^{k+p} y \|_2^2 \leqq (p+1) \binom{2p}{p} \| \Delta^k y \|_2^2 .$$

This shows that $y \in l_2^{(k)}$ implies that $y \in l_2^{(k')}$ if $k' > k$. We have therefore established the set inclusions

$$l_2 \subset l_2^{(1)} \subset \cdots \subset l_2^{(k)} \subset l_2^{(k+1)} \subset \cdots. \tag{1.4}$$

Sometimes we shall use their union

$$(1.5) \qquad l_2^* = \bigcup_{k=0}^{\infty} l_2^{(k)}.$$

We are also concerned with functions $F(x)$, of the complex variable x, defined by

$$(1.6) \qquad F(x) = \int_{-\pi}^{\pi} g(u)e^{-iux}du, \text{ where } g(u) \in L_2(-\pi,\pi).$$

It is clear from the representation that $F(x)$ *is an entire function of exponential type* $\leqq \pi$. Moreover, Plancherel's theorem shows that the *restriction of* $F(x)$ *to the real axis* $\mathbb{R}$ *is in* $L_2(\mathbb{R})$. The classical theorem of Paley and Wiener states that the two italicized statements characterize the class of functions $F(x)$ defined by (1.6) (see [1]). Using the initial letters of the authors' names, w denote this class of functions by the symbol

$$PW_\pi.$$

We need also the space $PW_\pi^{(k)}$ of functions $F(x)$ that are obtained from the elements of PW_π by a succession of k integrations. A direct definition of this new class is

$$(1.7) \qquad PW_\pi^{(k)} = \{F(x);\ F(x) \text{ is entire and } F^{(k)}(x) \in PW_\pi\}.$$

If $F(x)$ belongs to this class, hence

$$F^{(k)}(x) = \int_{-\pi}^{\pi} g(u)e^{-iux}du, \quad g(u) \in L_2(-\pi,\pi),$$

then

$$F^{(k+p)}(x) = \int_{-\pi}^{\pi} (-iu)^p g(u)e^{-iux}du, \quad (p = 1,2,\cdots).$$

The integrand here being evidently in $L_2(-\pi,\pi)$, we conclude that

$$F(x) \in PW_\pi^{(k)} \text{ implies that } F(x) \in PW_\pi^{(k')} \text{ if } k' > k.$$

Therefore

(1.8) $$PW_\pi \subset PW_\pi^{(1)} \subset \cdots \subset PW_\pi^{(k)} \subset PW_\pi^{(k+1)} \subset \cdots.$$

We shall sometimes use their union defined by

(1.9) $$PW_\pi^* = \bigcup_{k=0}^{\infty} PW_\pi^{(k)}.$$

A polynomial $P(x)$, no matter of what degree k, is an element of PW_π^*, because $P^{(k+1)}(x) = 0 \in PW_\pi$.

2. The correspondence between the spaces $l_2^{(k)}$ and $PW_\pi^{(k)}$. There is a known correspondence between the spaces l_2 and PW_π which may be stated as follows: *If*

(2.1) $$(y_\nu) \in l_2$$

then there is a unique

(2.2) $$F(x) \in PW_\pi$$

such that

(2.3) $$F(\nu) = y_\nu \text{ for all } \nu.$$

Conversely, if (2.2) *holds and we define* (y_ν) *by* (2.3), *then* (2.1) *holds.*

The proofs follow easily as applications of the Riesz-Fischer theorem and Bessel's inequality, respectively (see [11]).

The following theorem shows that this correspondence by interpolation generalizes to the spaces $l_2^{(k)}$ and $PW_\pi^{(k)}$.

Theorem 1. *If*

(2.4) $$(y_\nu) \in l_2^{(k)},$$

then there is a unique

$$F(x) \in PW_\pi^{(k)} \tag{2.5}$$

such that

$$F(v) = y_v \text{ for all } v. \tag{2.6}$$

Conversely, if (2.5) *holds and we define* (y_v) *by* (2.6), *then* (2.4) *holds.*

The following beautiful proof is due to Richard A. Askey.

Proof of direct part. We assume that (2.4) holds, i.e. $(\Delta^k y_v) \in l_2$. By the Riesz-Fischer theorem we may write

$$\Delta^k y_v = \int_{-\pi}^{\pi} g(u) e^{-iuv} du, \text{ for all } v, \text{ where } g(u) \in L_2(-\pi, \pi). \tag{2.7}$$

We define the entire function $F_1^{(k)}(x)$ by

$$F_1^{(k)}(x) = \int_{-\pi}^{\pi} g(u) \left(\frac{iu}{1 - e^{-iu}}\right)^k e^{-iux} du, \tag{2.8}$$

which is visibly an element of PW_π. Here we denote by $F_1(x)$ any one of the k-fold integrals of $F_1^{(k)}(x)$, it does not matter which.

On both sides of (2.8) we perform the operation

$$\int_x^{x+1} (\cdot) dx. \tag{2.9}$$

Since

$$\int_x^{x+1} e^{-iux} dx = \frac{1 - e^{-iu}}{iu} e^{-iux}, \tag{2.10}$$

we obtain

$$F_1^{(k-1)}(x+1) - F_1^{(k-1)}(x) = \int_{-\pi}^{\pi} g(u) \left(\frac{iu}{1 - e^{-iu}}\right)^{k-1} e^{-iux} du.$$

Repeating the operation (2.9) a total of k times we obtain

$$\Delta^k F_1(x) = \int_{-\pi}^{\pi} g(u)e^{-iux}du\,. \tag{2.11}$$

Setting $x = \nu$ and using (2.7), we find that

$$\Delta^k F_1(\nu) = \Delta^k y_\nu \text{ for all } \nu\,,$$

whence we conclude that

$$y_\nu = F_1(\nu) + P(\nu)\,, \text{ for all } \nu\,, \tag{2.12}$$

where $P(x)$ is an appropriate element of π_{k-1}. This shows that the function

$$F(x) = F_1(x) + P(x) \tag{2.13}$$

satisfies (2.6), because of (2.12), and also $F^{(k)}(x) = F_1^{(k)}(x)$. Therefore (2.5) holds in view of (2.8).

We have yet to show that the function $F(x)$ just obtained is unique. For this we need to show that the assumptions

$$F(x) \in PW_\pi^{(k)} \tag{2.14}$$

and

$$F(\nu) = 0 \text{ for all } \nu\,, \tag{2.15}$$

imply that

$$F(x) = 0 \text{ for all } x\,. \tag{2.16}$$

Indeed, by (2.14) we may write

$$F^{(k)}(x) = \int_{-\pi}^{\pi} g(u)e^{-iux}du\,, \quad g \in L_2(-\pi, \pi) \tag{2.17}$$

Applying (2.9) k times, (2.10) shows that

$$\Delta^k F(x) = \int_{-\pi}^{\pi} g(u) \left(\frac{1 - e^{-iu}}{iu}\right)^k e^{-iux} du .$$

However, (2.15) shows that $\Delta^k F(\nu) = 0$ and therefore that

$$\int_{-\pi}^{\pi} g(u) \left(\frac{1 - e^{-iu}}{iu}\right)^k e^{-iu\nu} du = 0 \quad \text{for all } \nu .$$

But then $g(u) = 0$ a.e. and now (2.17) shows that $F(x) \in \pi_{k-1}$. Finally (2.15) implies (2.16). □

Proof of converse part. Assuming (2.5) we write

$$F^{(k)}(x) = \int_{-\pi}^{\pi} g(u) e^{-iux} du , \quad g(u) \in L_2(-\pi, \pi) , \tag{2.18}$$

which, as above, implies

$$\Delta^k F(x) = \int_{-\pi}^{\pi} g(u) \left(\frac{1 - e^{-iu}}{iu}\right)^k e^{-iux} du .$$

Setting $x = \nu$ and using (2.6) we obtain that

$$\Delta^k y_\nu = \int_{-\pi}^{\pi} g(u) \left(\frac{1 - e^{-iu}}{iu}\right)^k e^{-iu\nu} du .$$

The integrand being evidently in $L_2(-\pi, \pi)$, we obtain (2.4) by Bessel's inequality. □

Theorem 1 and the relations (1.4) and (1.8) show that the correspondence set up by the interpolation conditions (2.6) maps $l_2^{(k)} \backslash l_2^{(k-1)}$ onto $PW_\pi^{(k)} \backslash PW_\pi^{(k-1)}$ one-to-one, $(k = 0, 1, \cdots)$. This implies a correspondence between the classes l_2^* and PW_π^*.

3. Examples

1. Let us assume that

$$(y_\nu) \in l_2, \tag{3.1}$$

so that we are dealing with the case $k = 0$ of Theorem 1. Equation (2.7) reduces to

$$y_\nu = \int_{-\pi}^{\pi} g(u) e^{-iu\nu} du, \quad g \in L_2(-\pi, \pi), \tag{3.2}$$

and (2.8) becomes

$$F(x) = \int_{-\pi}^{\pi} g(u) e^{-iux} du. \tag{3.3}$$

A result of J. M. Whittaker [11] states that

$$F(x) = \sum_{-\infty}^{\infty} y_\nu \frac{\sin \pi(x-\nu)}{\pi(x-\nu)}, \tag{3.4}$$

the series converging absolutely and uniformly for all real x.

Proof. By (3.2)

$$g(u) \sim \frac{1}{2\pi} \sum_{-\infty}^{\infty} y_j e^{iuj}$$

is the Fourier series of $g(u)$. In terms of

$$g_N(u) = \frac{1}{2\pi} \sum_{-N}^{N} y_j e^{iju}$$

we find that

$$\int_{-\pi}^{\pi} g_N(u)e^{-iux}du = \sum_{-N}^{N} y_j \frac{1}{2\pi} \int_{-\pi}^{\pi} e^{-iu(x-j)}du = \sum_{-N}^{N} y_j \frac{\sin \pi(x-j)}{\pi(x-j)},$$

and Schwarz's inequality shows that

$$\left| \int_{-\pi}^{\pi} g(u)e^{-iux}du - \int_{-\pi}^{\pi} g_N(u)e^{-iux}du \right|^2 \leqq 2\pi \| g - g_N \|_2^2 \to 0$$

as $N \to \infty$. □

For the special sequence

$$y_\nu = \delta_\nu, \text{ where } \delta_0 = 1, \delta_\nu = 0 \text{ if } \nu \neq 0,$$

we obtain the following

Corollary 1. *The function*

$$\frac{\sin \pi x}{\pi x}$$

is the only element of the class PW_π^* that interpolates the unit-sequence (δ_ν).

Corollary 2. *If $F(x) \in PW_\pi^*$ and $F(\nu) = 0$ for all integers ν, then $F(x) \equiv 0$.*

2. In order to show that the proof of Theorem 1 may at times lead to an effective construction of $F(x)$, we consider the sequence

$$y_\nu = |\nu|, \text{ for all } \nu. \tag{3.5}$$

Since $\Delta^2 y_\nu = 0$ if $\nu \neq -1$ and $\Delta^2 y_{-1} = 2$, we seet that

$$(y_\nu) \in l_2^{(2)}. \tag{3.6}$$

By (2.7) we need the function $g(u)$ such that

$$\Delta^2 y_\nu = \int_{-\pi}^{\pi} g(u)e^{-iu\nu}du \text{ for all } \nu,$$

and evidently

$$g(u) = \frac{1}{\pi}e^{-iu},$$

and (2.8) shows that

$$F''(x) = \frac{1}{\pi}\int_{-\pi}^{\pi} e^{-iu}\left(\frac{iu}{1-e^{-iu}}\right)^2 e^{-iux}du. \tag{3.7}$$

Since (3.5) is an even sequence, we expect $F(x)$ to be even; hence $F(0)=F'(0)=0$. Applying the operation $\int_0^x(\cdot)dx$ twice on both sides of (3.7) we find by elementary operations that

$$F(x) = \frac{1}{\pi}\int_0^{\pi}\left(\frac{\sin(ux/2)}{\sin(u/2)}\right)^2 du. \tag{3.8}$$

This interpolant of the sequence (3.5) could have been guessed if we had thought of the Fejér kernel

$$K_{n-1}(u) = \frac{1}{2n}\left(\frac{\sin(nu/2)}{\sin(u/2)}\right)^2$$

which has the property

$$\frac{1}{\pi}\int_{-\pi}^{\pi} K_{n-1}(u)du = 1$$

so that

$$\frac{1}{\pi}\int_0^{\pi}\left(\frac{\sin(nu/2)}{\sin(u/2)}\right)^2 du = n \qquad (n = 0, 1, \cdots).$$

II. The convergence of the spline interpolants

4. The main result (Theorem 2). Let $\mathscr{S}_n$ denote the class of cardinal spline functions $S(x)$ of degree n having the integers as knots. This means that $S(x) \in C^{n-1}(\mathbb{R})$, and that the restriction of $S(x)$ to every unit interval $[\nu, \nu + 1]$ is a polynomial of degree not exceeding n. Furthermore, let

$$y = (y_\nu), \quad (\nu = 0, \pm 1, \cdots), \tag{4.1}$$

be a sequence such that

$$y \in l_2^{(k)}, \text{ which means that } (\Delta^k y_\nu) \in l_2. \tag{4.2}$$

This assumption has two consequences:

1. From the first part of this paper we know that there is a unique

$$F(x) \in PW_\pi^{(k)} \tag{4.3}$$

such that

$$F(\nu) = y_\nu, \text{ for all } \nu. \tag{4.4}$$

2. If we assume that

$$m \geqq k, \tag{4.5}$$

then we know [6, Th. 9] that there exists a unique

$$S_m(x) \in \mathscr{S}_{2m-1} \tag{4.6}$$

such that

$$S_m^{(m)}(x) \in L_2(\mathbb{R}) \tag{4.7}$$

and

$$S_m(\nu) = y_\nu, \text{ for all } \nu. \tag{4.8}$$

Indeed, (4.2) and (4.5) imply that $y \in l_2^{(m)}$ and this is the assumption of [6, Th. 9, p. 174].

Theorem 2. *The relations*

$$\lim_{m\to\infty} S_m^{(\nu)}(x) = F^{(\nu)}(x), \quad (\nu = 0,1,2,\cdots) \tag{4.9}$$

hold uniformly in x for all real values of x.

5. A few lemmas. We require a few tools that have proved to be of use in the study of spline functions with equidistant knots. We recall them briefly, stating some of their properties as lemmas and giving references.

Lemma 1. *Let n be an integer,* $n \geqq 2$, *and let*

$$\psi_n(u) = \left(\frac{\sin(u/2)}{u/2}\right)^n. \tag{5.1}$$

Then

$$M_n(x) = \frac{1}{2\pi}\int_{-\infty}^{\infty} \psi_n(u)e^{-iux}du \tag{5.2}$$

is a spline function of degree $n-1$, *which is positive in the interval* $(-n/2, n/2)$, *vanishes in its complements and has its knots in the points* $x = -(n/2)+\nu$, $(\nu = 0,1,\cdots,n)$ (see [4, p. 68], or [6, §1]).

Lemma 1 was known to Laplace. The function $M_n(x)$ and its translates are called *B*-splines, in view of the following.

Lemma 2. *If* $S(x) \in \mathscr{S}_{n-1}$, *then* $S(x)$ *admits a unique representation of the form*

$$S(x) = \sum_{-\infty}^{\infty} c_j M_n(x-j-\tfrac{1}{2}n). \tag{5.3}$$

Conversely, for assigned constants c_j, *the series* (5.3) *defines an element of* $\mathscr{S}_{n-1}$ (see [4, Th. 1, p. 72]).

We also need the function defined by

$$\phi_n(u) = \sum_{-\infty}^{\infty} \psi_n(u + 2\pi j). \tag{5.4}$$

This is evidently a periodic function of period 2π. In [6, §2] $\phi_n(u)$ is shown to be a cosine polynomial of order $[(n + 1)/2] - 1$ which is positive for all real u. Notice that $\phi_n(0) = 1$.

Lemma 3.

$$\lim_{m\to\infty} \frac{\psi_{2m}(u)}{\phi_{2m}(u)} = 1 \ \textit{if} \ -\pi < u < \pi. \tag{5.5}$$

Proof. The functions involved being even, we may assume that

$$0 \leqq u < \pi. \tag{5.6}$$

The definitions (5.1) and (5.4) show that

$$\frac{\psi_{2m}(u)}{\phi_{2m}(u)} = 1\Big/\left\{1 + \sum_{j\neq 0} \left(\frac{u}{u + 2\pi j}\right)^{2m}\right\} = 1/\{1 + S_1 + S_2\},$$

where

$$S_1 = \sum_{1}^{\infty} \left(\frac{u}{u + 2\pi j}\right)^{2m}, \quad S_2 = \sum_{1}^{\infty} \left(\frac{u}{2\pi j - u}\right)^{2m}.$$

Hence

$$0 < 1 - \frac{\psi_{2m}(u)}{\phi_{2m}(u)} = \frac{S_1 + S_2}{1 + S_1 + S_2} \leqq S_1 + S_2.$$

However,

$$S_1 < \sum_{1}^{\infty} \left(\frac{\pi}{\pi + 2\pi j}\right)^{2m} \to 0 \quad \text{as } m \to \infty,$$

and

$$S_2 < \left(\frac{u}{2\pi - u}\right)^{2m} + \sum_{2}^{\infty} \left(\frac{\pi}{2\pi j - \pi}\right)^{2m} \to 0 \text{ as } m \to \infty,$$

because $0 \leqq u < 2\pi - u$. □

Lemma 4. *If h is a fixed natural number, then there is a constant C_h such that*

(5.7) $$\frac{\phi_{2m-2h}(u)}{\phi_{2m}(u)} < C_h \text{ for all } u \text{ and } m \geqq h+1.$$

Proof. The functions involved being periodic and even, we may assume that

(5.8) $$0 \leqq u \leqq \pi$$

From

$$\phi_{2m-2h}(u) = \left(2\sin\frac{u}{2}\right)^{2m-2h} \sum_j (u + 2\pi j)^{-(2m-2h)}$$

$$\phi_{2m}(u) = \left(2\sin\frac{u}{2}\right)^{2m} \sum_j (u + 2\pi j)^{-2m}$$

we obtain that

$$\frac{\phi_{2m-2h}(u)}{\phi_{2m}(u)} = \frac{u^{2h}}{(2\sin u/2)^{2h}} \cdot \frac{\sum_j u^{2m-2h}/(u+2\pi j)^{2m-2h}}{\sum_j u^{2m}/(u+2\pi j)^{2m}}.$$

Since the first factor on the right side is bounded by $(\pi/2)^{2h}$ in the interval (5.8), it suffices to show that the second factor, *which we denote by* $R_{m,h}$, also admits an upper bound which is independent of u and m. Observe that we can write

$$R_{m,h} = \frac{1 + \sum_{1}^{\infty} \left(\frac{u}{u+2\pi j}\right)^{2m-2h} + \sum_{1}^{\infty} \left(\frac{u}{2\pi j - u}\right)^{2m-2h}}{1 + \sum_{1}^{\infty} \left(\frac{u}{u+2\pi j}\right)^{2m} + \sum_{1}^{\infty} \left(\frac{u}{2\pi j - u}\right)^{2m}},$$

and dropping both sums in the denominator we obtain

$$R_{m,h} \leqq 1 + \sum_1^\infty \left(\frac{u}{u+2\pi j}\right)^{2m-2h} + \sum_1^\infty \left(\frac{u}{2\pi j - u}\right)^{2m-2h}$$

Notice that all fractions

$$\frac{u}{u+2\pi j}, \frac{u}{2\pi j - u}, \quad (j = 1, 2, \cdots),$$

reach their maximal values in $[0, \pi]$ for $u = \pi$, and that all these maxima are < 1, except for the first fraction of the second set whose maximum is $= 1$. If $2m - 2h \geqq 2$, we conclude that

$$R_{m,h} \leqq 1 + \sum_1^\infty \frac{1}{(2j+1)^2} + \sum_1^\infty \frac{1}{(2j-1)^2} = K.$$

Therefore

$$C_h = K(\pi/2)^{2h}$$

satisfies the conditions of the lemma. □

6. Proof of the relations (4.9) if $\nu \geqq k$. We start with $\nu = k$ and want to show that

(6.1) $$\lim_{m \to \infty} S_m^{(k)}(x) = F^{(k)}(x) \text{ uniformly for all real } x.$$

From (4.2), (4.4), (2.7) and (2.8) we may write the relations

(6.2) $$\Delta^k y_\nu = \int_{-\pi}^{\pi} g(u) e^{-iu\nu} du, \quad g(u) \in L_2(-\pi, \pi)$$

and

(6.3) $$F^{(k)}(x) = \int_{-\pi}^{\pi} g(u) \left(\frac{iu}{1 - e^{-iu}}\right)^k e^{-iux} du.$$

Denoting the periodic extension of $g(u)$ by $\bar{g}(u)$, *we define the function*

$$(6.4) \qquad s(x) = \int_{-\infty}^{\infty} \frac{\psi_{2m}(u)}{\phi_{2m}(u)} \bar{g}(u) \left(\frac{iu}{1-e^{-iu}}\right)^k e^{-iux} du \quad (2m > k+2)$$

and wish to show that

$$(6.5) \qquad s(x) = S_m^{(k)}(x),$$

where $S_m(x)$ *is the spline function satisfying* (4.6), (4.7), *and* (4.8).
This will be done in several stages.

1. *The right side of* (6.4) *represents an element of* $\mathscr{S}_{2m-k-1} \cap L_2(\mathbb{R})$.

Proof. In (6.4) we have assumed that

$$(6.6) \qquad 2m > k+2,$$

and this easily implies the absolute convergence of the integral (6.4). Observe next that

$$(6.7) \qquad \psi_{2m}(u)\left(\frac{iu}{1-e^{-iu}}\right)^k = \psi_{2m-k}(u)\left(\frac{\sin(u/2)}{u/2}\right)^k \left(\frac{iu}{1-e^{-iu}}\right)^k$$
$$= \psi_{2m-k}(u) e^{iuk/2},$$

and this allows us to write (6.4) as

$$(6.8) \qquad s(x) = \int_{-\infty}^{\infty} \frac{\psi_{2m-k}(u)}{\phi_{2m}(u)} \bar{g}(u) e^{-iu(x-\frac{1}{2}k)} du.$$

Let

$$(6.9) \qquad \bar{g}(u)/\phi_{2m}(u) \sim \sum_{-\infty}^{\infty} c_j e^{iju}$$

be the Fourier series of the left side. Substituting this formally into (6.8) we obtain

$$s(x) = \sum_{-\infty}^{\infty} c_j \int_{-\infty}^{\infty} \psi_{2m-k}(u) e^{-iu(x-j-\frac{1}{2}k)} du,$$

which, in view of (5.2), becomes

$$s(x) = \sum_{-\infty}^{\infty} c_j M_{2m-k}(x - j - \tfrac{1}{2}k).$$

Since

$$(c_j) \in l_2, \tag{6.10}$$

our formal derivation is easily justified. By Lemma 2 we conclude that $s(x) \in \mathscr{S}_{2m-k-1}$. Moreover, by (6.10) and a known theorem (see [6, Th. 12] or [7, Th. 4] for a simpler proof), we conclude that $s(x) \in L_2(\mathbb{R})$. □

Let $S(x)$ denote a kth integral of $s(x)$ so that

$$S^{(k)}(x) = s(x). \tag{6.11}$$

In any case it is clear from the definition of $S(x)$ that

$$S(x) \in \mathscr{S}_{2m-1} \cap L_2^{(k)} \tag{6.12}$$

where $L_2^{(k)}$ denotes the class of functions whose kth derivatives are in $L_2(\mathbb{R})$.

2. *The solution* $S(x)$ *of the differential equation* (6.11) *can be selected so that*

$$S(x) = S_m(x). \tag{6.13}$$

Proof. From (6.4) and (6.11) we see that

$$S^{(k)}(s) = \int_{-\infty}^{\infty} \frac{\psi_{2m}(u)}{\phi_{2m}(u)} \bar{g}(u) \left(\frac{iu}{1-e^{-iu}}\right)^k e^{-iux} du. \tag{6.14}$$

Performing the operation (2.9) k times on both sides of (6.14) we obtain that

$$\Delta^k S(x) = \int_{\infty}^{\infty} \frac{\psi_{2m}(u)}{\phi_{2m}(u)} \bar{g}(u) e^{-iux} du \,.$$

Setting $x = \nu$ (an integer) we may write

$$\Delta^k S(\nu) = \sum_{-\infty}^{\infty} \int_{-\pi+2\pi j}^{\pi+2\pi j} \frac{\psi_{2m}(u)}{\phi_{2m}(u)} \bar{g}(u) e^{-iu\nu} du$$

$$= \sum_{j=-\infty}^{\infty} \int_{-\pi}^{\pi} \frac{\psi_{2m}(u + 2\pi j)}{\phi_{2m}(u)} \bar{g}(u) e^{-iu\nu} du$$

and finally

$$\Delta^k S(\nu) = \int_{-\pi}^{\pi} g(u) e^{-iu\nu} du$$

in view of the definition (5.4) of $\phi_{2m}(u)$. By (6.2) we obtain

$$\Delta^k S(\nu) = \Delta^k y_\nu, \text{ for all } \nu,$$

and now it is clear that we may choose the solution $S(x)$ of (6.11) (an appropriate additive element of π_{k-1} is involved) so as to have

$$S(\nu) = y_\nu = F(\nu), \text{ for all } \nu. \tag{6.15}$$

On the other hand, since the interpolating spline function satisfying (6.12) is unique, we have established (6.13). □

Using (6.4), (6.11) and (6.13), we may state our own result as

$$S_m^{(k)}(x) = \int_{-\infty}^{\infty} \frac{\psi_{2m}(u)}{\phi_{2m}(u)} \bar{g}(u) \left(\frac{iu}{1 - e^{-iu}}\right)^k e^{-iux} du \,. \tag{6.16}$$

This expression for $S_m^{(k)}(x)$ can now be compared to the form (6.3) of $F^{(k)}(x)$.

3. *The relation* (6.1) *holds.*

Proof. Using (6.3) and (6.16) we are to estimate the absolute value of the difference

$$|F^{(k)}(x) - S_m^{(k)}(x)| \leqq E_1 + E_2, \tag{6.17}$$

where

$$E_1 = \left| \int_{-\pi}^{\pi} \left(1 - \frac{\psi_{2m}(u)}{\phi_{2m}(u)}\right) g(u) \left(\frac{iu}{1-e^{-iu}}\right)^k e^{-iux} du \right|, \tag{6.18}$$

$$E_2 = \left| \left(\int_{-\infty}^{-\pi} + \int_{\pi}^{\infty} \right) \frac{\psi_{2m}(u)}{\phi_{2m}(u)} \bar{g}(u) \left(\frac{iu}{1-e^{-iu}}\right)^k e^{-iux} du \right|. \tag{6.19}$$

Lemma 3 and the dominated convergence theorem immediately show that

$$\lim_{m\to\infty} E_1 = 0 \text{ uniformly for all real } x. \tag{6.20}$$

Using (6.7) we obtain from (6.19) that

$$E_2 = \left| \left(\int_{-\infty}^{-\pi} + \int_{\pi}^{\infty} \right) \frac{\psi_{2m-k}(u)}{\phi_{2m}(u)} \bar{g}(u) e^{-iux} e^{iuk/2} du \right|$$

whence

$$E_2 \leqq \left(\int_{-\infty}^{-\pi} + \int_{\pi}^{\infty} \right) \frac{|\psi_{2m-k}(u)|}{\phi_{2m}(u)} |\bar{g}(u)| du, \tag{6.21}$$

and it remains to show that the right side $\to 0$ as $m \to \infty$. Notice that $|\psi_n(u)|$ is a nonincreasing function of n, for every fixed $u \leqq -\pi$, or $u \geqq \pi$, and also that $\psi_n(u) \geqq 0$ if n is even. It is therefore sufficient to consider the right side of (6.21) only for even values of $k = 2h$. Denoting the right side of (6.21) by E_2', we may write

$$E_2' = \left(\int_{-\infty}^{\pi} + \int_{\pi}^{\infty}\right) \frac{\psi_{2m-2h}(u)}{\phi_{2m-2h}(u)} \cdot \frac{\phi_{2m-2h}(u)}{\phi_{2m}(u)} \left|\bar{g}(u)\right| du$$

$$= \left(\int_{-\infty}^{\infty} - \int_{-\pi}^{\pi}\right) \frac{\psi_{2m-2h}(u)}{\phi_{2m-2h}(u)} \cdot \frac{\phi_{2m-2h}(u)}{\phi_{2m}(u)} \left|\bar{g}(u)\right| du.$$

If we represent the first integral of the last term (the one over $\mathbb{R}$) as a sum of integrals over the interval $(-\pi + 2\pi j, \pi + 2\pi j)$ and use the definition (5.4) of $\phi_{2m-2h}(u)$, we obtain

$$E_2' = \int_{-\pi}^{\pi} \frac{\phi_{2m-2h}}{\phi_{2m}} |g| \, du - \int_{-\pi}^{\pi} \frac{\psi_{2m-2h}}{\phi_{2m-2h}} \cdot \frac{\phi_{2m-2h}}{\phi_{2m}} |g| \, du$$

and finally

$$E_2' = \int_{-\pi}^{\pi} \left(1 - \frac{\psi_{2m-2h}(u)}{\phi_{2m-2h}(u)}\right) \frac{\phi_{2m-2h}(u)}{\phi_{2m}(u)} \left|g(u)\right| du. \tag{6.22}$$

Now, using Lemmas 3 and 4, we find that $E_2' \to 0$. This establishes (6.1).

We have yet to establish (4.9) if $\nu > k$. This is immediately done by the following remark. The basic assumption is (4.2), where we naturally think of k as assuming the least possible value. This, however, is not necessary, and in view of the set inclusions (1.4) we may replace k in (4.2) by any other value ν, provided that $\nu > k$. But then the previous proof applies and the relations (4.9) are established. □

7. **Proof of the relations** (4.9) **if** $0 \leqq \nu < k$. Let $f(x) \in C^k(\mathbb{R})$. We denote by $P(x;f)$ the polynomial of degree $\leqq k-1$ that interpolates $f(x)$ at $x = 0, 1, \cdots, k-1$. For each fixed x

$$\mathscr{L} f = f(x) - P(x;f) \tag{7.1}$$

is a linear functional having the property that $\mathscr{L}f = 0$ if $f \in \pi_{k-1}$. Let $I = [a, b]$ such that $a < 0$ and $k-1 < b$. By Peano's theorem we may write

$$f(x) = P(x;f) + \int_I K(x,t) f^{(k)}(t) dt \quad \text{if } x \in I. \tag{7.2}$$

The kernel $K(x,t)$ is a spline function of t of degree $k-1$ having its knots at the points $0, 1, \cdots, k-1$, and x. Actually $K(x,t)$ can be explicitly expressed by Peano's formula

$$K(x,t) = \frac{1}{(k-1)!} \mathscr{L}_u (u-t)_+^{k-1} \tag{7.3}$$

by using Lagrange's expression for $P(x;f)$. In (7.3) the subscript "u" indicates that the operation (7.1) is to be performed on the variable u.

We apply (7.2) twice, for $f(x) = S_m(x)$ as well as for $f(x) = F(x)$. Observing that $P(x; S_m) = P(x; F)$, because $S_m(v) = y_v = F(v)$, and subtracting the second relation from the first, we obtain that

$$S_m(x) - F(x) = \int_a^b K(x,t)(S_m^{(k)}(t) - F^{(k)}(t)) dt. \tag{7.4}$$

On the other hand the explicit expression for $K(x,t)$ derived from (7.3) will easily show the existence of a constant H_k such that

$$\left| \frac{\partial^v}{\partial x^v} K(x,t) \right| < H_k \quad \text{if } (x,t) \in I \times I, \quad (v = 0, 1, \cdots, k-1). \tag{7.5}$$

Differentiating (7.4) v times, and using (7.5) we obtain that

$$|S_m^{(v)}(x) - F^{(v)}(x)| \leqq (b-a) H_k \| S_m^{(k)} - F^{(k)} \|_\infty \quad (v = 0, \cdots, k-1). \tag{7.6}$$

Applying here the relation (6.1), we have established (4.9) for $v = 0, \cdots, k-1$, if $x \leqq I$.

Now let $I_n = [a+n, b+n]$, where n is any integer $\lesseqgtr 0$. Using the polynomial that interpolates $F(x)$ at the k points $n, n+1, \cdots, n+k-1$, we find that

$$S_m(x) - F(x) = \int_{a+n}^{b+n} K_n(x,t)(S_m^{(k)}(t) - F^{(k)}(t))dt$$

where

$$K_n(x,t) = K(x-n, t-n).$$

The estimate (7.5) therefore holds for K replaced by K_n and $(x,t) \in I_n \times I_n$. It follows that the inequalities (7.6) hold for all real values of x. □

8. A summation method for the cardinal series. Let

$$f(x) = \sum_{-\infty}^{\infty} f(v)L_m(x-v) + R(x) \tag{8.1}$$

be the cardinal spline interpolation formula of degree $2m-1$ (see [7, Ths. 1, 2, and 3]). The fundamental function $L_m(x)$ is known to be characterized by the two conditions

$$L_m(x) \in \mathscr{S}_{2m+1} \cap L_2^{(m)}, \quad L_m(j) = \delta_j \text{ for all } j. \tag{8.2}$$

In fact we still obtain the same $L_m(x)$ if we replace the first condition (8.2) by

$$L_m(x) \in \mathscr{S}_{2m-1}$$

plus the much weaker requirement that $L_m(x)$ be at most of power growth.

A very special case of Theorem 2 is

Corollary 3. *The relation*

$$\lim_{m\to\infty} L_m(x) = \frac{\sin \pi x}{\pi x} \tag{8.3}$$

holds uniformly for all real x, as well as all differentiated relations.

Now observe the following two facts:

A. *The interpolating spline function $S_m(x)$ of §4 may be represented by*

$$S_m(x) = \sum_{-\infty}^{\infty} y_v L_m(x-v). \tag{8.4}$$

Indeed, the sequence (y_ν) satisfies (4.2) and is therefore of power growth, while for such sequences, (8.4) holds [7, Th. 3].

B. *By* (8.3) *the series on the right side of* (8.4) *goes over formally into*

$$\sum_{-\infty}^{\infty} y_\nu \frac{\sin \pi(x-\nu)}{\pi(x-\nu)}$$

as $m \to \infty$.

These facts seem to justify the following definition.

Definition 1. If $(y_\nu) \in l_2^*$, and if $F(x) \in PW_\pi^*$ is the interpolating function of Theorem 1, then we write

$$(S) \sum_{-\infty}^{\infty} y_\nu \frac{\sin \pi(x-\nu)}{\pi(x-\nu)} = F(x), \tag{8.5}$$

and call $F(x)$ the (S)-sum of the cardinal series on the left.

It follows that the relation

$$(S) \sum_{-\infty}^{\infty} F(\nu) \frac{\sin \pi(x-\nu)}{\pi(x-\nu)} = F(x)$$

holds for every $F(x) \in PW_\pi^*$, for instance if $F(x)$ is a polynomial. The example of §3, No. 2, may now be written as

$$(S) \sum_{-\infty}^{\infty} |\nu| \frac{\sin \pi(x-\nu)}{\pi(x-\nu)} = \frac{1}{\pi} \int_0^\pi \left(\frac{\sin(ux/2)}{u/2}\right)^2 du.$$

The name "(S)-sum" or "spline sum" is due to Theorem 2 which shows that $F(x)$ is the limit of the interpolating spline function $S_m(x)$, as $m \to \infty$, for real values of x.

Beyond merely restating Theorem 1, Definition 1 combines it with Theorem 2, the Lagrange expansion (8.4) and the limit relation (8.3).

References

1. R. P. Boas, Jr., *Entire functions*, Academic Press, New York, 1954.

2. P. R. Lipow and I. J. Schoenberg, *Cardinal interpolation and spline functions III. Cardinal Hermite interpolation*, J. Linear Algebra Appl. 6 (1973), 273–304. Also MRC T. S. Rep. #1113

3. W. Quade and L. Collatz, *Zur Interpolationstheorie der reelen periodischen Funktionen*, Sitzungsber. der Preuss. Akad. Wiss., Phys.-Math. Kl. XXX (1938), 383–429.

4. I. J. Schoenberg, *Contributions to the problem of approximatiion of equidistant data by analytic functions*, Quart. Appl. Math. **4** (1946), 45–99, 112–141.

5. I. J. Schoenberg, *On spline interpolation at all integer points of the real axis*, Mathematica (Cluj) **10** (1968), 151–170.

6. I. J. Schoenberg, *Cardinal interpolation and spline functions*, J. Approximation Theory **2** (1969), 167–206.

7. I. J. Schoenberg, *Cardinal interpolation and spline functions II. Interpolation of data of power growth*, J. Approximation Theory **6** (1972), 404–420. Also MRC T.S. Rep. #1104.

8. I. J. Schoenberg, *Cardinal interpolation and spline functions IV. The exponential Euler splines*, in: Linear Operators and Approximation (Proc. of Oberwolfach Conference, August 14–22, 1971), ISNM **20** (1972), 382–404. Also MRC T.S. Rep. #1153,

9. I. J. Schoenberg, *Cardinal interpolation and spline functions VI. Semi-cardinal interpolation and quadrature formulae*, MRC T.S.R. #1180, J. Analyse Math. **27** (1974), 159–204.

10. I. J. Schoenberg and A. Sharma, *Cardinal interpolation and spline functions V. The B-splines for cardinal Hermite interpolation*, Linear Algebra Appl. **7** (1973), 1–42. Also MRC T.S. Rep. # 1150,

11. J. M. Whittaker, *On the cardinal function of interpolation theory*, Proc. Edinburgh Math. Soc. **1** (1929), 41–46.

Mathematics Research Center
University of Wisconsin
Madison, Wisconsin, U.S.A.

(Received November 28, 1971)

Reprinted from
J. d'Analyse Math. **XXVII** (1974), pp. 205–229

ON THE REMAINDERS AND THE CONVERGENCE OF CARDINAL SPLINE INTERPOLATION FOR ALMOST PERIODIC FUNCTIONS

I. J. Schoenberg

1. INTRODUCTION

Remainders in terms of high-order derivatives might at times seem rather useless for numerical applications. However, they are often effective in theoretical problems of convergence. Our present topic is the remainder of cardinal spline interpolation (C.S.I.) of the odd degree $2m-1$, in its customary Peano form. Its kernel $K_{2m-1}(x,t)$ appears to be endowed with numerous worthwhile properties some of which are described in the first half (Part I) of this paper. The remainder of C.S.I. allows us to discuss the behavior of the interpolant, as $m \to \infty$, of entire functions of exponential type (Theorem 6 of §6).

This work was done during the winter of 1973-74 while the author, on leave from Madison, was at the Weizmann Institute of Science in Rehovot, Israel. Speaking on this subject in February 1973 at the Hebrew University in Jerusalem, Yitzhak Katznelson immediately noticed its relevance for the C.S.I. of almost periodic functions. His remarks, made orally to the author, are elaborated in §§8 and 9 below, with the originator's kind permission.

The paper concludes with a conjecture suggested by D.J. Newman's recent note [4]. An affirmative answer to this conjecture would show that C.S.I. is particularly effective for the approximation of entire functions of exponential type π and bounded on the real axis.

I. *THE REMAINDER OF CARDINAL SPLINE INTERPOLATION*

2. THE KERNEL OF THE REMAINDER AND SOME OF ITS PROPERTIES

Let m be a natural number and let

$$\mathcal{S}_{2m-1} = \{S(x)\} \tag{2.1}$$

denote the class of functions $S(x)$ $(\mathbb{R} \to \mathbb{C})$, which are piecewise polynomials of degree $\leq 2m-1$ in each unit interval $(\nu,\nu+1)$ for each integer ν, and such that $S(x) \in C^{2m-2}(\mathbb{R})$. It is known (see e.g. [6, Lecture 4]) that for every function $f(x)$ of power growth, i.e., belonging to the class

$$\mathcal{PG} = \{f(x); f(x) = O(|x|^{\gamma}) \text{ as } x \to \pm\infty, \text{ for some } \gamma \geq 0\}, \tag{2.2}$$

there exists a unique $S(x) \in \mathcal{S}_{2m-1}$, such that $S(x) \in \mathcal{PG}$, and which interpolates $f(x)$ at all integers, hence

$$S(\nu) = f(\nu) \quad \text{for all } \nu. \tag{2.3}$$

We denote this uniquely defined interpolant by $S_{2m-1}(f)$, writing $S_{2m-1}(f,x)$ to denote its value at x. We also know that the analogue of Lagrange's interpolation formula holds and allows us to represent the interpolant in the form

$$(2.4)\qquad S_{2m-1}(f,x) = \sum_{-\infty}^{\infty} f(\nu)L_{2m-1}(x-\nu) \quad ,$$

where $L_{2m-1}(x) = L(x)$ is the fundamental function of the process. The uniqueness of the interpolant implies that $S_{2m-1}(f,x) = f(x)$ for all x , if $f(x) \in \mathcal{S}_{2m-1} \cap \mathcal{PG}$. We say that C.S.I. is <u>exact</u> for elements of the class $\mathcal{S}_{2m-1} \cap \mathcal{PG}$.

We apply this interpolation process to the function

$$(2.5)\qquad f(x) = (x-t)_{+}^{2m-1}/(2m-1)! \quad , \text{ (t is a real parameter).}$$

Using (2.4) we find the remainder $K_{2m-1}(x,t) = K(x,t)$ of this interpolation to be given by

$$(2.6)\qquad (2m-1)!K_{2m-1}(x,t) = (x-t)_{+}^{2m-1} - \sum_{-\infty}^{\infty} (\nu-t)_{+}^{2m-1}L_{2m-1}(x-\nu) .$$

A study of this function from $R^2 \to R$, its properties and applications, is the topic of this paper.

The most obvious properties of $K(x,t)$ are as follows: $K(x,t) = 0$ if x is an integer, because the integers are the nodes of interpolation. However, $K(x,t) = 0$ also if t is an integer, because then $(x-t)_{+}^{2m-1} \in \mathcal{S}_{2m-1} \cap \mathcal{PG}$ and the interpolation is exact. We have found that

(2.7) $K_{2m-1}(x,t) = 0$ <u>if</u> x , <u>or</u> t , <u>are integers</u>.

We write

$$(2.8) \qquad K_0(x,t) = (2m-1)!K_{2m-1}(x,t)$$

and observe that by (2.6)

$$K_0(x+1,t+1) = (x-t)_+^{2m-1} - \sum(\nu-t-1)_+^{2m-1} L(x+1-\nu)$$

$$= (x-t)_+^{2m+1} - \sum(\nu-t)_+^{2m-1} L(x-\nu) = K_0(x,t) \quad ,$$

and therefore

$$(2.9) \qquad K_{2m-1}(x+1,t+1) = K_{2m-1}(x,t) \quad .$$

In studying the behavior of $K_{2m-1}(x,t)$ for large x it suffices therefore to assume that the point (x,t) is in the strip

$$(2.10) \qquad J = \{(x,t) \ ; \ 0 \leqslant t \leqslant 1\} \quad .$$

It is also clear from its definition (2.6) that for each fixed t , $K_{2m-1}(x,t) \in \mathcal{PL}$ as a function of x . A much stronger statement is provided by

<u>Lemma 1.</u> <u>For appropriate positive constants</u> A <u>and</u> α <u>we have that</u>

$$(2.11) \qquad |K_{2m-1}(x,t)| \leqslant A\, e^{-\alpha|x|} \quad \underline{\text{in the strip}}\ (2.10).$$

<u>Proof.</u> As the proof is somewhat long, if given with all details, we only sketch its main points. We assume $0 \leqslant t \leqslant 1$ and perform some surgery on $K_0(x,t)$. Looking at $K_0(x,t)$ only for $x \geqslant 1$, we obtain from (2.6) that

$$(2.12) \quad K_0(x,t) = (x-t)^{2m-1} - \sum_{\nu=1}^{\infty} (\nu-t)_+^{2m-1} L_{2m-1}(x-\nu)$$

$$\text{if } x \geqslant 1 \ .$$

The right side defines an element of $\mathcal{S}_{2m-1}[x]$ that we denote by

$$(2.13)\quad k_0(x,t) = (x-t)^{2m-1} - \sum_{\nu=1}^{\infty} (\nu-t)_+^{2m-1} L_{2m-1}(x-\nu) \ ,$$

$$-\infty < x < \infty \ ,$$

observing that

$$(2.14)\quad k_0(x,t) = K_0(x,t) \quad \text{if} \quad x \geq 1 \ .$$

From $k_0(x,t)$ we now derive the new cardinal spline (see [6, Lecture 4, §3])

$$(2.15)\quad \tilde{K}_0(x,t) = k_0(x,t)+b_1(t)(1-x)_+^{2m-1}+b_0(t)(-x)_+^{2m-1} + b_{-1}(t)(-1-x)_+^{2m-1} + \ldots$$

such that it becomes an element of the space

$$(2.16)\quad \mathcal{S}_{2m-1}^0 = \{S(x);S(x) \in \mathcal{S}_{2m-1}, S(\nu) = 0 \text{ for all } \nu\} \ .$$

This is possible because, by (2.14), $k_0(x,t)$ already vanishes if $x = 1,2,\ldots,$ and

$$(2.17)\quad \tilde{K}_0(x,t) \in \mathcal{S}_{2m-1}^0$$

only requires that $\tilde{K}_0(0,t) = 0$, $\tilde{K}_0(-1,t) = 0,\ldots$, and these equations determine the coefficients $b_1(t), b_0(t),\ldots$ successively and uniquely. Finally, as for all cardinal splines we have the standard representation

$$(2.18)\quad \tilde{K}_0(x,t) = \sum_{-\infty}^{\infty} a_j(t)\, Q_{2m}(x-j) \ ,$$

where $Q_{2m}(x)$ is the B-spline of $\mathcal{S}_{2m-1}$. The coefficients of (2.18) enjoy the property that

$$(2.19)\quad a_j(t) \in C[0,1] \quad \text{for all} \quad j \ .$$

This is obtained by showing first that the coefficients $b_i(t)$

in (2.15) are in $C[0,1]$, and then using the fact that the coefficients $a_j(t)$ are expressible as certain linear combinations with numerical coefficients of the even order derivatives (of orders $\leq 2m-2$) of $\tilde{K}_0(x,t)$ with respect to x , at the point $x = j$ (see [6, Lecture 2, §3]). We omit further details.

We recall next (see [6, Lecture 4, §3]) that the space (2.16) is apanned by the $2m-2$ eigensplines

$$(2.20)\qquad S_\nu(x) = \sum_{-\infty}^{\infty} \lambda_\nu^j \, Q_{2m}(x-j) \quad , \quad (\nu = 1,\ldots,2m-2)$$

of $\mathcal{S}_{2m-1}$, and therefore (2.17) implies a representation

$$(2.21)\qquad \tilde{K}_0(x,t) = \sum_{1}^{2m-2} c_\nu(t) S_\nu(x) \quad .$$

Now (2.18), (2.20) and (2.21) show that

$$(2.22)\qquad a_j(t) = \sum_{\nu=1}^{2m-2} c_\nu(t) \lambda_\nu^j \quad \text{for all } j \text{ .}$$

By solving for the $c_\nu(t)$ the system of $2m-2$ equations obtained from (2.22) for $j = 0,1,\ldots,2m-3$, we find, on using (2.19), that

$$(2.23)\qquad c_\nu(t) \in C[0,1] \quad , \quad \nu = 1,\ldots,2m-2 \quad .$$

Finally, the fact that $\tilde{K}_0(x,t) = k_0(x,t) = K_0(x,t)$, if $x \geq 1$, and that $K_0(\nu,t) \in \mathcal{PG}$ as $x \to +\infty$, for each fixed $t \in [0,1]$, shows that in the representation (2.21) only the "decreasing" eigensplines $S_1(x),\ldots,S_{m-1}(x)$ may occur. Hence, $c_\nu(t) = 0$ in $[0,1]$ if $\nu = m,m+1,\ldots,2m-2$, and (2.21) reduces to

$$(2.24)\qquad \tilde{K}_0(x,t) = \sum_{\nu=1}^{m-1} c_\nu(t) S_\nu(x) \quad .$$

Lemma 1 now follows readily: Firstly, from (2.23) we conclude that $|c_\nu(t)| < K$. Secondly, among the $m-1$ eigensplines of (2.24) the last eigenspline $S_{m-1}(x)$ has the slowest rate of exponential decay, so that $|S_\nu(x)| < B\exp(-\alpha x)$ if $x \geq 1$ for $\nu = 1,\ldots,m-1$. Finally, $\tilde{K}_0(x,t) = K_0(x,t)$ if $x \geq 1$, and (2.24) evidently implies (2.11).

From Lemma 1 we easily derive

Theorem 1. For appropriate positive constants A_1 and α we have

$$(2.25) \qquad |K_{2m-1}(x,t)| \leq A_1 e^{-\alpha|x-t|} \quad \text{for all} \quad (x,t) \in R^2 .$$

Proof. From (2.9) and (2.11) we obtain that

$$|K(x,t)| = |K(x-[t],t-[t])| \leq A\, e^{-\alpha|x-[t]|}$$

and (2.25) follows, because $|x-[t]| > |x-t| - 1$.

3. FURTHER PROPERTIES OF THE KERNEL $K_{2m-1}(x,t)$

Writing

$$(3.1) \qquad S_t(x) = \sum_{-\infty}^{\infty} (\nu-t)_+^{2m-1} L_{2m-1}(x-\nu) ,$$

we may express (2.6) as $K_0(x,t) = (x-t)_+^{2m-1} - S_t(x)$. If we now think of x as being fixed, and regard $(x-t)_+^{2m-1}$ as a function of t, we may apply to it cardinal spline interpolation in t and find its interpolant to be

$$(3.2) \qquad s_x(t) = \sum_{-\infty}^{\infty} (x-\nu)_+^{2m-1} L_{2m-1}(t-\nu) .$$

However, also $s_x(t) \in \mathcal{S}_{2m-1}[x]$. We therefore find both $S_t(x)$ and $s_x(t)$ to belong to the class (2.1) for t fixed. From (3.1) we obtain

$$(3.3) \qquad S_t(\nu) = (\nu-t)_+^{2m-1} .$$

Let us show that also

$$(3.4) \qquad s_\nu(t) = (\nu-t)_+^{2m-1} .$$

283

For indeed, $(\nu-t)_+^{2m-1} \in \mathcal{S}_{2m-1}[t]$ and the spline interpolation in t is exact and implies that (3.4) holds. Moreover, we easily find that both $S_t(x)$ and $s_x(t)$ are both elements of $\mathcal{P}\mathcal{S}$ as functions of x. By (3.3) and (3.4) they agree at all integers and from the unicity of C.S.I. we conclude that

(3.5) $$S_t(x) = s_x(t) \quad \text{for all} \quad (x,t) .$$

It follows that we can also write that

(3.6) $$K_0(x,t) = (x-t)_+^{2m-1} - \sum_{-\infty}^{\infty} (x-\nu)_+^{2m-1} L_{2m-1}(t-\nu) ,$$

while (2.6) gave

(3.7) $$K_0(x,t) = (x-t)_+^{2m-1} - \sum_{-\infty}^{\infty} (\nu-t)_+^{2m-1} L_{2m-1}(x-\nu) .$$

We now use the relation $u_+^{2m-1} - u^{2m-1} = (-u)_+^{2m-1}$, $(u \in R)$. It implies that if we subtract from (3.7) the identity

$$0 = (x-t)^{2m-1} - \sum_{-\infty}^{\infty} (\nu-t)^{2m-1} L_{2m-1}(x-\nu)$$

we obtain that

(3.8) $$K_0(x,t) = (t-x)_+^{2m-1} - \sum_{-\infty}^{\infty} (t-\nu)_+^{2m-1} L_{2m-1}(x-\nu) .$$

If we interchange here x and t and compare the result with (3.6), we find that we have established

<u>Theorem 2.</u> <u>The symmetry relation</u>

(3.9) $K_{2m-1}(x,t) = K_{2m-1}(t,x)$ holds for all real x and t.

Perhaps the most apt definition of our kernel is obtained if we define it as the remainder of the bivariate C.S.I. as discussed in [6, Lecture 4, §7], to obtain that

(3.10) $$(2m-1)!K_{2m-1}(x,t) = (x-t)_+^{2m-1} - \sum_{\mu} \sum_{\nu} (\mu-\nu)_+^{2m-1} \times L(x-\mu)L(t-\nu) .$$

That this defines the same kernel as before we see from

$$\sum_\mu \sum_\nu (\mu-\nu)_+^{2m-1} L(x-\mu)L(t-\nu) = \sum_\mu L(x-\mu)\sum_\nu (\mu-\nu)_+^{2m-1} L(t-\nu)$$

$$= \sum_\mu L(x-\mu)(\mu-t)_+^{2m-1} \quad ,$$

because $(\mu-t)_+^{2m-1} \in \mathcal{S}_{2m-1}[t]$, and now (3.10) reduces to (3.7).

Also Theorem 2 follows very easily from (3.10).

We now state the fundamental

Theorem 3. *Let* $m > 1$. *If* t *is fixed such that* $0 < t < 1$, *then* $K_{2m-1}(x,t)$ *has simple zeros at all integer values of* x *and vanishes nowhere else. It follows that*

(3.11) $$\operatorname{sgn} K_{2m-1}(x,t) = \varepsilon_m \operatorname{sgn} \sin \pi x , \text{ where } \varepsilon_m = \pm 1 .$$

A proof of this theorem will be given elsewhere in a study of the theorem of Budan and Fourier for splines and its applications.

An easy consequence is the

Corollary 1. *The sign of the kernel* $K_{2m-1}(x,t)$ $(m > 1)$ *in the* (x,t)*-plane exhibits a strict checker-board design of* $+1$ *and* -1 . In symbols: *If we write*

(3.12) $$\omega(x) = \operatorname{sgn} \sin \pi x$$

then

(3.13) $$\operatorname{sgn} K_{2m-1}(x,t) = \varepsilon_m \omega(x)\omega(t) \quad \text{for all } (x,t) .$$

Proof. (3.13) is trivially true if x , or t , is an integer by (2.7). We may therefore assume that neither x , nor t , is an integer. From (2.9) and (3.11) we obtain that

$$\operatorname{sgn} K_{2m-1}(x,t) = \operatorname{sgn} K(x-[t],t-[t])$$
$$= \varepsilon_m \omega(x-[t]) = \varepsilon_m \omega(x)(-1)^{[t]}$$
$$= \varepsilon_m \omega(x)\omega(t) \quad .$$

We will find below that $\varepsilon_m = (-1)^m$.

We have excluded above the case $m=1$ of linear C.S.I. Actually, Theorem 3 and Corollary 1 do then not hold as stated. Rather from (2.6) we easily find the following situation. We consider the Green's function

$$G(x,t) = \begin{cases} t(1-x) & \text{if} \quad 0 \leqslant t \leqslant x \leqslant 1 \quad , \\ x(1-t) & \text{if} \quad 0 \leqslant x \leqslant t \leqslant 1 \quad , \end{cases}$$

of the differential equation $y'' = g(x)$ for the boundary conditions $y(0) = y(1) = 0$. In terms of it we find that

$$K_1(x,t) = \begin{cases} -G(x-[x],t-[t]) & \text{if} \quad [x] = [t] \quad , \\ 0 & \text{if} \quad [x] \neq [t] \quad . \end{cases}$$

4. THE CARDINAL SPLINE INTERPOLATION FORMULA WITH REMAINDER

From Peano's well known theorem we expect the kernel $K_{2m-1}(x,t)$ to play the usual role of a Peano kernel in the expression of the remainder. However, as we deal with the entire axis, a detailed discussion seems to be called for.

Theorem 4. We assume that $f(x) \in C^{2m-1}(R)$, $f^{(2m-1)}(x)$ *is locally absolutely continuous and therefore* $f^{(2m)}(x)$ *is locally integrable, and also that*

(4.1) $$f^{(2m)}(x) \in \mathcal{P}\mathcal{G} \quad \text{as} \quad x \to \pm\infty \quad .$$

Then

(4.2) $$f(x) = \sum_{-\infty}^{\infty} f(\nu) L_{2m-1}(x-\nu) + \int_{-\infty}^{\infty} K_{2m-1}(x,t) f^{(2m)}(t)\,dt, \quad (x \in R).$$

286

<u>Proof.</u> It follows (4.1) that

$$f^{(\nu)}(x) \in \mathcal{PG} \text{ as } x \to \pm\infty \quad \text{for} \quad \nu = 0,1,\ldots,2m \,. \tag{4.3}$$

Since (4.2) is trivially correct if x is an integer, by (2.7), we may assume without loss of generality that

$$0 < x < 1 \,. \tag{4.4}$$

From (3.8) we know that

$$(2m-1)!K_{2m-1}(x,t) = (t-x)_+^{2m-1} - \sum_{-\infty}^{\infty} (t-\nu)_+^{2m-1} L_{2m-1}(x-\nu) \,. \tag{4.5}$$

We integrate repeatedly by parts the remainder term of (4.2), as can be done in view of (2.25) and (4.3). Omitting the subscript $2m-1$ we obtain

$$\int_{-\infty}^{\infty} Kf^{(2m)}dt = \int_{-\infty}^{\infty} Kdf^{(2m-1)} = - \int_{-\infty}^{\infty} K'f^{(2m-1)}dt = \ldots \tag{4.6}$$

$$= - \int_{-\infty}^{\infty} K^{(2m-1)} f'dt \,,$$

where $K^{(\nu)}(x,t)$ are derivatives with respect to t. At this point we must stop, because $K^{(2m-1)}$ is a step-function in t with discontinuities at the points

$$\ldots, -2, -1, 0, x, 1, 2, \ldots \quad . \tag{4.7}$$

Denoting by (α,β) a generic interval with its endpoints at consecutive points of (4.7), we may rewrite the last result of (4.6) as

$$\int_{-\infty}^{\infty} Kf^{(2m)}dt = \sum_{(\alpha,\beta)} \{\text{value of } K^{(2m-1)} \text{ in } (\alpha,\beta)\}(f(\alpha)-f(\beta))$$

$$= \sum_{\alpha} f(\alpha)\{\text{jump of } K^{(2m-1)} \text{ at } t = \alpha\} \,.$$

From (4.5) we find that

$$\text{the jump of } K^{(2m-1)} \text{ at } t=\alpha \text{ is } = \begin{cases} 1 & \text{if } \alpha = x \,, \\ -L_{2m-1}(x-\nu) & \text{if } \alpha = \nu \,, \end{cases}$$

and the relation (4.2) follows.

Generalizing the remark of the last paragraph of §3, let us show that $K_{2m-1}(x,t)$ <u>is the Green's function for the differential equation</u> $y^{(2m)} = g(x)$ <u>for the multi-boundary value problem</u>

(4.8) $$y(\nu) = 0 , \quad \text{for all integers } \nu .$$

Let $g(x)$ be locally integrable, $g(x) \in \mathcal{PG}$, and let us find $G(x)$ such as to satisfy the following conditions:

1. $G(x)$ is a solution of $y^{(2m)} = g(x)$ in each open interval $(\nu,\nu+1)$, for all ν ,

(4.9) 2. $G(x) \in C^{2m-2}(R)$,

3. $G(\nu) = 0$ for all ν ,

4. $G(x) \in \mathcal{PG}$.

<u>The solution</u> $G(x)$ <u>of this problem is unique and given by</u>

(4.10) $$G(x) = \int_{-\infty}^{\infty} K_{2m-1}(x,t)g(t)dt .$$

<u>Proof.</u> Let $G_0(x)$ be any fixed solution of $y^{(2m)} = g(x)$ on R , and let us define

(4.11) $$G(x) = G_0(x) - S_{2m-1}(G_0,x) .$$

Since $G_0(x)$ satisfies all conditions of Theorem 4, we see that on applying (4.2) to $f(x) = G_0(x)$ we obtain (4.10). This also proves the uniqueness of $G(x)$. Incidentally, the uniqueness of $G(x)$ was clear in the first place, because the difference of two solutions would have to be an element of the class (2.16) which is also of power growth, and we know that it must therefore vanish identically.

5. A FEW SPECIAL CHOICES OF f(x)

A. In (3.13) of Corollary 1 we found the nice checker-board design

(5.1) $\qquad \operatorname{sgn} K_{2m-1}(x,t) = \varepsilon_m \, \omega(x)\omega(t) \quad .$

To find ε_m we apply the C.I.F. (4.2) to the function

$$f(x) = \sin \pi x \ .$$

From $f^{(2m)}(t) = (-1)^m \pi^{2m} \sin\pi x$, and because $f(\nu) = 0$ for all ν , (4.2) reduces to

$$\sin \pi x = (-1)^m \pi^{2m} \int_{\infty}^{\infty} K(x,t) \sin\pi t \, dt^{\dagger}$$

$$= (-1)^m \pi^{2m} \int_{-\infty}^{\infty} \varepsilon_m \omega(x)\omega(t) |K(x,t)| \cdot \omega(t) |\sin\pi t| dt \ .$$

Since $\omega(t)\omega(t) = 1$ for non-integral t , we obtain

$$\omega(x) |\sin\pi x| = (-1)^m \pi^{2m} \varepsilon_m \omega(x) \int_{-\infty}^{\infty} |K(x,t) \sin\pi t| dt$$

and also, if x is not an integer, that

$$|\sin\pi x| = (-1)^m \varepsilon_m \pi^{2m} \int_{-\infty}^{\infty} |K(x,t) \sin\pi t| dt \ .$$

Because of (2.7), this holds for all x and shows that

(5.2) $\qquad \varepsilon_m = (-1)^m$

and

(5.3) $\qquad |\sin\pi x| = \pi^{2m} \int_{-\infty}^{\infty} |K(x,t) \sin\pi t| dt \quad .$

We have therefore obtained the

<u>Corollary 2.</u> <u>For the sign of</u> $K_{2m-1}(x,t)$ <u>we have</u>

(5.4) $\qquad \operatorname{sgn} K_{2m-1}(x,t) = (-1)^m (\operatorname{sgn} \sin \pi x)(\operatorname{sgn} \sin \pi t) \ .$

---

† Similarly $\sin(k\pi t)$ is an eigenfunction of $K(x,t)$ for the eigenvalue $(-1)^m (k\pi)^{2m}$.

This includes the results (2.7), (3.11) and (3.13).

B. We obtain additional information on $K_{2m-1}(x,t)$ by using appropriate <u>Euler splines</u> (See [6, Lecture 4, §6B]). The even degree Euler spline $\mathcal{E}_{2m}(x)$ satisfies $\mathcal{E}_{2m}(\nu) = (-1)^{\nu}$ for all ν, $\mathcal{E}_{2m}(x+1) = -\mathcal{E}_{2m}(x)$, and the knots of the spline are at the points $\nu+\frac{1}{2}$. It follows that $\mathcal{E}_{2m}(x)$ is a so-called perfect spline, i.e., such that

$$|\mathcal{E}^{(2m)}(x)| = \gamma_{2m} \qquad (x \neq \nu+\tfrac{1}{2}) \tag{5.5}$$

has the same value for all $x \neq \nu+\frac{1}{2}$. Therefore the function

$$f(x) = \mathcal{E}_{2m}(x-\tfrac{1}{2}) \qquad (x \in R) \tag{5.6}$$

vanishes at all integers and

$$\operatorname{sgn} \mathcal{E}^{(2m)}(x-\tfrac{1}{2}) = \eta_m \gamma_{2m} \operatorname{sgn} \sin \pi x \qquad (x \in R) , \tag{5.7}$$

where $\eta_m = \pm 1$. Applying the C.S.I.F. (4.2) to the function (5.6) we obtain the identity

$$\mathcal{E}_{2m}(x-\tfrac{1}{2}) = \int_{-\infty}^{\infty} K_{2m-1}(x,t)\mathcal{E}_{2m}^{(2m)}(t-\tfrac{1}{2})dt$$

and therefore

$$\mathcal{E}_{2m}(x-\tfrac{1}{2}) = \int_{-\infty}^{\infty} |K_{2m-1}(x,t)|(-1)^m \omega(x)\omega(t)\eta_m \gamma_{2m}\omega(t)dt ,$$

which simplifies to

$$\mathcal{E}_{2m}(x-\tfrac{1}{2}) = (-1)^m \eta_m \gamma_{2m} \omega(x) \int_{-\infty}^{\infty} |K_{2m-1}(x,t)|dt .$$

For $x = \frac{1}{2}$ we obtain that $\eta_m = (-1)^m$ and finally

$$\int_{-\infty}^{\infty} |K(x,t)|dt = \frac{1}{\gamma_{2m}} |\mathcal{E}_{2m}(x-\tfrac{1}{2})| \qquad (x \in R) . \tag{5.8}$$

Of this we make some applications below.

C. For another example in the same spirit we let

$$f(x) = x^{2m}/(2m)! \tag{5.9}$$

Here $f^{(2m)}(t) = 1$ and the remainder of (4.2) is known to be identical with the Bernoulli monospline $(\bar{B}_{2m}(x)-B_{2m})/(2m)!$ (See [6, Lecture 4, §6C]). Therefore (4.2) becomes

$$\int_{-\infty}^{\infty} K(x,t)\,dt = \frac{1}{(2m)!}(\bar{B}_{2m}(x)-B_{2m}) \;, \quad (x \in R) \;. \tag{5.10}$$

6. APPLICATIONS

A. For the applications of (4.2) and (5.8) that we have in mind we consider the class

$$L_\infty^{2m} = \{f(x); f^{(2m-1)} \text{ locally abs. continuous, } ||f^{(2m)}||_\infty < \infty\} \;. \tag{6.1}$$

Assuming that $f(x) \in L_\infty^{2m}$ and denoting by $S_{2m-1}(f,x)$ the interpolating cardinal spline we obtain from (4.2) and (5.8) that

$$|f(x)-S_{2m-1}(f,x)| \leq \frac{1}{\gamma_{2m}} |\mathcal{E}_{2m}(x-\tfrac{1}{2})| \cdot ||f^{(2m)}||_\infty \;, \quad (x \in R). \tag{6.2}$$

<u>Theorem 5.</u> <u>In the inequality</u> (6.2) <u>the constant</u>

$$\frac{1}{\gamma_{2m}} |\mathcal{E}_{2m}(x-\tfrac{1}{2})| \tag{6.3}$$

<u>is best. Moreover, if</u> x_0 <u>is not an integer, while</u>

$$|f(x_0)-S_{2m-1}(f;x_0)| = \frac{1}{\gamma_{2m}} |\mathcal{E}_{2m}(x_0-\tfrac{1}{2})| \tag{6.4}$$

and

$$||f^{(2m)}||_\infty \leq 1 \;, \tag{6.5}$$

<u>then</u> $f(x)$ <u>must be of the form</u>

$$f(x) = \pm \frac{1}{\gamma_{2m}} \mathcal{E}_{2m}(x-\tfrac{1}{2}) + P(x) \;, \text{ where } P(x) \in \pi_{2m-1} \;. \tag{6.6}$$

<u>Proof.</u> We assume (6.4) and (6.5) to hold. Then, using (5.8),

$$\gamma_{2m}^{-1}|\mathcal{E}_{2m}(x_0-\tfrac{1}{2})| = |f(x_0)-S_{2m-1}(f;x_0)| = |\int_{-\infty}^{\infty} K_{2m-1}(x_0,t)f^{(2m)}(t)dt|$$

$$\leq \int_{-\infty}^{\infty} |K_{2m-1}(x_0,t)|dt\cdot||f^{(2m)}||_\infty \leq \int_{-\infty}^{\infty} |K_{2m-1}(x_0,t)|dt$$

$$= \gamma_{2m}^{-1}\ |\mathcal{E}_{2m}(x_0-\tfrac{1}{2})|\quad ,$$

where the first and last term are equal. It follows that the first integral is equal to the third integral. Since x_0 is not an integer, we have $\operatorname{sgn} K_{2m-1}(x_0,t) = \varepsilon\ \omega(t)$ $(\varepsilon = \pm 1)$, and we must therefore have that

$$f^{(2m)}(t) = \eta \operatorname{sgn} \sin \pi t \quad \text{for almost all} \quad t\ .$$

From this we obtain (6.6) by $2m$ integrations.

B. If $f(x)$, defined on R , is the restriction to R of an entire function, then we denote this entire function by the symbol $f(z)$. Conversely, if $f(z)$ is entire, then $f(x)$ will denote its restriction to R .

We consider the well-known classes (see e.g. [1, Chapter 4])

(6.7) $\quad E_\sigma = \{f(z);f(z) \text{ entire of exponential type } \leq \sigma\}$,

and

(6.8) $\quad B_\sigma = \{f(z);f(z) \in E_\sigma\ ,\ \sup|f(x)| = ||f||_\infty < \infty\}$.

It is known that if $f(z) \in B_\sigma$, then not only does $|f(z)| < A \exp(\sigma|z|)$ $(z \in \mathbb{C})$ hold, but even

$$|f(z)| \leq A \exp(\sigma|\operatorname{Im} z|)\ .$$

If $f(z) \in B_\sigma$, then we write $\sup_{x\in R} |f(x)| = ||f||_\infty$.

We also need S. Bernstein's theorem that if $f(z) \in B_\sigma$, then also for each integer n $f^{(n)}(z) \in B_\sigma$ and we have the inequality

(6.9) $$||f^{(n)}||_\infty \leq \sigma^n||f||_\infty\ .$$

For further applications of (6.2) we need a good estimate for the constant $1/\gamma_{2m}$. This is easily obtained from the Fourier series

$$(6.10)\quad \mathcal{E}_n(x) = \sum_0^\infty \frac{(-1)^{\nu(n+1)}}{(2\nu+1)^{n+1}} \cos(\pi(2\nu+1)x) \Big/ \sum_0^\infty \frac{(-1)^{\nu(n+1)}}{(2\nu+1)^{n+1}} .$$

If $n=2m$ this becomes

$$\mathcal{E}_{2m}(x) = \sum_0^\infty \frac{(-1)^\nu}{(2\nu+1)^{2m+1}} \cos(\pi(2\nu+1)x) \Big/ \sum_0^\infty \frac{(-1)^\nu}{(2\nu+1)^{2m+1}}$$

and by differentiation we obtain

$$\mathcal{E}_{2m}^{(2m)}(x) = (-1)^m \pi^{2m} \sum_0^\infty \frac{(-1)^\nu}{2\nu+1} \cos(\pi(2\nu+1)x) \Big/ \sum_0^\infty \frac{(-1)^\nu}{(2\nu+1)^{2m+1}} .$$

For $x=0$ this shows that

$$\gamma_{2m} = |\mathcal{E}_{2m}^{(2m)}(0)| = \pi^{2m}\left(1-\frac{1}{3}+\frac{1}{5}-\ldots\right) \Big/ \left(1-\frac{1}{3^{2m+1}}+\frac{1}{5^{2m+1}}-\ldots\right) ,$$

whence

$$(6.11)\quad \frac{\pi^{2m}}{\gamma_{2m}} = \frac{4}{\pi}\left(1-\frac{1}{3^{2m+1}}+\frac{1}{5^{2m+1}}-\ldots\right) < \frac{4}{\pi} .$$

We now derive from (5.8), observing that $||\mathcal{E}_{2m}||_\infty = 1$, the inequality

$$(6.12)\qquad \int_{-\infty}^{\infty} |K(x,t)|\,dt < \frac{4}{\pi}\,\pi^{-2n} \qquad (x \in R) .$$

Assuming that $f(x) \in B_\sigma$, and using Bernstein's inequality (6.9), we derive immediately the following†

Theorem 6. *If*

$$(6.13)\qquad f(x) \in B_\sigma ,$$

then

---

† See bottom of next page.

(6.14) $$||f - S_{2m-1}(f)||_\infty \leq \frac{4}{\pi}(\frac{\sigma}{\pi})^{2m} ||f||_\infty .$$

The following are immediate consequences.

<u>Corollary 3.</u> <u>If</u>

(6.15) $$f(x) \in B_\sigma \quad \underline{\text{and}} \quad \sigma < \pi ,$$

<u>then</u>

(6.16) $$\lim_{m\to\infty} S_{2m-1}(f,x) = f(x) \quad \underline{\text{uniformly on}} \quad R .$$

<u>Corollary 4.</u> <u>If</u>

(6.17) $$f(x) \in B_\pi ,$$

<u>then</u>

(6.18) $$|f(x) - S_{2m-1}(f,x)| \leq \frac{4}{\pi} ||f||_\infty \quad (x \in R) .$$

From (6.17) we can no longer conclude that we have convergence to $f(x)$, but at least we see that the interpolating splines are equibounded on R. Actually, if we assume only (6.17), then (6.16) need not hold, as we see if we take

---

†If we apply Theorem 6 to $f(x) = e^{iux} (-\pi \leq u \leq \pi)$ we obtain

(*) $$||f-S_{2m-1}(f)||_\infty \leq \frac{4}{\pi} (\frac{|u|}{\pi})^{2m}$$

because $||f||_\infty = 1$. In this case the interpolant $S_{2m-1}(f)$ is identical with what we denote by $S_{2m-1}(x;e^{iu})$ in [6, Lecture 3, §§5 and 6] and called the <u>exponential Euler spline</u>. In Theorem 8 of [6, p.30] it is stated that in our case when $f(x) = e^{iux}$, the left side of (*) is $\leq C_m(|u|/\pi)^{2m}$, where $C_m < 3$. We see that (*) is a slight improvement. It is also stated in [6, p.30] that $C_2 = C_3 = 1$ and it is there conjectured that $C_m = 1$ if $m > 3$. An oral communication from C. de Boor seems to throw doubt on these statements and conjecture. See the important paper: C. de Boor, On the cardinal spline interpolant to e^{int}, to appear in SIAM J. of Num. Analysis.

$f(x) = \sin \pi x$, when $S_{2m-1}(f,x) = 0$ for all real x .

II. THE BEHAVIOR OF THE CARDINAL SPLINE INTERPOLANT AS $m \to \infty$

7. DEFINITIONS AND KNOWN RESULTS

The theme of the present second part is as follows: $f(x)$ being a function from $\mathbb{R}$ to $\mathbb{C}$ and $S_{2m-1}(f,x)$ denoting its cardinal spline interpolant, we are interested in the behavior of $S_{2m-1}(f,x)$ as $m \to \infty$, the expectation being that in some sense $S_{2m-1}(f,x)$ will converge to $f(x)$. However, since C.S.I. samples $f(x)$ at the integers only, it cannot distinguish between $f(x)$ and $f(x) - C \sin \pi x$, where C is a constant. We might therefore expect that $S_{2m-1}(f,x) \to f(x) - C \sin \pi x$, rather than $f(x)$.

Let

$$\mathcal{B} = \{f(x)\}$$

denote the class of functions $f(x)$ which are bounded on $\mathbb{R}$.

<u>Definition 1.</u> Let

$$(7.1) \qquad f(x) \in \mathcal{B} , \quad f_n(x) \in \mathcal{B} \quad (n = 1,2,\ldots) .$$

<u>We say that</u> $f_n(x)$ <u>converges strongly to</u> $f(x)$ <u>and write</u>

$$(7.2) \qquad \operatorname*{strong\,lim}_{n \to \infty} f_n(x) = f(x) ,$$

<u>provided that</u> $||f - f_n||_\infty \to 0$ <u>as</u> $n \to \infty$.

A different topology in $\mathcal{B}$ was recently introduced by D.J. Newman [4] as follows.

<u>Definition 2.</u> <u>We assume</u> (7.1) <u>and say that</u> $f_n(x)$ <u>converges weakly to</u> $f(x)$ <u>and write</u>

$$(7.3) \qquad \operatorname*{weak\,lim}_{n \to \infty} f_n(x) = f(x) ,$$

<u>provided that</u>

(7.4) $|f_n(x)| < K$ <u>for all</u> n <u>and</u> x ,

(7.5) $\lim_{n\to\infty} f_n(x) = f(x)$ <u>locally uniformly</u>.

We use weak convergence only in our last §10.

Let us review the main known facts about the asymptotic behavior of $S_{2m-1}(f)$. The simplest example is $f(x) = \cos \pi x$ when

(7.6) $S_{2m-1}(\cos \pi x) = \mathcal{E}_{2m-1}(x)$

is the Euler spline. By taking only the first term of the Fourier series (6.10), for $n = 2m-1$, we find that

(7.7) strong $\lim S_{2m-1}(\cos \pi x) = \cos \pi x$.

It is also known (see [6, Lecture 3, §6]) that if $u \in \mathbb{C}$, but e^{iu} is not a negative number, then

(7.8) $\lim_{m\to\infty} S_{2m-1}(e^{iux}) = e^{iux}$ locally uniformly on $\mathbb{R}$.

In the opposite direction (see [5]) we have

<u>Theorem 7.</u> (Richards and Schoenberg). <u>If</u> $f(x) \in \mathcal{B}$ <u>and</u>

(7.9) strong $\lim_{m\to\infty} S_{2m-1}(f,x) = f(x) - C \sin \pi x$ (C <u>is a constant</u>)

<u>then</u>

(7.10) $f(x) \in B_\pi$.

By (7.10) we mean that $f(x)$ is the restriction to $\mathbb{R}$ of a function $f(z) \in B_\pi$. Corollaries 3,4, and especially Theorem 7 suggest that the class B_π if of particular relevance in our problem. A direct result is (see [7])

<u>Theorem 8.</u> (Schoenberg). <u>If</u> $f(x)$ <u>is of the form</u>

(7.11) $f(x) = \int_{-\pi}^{\pi} e^{iux}\, d\alpha(u)$, $\alpha(u)$ <u>of B.V. in</u> $[-\pi,\pi]$,

<u>then</u>

(7.12) $$\text{strong}\lim S_{2m-1}(f) = f(x) - C\sin\pi x \quad ,$$

<u>where</u>

(7.13) $$C = i\{\alpha(\pi)-\alpha(\pi-0) + \alpha(-\pi)-\alpha(-\pi+0)\} \quad .$$

In §§8 and 9 of this paper we do the following:
1. We describe a definition and a lemma due to Y. Katznelson (oral communication) in §8. 2. We apply Katznelson's lemma in §9 to obtain a new proof of Theorem 8, and, what is more important, we show following Katznelson, that C.S.I. converges for the case of almost periodic functions.

8. KATZNELSON'S DEFINITION AND LEMMA

<u>Definition 3.</u> <u>We denote by</u> B_π^* <u>the class of functions which are uniform limits on</u> R <u>of functions belonging to</u> B_p <u>for</u> $p < \pi$. <u>More precisely</u>

(8.1) $$f(x) \in B_\pi^*$$

<u>means the existence of a sequence of functions</u> $f_j(x)$, $(j = 1,2,\ldots)$, <u>such that</u>

(8.2) $$f_j(x) \in B_{p_j} \quad \text{with} \quad p_j < \pi \quad ,$$

(8.3) $$||f-f_j||_\infty \to 0 \quad \text{as} \quad j \to \infty \quad .$$

<u>Lemma 2.</u> (Katznelson). <u>If</u> $f(x)$ <u>satisfies</u> (8.1) <u>then</u>

(8.4) $$\text{strong}\lim_{m\to\infty} S_{2m-1}(f) = f(x) \quad .$$

<u>Proof.</u> We assume that (8.2), (8.3), hold. Observe that

$$\begin{aligned} f-S_{2m-1}(f) &= f-f_j+f_j-S_{2m-1}(f_j)+S_{2m-1}(f_j)-S_{2m-1}(f) \\ &= f_j-S_{2m-1}(f_j)+\{f-f_j-S_{2m-1}(f-f_j)\} \end{aligned}$$

because we deal with a linear interpolation process, hence $S(f)-S(f_j) = S(f-f_j)$. Therefore

(8.5) $||f-S_{2m-1}(f)||_\infty \leqslant ||f_j-S_{2m-1}(f_j||_\infty+||f-f_j-S_{2m-1}(f-f_j)||_\infty .$

From (8.2) and the inequality (6.14) of Theorem 6 we obtain

$$||f_j-S_{2m-1}(f_j)||_\infty \leqslant \frac{4}{\pi}\left(\frac{p_j}{\pi}\right)^{2m} ||f_j||_\infty < C\left(\frac{p_j}{\pi}\right)^{2m} \quad \text{for some } C > 0 .$$

Also $f-f_j \in B_\pi$, and so by (6.18) of Corollary 4

$$||f-f_j-S_{2m-1}(f-f_j)||_\infty \leqslant \frac{4}{\pi}||f-f_j||_\infty .$$

From (8.5) and both estimates we have

$$||f-S_{2m-1}(f)||_\infty < C\left(\frac{p_j}{\pi}\right)^{2m} + \frac{4}{\pi}||f-f_j||_\infty .$$

This establishes (8.4), because by (8.3) the second term on the right side is $< \varepsilon/2$ if j is sufficiently large. We select one such j , and then $p_j < \pi$ implies that also the first term is $< \varepsilon/2$, if only m is large enough.

9. TWO APPLICATIONS OF KATZNELSON'S LEMMA

A. Stieltjes integrals. We give a new proof of Theorem 8. In terms of the function $\alpha(u)$ of (7.11) we define the new function

$$(9.1) \qquad \alpha_0(u) = \begin{cases} \alpha(-\pi+0) & \text{if } u = -\pi , \\ \alpha(u) & \text{if } -\pi < u < \pi , \\ \alpha(\pi-0) & \text{if } u = \pi . \end{cases}$$

In words: $\alpha_0(u)$ is identical with $\alpha(u)$ in $-\pi < u < \pi$, and $\alpha_0(u) \in C[-\pi,\pi]$. Let

$$(9.2) \qquad f_0(x) = \int_{-\pi}^{\pi} e^{iux}\, d\alpha_0(u) .$$

Setting

$$(9.3) \qquad A_1 = \alpha(-\pi+0)-\alpha(-\pi) , \quad A_2 = \alpha(\pi)-\alpha(\pi-0) ,$$

we can rewrite (7.11) in terms of $f_0(x)$ as

$$f(x) = f_0(x) + A_1 e^{-i\pi x} + A_2 e^{i\pi x} .$$

Setting

(9.4) $$B = A_2+A_1 \ , \quad C = i(A_2-A_1) \ ,$$

we obtain for (7.11) the final form

(9.5) $$f(x) = f_0(x) + B \cos \alpha\pi + C \sin \pi x \ .$$

We choose a sequence $\{p_j\}$ such that

(9.6) $$0 < p_j < \pi \ , \quad p_j \to \pi \quad \text{as} \quad j \to \infty \ ,$$

and define

(9.7) $$f_j(x) = \int_{-p_j}^{p_j} e^{iux} \, d\,\alpha_0(u) \ .$$

Evidently

$$f_j(x) \in B_{p_j} \quad \text{and} \quad ||f_0-f_j||_\infty \to 0 \quad \text{as} \quad j \to \infty \ ,$$

because $\alpha_0(u)$ is continuous at the endpoints $u = \pm\pi$. By Katznelson's Lemma 2 we conclude that

(9.8) $$\lim_{m\to\infty} S_{2m-1}(f_0) = f_0(x) \quad \text{uniformly on} \quad R \ .$$

By (7.6) and (7.7) we already know that

(9.9) $$\lim_{m\to\infty} S_{2m-1}(B \cos \pi x) = B \cos \pi x \quad \text{uniformly on} \quad R \ .$$

From (9.5) we derive

$$\begin{aligned} S_{2m+1}(f) &= S_{2m-1}(f-C \sin \pi x) \\ &= S_{2m-1}(f_0+B \cos \pi x) \\ &= S_{2m-1}(f_0) + S_{2m-1}(B \cos \pi x) \end{aligned}$$

and, by (9.8), (9.9), this converges strongly as $m \to \infty$, to the limit

$$f_0(x) + B \cos \pi x = f(x) - C \sin \pi x \ .$$

Since (9.3) and (9.4) imply (7.13), the proof is completed.

B. <u>Almost periodic functions.</u> Let

$$\mathcal{AP} = \{f(x)\}$$

denote the class of almost periodic functions $f(x)$ in the

sense of H. Bohr [2, Article C3, p.30]. To every $f(x) \in \mathcal{AP}$ corresponds a Fourier series

$$(9.10) \qquad f(x) \sim \sum_{\nu=1}^{\infty} A_\nu e^{i\lambda_\nu x} \quad , \quad \text{finite or infinite.}$$

The λ_ν , all real, are the Fourier exponents, or frequencies, of $f(x)$. We also define the subclass

$$\mathcal{AP}[-\sigma,\sigma] = \{f(x);f(x) \in \mathcal{AP} \quad , -\sigma \leq \lambda_\nu \leq \sigma\} \quad .$$

To Bohr is also due the result that

$$(9.11) \qquad \mathcal{AP} \cap B_\sigma = \mathcal{AP}[-\sigma,\sigma]$$

(see [2, Article C10] and [1, §6.10]).

Let $f(x) \in \mathcal{AP}$ and let us investigate conditions which will insure that $S_{2m-1}(f) \to f(x)$ as $m \to \infty$. Our Theorem 7 and (9.11) show that in studying this problem we should consider only functions such that

$$(9.12) \qquad f(x) \in \mathcal{AP}[-\pi,\pi] \quad .$$

Accordingly, we assume that the frequencies λ_ν of the Fourier series (9.10) satisfy

$$(9.13) \qquad -\pi \leq \lambda_\nu \leq \pi \quad \text{for all} \quad \nu \quad .$$

Let us so choose our notation that

$$(9.14) \qquad \lambda_1 = -\pi \quad , \quad \lambda_2 = \pi \quad ,$$

with the understanding that $A_1 = 0$ if the frequency $-\pi$ is absent, and similarly that $A_2 = 0$ if π does not occur.

<u>Theorem 9.</u> (Katznelson). <u>If</u> $f(x)$ <u>is an almost periodic function satisfying</u> (9.10) <u>and</u> (9.13), <u>then</u>

$$(9.15) \qquad \text{strong}\lim_{m\to\infty} S_{2m-1}(f) = f(x) - C \sin \pi x$$

<u>where</u>

$$(9.16) \qquad C = 2 \lim_{T\to\infty} \frac{1}{T} \int_0^T f(x) \sin \pi x \, dx \quad .$$

<u>Proof.</u> Let $A_1 e^{-i\pi x} + A_2 e^{i\pi x} = B \cos \pi x + C \sin \pi x$, where $B = A_2 + A_1$, $C = i(A_2 - A_1)$. It follows from (9.10), (9.14),

that the new function

$$g(x) = f(x) - B\cos\pi x - C\sin\pi x \sim \sum_{\nu=3}^{\infty} A_\nu e^{i\lambda_\nu x} \tag{9.17}$$

is almost periodic and has the Fourier series as indicated.

The crucial observation is that $g(x)$ is in Katznelson's class B_π^* of Definition 3:

$$g(x) \in B_\pi^* \ . \tag{9.18}$$

Proof. By our assumption (9.14) all Fourier exponents λ_ν of $g(x)$ are in the open interval $(-\pi,\pi)$, because

$$-\pi < \lambda_\nu < \pi \quad \text{if} \quad \nu \geq 3 \ . \tag{9.19}$$

From a theorem of Bohr's (see e.g., [2, Article C 28, p.69]) we know that $g(x)$ is the strong limit of an appropriate sequence of trigonometric polynomials

$$f_j(x) = \sum_{\nu=3}^{j} a_{\nu,j} e^{i\lambda_\nu x} \ , \tag{9.20}$$

for instance the Bochner-Fejér means of the Fourier series of $g(x)$ enjoy this property. However,

$$f_j(x) \in B_{p_j} \ , \text{ where } p_j = \max_{\nu=3,\dots,j} |\lambda_\nu| < \pi \ ,$$

the last inequality following from (9.19). This establishes (9.18).

By Lemma 2 we conclude that

$$\text{strong}\lim_{m\to\infty} S_{2m-1}(g) = g(x) \ . \tag{9.21}$$

From (9.17) we conclude that

$$S_{2m-1}(f) = S_{2m-1}(f - C\sin\pi x) = S_{2m-1}(g + B\cos\pi x)$$
$$= S_{2m-1}(g) + B\,S_{2m-1}(\cos\pi x) \ ,$$

and now (9.21) and (7.7) show that

$$\text{strong}\lim_{m\to\infty} S_{2m-1}(f) = g(x) + B\cos\pi x = f(x) - C\sin\pi x \ ,$$

using (9.17) again. The relations

$$C = i(A_2 - A_1) = i\, M\{f(x)e^{-i\pi x} - e^{i\pi x})\} = 2M\{f(x)\ \sin \pi x\}$$

complete a proof of the theorem.

Remark. Theorems 8 and 9 are related, but none implies the other. However, if the $f(x)$ of Theorem 9 has an absolutely convergent Fourier series, $\sum |A_\nu| < \infty$, then $f(x)$ can be expressed as a Stieltjes integral (7.11) and then Theorem 8 implies (9.15).

10. A CONJECTURE

Let T_π denote the class of trigonometric polynomials $\sum A_\nu e^{i\lambda_\nu x}$ with frequencies λ_ν in the interval $[-\pi,\pi]$. H. Bohr has shown that $\mathcal{AP}[-\pi,\pi]$ is identical with the closure of T_π in the sense of the strong limit. In terms of the Definition 2 (§7), D.J. Newman has recently established in [4] the following interesting theorem.

Theorem 10. (Newman). The class B_π is the closure of T_π in the sense of the weak limit.

If we analyze the proof of Theorem 7 as given in [5] we find that it actually proves the following stronger version (stronger, because its assumption is weaker).

Theorem 11. (Richards and Schoenberg). If $f(x) \in \mathcal{B}$ and

(10.1) $$\operatorname*{weak\ lim}_{m \to \infty} S_{2m-1}(f) = f(x) - C \sin \pi x \quad (C \text{ constant})$$

then

(10.2) $$f(x) \in B_\pi \ .$$

Theorems 10 and 11 suggest a converse of Theorem 11 which we state as

Conjecture 1. If (10.2) holds, then also (10.1) holds for an appropriate constant C.

This, if true, would be a fitting conclusion of the study

of the asymptotic behavior of cardinal spline interpolation for bounded functions.

REFERENCES

1. Boas, R.P., Jr. Entire Functions. Academic Press, New York, 1954.

2. Bohr, H. Collected Mathematical Works, Vol. II. Danish Math. Society, Copenhagen, 1952.

3. Katznelson, Y. Oral communication.

4. Newman, D.J. Finite type functions as limits of exponential sums. MRC T.S.R. #1477, 4 pp. September, 1974.

5. Richards F.B. and I.J. Schoenberg. Notes on spline functions IV. A cardinal spline analogue of the theorem of the brothers Markov. Israel J. Math. 16: 94-102. 1973.

6. Schoenberg, I.J. Cardinal spline interpolation. CBMS Regional Conf. Monograph No.12. SIAM, Philadelphia. 1973.

7. Schoenberg, I.J. Notes on spline functions III. On the convergence of the interpolating cardinal splines as their degree tends to infinity. Israel J. Math. 16: 87-93. 1973.

Reprinted from
Studies in Spline Functions and Approximation Theory
S. Karlin et al., eds., Academic Press
1976, pp. 277–303

Cardinal Splines

Splines came out of the barrel of a cannon. Papers on what we now call splines had appeared before Schoenberg's basic spline paper [**31***] on techniques for interpolation and smoothing of ballistic tables. Two of the most interesting ones, [**Ea**] and [**QC**], even share with [**31***] the subject matter, namely cardinal spline interpolation, and approach it in the same way, namely with the aid of the Fourier transform. [**31***] is nevertheless the foundation of spline theory since it is the first paper which treats smooth piecewise polynomials not just as useful auxiliary constructs but as objects deserving of study in their own right, in token of which it provided them with a name, **splines**. More than that, [**31***] bases its analysis of splines on that most useful of spline functions, the **B-spline**, which is introduced there in terms of its Fourier transform.

The papers [**31*,32***] are the first in a series ([**31***], [**32***], [**94**], [**98,108,109,110*, 111,112,113*,134**], [**99**], [**102**], [**107,115,123,124**], [**116**], [**126**], [**128**], [**129***], [**139**], [**159**], [**160**], [**161**]) on what Schoenberg eventually (in [**108**]) termed **cardinal** splines. Of these, relatively few have been included in these *Selecta* since many of their results (though far from all the details) have been incorporated in Schoenberg's masterful monograph [**114**] on *Cardinal Spline Interpolation.*

The term "cardinal" refers to Whittaker's cardinal series

$$\sum_{n\in\mathbb{Z}} \operatorname{sinc}(\cdot - n)y_n$$

with

$$\operatorname{sinc}(x) := \sin(\pi x)/(\pi x),$$

which serves Schoenberg in [**31***] as a model for "interpolation" formulae of "cardinal type". These are formulae of the form

$$\sum_{n\in\mathbb{Z}} L(\cdot - n)y_n. \tag{1}$$

They give rise to the "cardinal" approximation

$$Cf := \sum_{n\in\mathbb{Z}} L(\cdot - n)f(n)$$

to the function f in terms of its values at the integers. [**31***] provides a characterization, in terms of the Fourier transform $\hat{L}$ of the "basic" L, of various desirable properties such a formula of cardinal type might have. For example,

(i) C reproduces all polynomials of degree $< k$ iff $\hat{L}(2\pi n) = \delta_n$ k–fold, i.e., for all $n \in \mathbb{Z}$,

$$\hat{L}(2\pi n) = \delta_n, \tag{2a}$$

$$D^j\hat{L}(2\pi n) = 0 \text{ for } j = 1,\ldots,k-1. \tag{2b}$$

(ii) C "preserves the degree" of all polynomials of degree $< k$ (in the sense that, for all such polynomials f, $Cf - f$ is a polynomial of degree $< \deg f$) iff (2a) holds for all integers and (2b) holds for all nonzero integers.

(iii) C provides interpolation (in the sense that $Cf = f$ on $\mathbb{Z}$) iff

$$\sum_{n \in \mathbb{Z}} \hat{L}(\cdot + 2\pi n) = 1. \tag{3}$$

The simplest analytic function $\hat{L}$ satisfying (2a-b) is the appropriately scaled k-fold power of the sinc-function. The corresponding "basic" function L is the k-order centered cardinal B-spline, i.e., the B-spline

$$M_k(t) := M(t| - k/2, -k/2+1, \ldots, k/2) = \frac{1}{2\pi} \int_{-\infty}^{\infty} \left(\text{sinc}\left(\frac{\eta}{2\pi}\right)\right)^k e^{-it\eta} d\eta$$

with knots $-k/2, -k/2+1, \ldots, k/2$, whose alternate definition as the k-th convolution power of the characteristic function of the interval $[-1/2, 1/2]$ becomes evident thereby.

These characterizations surfaced again in the late sixties and early seventies when approximation by integer shifts $L(\cdot - n)$ of a compactly supported function L was studied in connection with the Finite Element Method, by Aubin, Di Guglielmo, Babuška, Fix & Strang, and others. Their results, described in final generality in [**SF**], concern functions on $\mathbb{R}^d$ and show that degree preservation is the crucial property. Precisely, they show that having the distance of any smooth function from

$$S_h := \text{span}\{L(\frac{\cdot - n}{h}) : n \in \mathbb{Z}^d\}$$

be $O(h^k)$ is equivalent to having C preserve degree for all polynomials of degree $< k$.

This theory, in turn, made new contact with cardinal splines around 1980 when Schoenberg's characterization [**86***] of the B-spline $M(\cdot|t_0, \ldots, t_k)$ with arbitrary knots $t_0, \ldots, t_k$ as the "shadow" of a k-dimensional simplex had led to multivariate B-splines ([**B**]) and eventually to box splines ([**BH**]). The box spline $M(\cdot|\Xi)$ associated with the $d \times n$-matrix Ξ is, by definition, the distribution

$$M(\cdot|\Xi) : C_0^\infty(\mathbb{R}^d) \to \mathbb{R} : \varphi \mapsto \int_B \varphi(\Xi y) dy,$$

with the domain of integration the box $B := [-1/2, 1/2]^n$, hence the name. Its close connection to Schoenberg's original cardinal B-spline becomes evident from its Fourier transform:

$$\hat{M}(\eta|\Xi) = \prod_{\xi = \Xi(\cdot, j)} \text{sinc}(\frac{\eta \cdot \xi}{2\pi}).$$

Box splines have provided nontrivial and surprising examples for the Strang-Fix theory. In addition, particularly in the hands of Dahmen and Micchelli (e.g., [**DM**]), they have lived up to the cardinal spline tradition of leading to exceptionally beautiful mathematics.

Schoenberg pursued the cardinal idea in a different direction, that of interpolation. This brought him to a study of the cardinal spline space

$$S := \text{span}\{M_k(\cdot - n) : n \in \mathbb{Z}\}$$

and its intersection with L_p, L_p^m, etc. Starting point (in [**98**]) is the **cardinal interpolation problem**: Determine choices for the pair (X, x) of normed linear function (resp. sequence) spaces for which the restriction map

$$S \cap X \to \mathbb{R}^{\mathbb{Z}} \cap x : f \mapsto f_{|\mathbb{Z}}$$

is boundedly invertible. He shows for example that, for any $p \in [1, \infty]$, the pair (L_p, ℓ_p) is such a choice. Ultimately, such questions are resolved with the aid of the fact that the inverse of the biinfinite band matrix $\big(M_k(i-j)\big)_{i,j\in\mathbb{Z}}$ decays exponentially away from the diagonal. This fact is relevant since B-splines are well-conditioned, i.e., the coordinate map

$$\mathbb{R}^{\mathbb{Z}} \cap x \to S \cap X : a \mapsto \sum_{n\in\mathbb{Z}} M_k(\cdot - n)a_n$$

is bounded and boundedly invertible for the same pairs (X, x).

Subsequent papers deal with the nature of the cardinal spline interpolant for specific data([**110***],[**161**]) and spin-offs ([**126**],[**139**],[**160**]), its limit as the order k goes to infinity ([**107**],[**113***],[**115**]), the error in cardinal spline interpolation ([**129***],[**134**]), Hermite interpolation ([**109**],[**111**]), quadrature rules based on data at the integers ([**112**],[**117**]), and on certain cardinal splines in extremum problems ([**99**],[**143**]).

Schoenberg's Cardinal Spline Theory shows splines at their best. It continues to serve as a laboratory in which spline questions can be tried out and examined in careful detail before the same questions are attacked for splines with arbitrary knots.

References

[**B**] C. de Boor (1976), Splines as Linear Combinations of B-Splines, in *Approximation Theory II*, G. G. Lorentz, C. K. Chui and L. L. Schumaker, eds., Academic Press, 1–47.

[**BH**] C. de Boor and K. Höllig (1982/83), B-splines from Parallelepipeds, *J. d'Anal. Math.* **42**, 99–115.

[**DM**] W. Dahmen and C.A. Micchelli, Some results on box splines, *Bull. Amer. Math. Soc.* **11** (1984), 147–150.

[**Ea**] A. Eagle (1928), On the relation between Fourier constants of a periodic function and the coefficients determined by harmonic analysis, *Phil.Mag.* **5**, 113–132.

[**QC**] W. Quade & L. Collatz (1938), Zur Interpolationstheorie der reellen periodischen Funktionen, *Akad. Wiss., Math.-Phys. Klasse* **30**, 383–429.

[**SF**] G. Strang and G. Fix (1973), A Fourier Analysis of the Finite Element Variational Method, C.I.M.E., II Ciclo 1971, in *Constructive Aspects of Functional Analysis*, G. Geymonat ed., 793–840.

Carl de Boor
University of Wisconsin, Madison

SMOOTHING

ON SMOOTHING OPERATIONS AND THEIR GENERATING FUNCTIONS[1]

I. J. SCHOENBERG

INTRODUCTION

In this paper we are mainly concerned with two kinds of linear transformations, the sequence convolution transformation

$$(1) \qquad y_n = \sum_{\nu=-\infty}^{\infty} a_{n-\nu} x_\nu$$

and the integral convolution transformation

$$(2) \qquad g(x) = \int_{-\infty}^{\infty} \Lambda(x-t) f(t) dt,$$

where the sequence $\{a_n\}$ and the function $\Lambda(x)$ are thought of as given. In §2 we also consider the ordinary linear transformation

$$(3) \qquad y_i = \sum_{k=1}^{n} a_{ik} x_k \qquad (i = 1, \cdots, m).$$

The loosely connected topics to be discussed concerning these transformations are perhaps best brought together under the general subject of smoothing operations.

In §1 we are concerned with transformations (1) of the kind used for the purpose of smoothing numerical data. Erastus L. De Forest proposed long ago the problem of describing the asymptotic behavior of the coefficients of high-order iterates of (1) (see Wolfenden [**15**]). This question, as well as the question of when a formula (1) may rightly be called a smoothing formula, was answered by the author in [**38**] and [**31**]. It is shown in §1 that the author's criterion for a smoothing formula is essentially of the nature of a stability condition of the kind required of difference methods for the numerical integration of partial differential equations. From this point of view De Forest's problem amounts to constructing by difference methods the fundamental solutions of certain parabolic differential equations. These remarks are merely special cases of Fritz John's recent work

An address delivered before the Fresno meeting of the Society on May 3, 1952 by invitation of the Committee to Select Hour Speakers for Far Western Sectional Meetings; received by the editors June 30, 1952.

[1] This work was performed on a National Bureau of Standards contract with the University of California, Los Angeles and was sponsored (in part) by the Office of Scientific Research, USAF.

[49] on the integration of parabolic differential equations.

The remainder of the paper discusses those transformations (1) and (2) which are variation-diminishing. These special classes of smoothing operations, while perhaps too narrow for practical smoothing purposes (because, in actuarial terms, they preserve only linear functions), seem to deserve particular attention because of the interesting problems which arise if we attempt to characterize their respective generating functions

$$\sum_{-\infty}^{\infty} a_n z^n \quad \text{and} \quad \int_{-\infty}^{\infty} e^{-xs}\Lambda(x)dx.$$

§§3 and 4 are devoted to these problems concerning (2) and (1), respectively. A discussion of (2), initiated by Pólya in 1915, has recently received considerable attention by Widder, Hirschman, and the author. The transformation (1), studied by M. Aissen, Anne Whitney and the author, is the subject of very significant as yet unpublished contributions at the hands of Albert Edrei (§4). Edrei's results are used to establish (Article 4.3) a conjecture of Schoenberg [31, Part B, pp. 131–132].

Our discussion in §2 of the transformation (3) is mainly intended to provide a background for the transcendental cases of (1) and (2). Attention is called to a recent theorem of R. Gantmakher and M. Krein which is of interest in the present connection and is also helpful in a discussion of curves which are convex in higher-dimensional spaces.[2]

1. The stability of smoothing formulae

1.1. What is a smoothing formula? Let

$$y_m = \sum_{\nu=-\infty}^{\infty} a_{m-\nu} x_\nu \tag{1.1}$$

be a given "moving average" formula which is such that the function

$$F(z) = \sum_{-\infty}^{\infty} a_\nu z^\nu \qquad (r < |z| < r^{-1}), \tag{1.2}$$

which may be called its *generating function*, is regular in an open ring

[2] This remark explains the presence of several geometrical papers in our bibliography. As seen from this brief description of contents, the paper is concerned with the most general smoothing operations on the one hand (§1), with a very restricted class on the other (§§3, 4). For the middle-range of operations which are useful in the practice of smoothing data we refer to the excellent expositions by Whittaker-Robinson [17], and Wolfenden [15]. For the related subject of osculatory interpolation see Greville [29] and Schoenberg [31].

containing the circle $|z|=1$. Let us assume, as is generally the case in practice, that the given sequence of "weights" a_m is symmetric about a_0:

$$a_m = a_{-m} \qquad (m = 1, 2, \cdots). \tag{1.3}$$

It was pointed out in [31] that the generating function (1.2) or, equivalently, the *characteristic function*

$$\phi(u) = F(e^{iu}) = a_0 + 2a_1 \cos u + 2a_2 \cos 2u + \cdots \tag{1.4}$$

may be used to advantage in describing some of the useful attributes of a moving average formula (1.1). Thus the formula (1.1) will leave unchanged, or invariant, sequences $\{x_m\}$ which are equidistant ordinates of polynomials of degree $2k-1$, but of no higher degree, provided $\phi(u)$ has an expansion in powers of u which is of the form

$$\phi(u) = 1 - \lambda u^{2k} + \cdots \qquad (\lambda \neq 0). \tag{1.5}$$

Good arguments were presented in [31] and still better ones in [38] to the effect that a moving average formula (1.1) could be called a *smoothing formula* only if (perhaps also if) its characteristic function satisfies the condition

$$|\phi(u)| < 1 \qquad \text{if } 0 < u < 2\pi. \tag{1.6}$$

This condition implies that the coefficient λ of (1.5) is positive.[3]

1.2. **Behavior of the iterates of (1.1): De Forest's problem.** A conclusive argument in support of the necessity of our condition (1.6) is furnished by the solution of the following problem first stated and attacked by Erastus L. De Forest.[4] If we subject the given se-

[3] During his connection in the last war with the Data Analysis Section of the Boeing Aircraft Company, Dr. C. Lanczos has reached the same criterion (1.6) from the following point of view: If the formula (1.1) is applied to the simple vibration of frequency u, $x_\nu = e^{iu\nu}$ $(-\infty<\nu<\infty)$, we obtain the transformed sequence $y_m = e^{ium}\phi(-u)$ $(-\infty<m<\infty)$, which is again a simple vibration of frequency u, of amplitude $|\phi(-u)|$. Requiring that the amplitude of every simple vibration with frequency in the range $0<u<2\pi$ be *diminished* in the process, Dr. Lanczos reached our condition (1.6). This interesting remark shows that the values of $\phi(u)$ play the role of characteristic values of the matrix of the linear operation (1.1). The condition (1.6) is now seen to be related to the conditions found by R. Oldenburger and A. Dresden (for references see [47]) which insure that the nth power of an ordinary matrix converges to a limit as n tends to infinity.

[4] Erastus L. De Forest (1834–1888) contributed numerous original and fundamental papers on various aspects of the problem of fitting and smoothing irregular series by means of polynomials. His pioneer work remained practically unknown, however, until 1924, when its importance in relation to the graduation processes of actuaries was described by Hugh H. Wolfenden in his paper [15]. A further summary of De Forest's contributions can be found in H. H. Wolfenden, *The fundamental principles of mathematical statistics*, Toronto, Macmillan, 1942.

quence $\{x_m\}$ n times in succession to the same transformation (1.1), we obtain a linear transformation

$$y_m^{(n)} = \sum_{\nu=-\infty}^{\infty} a_{m-\nu}^{(n)} x_\nu \tag{1.7}$$

which is the n-fold iterate of (1.1). *What is the asymptotic behavior of the coefficients of* (1.7) *as* $n\to\infty$? This question was answered for the case when all coefficients of (1.1) are non-negative, hence necessarily that $k=1$ in (1.5), by De Forest and by G. B. Dantzig [26].[5] A general solution is as follows:

Let (1.1) be such that (1.3), (1.6), and (1.5) are satisfied, hence $\lambda>0$. Let

$$G_k(x) = \frac{1}{2\pi}\int_{-\infty}^{\infty} e^{-v^{2k}} \cos vx dv, \tag{1.8}$$

which is the normal function

$$G_1(x) = \frac{1}{2\pi^{1/2}} e^{-x^2/4} \tag{1.9}$$

if $k=1$, otherwise $(k=2, 3, \cdots)$ an entire even function having infinitely many zeros, all real. *The coefficients of* (1.7) *satisfy the asymptotic relations*

$$a_\nu^{(n)} = (\lambda n)^{-1/2k} G_k(\nu(\lambda n)^{-1/2k}) + o(n^{-1/2k}), \qquad as\ n\to\infty, \tag{1.10}$$

where the "little o" symbol holds uniformly for all integers ν (see Schoenberg [38]). It was also shown there by examples that (1.10) no longer holds if the equality sign is allowed in (1.6), and that the coefficient $a_0^{(n)}$ diverges exponentially to $+\infty$, as $n=2m$ tends to infinity through even values, if the inequality (1.6) is reversed anywhere in the range $(0, 2\pi)$.

[5] The following additional references, which I owe to T. S. Motzkin, came too late to my attention for inclusion in our bibliography: The case of non-negative weights $(k=1)$ in the more general form where the convoluted formulae need not be identical was treated by Eugen Slutzky in a Russian paper of 1927. See its English translation *The summation of random causes as the source of cyclic processes*, Econometrica vol. 5 (1937) pp. 105–146, in particular pages 134–138. Concerning the same problem see also R. v. Mises, *Wahrscheinlichkeitsrechnung und ihre Anwendungen in der Statistik und theoretischen Physik*, Leipzig and Wien, 1931, and R. v. Mises, *Generalizzazione di un teorema sulla probabilità della somma di un numero illimitato di variabili casuali*, Giornale dell' Istituto Italiano degli Attuari vol. 5 (1934) pp. 483–495.

A corresponding discussion of smoothing of periodic data is related to the so-called finite Fourier series (see Schoenberg [**47**]). Some unpublished work by Fritz John deals with the analogue of (1.10) for the periodic case.

1.3. **A differential equation.** The connection with differential equations is briefly as follows. Observe that (1.8) implies

$$U(x, t) = t^{-1/2k}G_k(xt^{-1/2k}) = \frac{1}{2\pi}\int_{-\infty}^{\infty} e^{-tv^{2k}+ixv}dv \qquad (t > 0). \tag{1.11}$$

The function under the integral sign is immediately seen to satisfy, for all v, the differential equation

$$\frac{\partial u}{\partial t} = (-1)^{k+1}\frac{\partial^{2k}u}{\partial x^{2k}}, \tag{1.12}$$

which reduces to the familiar heat-equation if $k=1$. It follows that also $U(x, t)$, defined by (1.11), is a solution of (1.12) in the upper half-plane $t>0$. On the other hand, applying to (1.8) Fourier's inversion formula and setting $v=0$, we find that

$$\int_{-\infty}^{\infty} G_k(x)dx = 1.$$

These remarks imply the following: If $f(x)$ is continuous and $o(|x|^{-2})$, say, as $|x|\to\infty$, then

$$u(x, t) = t^{-1/2k}\int_{-\infty}^{\infty} G_k((x-v)t^{-1/2k})f(v)dv \qquad (t > 0) \tag{1.13}$$

is a solution of the differential equation (1.12) satisfying the boundary condition

$$\lim_{t\to+0} u(x, t) = f(x).$$

This particular solution $u(x, t)$ may now also be constructed as follows: Draw in the (x, t)-plane the lattice of points

$$(\nu\Delta x,\ n\Delta t) \qquad (\nu = 0, \pm 1, \cdots ; n = 0, 1, \cdots).$$

Define in it a lattice-function $u_{\nu,n}$ by starting with

$$u_{\nu,0} = f(\nu\Delta x)$$

and computing the values along each horizontal line from those on the line below by means of (1.1). This evidently amounts to *iterating* (1.1), so that we have

$$u_{\nu,n} = \sum_{\mu=-\infty}^{\infty} a_{\nu-\mu}^{(n)} f(\mu\Delta x). \tag{1.14}$$

For any given x and $t>0$, (1.14) goes over into (1.13) under the following circumstances: We first connect the mesh-sizes Δx and Δt by the relation

$$\Delta t = \lambda(\Delta x)^{2k}. \tag{1.15}$$

If the integers ν and n are such that

$$\nu\Delta x \to x \quad \textit{and} \quad n\Delta t \to t, \quad \textit{as} \quad \Delta x \to +0,$$

then

$$u_{\nu,n} \to u(x, t).$$

This follows readily from (1.14) and (1.13), in view of the asymptotic relations (1.10): (1.14) differs from a Cauchy-Riemann sum for the integral (1.13) by a quantity which tends to zero because of the uniformity in ν of the error term of (1.10).

It is interesting to note that it matters little which formula (1.1) we use in this construction, as long as it preserves polynomials of degree $2k-1$ and of no higher degree, i.e. satisfies (1.5), and above all that it satisfy the *stability condition* (1.6), the term "stability" meaning here stability on iteration. For the general theory of Fritz John of which the above situation (for $k=1$) is merely a very special example, see [**49**].

2. Variation-diminishing linear transformations

2.1. Fundamental criteria. Let

$$y_i = \sum_{k=1}^{n} a_{ik}x_k \qquad (i = 1, \cdots, m) \tag{2.1}$$

be a linear transformation where coefficients and variables are all real. Let $v(x)$ denote the number of changes of sign in the sequence x_k and $v(y)$ the similar number for y_i. F. Klein discussed the problem of comparing the upper bounds for the number of real roots of an algebraic equation in a given interval, as furnished by the Descartes rule of signs and similar theorems. These bounds are usually equal to the number of changes of sign of sequences arising from the coefficients of the given equation by linear transformations. For this reason, Klein's problem led Schoenberg [**19; 21**] to study transformations (2.1) which have the property that

$$v(y) \leqq v(x), \tag{2.2}$$

for all values of the variables x_k. Such transformations were called *variation-diminishing* by Pólya.

The matrix $A=\|a_{ik}\|$ is called *totally positive*, provided all its minors of all orders are non-negative [**19**, p. 322]. A useful criterion for such matrices, based on a lemma of M. Fekete [**4**, pp. 4–6], was recently given by Anne Whitney [**58**]. Totally positive square-shaped matrices were thoroughly investigated by Gantmakher-Krein [**24**; **44**] in connection with boundary-value problems arising in vibration problems. Schoenberg [**19**] showed that (1.1) is variation-diminishing if A is totally positive. Assuming that the matrix A is of rank n, he also proved the following: (1.1) is variation-diminishing if and only if A enjoys the property that two minors of A, of equal orders, are never of opposite signs. If the rank r of A is unrestricted, Motzkin [**22**] proved that (1.1) is variation-diminishing if and only if two minors of A, of the same order s, are never of opposite sign if $s<r$, while if $s=r$ they should never be of opposite signs if they belong to the same set of r columns of A. For a proof of this theorem, differing from Motzkin's, see Schoenberg-Whitney [**51**]. There also cyclic variation-diminishing transformations are characterized: (1.1) is said to be *cyclic variation-diminishing* provided it is such that the inequality

$$v_c(y) \leqq v_c(x) \tag{2.3}$$

always holds, where $v_c(x)$ is the number of changes of sign in the sequence $x_1, \cdots, x_n$ counted after these variables are placed cyclically along a circle, with a similar definition for $v_c(y)$.

2.2. **A theorem of Gantmakher-Krein and convex curves.** Let us assume that the rank of the matrix A, of (2.1), is equal to n. Choose n among the linear forms y_i which are linearly independent. As their values may be chosen at wish, for instance alternating in sign, it is clear that $\sup_x v(y) \geqq n-1$. The theorem of Gantmakher-Krein answers the question as to when we have the equality sign in this relation and may be stated as follows:

The system (1.1), *of rank equal to n, is such that we always have*

$$v(y) \leqq n - 1 \tag{2.4}$$

if and only if all nonvanishing minors of order n, of A, are of the same sign. See Gantmakher-Krein [**44**, Theorem 3 on p. 297], and also Schoenberg-Whitney [**51**, Theorem 1 on p. 142].[6]

[6] Anne Whitney and the author were unaware of the book by Gantmakher and Krein when [**51**] was published. However, the priority clearly belongs to the Russian authors.

This theorem seems fundamental for two reasons: Firstly, it implies easily the results of Schoenberg and Motzkin concerning variation-diminishing transformations and plays a similar role in a discussion of the related problem (2.3). Secondly, it seems indispensable in a discussion of curves which are convex in higher-dimensional spaces. The use of the term "convex" in this connection requires some explanation. We mention first the concept of an arc in E_m of (linear) order m, due to C. Juel, intensively studied and generalized by O. Haupt, A. Marchaud **[18]**, P. Scherk **[25]**, and others: A Jordan arc in E_m is of order m provided it is intersected by every hyperplane in at most m points. A related but distinct notion is as follows: A continuous arc

$$\Gamma: x_i = x_i(t) \qquad (i = 1, \cdots, m; \alpha \leqq t \leqq \beta) \tag{2.5}$$

in the euclidean space E_m is said to be *convex in* E_m if it crosses every hyperplane of E_m at most m times. If we also require that Γ should not be contained in a lower-dimensional flat space, then we say that Γ is *convex on* E_m (see Schoenberg **[56]**.)

The following criterion is an almost immediate corollary of the theorem of Gantmakher-Krein: *The arc* Γ, *defined by* (2.5), *is convex on* E_m *if and only if the determinants*

$$\begin{vmatrix} 1 & x_1(t_0) & \cdots & x_m(t_0) \\ 1 & x_1(t_1) & \cdots & x_m(t_1) \\ \cdot & \cdot & & \cdot \\ \cdot & \cdot & & \cdot \\ \cdot & \cdot & & \cdot \\ 1 & x_1(t_m) & \cdots & x_m(t_m) \end{vmatrix} \qquad (\alpha \leqq t_0 < t_1 < \cdots < t_m \leqq \beta) \tag{2.6}$$

do not all vanish and the nonvanishing among them are all of the same sign. The case when $m=2$ seems intuitively obvious, for the theorem says that the plane arc $P=P(t)$ is convex on E_2 provided the nondegenerate ones among the triangles $P(t_0)P(t_1)P(t_2)$ have all the same orientation. All convex curves are rectifiable. If Γ is convex *on* E_m and m is *odd*, then the arc Γ can never close. In even-dimensional spaces E_{2n} we may well have closed curves convex on E_{2n} as seen by the example of the special curve

$$\begin{aligned} C_0:\quad & x_1 = \cos t,\ x_3 = \frac{1}{2}\cos 2t, \cdots, x_{2n-1} = \frac{1}{n}\cos nt, \\ & x_2 = \sin t,\ x_4 = \frac{1}{2}\sin 2t, \cdots, x_{2n} \ \ = \frac{1}{n}\sin nt \end{aligned} \tag{2.7}$$

$$(0 \leqq t \leqq 2\pi).$$

That C_0 spans E_{2n} is clear and equally so that C_0 is convex in E_{2n}: If $l(x_i)$ is a linear function and if we substitute the x_i from (2.7), then $l(x_i)$ becomes a real trigonometric polynomial of order n which is known to change sign within a period at most $2n$ times.

Let

$$(2.8) \qquad C: x_i = x_i(t) \qquad (i = 1, \cdots, 2n; 0 \leqq t \leqq 2\pi),$$

be a closed curve in E_{2n} defined by continuous functions of period 2π. C is found to be convex on E_{2n} provided the above restriction on the signs of the determinants (2.6) (with $m=2n$) holds for parameter values restricted by the inequalities $t_0<t_1< \cdots <t_n<t_0+2\pi$. This criterion leads naturally to a useful expression for the $2n$-dimensional volume $V=V(K)$ of the convex hull $K=K(C)$ of the curve C. Indeed, assuming the $x_i(t)$ to be absolutely continuous, as is the case if t is proportional to the arc-length, then V can be expressed by the following Lebesgue integral

$$V = \frac{\epsilon}{n!(2n)!}\int_0^{2\pi} \cdots \int_0^{2\pi} \det \| x_i(t_1), x_i'(t_1), x_i(t_2), x_i'(t_2), \cdots, x_i(t_n), x_i'(t_n) \| dt_1 \cdots dt_n,$$

where $\epsilon=1$, or -1, depending on the common sign of the determinants (2.6). This expression for V, which reduces to the classical area-integral if $n=1$, allows one to generalize, by A. Hurwitz's method of Fourier series [**3**], the classical isoperimetric inequality to closed curves in E_{2n}: *If the closed curve C is convex in E_{2n}, of length L, then the following inequality holds*:

$$(2.9) \qquad L^{2n} \geqq (2\pi n)^n n!(2n)!V,$$

with the equality sign if and only if C is similar to the curve C_0, defined by (2.7), *or to its reflexion* (see [**56**]). If $n=1$, (2.9) reduces to the well known inequality $L^2 \geqq 4\pi V$ for convex curves in E_2, V being the area enclosed by C, and where equality holds only if C is a circle.

It is to be expected that this new approach to curves convex in E_m will be fruitful and work in collaboration with T. S. Motzkin is in progress. In concluding it should be pointed out that the notion of convexity of curves is also related to the Chebyshev systems of functions (Kellogg [**9**; **10**; **11**], S. Bernstein [**16**], Laasonen [**42**]) and to the systems satisfying Descartes' rule of signs (Pólya-Szegö [**14**]). Especially close is Pólya's discussion [**13**] of linear homogeneous differential equations. Indeed, the notion of a curve convex on E_m being affine invariant, such differential equations would seem to be the appropriate tool.

3. Variation-diminishing integral transformations

3.1. Pólya's work of 1915. An entire function $\Psi(s)$ may obviously be represented as a limit of a sequence of polynomials, uniformly convergent in every bounded domain of the complex plane: The sections of its expansion in powers of s have this property. Laguerre [1] and Pólya [5] have investigated the following problem: Which entire $\Psi(s)$ ($\not\equiv 0$) are limits of a sequence of real polynomials having only real zeros? They found this to be the case if and only if $\Psi(s)$ is of the form

$$\Psi(s) = Ce^{-\gamma s^2+\delta s}s^m\prod_1^\infty (1+\delta_\nu s)e^{-\delta_\nu s} \tag{3.1}$$

$$(C \gtrless 0,\ \gamma \geqq 0,\ \delta,\ \delta_\nu \text{ real},\ \sum \delta_\nu^2 < \infty).$$

They also showed that the subclass of those functions $\Phi(s)$ which are limits of polynomials with only real zeros and non-negative coefficients is characterized by the representation

$$\Phi(s) = Ce^{\gamma s}s^m\prod_1^\infty (1+\delta_\nu s)\quad (C > 0,\ \gamma \geqq 0,\ \delta_\nu \geqq 0,\ \sum \delta_\nu < \infty). \tag{3.2}$$

See also Szász [28]. For an extension of these results to entire functions of n variables see Motzkin-Schoenberg [50].

Pólya and Schur [7] have described these classes in terms of their Taylor expansions: A formal real power series

$$a_0 + (a_1/1!)\, s + (a_2/2!)\, s^2 + \cdots$$

is the expansion of a function of the form (3.1) if and only if the polynomials

$$P_n(x) = a_0 + \binom{n}{1} a_1 x + \cdots + a_n x^n \qquad (n = 1, 2, \cdots) \tag{3.3}$$

have only real zeros; also that it is the expansion of a function of the form (3.2) if and only if the polynomials (3.3) have only real zeros and non-negative coefficients.

In [8], Pólya considered the subclass

$$\Psi(s) = Ce^{-\gamma s^2+\delta s}\prod_1^\infty (1+\delta_\nu s)e^{-\delta_\nu s} \tag{3.4}$$

$$(C > 0,\ \gamma \geqq 0,\ \delta,\ \delta_\nu \text{ real},\ 0 < \gamma + \sum \delta_\nu^2 < \infty),$$

of those functions of the class (3.1), with $\Psi(0) > 0$, which are not exponentials $C\exp\{\delta s\}$, and investigated the power series expansions of their reciprocals

$$\frac{1}{\Psi(s)} = \sum_{0}^{\infty} \frac{(-1)^{\nu}\mu_{\nu}}{\nu!} s^{\nu}. \tag{3.5}$$

He discovered the following properties:

1. The Hankel determinants

$$\begin{vmatrix} \mu_0 & \mu_1 & \cdots & \mu_n \\ \mu_1 & \mu_2 & \cdots & \mu_{n+1} \\ \cdot & & & \\ \cdot & & & \\ \cdot & & & \\ \mu_n & \mu_{n+1} & \cdots & \mu_{2n} \end{vmatrix} \quad \text{are all positive } (n = 1, 2, \cdots). \tag{3.6}$$

2. If $f(x)$ is a real polynomial, then

$$g(x) = \sum_{0}^{\infty} \frac{(-1)^{\nu}\mu_{\nu}}{\nu!} f^{(\nu)}(x) = \frac{1}{\Psi(D)} f(x) \tag{3.7}$$

is a polynomial having no more real zeros than $f(x)$. Conversely, if a real formal power series $\sum(-1)^{\nu}\mu_{\nu}s^{\nu}/\nu!$ $(\mu_0>0)$ is such that $g(x)$, defined by (3.7), has never more zeros than the arbitrary polynomial $f(x)$, then it is the expansion (3.5) of the reciprocal of a function $\Psi(s)$ of the form (3.4).

In 1920 Hamburger [12], using his then recent solution of the moment problem bearing his name, concluded from (3.6) that the coefficients μ_{ν}, of (3.5), are the moments of a nondecreasing function $\psi(x)$ and that (3.5) may be represented as a Laplace-Stieltjes integral

$$\frac{1}{\Psi(s)} = \int_{-\infty}^{\infty} e^{-xs} d\psi(x)$$

within the vertical strip of regularity of the left-hand function, containing the origin of the s-plane. More recently Hirschman-Widder [40] and Schoenberg [34; 55] have established a representation of (3.5) in terms of a bilateral Laplace integral

$$\frac{1}{\Psi(s)} = \int_{-\infty}^{\infty} e^{-xs}\Lambda(x)dx. \tag{3.8}$$

Here $\Lambda(x)$ is a frequency function,[7] i.e. $\Lambda(x)\geqq 0$, which may be repre-

[7] Widder [35] interprets $\Lambda(x-y)$ as the Green's function of a differential system for the whole real axis. For a thorough investigation of differential systems in a *finite* range whose Green's functions are totally positive (Kellogg kernels) see Gantmakher-Krein [44].

sented by the familiar inversion formula

$$(3.9) \qquad \Lambda(x) = \frac{1}{2\pi i}\int_{-i\infty}^{i\infty} \frac{e^{xs}}{\Psi(s)} ds \qquad (-\infty < x < \infty),$$

where the integral is defined as usual as a principal value.

Since

$$\mu_\nu = \int_{-\infty}^{\infty} x^\nu \Lambda(x) dx,$$

we may now write the differential operation (3.7) as an integral transformation. Indeed, if $f(x)$ is a polynomial, we find

$$\int_{-\infty}^{\infty} \Lambda(t) f(x-t) dt = \int_{-\infty}^{\infty} \Lambda(t) \left\{ \sum_0^\infty \frac{1}{\nu!} f^{(\nu)}(x)(-t)^\nu \right\} dt$$

$$= \sum_0^\infty \frac{(-1)^\nu \mu_\nu}{\nu!} f^{(\nu)}(x) = g(x)$$

or

$$(3.10) \qquad g(x) = \int_{-\infty}^{\infty} \Lambda(x-t) f(t) dt.$$

Hence for polynomials, (3.10) being equivalent to (3.7), this integral transformation is inverted by

$$(3.11) \qquad f(x) = \Psi(D) g(x),$$

where the operator $\Psi(D)$ is to be applied after its expansion in powers of D. This theory goes back to Pólya's paper [8] of 1915, except for the integral representations (3.8) and (3.10), the transformation (3.10), for polynomials, appearing in Pólya's paper in terms of the equivalent differential operation (3.7).

3.2. **Pólya frequency functions.** The determining functions $\Lambda(x)$, of (3.8), may be described in terms of the following definition: A frequency function $\Lambda(x)$, i.e. a measurable non-negative function such that

$$0 < \int_{-\infty}^{\infty} \Lambda(x) dx < \infty,$$

is said to be a *Pólya frequency function* provided that for any two sets of numbers

$$(3.12) \qquad x_1 < x_2 < \cdots < x_n, \quad t_1 < t_2 < \cdots < t_n \qquad (n = 1, 2, \cdots)$$

we have the inequality

$$\det \|\Lambda(x_i - t_j)\|_{1,n} \geqq 0. \tag{3.13}$$

In terms of this definition, the following theorem holds: If $\Psi(s)$ is of the form (3.4), then its reciprocal may be represented in the form (3.8) within the vertical strip of regularity of $1/\Psi$ which contains the origin, and where $\Lambda(x)$ is a Pólya frequency function. Conversely, given a Pólya frequency function $\Lambda(x)$, then its Laplace transform (3.8) converges within such a vertical strip and represents the reciprocal of an entire function of the form (3.4). See Schoenberg [**34**; **55**]; also Wintner [**23**, Theorem 1].

Similar results concerning the ordinary Laplace transform are connected with entire functions of the form

$$\Phi(s) = Ce^{\gamma s}\prod_{1}^{\infty} (1 + \delta_\nu s) \tag{3.14}$$

$$(C > 0,\ \gamma \geqq 0,\ \delta_\nu \geqq 0,\ 0 < \gamma + \textstyle\sum \delta_\nu < \infty).$$

These functions forming a subclass of the class of functions (3.4), the previous theorem applies to their reciprocals. However, the function $\Psi(s)$ of (3.8) is of the special form (3.14) if and only if the Pólya frequency function $\Lambda(x)$, of (3.8), vanishes for all negative values of x (see [**34**; **55**]).

A related question is the following: The meromorphic function on the left-hand side of (3.8) has in general several vertical strips of regularity separated by vertical lines through its poles. What is the nature of the Laplace integral representations of $1/\Psi$ in the strips which do not contain the origin? An answer requires the following definition: A real measurable function $\Lambda(x)$ $(-\infty < x < \infty)$ is said to be *totally positive* provided it satisfies the following three conditions: 1. (3.12) implies (3.13), 2. $\Lambda(x)$ is different from zero for at least two distinct values of x, 3. $\Lambda(x) \not\equiv \exp\{ax+b\}$. It is clear that the Pólya frequency functions are totally positive. Moreover, it can be shown that every totally positive function turns into a Pólya frequency function, upon multiplication by appropriate exponentials. As a result the above stated correspondence between functions $\Psi(s)$ and Pólya frequency functions now extends to a one-to-one correspondence between functions $\Psi(s)$ and the totally positive functions $\Lambda(x)$, by the relation (3.8) with the sign changed if necessary, in vertical strips of regularity for $1/\Psi$ which are no longer required to contain the origin. In particular, a given $1/\Psi(s)$ allows a representation (3.8) in each of its strips of regularity with a different totally positive $\Lambda(x)$ corresponding to each of the strips (see [**34**; **55**]).

As particular examples of Pólya frequency functions we mention

$$(3.15)\qquad \Lambda_1(x) = e^{-x^2}, \qquad \Lambda_2(x) = e^{-x-e^{-x}}, \qquad \Lambda_3(x) = \frac{1}{\cosh x},$$

whose respective transforms $\pi^{1/2}e^{s^2/4}$, $\Gamma(s+1)$, $\pi/\cos(\pi s/2)$ are seen to be reciprocals of functions of the form (3.4). As a nontrivial example of Pólya frequency functions vanishing for $x<0$ we mention

$$\Lambda_4(x) = \begin{cases} \sum_{-\infty}^{\infty} (-1)^{\nu+1}\nu^2 e^{-x\nu^2} & \text{if } x > 0, \\ 0 & \text{if } x \leqq 0, \end{cases}$$

whose transform $\prod_1^\infty (1+s^2/\nu^2)^{-1}$ is indeed the reciprocal of a function of the form (3.14). Its cumulative distribution function

$$\int_{-\infty}^{x} \Lambda_4(x)dx = \begin{cases} \sum_{-\infty}^{\infty} (-1)^{\nu} e^{-x\nu^2} & \text{if } x > 0, \\ 0 & \text{if } x \leqq 0 \end{cases}$$

is the subject of the Kolmogorov-Smirnov limit theorem (see Feller [**36**]).

Given a Pólya frequency function $\Lambda(x)$ in terms of its transform (3.8) and two sets of numbers (3.12), Schoenberg-Whitney [**43**; **57**] have answered the question as to when the determinant (3.13) is actually positive. For a discussion of the differentiability properties of Pólya frequency functions see [**55**, Corollary 2], Widder [**35**, Theorems 7 and 12], Hirschman-Widder [**40**]. For a new characterization of Pólya frequency functions as the only possible limits of so-called spline frequency functions see Curry-Schoenberg [**33**]. Concerning the interesting property of Pólya frequency functions of being "bell-shaped" see Hirschman-Widder [**40**], Schoenberg [**34**], and a related note [**45**] by Hirschman.

3.3. **Smoothing properties of Pólya frequency functions.** The remarkable smoothing property of the transformation (3.7), or (3.10), to the effect that the polynomial $g(x)$ has no more real zeros than $f(x)$, suggested the problem of finding the most general transformation of the form (3.10) which is *variation-diminishing*. By this we mean, obviously enough, the following: Let $v(f)$ (the number of changes of sign of $f(x)$) denote the supremum of the number of changes of sign in the sequence $f(x_1), \cdots, f(x_n)$, where $-\infty < x_1 < \cdots < x_n < +\infty$, $n=2, 3, \cdots$. Let $\Lambda(x)$ be summable, not

$=0$ p.p., and such that, for every bounded $f(x)$, (3.10) should imply the inequality

$$v(g) \leqq v(f).$$

It turns out that this is the case if and only if $\Lambda(x)$ or $-\Lambda(x)$ is a Pólya frequency function. The variation-diminishing property of (3.10) therefore characterizes the Pólya frequency functions. For a somewhat stronger result in terms of Stieltjes integrals see Schoenberg [**46**]. The author's old theorem [**19**] to the effect that the linear transformation (2.1) is variation-diminishing if the matrix A is totally positive is the origin of the peculiar property (3.13) of Pólya frequency functions. Indeed, (3.10) may be regarded as a linear transformation defined by the matrix $\Lambda(x-t)$, while (3.13) expresses its total positivity.

3.4. Inversion and representation problems. The prototype of the functional transformations to be now discussed is the Laplace transformation

$$\gamma(z) = \int_0^\infty e^{-zu}\phi(u)du. \tag{3.16}$$

Well known results are its inversion by means of the Post-Widder inversion operator (Widder [**32**, Chap. VII]) and the Bernstein-Widder theorem which characterizes the functions representable in the form

$$\gamma(z) = \int_0^\infty e^{-zu}d\alpha(u) \qquad (d\alpha(u) \geqq 0)$$

by the condition of complete monotoneity [**32**, Chap. VII, §14]. Following Widder we now set $u=\exp(t)$, $z=\exp(-x)$, whereupon (3.16) changes into

$$g(x) = \int_{-\infty}^\infty \Lambda_2(x-t)f(t)dt, \tag{3.17}$$

where $g(x)=\gamma(e^{-x})e^{-x}$, $f(t)=\phi(e^t)$, while $\Lambda_2(x)=\exp(-x-e^{-x})$ is one of the Pólya frequency functions (3.15). The Laplace transformation (3.16) is thus seen to be equivalent to a special instance of the convolution transformation (3.10).[8] For wide classes of Pólya fre-

[8] This explains, in view of Article 3.3, the variation-diminishing properties of the Laplace transformation discovered by Laguerre and Pólya. See Pólya-Szegö [**14**, Problem 80, p. 50].

quency functions $\Lambda(x)$, Hirschman and Widder have extended the above results concerning the Laplace transformation to the general transformation

$$g(x) = \int_{-\infty}^{\infty} \Lambda(x - t)f(t)dt, \tag{3.18}$$

obtaining an inversion theory (Widder [**35**; **52**; **53**], Hirschman-Widder [**37**; **40**]) and a representation theory (Hirschmann-Widder [**41**]). Their results include as special cases earlier work by R. P. Boas and H. Pollard on the so-called Stieltjes transformation (which again may be reduced to (3.18) if $\Lambda(x) = \Lambda_3(x) = 1/\cosh x$) and its iterates. In case $f(x)$ is a polynomial we have seen that (3.18) is inverted by the differential operation (3.11). However, for arbitrary $f(x)$ the relation (3.11) is meaningless.

In order to illustrate the Hirschman-Widder inversion theory for (3.18) and to see how it goes beyond Pólya's formal inversion by (3.11), let us consider briefly the following special case discussed by Widder, *The inversion of a generalized Laplace transform*, Proc. Nat. Acad. Sci. U.S.A. vol. 33 (1947) pp. 295–297: Let $\Lambda(x)$ be a Pólya frequency function defined by the relations (3.8) and (3.4), with $\gamma = 0$ and all $\delta_\nu \neq 0$. If the given function $f(x)$ is continuous and summable ($f \in C \cdot L$), then (3.18) is inverted by the relation

$$f(x) = \lim_{n\to\infty} \prod_{\nu=1}^{n} (1 + \delta_\nu D) g\left(x - \sum_1^n \delta_\nu\right) \qquad (-\infty < x < \infty). \tag{3.19}$$

A proof follows readily from the relation

$$(1 + s\delta)e^{-\delta s}/\Psi(s) = \int_{-\infty}^{\infty} e^{-sx}(1 + \delta D)\Lambda(x - \delta)dx, \tag{3.20}$$

where δ is a real constant, and which follows from (3.8) by integration by parts. Applying the transformation (3.20) to (3.8), n times in succession with $\delta = \delta_1, \delta_2, \cdots, \delta_n$, we obtain the relation

$$\prod_{\nu=n+1}^{\infty} \frac{e^{\delta_\nu s}}{1 + \delta_\nu s} = \int_{-\infty}^{\infty} e^{-xs}\Lambda_n(x)dx, \tag{3.21}$$

where $\Lambda_n(x)$ is evidently a Pólya frequency function explicitly given by

$$\Lambda_n(x) = \prod_{\nu=1}^{n} (1 + \delta_\nu D)\Lambda\left(x - \sum_1^n \delta_\mu\right). \tag{3.22}$$

If we now operate with

$$\prod_{\nu=1}^{n} (1 + \delta_\nu D)e^{-\delta_\nu D}$$

on both sides of the relation (3.18), we obtain, by (3.22),

$$f_n(x) \equiv \prod_{1}^{n} (1 + \delta_\nu D) g\left(x - \sum_{1}^{n} \delta_\mu\right) = \int_{-\infty}^{\infty} \Lambda_n(x - t)f(t)dt,$$

and we are to show that

$$f_n(x) = \int_{-\infty}^{\infty} \Lambda_n(x - t)f(t)dt \to f(x), \qquad \text{as } n \to \infty. \tag{3.23}$$

This follows from the fact that (3.23) is a singular integral satisfying the classical conditions of Jordan:

(i) $$\int_{-\infty}^{\infty} \Lambda_n(x)dx = 1 \qquad \text{for all } n,$$

(ii) $$\lim_{n\to\infty} \int_{|x|\geqq\delta} \Lambda_n(x)dx = 0, \qquad \text{for every } \delta > 0.$$

Indeed, on expanding both sides of (3.21) in powers of s and comparing coefficients we find

$$\int_{-\infty}^{\infty} \Lambda_n(x)dx = 1, \qquad \int_{-\infty}^{\infty} x^2\Lambda_n(x)dx = \sum_{n+1}^{\infty} \delta_\nu^2 = \sigma_n^2.$$

The fact that the variance σ_n^2 converges to zero, as $n \to \infty$, implies the property (ii) by a familiar Chebyshev argument. See Widder, loc. cit., where it is shown that the Post-Widder inversion operator for the Laplace transformation (3.16) is a special case of (3.19).

See Widder [52; 53] for the particularly interesting case of $\Lambda(x) = \Lambda_1(x) = \exp(-x^2)$ when (3.18) becomes the Weierstrass transform.

4. Variation diminishing sequence transformations

4.1. **Totally positive sequences.** The sequence transformation

$$y_n = \sum_{\nu=-\infty}^{\infty} a_{n-\nu}x_\nu \tag{4.1}$$

gives rise to a theory in some respects analogous to the theory of §3. Assuming that $\sum |a_\nu| < \infty$, we say that the transformation (4.1) is variation-diminishing, provided that for every bounded sequence $\{x_n\}$, (4.1) implies the inequality

$$v(y) \leqq v(x). \tag{4.2}$$

Here, as before, $v(x)$ and $v(y)$ denote the numbers of changes of sign in their respective sequences and may well be infinite. Since $x_0=1$, $x_n=0$ (if $n\neq 0$) implies that $y_n=a_n$, for all n, we see that (4.2) implies that $v(a_n)=0$. We may therefore assume all a_n non-negative.

A sequence $\{a_n\}$ is called *totally positive* provided that the matrix

$$A = \|a_{i-j}\| \qquad (-\infty < i, j < \infty), \tag{4.3}$$

is totally positive, i.e. has only non-negative minors, and that $\{a_n\}$ is *not* a geometric progression of the form $a_n = Cr^n (C \geqq 0,\ r > 0)$. We say that the sequence $\{a_n\}$ is *totally positive normalized* provided $\{a_n\}$ is totally positive and the series $\sum a_n$ converges. The totally positive normalized sequences are the discrete analogues of the Pólya frequency functions of §3. A restricted kind of total positivity (k-times positive sequences) was discussed as early as 1912 by M. Fekete [4].

We saw in §2 that the transformation (2.1) is variation-diminishing if its matrix $\|a_{ik}\|$ is totally positive. In our case of the convolution transformation (4.1) this condition is not only sufficient but also necessary as stated by the following theorem: *The transformation* (4.1) *is variation-diminishing if and only if the sequence* $\{a_n\}$ *is totally positive normalized* (see [38]).

4.2. **Generation of totally positive sequences: Edrei's theorem.** In view of the last theorem, the problem of constructing variation-diminishing transformations (4.1) is equivalent to the problem of constructing sequences $\{a_n\}$ which are totally positive and normalized. A first step in this direction was made in [38] where the following was proved: If

$$F(z) = Ce^{az+bz^{-1}}z^m \frac{\prod_1^\infty (1+\alpha_\nu z)\prod_1^\infty (1+\beta_\nu z^{-1})}{\prod_1^\infty (1-\gamma_\nu z)\prod_1^\infty (1-\delta_\nu z^{-1})} \tag{4.4}$$

is a meromorphic function in the ring $0<|z|<\infty$, such that its constants satisfy the conditions

$$\begin{aligned} &C>0,\ a\geqq 0,\ b\geqq 0,\ m \text{ integer},\ \alpha_\nu \geqq 0,\ \beta_\nu \geqq 0,\\ &0\leqq \gamma_\nu <1,\ 0\leqq \delta_\nu <1,\ \textstyle\sum (\alpha_\nu+\beta_\nu+\gamma_\nu+\delta_\nu)<\infty, \end{aligned} \tag{4.5}$$

and if we expand $F(z)$ in a Laurent series

$$F(z) = \sum_{-\infty}^{\infty} a_n z^n \qquad (r < |z| < r^{-1}) \tag{4.6}$$

in the ring containing the unit circle inside, then the sequence $\{a_n\}$ so generated is totally positive normalized.[9]

This was easily established, the real difficulties arising as soon as we raise the following inverse problem: Given a totally positive normalized sequence $\{a_n\}$, it is not hard to show that the corresponding Laurent series (4.6) converges in an open ring containing the unit circle and defining an analytic function $F(z)$. *Is it true, then, that* $F(z)$ *is meromorphic in* $0<|z|<\infty$ *and of the form* (4.4), (4.5)?

Let us look at some examples. The following remark is obvious: If $\Lambda(x)$ is a Pólya frequency function (Article 3.2) then $a_n=\Lambda(n)$ $(-\infty<n<\infty)$ is a totally positive sequence which is normalized; for, indeed, $\|a_{i-j}\|=\|\Lambda(i-j)\|$ is a "submatrix" of $\Lambda(x-t)$. Using the functions (3.15) we now obtain sequences $\{a_n\}$ for exploratory purposes. Choosing $\Lambda(x)=\exp(-hx^2)$, where h is a positive constant and setting $q=\exp(-h)$, we obtain the sequence $a_n=\Lambda(n)=q^{n^2}$ which is totally positive and normalized. That the corresponding generating function (4.6) is of the form (4.4), (4.5) is shown by the famous Jacobian identity

$$\sum_{-\infty}^{\infty} q^{n^2} z^n = C_1 \prod_{n=0}^{\infty} (1+q^{2n+1}z)(1+q^{2n+1}z^{-1}) \quad (0<|z|<\infty), \tag{4.7}$$

where C_1 does not depend on z. For the same reason (see (3.15)) also the sequence

$$a_n = \Lambda_3(hn) = 1/\cosh(hn)$$

is totally positive normalized. The following identity which is equivalent to the Fourier expansion of the elliptic function $\mathrm{dn}\,u$:

$$\sum_{-\infty}^{\infty} \frac{1}{\cosh(hn)} z^n = C_2 \frac{\prod_0^{\infty} (1+q^{2n+1}z)(1+q^{2n+1}z^{-1})}{\prod_0^{\infty} (1-q^{2n+1}z)(1-q^{2n+1}z^{-1})} \qquad (q<|z|<q^{-1})$$

[9] In Rademacher-Schoenberg [30, pp. 156–159], the function $-1/J_0((13z)^{1/2})$ is expanded, as a numerical illustration, in a Laurent series $\sum \omega_n z^n$ in a ring between its first and second pole. The extremely smooth behavior of the coefficients ω_n there obtained is explained by the fact that the sequence $\{\omega_n\}$ is totally positive normalized, in view of the result just stated.

shows that its generating function is of the form (4.4), (4.5).

It was conjectured in [38] that the answer to the question is affirmative. This was recently established by A. Edrei [62] who proved the following theorem: *If* $\{a_n\}$ *is a sequence which is totally positive and normalized, then its generating function* (4.6) *is necessarily of the form* (4.4), (4.5).[10]

Three noteworthy special cases, stated with their converses, are as follows:

1. If the power series

$$\sum_0^\infty a_n z^n \qquad (a_0 = 1)$$

converges for all z, then it represents a function of the form

$$e^{\gamma z}\prod_1^\infty (1+\delta_\nu z) \qquad (\gamma \geqq 0,\ \delta_\nu \geqq 0,\ \sum \delta_\nu < \infty) \tag{4.8}$$

if and only if the matrix

$$A = \left\| \begin{matrix} a_0 & 0 & 0 & \cdots \\ a_1 & a_0 & 0 & \cdots \\ a_2 & a_1 & a_0 & \cdots \\ \vdots & \vdots & \vdots & \end{matrix} \right\|$$

is totally positive.

2. If the Laurent series

$$\sum_{-\infty}^\infty a_n z^n$$

converges for all $z \neq 0$, then it represents a function of the form

$$Ce^{az+bz^{-1}}z^m\prod_1^\infty (1+\alpha_\nu z)\prod_1^\infty (1+\beta_\nu z^{-1}) \tag{4.9}$$

$$(C \geqq 0,\ a \geqq 0,\ b \geqq 0,\ m \text{ integer},\ \alpha_\nu \geqq 0,\ \beta_\nu \geqq 0,\ \sum (\alpha_\nu + \beta_\nu) < \infty)$$

---

[10] Notice the interesting symmetry of (4.4) with respect to reciprocation: Also $1/F(-z)$ is of the form (4.4). An easy corollary of Edrei's theorem is as follows: If the identity $(\sum_{-\infty}^\infty a_n z^n)(\sum_{-\infty}^\infty (-1)^n b_n z^n) = 1$ holds in a ring $r_1 < |z| < r_2$, and if $\{a_n\}$ is totally positive, then either $\{b_n\}$ or else $\{-b_n\}$ is a totally positive sequence. As it happens, this proposition is actually a lemma in Edrei's proof of his theorem.

It should also be remarked that in analogy with the situation of §3, the Laurent expansions of (4.4) in every ring of regularity of $F(z)$ will generate, up to a common negative sign, totally positive sequences.

if and only if the sequence $\{a_n\}$ is totally positive.

3. If the symmetric Laurent series

$$\sum_{-\infty}^{\infty} a_n z^n \qquad (a_n = a_{-n})$$

converges for all $z\neq 0$, then it represents a function of the form

$$Ce^{a(z+z^{-1})}\prod_1^{\infty}(1+\alpha_\nu z)(1+\alpha_\nu z^{-1}) \tag{4.10}$$

$$(C \geqq 0, a \geqq 0, 1 \geqq \alpha_1 \geqq \alpha_2 \geqq \cdots \geqq \alpha_n \geqq \cdots \geqq 0, \sum \alpha_\nu < \infty)$$

if and only if the sequence $\{a_n\}$ is totally positive.

For a different characterization of the power series expansion of entire functions of the form (4.8) see J. Grommer [6, pp. 157–158]. A third characterization of the same expansion in terms of the zeros of the polynomials (3.3) was stated in Article 3.1. Edrei's general theorem should be regarded as a notable contribution to the general type of problem initiated by Hadamard in his classical dissertation [2]. Of course, the condition of Edrei's theorem that the sequence $\{a_n\}$ be totally positive is very strong, but then (4.4), (4.5) show the precise nature of the information on $F(z)$ which the theorem furnishes. Moreover, the condition is not only sufficient but also necessary.

Let us briefly review the work leading up to Edrei's theorem. Aissen, Schoenberg, and Whitney [59] investigated totally positive sequences $\{a_n\}$ with the property that $a_n=0$ if $n<0$, $a_0=1$, and showed that their generating function is necessarily of the form

$$\sum^{\infty} a_n z^n = e^{g(z)}\frac{\prod_1^{\infty}(1+\alpha_\nu z)}{\prod_1^{\infty}(1-\gamma_\nu z)},$$

where $\alpha_\nu \geqq 0$, $\gamma_\nu \geqq 0$, $\sum(\alpha_\nu+\gamma_\nu)<\infty$, and where $g(z)$ is entire and such that the exponential factor $\exp\{g(z)\}$, if expanded in powers of z, will also generate a totally positive sequence. Combining this result with a refinement of Picard's theorem due to R. Nevannlina, Edrei [60] showed that necessarily $g(z)=\gamma z$ $(\gamma\geqq 0)$, thereby proving his theorem for the case when (4.6) reduces to a power series. A different approach to this case was given by Edrei in [61]. For the general case see Edrei [62].

4.3. **An application of Edrei's theorem.** With the problem to be now discussed we return to the stability of smoothing formulae which is the theme of §1. In 1946 the author constructed certain classes of analytic approximation formulae to given equidistant ordinates $\{y_n\}$ $(-\infty < n < \infty)$, which we need not describe here in detail; they are of the "cardinal" type

$$f(x) = \sum_{\nu=-\infty}^{\infty} y_\nu L(x - \nu),$$

$L(x)$ being a certain even entire function, accurate tables of values for $L(x)$ and its derivatives being available. Moreover, if $y_\nu = P(\nu)$, where $P(x)$ is a polynomial of degree $2k-1$ (k being a certain fixed number), then $f(x) \equiv P(x)$. If we now let $x = n$ be an integer, we are faced with the natural question whether the sequence transformation

$$f(n) = \sum_{\nu=-\infty}^{\infty} L(n - \nu) y_\nu$$

satisfies our stability (or smoothing) condition (1.6) (see Schoenberg [**31**, Part B, pp. 131–133]). This question was there shown to depend on a conjecture which the author was unable to prove at the time. The conjecture there formulated is an easy corollary of a theorem to be stated and proved below by means of Edrei's theorem.

We start with the "rectangular" frequency function

$$M_1(x) = \begin{cases} 1 & \text{if } |x| \leqq 1/2, \\ 0 & \text{if } |x| > 1/2 \end{cases} \tag{4.11}$$

having the transform

$$\int_{-\infty}^{\infty} e^{-xs} M_1(x)dx = \frac{2 \sinh (s/2)}{s} \tag{4.12}$$

and the normal frequency function

$$M_0(x, t) = \frac{1}{(\pi t)^{1/2}} e^{-x^2/t} \qquad (t > 0) \tag{4.13}$$

of transform

$$\int_{-\infty}^{\infty} e^{-xs} M_0(x, t)dx = e^{ts^2/4}. \tag{4.14}$$

If we convolute $M_1(x)$ with itself $k-1$ times and convolute the result with $M_0(x, t)$, we obtain a distribution function $M_k(x, t)$ whose transform, by (4.12) and (4.14), is evidently

$$\int_{-\infty}^{\infty} e^{-xs} M_k(x, t)dx = e^{ts^2/4}\left(\frac{2 \sinh s/2}{s}\right)^k. \tag{4.15}$$

The theorem to be proved is as follows: *The Laurent series*

$$F(z) = \sum_{-\infty}^{\infty} M_k(n, t)z^n \qquad (0 < |z| < \infty) \tag{4.16}$$

admits a product representation

$$F(z) = C\cdot \prod_{\nu=1}^{\infty} (1 + \alpha_\nu z)(1 + \alpha_\nu z^{-1}) \tag{4.17}$$

where the $\alpha_\nu = \alpha_\nu(t)$ *are real and satisfy the conditions* $0 < \alpha_\nu < 1$, $\sum \alpha_\nu < \infty$, *while* $C = \prod(1+\alpha_\nu)^{-2}$.

REMARKS.[11] 1. The conclusion (4.17) "implies" that the sequence $a_n = M_k(n, t)$ is totally positive. This, naturally, will turn out to be the case. However, if $k>0$, the transform (4.15), of $M_k(x, t)$, has zeros. It is therefore not of the form (3.8) and we conclude that $M_k(x, t)$ is not a Pólya frequency function. We can therefore not apply the convenient argument used in Article 4.2 to show that the sequence $M_k(n, t)$ is totally positive.

2. The argument is of course valid if $k=0$ since $M_0(x, t)$ is the normal function (4.13). In this case our conclusion (4.17) is evidently true since it reduces to Jacobi's identity (4.7). If $k>0$, the representation (4.17) may be regarded as a generalization of Jacobi's relation.

3. Quite recently, J. Berghuis [54] succeeded in proving (4.17) for the two values $k=1$ and $k=2$ by standard methods of the theory of functions.

A general proof of (4.17) is as follows: We define the successive integrals of (4.13) by the relations

$$g_0(x) = M_0(x, t)$$

and

$$g_\nu(x) = \int_{-\infty}^{x} g_{\nu-1}(x)dx \qquad (\nu = 1, 2, \cdots).$$

The Laplace transform of $g_k(x)$ is found from (4.14), by successive

[11] The conjecture stated in [31, Part B, pp. 131–132] is as follows: *The reciprocal of the function* $F(e^{iu})$ *admits a representation* (*) $1/F(e^{iu}) = \sum_0^\infty C_{2n}(t)(2 \sin u/2)^{2n}$ *whose coefficients are positive and which converges for all real u.* This is implied by (4.17) as follows: $F(e^{iu}) = C\prod(1+\alpha_\nu^2+2\alpha_\nu \cos u) = C\prod\{(1+\alpha_\nu)^2-4\alpha_\nu \sin^2 u/2\}$ and (*) follows by reciprocation and expansion in powers of $\sin^2 u/2$.

integrations by parts, to be

$$(4.18)\qquad \int_{-\infty}^{\infty} e^{-xs} g_k(x)\,dx = \frac{1}{s^k}\, e^{ts^2/4} \qquad (\mathrm{R}s > 0),$$

whence

$$(4.19)\qquad \int_{-\infty}^{\infty} e^{-xs} g_k(x+h)\,dx = \frac{1}{s^k}\, e^{ts^2/4} e^{hs} \qquad (\mathrm{R}s > 0).$$

From these relations we draw two conclusions: 1. Replacing s by $s+1$ in (4.18) we obtain

$$\int_{-\infty}^{\infty} e^{-xs} e^{-x} g_k(x)\,dx = \frac{1}{(s+1)^k}\, e^{(ts^2+2ts+t)/4} \quad (\mathrm{R}s > -1).$$

The right-hand side being a reciprocal $1/\Psi(s)$, where $\Psi(s)$ is of the form (3.4), we conclude (Article 3.2) that $g_k(x)e^{-x}$ is a Pólya frequency function and hence:

(4.20) *The function $g_k(x)$ is totally positive.*

2. Writing $h=1/2$, $h=-1/2$ in (4.19) and subtracting one of the equations so obtained from the other, we find

$$\int_{-\infty}^{\infty} e^{-xs} \delta g_k(x)\,dx = \frac{1}{s^k}\, e^{ts^2/4} 2 \sinh s/2,$$

where δ is the symbol of central differencing with unit step. On repeating the operation altogether k times we find that

$$\int_{-\infty}^{\infty} e^{-xs} \delta^k g_k(x)\,dx = \frac{1}{s^k}\, e^{ts^2/4} (2 \sinh s/2)^k.$$

The transform so obtained agrees with (4.15) and we obtain the relation[12]

$$(4.21)\qquad M_k(x, t) = \delta^k g_k(x).$$

For integral $x=n$ we obtain from (4.21)

$$M_k(n, t) = \delta^k g_k(n) = \Delta^k g_k(n - k/2),$$

a relation which is equivalent to the identity

$$\sum_{-\infty}^{\infty} M_k(n, t) z^n = (1-z)^k \sum_{n=-\infty}^{\infty} g_k\left(n + \frac{k}{2}\right) z^n \qquad (0 < |z| < 1).$$

[12] See [31, Part A, p. 85] for a different derivation of the relation (4.21) which is used there to construct tables of values of $M_k(x,t)$.

Replacing the left-hand side by $F(z)$, in view of (4.16), and replacing z by $z/2$ we obtain

$$\sum_{-\infty}^{\infty} \frac{1}{2^n} g_k\left(n+\frac{k}{2}\right) z^n = \frac{F(z/2)}{(1-z/2)^k} \qquad (0<|z|<2). \tag{4.22}$$

From (4.20) we conclude that $\{g_k(n+k/2)\}$ is a totally positive sequence. Therefore $\{2^{-n}g_k(n+k/2)\}$ is a totally positive sequence which is also normalized. By Edrei's theorem the left-hand side of (4.22) can be continued into the domain $0<|z|<\infty$ as a function of the form (4.4), (4.5). On the other hand, the right-hand side of (4.22) is another such representation valid throughout the region $0<|z|<\infty$, having as only singularity the pole $z=2$ of exact order k. On comparing these two analytic continuations of the left-hand side of (4.22) we conclude that $F(z/2)$ is of the form (4.9) and therefore so is $F(z)$. However, the Laurent expansion (4.16) of $F(z)$ being symmetric, we conclude that $F(z)$ allows a representation of the form (4.10), hence

$$F(z) = Ce^{a(z+z^{-1})}\prod_{1}^{\infty}(1+\alpha_\nu z)(1+\alpha_\nu z^{-1}), \tag{4.23}$$

where $C>0$, $a\geqq 0$, $1\geqq\alpha_1\geqq\alpha_2\cdots\geqq 0$, $\sum\alpha_\nu<\infty$. A proof of (4.17) will be complete as soon as we show that the constants satisfy the following conditions:

1. $a=0$.
2. $\alpha_1<1$.
3. $\alpha_\nu>0$ for $\nu=1, 2, \cdots$.

PROOF OF 1. If we denote by $M_k(x)$ the convolution of k functions all identical with $M_1(x)$, defined by (4.11), then

$$M_k(x,t) = \frac{1}{(\pi t)^{1/2}}\int_{-\infty}^{\infty} e^{-(x-u)^2t^{-1}}M_k(u)du.$$

However, $M_k(u)=0$ if $|u|>k/2$, hence if $x>k/2$ we find

$$\begin{aligned} M_k(x,t) &= \frac{1}{(\pi t)^{1/2}}\int_{-k/2}^{k/2} e^{-(x-u)^2t^{-1}}M_k(u)du \\ &< \frac{1}{(\pi t)^{1/2}}\int_{-k/2}^{k/2} M_k(u)du\cdot e^{-(x-k/2)^2t^{-1}} \end{aligned}$$

or

$$M_k(x,t) < \frac{1}{(\pi t)^{1/2}} e^{-(x-k/2)^2t^{-1}} \qquad \text{if } x>k/2.$$

For appropriate positive constants C_1 and γ ($\gamma < t^{-1}$) we therefore have $M_k(n, t) < C_1 e^{-\gamma n^2}$, for all n, and (4.16) implies the estimate

$$F(z) < 2C_1 \sum_0^\infty e^{-\gamma n^2} z^n, \qquad \text{if } z > 1.$$

However, if $z > 1$, all factors of (4.23) are $\geqq 1$ so that a fortiori

$$e^{az} < 2C_1 \sum_0^\infty e^{-\gamma n} z^n, \qquad \text{if } z > 1.$$

Since on the right side we have an entire function of order $\rho = 0$, we must indeed have that $a = 0$.

PROOF OF 2. It suffices to show that $F(-1) > 0$. This last inequality can be derived as follows: Setting $s = iu$ in (4.15) we obtain

$$\int_{-\infty}^{\infty} e^{-ixu} M_k(x, t) dx = e^{-tu^2/4} \left(\frac{2 \sin u/2}{u} \right)^k \equiv \psi_k(u)$$

and by Poisson's summation formula we obtain the relation

$$\sum_{n=-\infty}^{\infty} M_k(n, t) e^{inu} = \sum_{\nu=-\infty}^{\infty} \psi_k(u + 2\pi\nu).$$

Setting $u = \pi$, we obtain

$$F(-1) = \left(\frac{2}{\pi} \right)^k \sum_{\nu=-\infty}^{\infty} (-1)^{\nu k} \frac{1}{(2\nu + 1)^k} e^{-t\pi (2\nu+1) /4}.$$

This is visibly positive if k is even. For odd values of k we write it in the form

$$\left(\frac{\pi}{2} \right)^k F(-1) = \left(\sum_{\nu=0}^{\infty} + \sum_{\nu=-1}^{-\infty} \right) (-1)^\nu \frac{1}{(2\nu + 1)^k} e^{-t\pi^2 (2\nu+1)^2/4}$$

and find that each of the two ordinary infinite series indicated has terms decreasing in absolute values and alternating sign. Thus the sums of these series are positive and again $F(-1) > 0$.

PROOF OF 3. If only a finite number of α_ν in (4.17) were positive, then $F(z)$ would be rational, hence $z = 0$ would be a polar singularity for the right-hand side of (4.22). However, the Laurent expansion on the left-hand side of (4.22) shows that $z = 0$ is actually an essential singularity. This contradiction completes a proof of the factorization (4.17).

A concluding remark on this subject is as follows. In [**31**, p. 79], it was pointed out that

$$f(x) = \sum_{\nu=-\infty}^{\infty} y_\nu M_k(x - \nu, t)$$

is an analytic approximation to the given sequence of ordinates $\{y_n\}$ such that

$$f(n) = \sum_{-\infty}^{\infty} y_\nu M_k(n - \nu, t)$$

is a smoothing operation in the sense of §1. We can now make the stronger assertion that this last operation is variation-diminishing. Indeed, as already pointed out, the sequence $a_n = M_k(n, t)$ is totally positive, a fact equivalent to our assertion (see Article 4.1).

4.4. **A few unsolved problems.** 1. Our first question concerns the proof of Edrei's theorem. The ties between the totally positive functions (§3) and totally positive sequences (§4) are apparently not as close as to allow a proof of Edrei's theorem along the lines used in characterizing Laplace transforms of Pólya frequency functions. Is this really in the nature of things or could an "elementary" proof of Edrei's theorem be devised, at least for the case when (4.6) is a power series?

2. It was pointed out in Article 4.2 that if $\Lambda(x)$ is a Pólya frequency function, then

$$a_n = \Lambda(n) \qquad (-\infty < n < \infty) \tag{4.24}$$

is a totally positive normalized sequence. However, not all such sequences can be obtained in this way. Indeed, let $\{a_n\}$ be a totally positive sequence such that its generating Laurent series converges in $0<|z|<\infty$, and hence is of the form (4.9). However, let it be different from the series (4.7). Such a sequence can *not* be obtained by (4.24), by interpolating a Pólya frequency function $\Lambda(x)$. The reason for this is very simple: Assuming (4.24) to hold, $\Lambda(x)$ would have a Laplace transform (3.8) which is entire. However, (3.8) and (3.4) show that the only Pólya frequency function having an entire transform is the normal frequency function. This situation raises the question: *What is the nature of totally positive sequences obtained by* (4.24) *from Pólya frequency functions?*

3. Let us apply the integral transformation (3.10) to a bounded function $f(x)$ of period 2π. Setting

$$\chi(x) = \sum_{n=-\infty}^{\infty} \Lambda(x + 2\pi n), \tag{4.25}$$

the transformation (3.10) becomes

$$g(x) = \int_0^{2\pi} \chi(x - t)f(t)dt, \tag{4.26}$$

which shows that $g(x)$ is continuous and of period 2π. Assuming $\Lambda(x)$ to be a Pólya frequency function, (4.26) implies that $v(g) \leqq v(f)$ (see Article 3.3); however, this relation is now meaningless since $v(g)$ and $v(f)$ either vanish or else are infinite. It is not difficult to see that (4.26) now enjoys the variation-diminishing property in the cyclic sense, i.e. it implies

$$v_c(g) \leqq v_c(f), \tag{4.27}$$

where these numbers of changes of sign are counted on the circle, or within a period. The special case of (4.26) when

$$\Lambda_\tau(x) = \frac{1}{(\pi\tau)^{1/2}} e^{-x^2/\tau} \qquad (\tau > 0), \tag{4.28}$$

was already noticed by Pólya [**20**], not in its integral form (4.26) but rather in terms of the equivalent convolution of the respective Fourier series;[13] $\chi(x)$ then becomes the Fourier kernel for a ring known from heat-conduction.

The following question arises: *What is the nature of the periodic functions* $\chi(x)$ *with the property that the transformation* (4.26) *is cyclically variation-diminishing, i.e. implies* (4.27)? Our formula (4.25) furnishes such functions but does not produce all of them. Indeed

$$\chi(x) = a + \cos t \qquad (a \text{ constant} \geqq 1)$$

is readily shown to have the property (4.27), without, however, being of the form (4.25).

Concluding, let us briefly mention a connection between the transformations (4.26) and the convex curves of Article 2.2. Let us assume

$$\int_0^{2\pi} \chi(t)dt = 1, \tag{4.29}$$

and let C, defined by (2.8), be convex in E_{2n}. *Then the new curve*

$$\overline{C}: \quad x_i = \bar{x}_i(x) = \int_0^{2\pi} \chi(x - t)x_i(t)dt \tag{4.30}$$

$$(i = 1, \cdots, 2n;\ 0 \leqq x \leqq 2\pi)$$

[13] For yet another example of a special function $\chi(x)$ see Pólya-Wiener [**27**, Lemma 1].

is also convex in E_{2n}. Indeed, (4.29) and (4.30) imply the relation

$$A_0 + \sum_1^{2n} A_i \bar{x}_i(x) = \int_0^{2\pi} \chi(x-t)\left(A_0 + \sum_1^{2n} A_i x_i(t)\right) dt,$$

and by (4.27):

$$v_c(A_0 + \sum A_i \bar{x}_i(t)) \leqq v_c(A_0 + \sum A_i x_i(t)).$$

This relation means that $\overline{C}$ crosses a hyperplane not more often than C does, hence at most $2n$ times.

If we choose in particular the Fourier kernel $\chi_\tau(x)$ arising from (4.25) and $\Lambda(x) = \Lambda_\tau(x)$, defined by (4.28), then we see that the $2n$ functions (4.30) are analytic (even entire) functions of the parameter x. As $\tau \to +0$, we evidently have $\bar{x}_i(x) \to x_i(x)$. This remark proves the following: *A curve convex in* E_{2n} *is the limit of a family of curves convex in* E_{2n} *which are analytic.*

Added in proof, April 1953. Professor W. Fenchel kindly pointed out to me that also Gantmakher and Krein have been essentially anticipated in the discovery of their theorem of Article 2.2 by the late J. Hjelmslev in his paper *Introduction à la théorie des suites monotones*, Oversigt over det Kgl. Danske Videnskabernes Selskabs Forhandlinger, 1914, pp. 1–74. Indeed Hjelmslev gives analytic conditions in order that the arc Γ in E_m, defined by (2.5), be of order m. The precise statement of his result (loc. cit. p. 62) is as follows: *If no subarc of* Γ *is in* E_{m-1}, *then* Γ *is of order* m *if and only if all determinants* (2.6) *are positive, or all these determinants are negative.* Arcs satisfying these determinant conditions are called *monotone arcs* by Hjelmslev.

Bibliography

1. E. Laguerre, *Sur les fonctions de genre zéro et du genre un*, C. R. Acad. Sci. Paris vol. 95 (1882) pp. 828–831 [also *Oeuvres*, vol. I, pp. 174–177].

2. J. Hadamard, *Essai sur l'étude des fonctions données par leur développement de Taylor*, Journal de Mathématiques (4) vol. 8 (1892) pp. 101–186.

3. A. Hurwitz, *Sur le problème des isopérimetres*, C. R. Acad. Sci. Paris vol. 132 (1901) pp. 401–403 [also *Werke*, vol. I, pp. 490–491].

4. M. Fekete, *Über ein Problem von Laguerre*, Rendiconti di Palermo vol. 34 (1912) pp. 1–12.

5. G. Pólya, *Über Annäherung durch Polynome mit lauter reellen Wurzeln*, Rendiconti di Palermo vol. 36 (1913) pp. 1–17.

6. J. Grommer, *Ganze transzendente Funktionen mit lauter reellen Nullstellen*, Journal für Mathematik vol. 144 (1914) pp. 114–165.

7. G. Pólya and I. Schur, *Über zwei Arten von Faktorenfolgen in der Theorie der algebraischen Gleichungen*, Journal für Mathematik vol. 144 (1914) pp. 89–113.

8. G. Pólya, *Algebraische Untersuchungen über ganze Funktionen vom geschlechte Null und Eins*, Journal für Mathematik vol. 145 (1915) pp. 224–249.

9. O. D. Kellogg, *The oscillation of functions of an orthogonal set*, Amer. J. Math. vol. 38 (1916) pp. 1–5.

10. ———, *Orthogonal function sets arising from integral equations*, Amer. J. Math. vol. 40 (1918) pp. 144–154.

11. ———, *Interpolation properties of orthogonal sets of solutions of differential equations*, Amer. J. Math. vol. 40 (1918) pp. 225–234.

12. H. Hamburger, *Bemerkungen zu einer Fragestellung des Herrn Pólya*, Math. Zeit. vol. 7 (1920) pp. 302–322.

13. G. Pólya, *On the mean-value theorem corresponding to a given linear homogeneous differential equation*, Trans. Amer. Math. Soc. vol. 24 (1922) pp. 312–324.

14. G. Pólya and G. Szegö, *Aufgaben und Lehrsätze aus der Analysis*, vol. II, Berlin, 1925.

15. H. H. Wolfenden, *On the development of formulae for graduation by linear compounding, with special reference to the work of Erastus L. De Forest*, Transactions of the Acturial Society of America vol. 26 (1925) pp. 81–121.

16. S. Bernstein, *Leçons sur les propriétés extrémales des fonctions analytiques*, Paris, 1926.

17. E. T. Whittaker and G. Robinson, *The calculus of observations*, 2d ed., London and Glasgow, 1926.

18. A. Marchaud, *Sur les continus d'ordre borné*, Acta Mathematica vol. 55 (1930) pp. 67–115.

19. I. J. Schoenberg, *Über variationsvermindernde lineare Transformationen*, Math. Zeit. vol. 32 (1930) pp. 321–328.

20. G. Pólya, *Qualitatives über Wärmeausgleich*, Zeitschrift für Angewandte Mathematik und Mechanik vol. 13 (1933) pp. 125–128.

21. I. J. Schoenberg, *Zur Abzählung der reellen Wurzeln algebraischer Gleichungen*, Math. Zeit. vol. 38 (1934) pp. 546–564.

22. T. S. Motzkin, *Beiträge zur Theorie der linearen Ungleichungen*, dissertation, Basel, 1933, Jerusalem, 1936.

23. A. Wintner, *On a class of Fourier transforms*, Amer. J. Math. vol. 58 (1936) pp. 45–90.

24. F. Gantmakher and M. Krein, *Sur les matrices complètement non négatives et oscillatoires*, Compositio Math. vol. 4 (1937) pp. 445–476.

25. P. Scherk, *Über differenzierbare Kurven und Bögen*, Casopis pro pěstovani Matematiky a Fysiky vol. 66 (1937) pp. 165–191.

26. G. B. Dantzig, *On a class of distributions that approach the normal distribution function*, Ann. Math. Statist. vol. 10 (1939) pp. 247–253.

27. G. Pólya and N. Wiener, *On the oscillation of the derivatives of a periodic function*, Trans. Amer. Math. Soc. vol. 52 (1942) pp. 249–256.

28. O. Szász, *On sequences of polynomials and the distribution of their zeros*, Bull. Amer. Math. Soc. vol. 49 (1943) pp. 377–383.

29. T. N. E. Greville, *The general theory of osculatory interpolation*, Transactions of the Actuarial Society of America vol. 45 (1944) pp. 202–265.

30. H. Rademacher and I. J. Schoenberg, *An iteration method for calculation with Laurent series*, Quarterly of Applied Mathematics vol. 4 (1946) pp. 142–159.

31. I. J. Schoenberg, *Contributions to the problem of approximation of equidistant data by analytic functions*, Quarterly of Applied Mathematics vol. 4 (1946) Part A pp. 45–99, Part B, pp. 112–141.

32. D. V. Widder, *The Laplace transform*, Princeton, 1946.

33. H. B. Curry and I. J. Schoenberg, *On Pólya frequency functions* IV: *The spline*

functions and their limits, as yet unpublished; see Bull. Amer. Math. Soc. Abstract 53-11-380.

34. I. J. Schoenberg, *On totally positive functions, Laplace integrals and entire functions of the Laguerre-Pólya-Schur type*, Proc. Nat. Acad. Sci. U.S.A. vol. 33 (1947) pp. 11–17.

35. D. V. Widder, *Inversion formulas for convolution transforms*, Duke Math. J. vol. 14 (1947) pp. 217–249.

36. W. Feller, *On the Kolmogorov-Smirnov limit theorems for empirical distributions*, Ann. Math. Statist. vol. 19 (1948) pp. 177–189.

37. I. I. Hirschman and D. V. Widder, *The inversion of convolution transform with totally positive kernels*, Proc. Nat. Acad. Sci. U.S.A. vol. 34 (1948) pp. 152–156.

38. I. J. Schoenberg, *Some analytical aspects of the problem of smoothing*, Courant Anniversary volume "Studies and Essays," New York, 1948, pp. 351–370.

39. ———, *On variation-diminishing integral operators of the convolution type*, Proc. Nat. Acad. Sci. U.S.A. vol. 34 (1948) pp. 164–169.

40. I. I. Hirschman and D. V. Widder, *The inversion of a general class of convolution transforms*, Trans. Amer. Math. Soc. vol. 66 (1949) pp. 135–201.

41. ———, *A representation theory for a general class of convolution transforms*, Trans. Amer. Math. Soc. vol. 67 (1949) pp. 69–97.

42. P. Laasonen, *Einige Sätze über Tschebycheffsche Systeme*, Ann. Acad. Sci. Fennicae, Ser. A. I. Math. Phys. no. 52, 24 pp., 1949.

43. I. J. Schoenberg and Anne Whitney, *Sur la positivité des determinants de translations des fonctions de fréquence de Pólya . . .* , C. R. Acad. Sci. Paris vol. 228 (1949) pp. 1996–1998.

44. F. Gantmakher and M. Krein, *Oscillatory matrices and kernels and small vibrations of mechanical systems* (in Russian), 2d ed., Moscow, 1950.

45. I. I. Hirschman, *Proof of a conjecture of I. J. Schoenberg*, Proc. Amer. Math. Soc. vol. 1 (1950) pp. 63–65.

46. I. J. Schoenberg, *On Pólya frequency functions* II: *Variation-diminishing integral operators of the convolution type*, Acta Sci. Math. Szeged vol. 12 (1950) pp. 97–106.

47. ———, *The finite Fourier series and elementary geometry*, Amer. Math. Monthly vol. 57 (1950) pp. 390–404.

48. M. Aissen, A. Edrei, I. J. Schoenberg, and Anne Whitney, *On the generating functions of totally positive sequences*, Proc. Nat. Acad. Sci. U.S.A. vol. 37 (1951) pp. 303–307.

49. F. John, *On integration of parabolic equations by difference methods*, Communications on Pure and Applied Mathematics vol. 5 (1952) pp. 155–211.

50. T. S. Motzkin and I. J. Schoenberg, *On lineal entire functions of n complex variables*, Proc. Amer. Math. Soc. vol. 3 (1952) pp. 517–526.

51. I. J. Schoenberg and Anne Whitney, *A theorem on polygons in n dimensions with applications to variation-diminishing and cyclic variation-diminishing linear transformations*, Compositio Math. vol. 9 (1951) pp. 141–160.

52. D. V. Widder, *Weierstrass transforms of positive functions*, Proc. Nat. Acad. Sci. U.S.A. vol. 37 (1951) pp. 315–317.

53. ———, *Necessary and sufficient conditions for the representation of a function by a Weierstrass transform*, Trans. Amer. Math. Soc. vol. 71 (1951) pp. 430–439.

54. J. Berghuis, *A class of entire functions used in analytic interpolation*, Proceedings Royal Neth. Acad. of Sciences. Ser. A vol. 55 (1952) pp. 468–473.

55. I. J. Schoenberg, *On Pólya frequency functions* I: *The totally positive functions*

and their Laplace transforms, Journal d'Analyse Mathématique vol. 1 (1951) pp. 331–374; actually appeared in 1952.

56. ———, *An isoperimetric inequality for closed curves convex in even-dimensional euclidean spaces*, to appear in the Acta Math.

57. I. J. Schoenberg and Anne Whitney, *On Pólya frequency functions* III: *The positivity of translation determinants with an application to the interpolation problem by spline curves*, Trans. Amer. Math. Soc. vol. 74 (1953) pp. 246–259.

58. Anne Whitney, *A reduction theorem for totally positive matrices*, Journal d'Analyse Mathématique vol. 2 (1952) pp. 88–92.

59. M. Aissen, I. J. Schoenberg, and A. Whitney, *On the generating functions of totally positive sequences* I, Journal d'Analyse Mathématique vol. 2 (1952) pp. 93–103.

60. A. Edrei, *On the generating functions of totally positive sequences* II, Journal d'Analyse Mathématique vol. 2 (1952) pp. 104–109.

61. ———, *Proof of a conjecture of Schoenberg on the generating function of a totally positive sequence*, Canadian Journal of Mathematics vol. 5 (1953) pp. 86–94.

62. ———, *On the generating function of doubly infinite, totally positive sequences*, submitted for publication in Trans. Amer. Math. Soc.

University of California, Los Angeles, and
University of Pennsylvania

Reprinted from the
Bulletin of the American Mathematical Society
Vol. 59, No. 3, pp. 199-230
May, 1953

Reprinted from the PROCEEDINGS OF THE NATIONAL ACADEMY OF SCIENCES
Vol. 52, No. 4, pp. 947–950. October, 1964.

SPLINE FUNCTIONS AND THE PROBLEM OF GRADUATION*

BY I. J. SCHOENBERG

UNIVERSITY OF PENNSYLVANIA AND INSTITUTE FOR ADVANCED STUDY

Communicated by S. Bochner, August 19, 1964

1. *Introduction.*—The aim of this note is to extend some of the recent work on spline interpolation so as to include also a solution of the problem of graduation of data. The well-known method of graduation due to E. T. Whittaker suggests how this should be done. Here we merely describe the idea and the qualitative aspects of the new method, while proofs and the computational side will be discussed elsewhere.

2. *Spline Interpolation.*—Let $I = [a, b]$ be a finite interval and let (x_ν, y_ν), $(\nu = 1, \ldots, n)$, be given data such that $a \leqq x_1 < x_2 < \ldots < x_n \leqq b$. The following facts are known:[1]

Let m be a natural number, $m \leqq n$. The problem of finding a function $f(x)$ $(x \in I)$ having a square integrable mth derivative and satisfying the two conditions

$$f(x_\nu) = y_\nu, \quad (\nu = 1, \ldots, n), \tag{1}$$

$$Jf \equiv \frac{1}{(m!)^2} \int_I (f^m(x))^2 dx = \text{minimum}, \tag{2}$$

has a unique solution which is the restriction to $[a, b]$ of the function $s(x) = S(x; y_1, y_2, \ldots, y_n)$ which is uniquely characterized by the three conditions

$$s(x_\nu) = y_\nu, \ (\nu = 1, \ldots, n) \tag{3}$$

$$s(x) \in C^{2m-2}(-\infty, \infty) \tag{4}$$

$$\begin{cases} s(x) \in \pi_{2m-1} \text{ in each of the intervals } (x_\nu, x_{\nu+1}),^2 \\ s(x) \in \pi_{m-1} \text{ in } (-\infty, x_1) \text{ and also in } (x_n, \infty). \end{cases} \tag{5}$$

The functions defined by the two conditions (4) and (5) are called *spline functions of order* $2m$ (or degree $2m - 1$), having the *knots* x_ν; we denote their class by the symbol $\mathcal{S}_m$.

We have assumed that $1 \leqq m \leqq n$. If $m = 1$, then $s(x)$ is obtained by linear interpolation between successive y_ν, while $s(x) = y_1$ if $x < x_1$ and $s(x) = y_n$ if $x > x_n$. If $m = n$, then $\mathcal{S}_m = \pi_{n-1}$ and $s(x)$ is the polynomial interpolating the y_ν.

3. *Whittaker's Method of Graduation.*—In 1923 E. T. Whittaker[3] proposed the following method of adjusting the ordinates y_ν if these are only imperfectly known and are in need of a certain amount of smoothing: he chooses m, $1 \leqq m < n$, and the (smoothing) parameter, ϵ, $\epsilon > 0$. The graduated sequence $y_\nu^* = y_\nu^*(\epsilon)$ is then obtained as the solution of the problem

$$\epsilon \sum_{\nu=1}^{n-m} (\Delta^m y_\nu^*)^2 + \sum_1^n (y_\nu^* - y_\nu)^2 = \text{minimum}, \tag{6}$$

where

$$\Delta^m y_\nu^* = \sum_{i=\nu}^{\nu+m} y_i^*/\omega'_\nu(x_i), \qquad \omega_\nu(x) = (x - x_\nu) \ldots (x - x_{\nu+m}),$$

are the divided differences.

We use throughout this note the notation

$$Ef = Ef(x) = \sum_{1}^{n} (f(x_\nu) - y_\nu)^2 \tag{7}$$

and define the familiar least squares polynomial $Q(x) \in \pi_{m-1}$ as the solution of the problem

$$Ef = \text{minimum}, \ (f \in \pi_{m-1}). \tag{8}$$

It is easily shown that Whittaker's graduated values $y_\nu^*(\epsilon)$ have the properties

$$\lim_{\epsilon \to 0} y_\nu^*(\epsilon) = y_\nu, \qquad \lim_{\epsilon \to \infty} y_\nu^*(\epsilon) = Q(x_\nu). \tag{9}$$

4. *Graduation by Spline Functions.*—In an attempt to combine the spline interpolation described in section 2 with Whittaker's idea, we propose the following

PROBLEM 1. *Let $m < n$ and $\epsilon > 0$. Among all $f(x)$, defined in I, having a square integrable mth derivative we propose to find the solution of the problem*

$$\epsilon Jf + Ef = \text{minimum}. \tag{10}$$

If the solution $Q(x)$ of the problem (8) is such that $EQ = 0$, then it is clear that $f = Q$ also solves the problem (10) for all $\epsilon > 0$. In order to exclude this trivial case we shall assume throughout that

$$EQ > 0, \text{ or equivalently } Js > 0. \tag{11}$$

THEOREM 1. *The minimum problem* (10) *has a unique solution* $f(x) = S(x, \epsilon)$ *which is a spline function of the family* $\mathcal{S}_m$.

We state the analogues of the relations (9) as

THEOREM 2. *The functions $s(x)$ and $Q(x)$ being as defined before, the following relations hold*

$$\lim_{\epsilon \to 0} S(x, \epsilon) = s(x), \qquad \lim_{\epsilon \to \infty} S(x, \epsilon) = Q(x). \tag{12}$$

5. *An Equivalent Approach.*—The quantity Jf evidently measures the departure of $f(x)$ from being an element of π_{m-1}; likewise Ef measures how well $f(x)$ describes the data (x_i, y_i). A sensible approach to the problem of graduation is as follows:

Assuming (11), *we choose u in the range $0 \leq u \leq Js$ and propose to find the solution of the problem*

$$Ef = \text{minimum}, \textit{ among functions } f(x) \textit{ subject to } Jf \leq u. \tag{13}$$

That this approach again leads to the solution of Problem 1, as described by Theorem 1, is stated as

THEOREM 3. *The solution* $f(x) = S_u(x)$ *of the problem* (13) *is unique and such that*

$$S_u(x) \in \mathcal{S}_m, \ JS_u = u.$$

The two families of spline functions

$$S_u(x), \ (0 \leq u \leq Js) \text{ and } S(x,\epsilon) \ (0 \leq \epsilon \leq \infty)$$

are identical. If we regard $v = ES_u$ as a function of $u = JS_u$ and express the dependence as

$$v = \Phi(u), \ (0 \leqq u \leqq Js), \tag{14}$$

then the graph of (14) *is a smooth and strictly convex arc with*

$$\Phi(0) = EQ, \ \Phi'(0) = -\infty, \ \Phi(Js) = 0, \ \Phi'(Js) = 0. \tag{15}$$

It follows that the function $\Phi(u)$ *is strictly decreasing in its domain of definition. Finally, the relation between u and the smoothing parameter* ϵ *of section* 4 *is described by the relation*

$$\epsilon = -\Phi'(u). \tag{16}$$

The convexity of the graph of (14) and (15) allows us to see readily on the graph why $S_u(x)$ is the solution of the problem

$$\epsilon JS + ES = \text{minimum, for } S \in \mathcal{S}_m,$$

and why therefore $S_u(x) = S(x, \epsilon)$. I add that (14) may be represented in parametric form and that u and v are rational functions of the parameter ϵ.

6. *A Formal Comparison with Whittaker's Method.*—We return to section 2 and wish to express $JS(x; y_1, \ldots, y_n)$ in terms of the y_ν. This can be done as follows: We denote by $M_i(x)$ the kernel in the integral representation of the divided difference

$$\Delta^m g(x_i) = \frac{1}{m!}\int_{x_i}^{x_{i+m}} M_i(x) g^{(m)}(x)dx \qquad (i = 1, \ldots, n - m)$$

and extending the definition of $M_i(x)$ to all x by setting $M_i(x) - 0$ if x is outside (x_i, x_{i+m}), we write

$$L_{ij} = \int_{-\infty}^{\infty} M_i(x)M_j(x)dx, \qquad (i,j = 1, \ldots, n - m).$$

The matrix $||L_{ij}||$ is positive definite, and if we introduce its inverse

$$||\Lambda_{ij}|| = ||L_{ij}||^{-1},$$

then[4]

$$JS(x; y_1, \ldots, y_n) = \sum_1^{n-m} \Lambda_{ij}\Delta^m y_i \Delta^m y_j. \tag{17}$$

Setting $S(x_i, \epsilon) = \eta_i$, hence $S(x, \epsilon) = S(x; \eta_1, \ldots, \eta_n)$ it follows from (17) that the solution of the problem (10) reduces to the solution of the algebraic problem

$$\epsilon \sum_1^{n-m} \Lambda_{ij}\Delta^m \eta_i \Delta^m \eta_j + \sum_1^n (\eta_\nu - y_\nu)^2 = \text{minimum}. \tag{18}$$

A comparison of the first sums in (6) and (18) shows that the new method arises if we replace in (6) the form $\sum_1^{n-m} \xi_i^2$ by the positive definite quadratic form $\sum \Lambda_{ij}\,\xi_i\xi_j$. This increase in complexity might be compensated by the new method furnishing also the approximating spline function $S(x, \epsilon)$, if such an approximation is desirable [e.g., compare the first relations (9) and (12)]. A further actual comparison of the two methods will require numerical experimentation.

7. *The Case of Periodic Data.*—In a recent paper[5] I introduced the method of trigonometric spline interpolation. The discussion of sections 4 and 5 carries over to the periodic case and need not be elaborated. The analogue of Problem 1 is as follows: assuming $2m + 2 \leqq n$, $\epsilon > 0$, we are seeking the function $f(x)$, of period 2π, having a square integrable $(2m + 1)$st derivative and which solves the problem

$$\epsilon\int(\Delta_m f)^2dx + \sum_1^n(f(x_\nu) - y_\nu)^2 = \text{minimum}, \ [\Delta_m = D(D^2 + 1^2)\ldots(D^2 + m^2)],$$

the integration being over an entire period while the x_ν are increasing with $x_n - x_1 < 2\pi$. The unique solution is a trigonometric spline function $S(x, \epsilon)$ having properties analogous to those stated in Theorems 1, 2, and 3. Naturally, the role of $Q(x)$ is now played by the trigonometric polynomial $T(x)$, of order m, which solves the problem

$$\sum_1^n (T(x_\nu) - y_\nu)^2 = \text{minimum}.$$

* Work done with partial support by a grant from the National Science Foundation contract GP-2442.

[1] For references, see *Proc. Roy. Netherl. Acad.*, **A67**, 155–163 (1964). A recent paper is by T. N.E. Greville (Math. Res. Center Tech. Report No. 450, Madison, Wis., January 1964).

[2] We denote by π_k the class of real polynomials of degree not exceeding k.

[3] Whittaker, E. T., *Proc. Edinburgh Math. Soc.*, **41**, 63–75 (1923).

[4] See these PROCEEDINGS, **51**, 28 (1964), formula (15), for a simplification occurring in the case when the x_ν are in arithmetic progression.

[5] To appear in the November 1964 issue of *J. Math. Mech.*

TOTAL POSITIVITY and VARIATION DIMINUTION

Über variationsvermindernde lineare Transformationen.

Von

Isac Schoenberg in Jassy (Rumänien).

Wir betrachten die reelle lineare Transformation

$$(1)\qquad \begin{aligned} y_1 &= a_{11}\, x_1 + a_{12}\, x_2 + \ldots + a_{1n}\, x_n \\ y_2 &= a_{21}\, x_1 + a_{22}\, x_2 + \ldots + a_{2n}\, x_n \\ &\vdots \\ y_m &= a_{m1}\, x_1 + a_{m2}\, x_2 + \ldots + a_{mn}\, x_n . \end{aligned}$$

und wollen dieselbe mit Herrn Pólya variationsvermindernd nennen, falls für beliebige reelle x die Zeichenwechselanzahl der Folge $y_1, y_2, \ldots, y_n$ diejenige der Folge $x_1, x_2, \ldots, x_n$ nie übertrifft, also immer $v_y \leqq v_x$ ist[1]). Derartige Transformationen in besonderer Gestalt treten zu Hilfszwecken in verschiedenen Abhandlungen aus dem Ideenkreise der Descartesschen Vorzeichenregel auf[2]).

In den folgenden Zeilen versuche ich die variationsvermindernde Eigenschaft der linearen Transformationen aus den Vorzeichen der Minoren ihrer Matrix direkt abzulesen. Für den Fall, daß der Rang der Transformation gleich n ist, gelingt diese Charakterisierung mit Hilfe der Minoren vollständig.

Bei anderer Gelegenheit werde ich Anwendungen des hier erhaltenen Satzes geben, insbesondere zum Vergleich der verschiedenen Sätze, welche Schranken für die Wurzelanzahl algebraischer Gleichungen liefern.

Ich möchte noch Herrn Professor Pólya für sein freundliches Interesse meinen Dank aussprechen.

---

[1]) Wir behalten diese Bezeichnung auch fernerhin bei.

[2]) Vgl. M. Fekete und G. Pólya, Über ein Problem von Laguerre, Rendiconti di Palermo **34** (1912), S. 89–120, insbesondere S. 91–95. Siehe ferner Fußnote [4]). — N. Obreschkoff, Über die Wurzeln von algebraischen Gleichungen, Jahresbericht der deutschen Math.-Ver. **33** (1925), S. 52–64, insbesondere die Hilfssätze des § 1. — G. Pólya und G. Szegö, Aufgaben und Lehrsätze aus der Analysis, Berlin 1925, 2, 5. Abschnitt, insbesondere die Aufgaben 4, 5 und 21.

1. Wir betrachten die reelle Matrix

$$A = \begin{pmatrix} a_{11} & a_{12} & \cdots & a_{1n} \\ a_{21} & a_{22} & \cdots & a_{2n} \\ \vdots & \vdots & & \vdots \\ a_{m1} & a_{m2} & \cdots & a_{mn} \end{pmatrix}$$

und alle ihre Unterdeterminanten der Ordnung ν für $1 \leqq \nu \leqq \mathrm{Min}(m,n)$. *Wenn sämtliche Minoren der Ordnung ν entweder Null sind oder alle dasselbe Vorzeichen besitzen, also keine zwei Minoren der Ordnung ν verschiedene Vorzeichen aufweisen und dies für alle in Frage kommenden ν gilt, so wollen wir A „minorendefinit" oder auch kurz „definit" nennen.* Es kann dabei das gemeinsame Vorzeichen sehr wohl von ν abhängen.

Wenn die Matrix A für alle ν bloß nichtnegative Minoren aufweist, so kann man A auch *minorenpositiv* oder auch *total positiv* nennen[3]).

2. Es gelten nun folgende Sätze:

Satz 1. *Wenn die Matrix A der linearen Transformation* (1) *minorendefinit ist, so ist diese Transformation auch variationsvermindernd*[4]).

Satz 2. *Es sei h der Rang von A. Falls $h = n$ und die Transformation* (1) *variationsvermindernd ist, so folgt daraus auch umgekehrt, daß die Koeffizientenmatrix A minorendefinit ist*[5]).

---

[3]) Interessante total positive Matrizen findet man in Pólya-Szegö, a. a. O., Aufgabe 48 und 76. Man kann folgenden Satz leicht beweisen: *Die Folge reeller Zahlen $\mu_0, \mu_1, \mu_2, \ldots$ ist dann und nur dann total monoton (im Sinne der Herren Schur und Hausdorff), wenn die mit den μ_ν geränderte Matrix der Binomialkoeffizienten*

$$\begin{pmatrix} \mu_0 & 1 & 0 & 0 & 0 & \cdots \\ \mu_1 & 1 & 1 & 0 & 0 & \cdots \\ \mu_2 & 1 & 2 & 1 & 0 & \cdots \\ \mu_3 & 1 & 3 & 3 & 1 & \cdots \\ \vdots & \vdots & \vdots & \vdots & \vdots & \end{pmatrix}$$

total positiv ist.

[4]) Herr Fekete beweist a. a. O. diesen Satz für lineare Transformationen mit rekurrenten total positiven Matrizen. Siehe a. a. O. auf S. 91 Satz 1 und Fußnote [6]).

[5]) Es sei $(y) = A(x)$ eine lineare Transformation mit nichtsingulärer quadratischer Matrix (also $m = n$, $|A| \neq 0$). Wir wollen annehmen, es sei immer $v_y \geqq v_x$, dann ist wegen $(x) = A^{-1}(y)$ und Satz 2 auch A^{-1} minorendefinit. Nach einem Hauptsatz über die Minoren der reziproken Matrix folgt für A folgende charakteristische Eigenschaft: *Alle Minoren derselben Ordnung ν $(\nu = 1, 2, \ldots, n-1)$, mit ihren Vorzeichen als algebraische Komplimente versehen, haben dasselbe Vorzeichen (oder Null).* Solche nichtsinguläre quadratische Matrizen könnte man *minoren-algebraisch-definit* oder kurz *algebraisch definit* nennen. Sie besitzen die Eigenschaft $v_y \geqq v_x$ und die entsprechende Transformation ist also variationsvermehrend.

Wir wollen sofort bemerken, daß die Zusatzbedingung $h = n$ für Satz 2 wesentlich ist. Wir betrachten als Beispiel das System

$$\begin{aligned} y_1 &= x_1 + x_2 + x_3 \\ y_2 &= \phantom{x_1 + {}} x_2 . \end{aligned}$$

Offenbar ist immer $v_y \leqq v_x$, aber die Matrix $\begin{pmatrix} 1 & 1 & 1 \\ 0 & 1 & 0 \end{pmatrix}$ ist nicht minorendefinit. Entsprechend ist auch die Zusatzbedingung $h = n$ von Satz 2 nicht erfüllt: $h = 2$ und $n = 3$.

3. Zum Beweis des Satzes 1 liefert uns folgender Hilfssatz eine erste obere Schranke für v_y.

Hilfssatz. *Es sei die Koeffizientenmatrix A des Systems* (1) *minorendefinit und es sei $h\,(> 0)$ ihr Rang.*

Für alle reellen Werte der $x_1, x_2, \ldots, x_n$ ist dann

$$v_y \leqq h - 1 . \tag{2}$$

Der Beweis erfolgt durch vollständige Induktion in bezug auf h. *Für $h = 1$ ist der Hilfssatz richtig,* denn alle Elemente der minorendefiniten Matrix A haben dasselbe Vorzeichen (oder sind gleich Null), und aus $h = 1$ folgt, daß alle Linearformen (15) sich von einer gewissen nicht identisch verschwindenden unter ihnen bloß durch konstante nichtnegative Faktoren unterscheiden.

Es ist also immer $v_y = 0 = h - 1$.

Wir greifen noch einen anderen Spezialfall vorweg und zwar $h = m - 1$.

Jedenfalls gilt $v_y \leqq m - 1$, also $v_y \leqq h$. Es wird also (2) bewiesen sein, wenn wir zeigen, daß $v_y = h$ ausgeschlossen ist. Wir wollen nun $v_y = h = m - 1$ voraussetzen und zu einem Widerspruch gelangen. Die Matrix A besitzt eine nichtverschwindende Unterdeterminante der Ordnung $h = m - 1$. Es seien

$$A_1, A_2, \ldots, A_m$$

die m Unterdeterminanten der Ordnung $m - 1$, welche sich aus den $m - 1$ Spalten jener nichtverschwindenden Unterdeterminante bilden lassen. Dieselbe ist mit einer von diesen $m - 1$ Minoren identisch.

Nach einer Eigenschaft des Ranges gilt identisch in $x_1, x_2, \ldots, x_n$

$$A_1 y_1 - A_2 y_2 + A_3 y_3 - \cdots + (-1)^{m-1} A_m y_m = 0 ,$$

und da für unsere speziellen x die Beziehung $v_y = h = m - 1$ gilt, also die Vorzeichen von $y_1, y_2, \ldots, y_m$ abwechseln, so folgt

$$|A_1| \cdot |y_1| + |A_2| \cdot |y_2| + \cdots + |A_m| \cdot |y_m| = 0 ,$$

weil alle A_i dasselbe Vorzeichen haben (A ist minorendefinit). Diese letzte Gleichung ist aber unmöglich, da kein y_i verschwindet und nicht alle A_i verschwinden.

Wir betrachten jetzt den allgemeinen Fall und wollen annehmen, daß der Satz (2) für kleinere Werte von h schon bewiesen ist.

Angenommen, es gäbe feste $x_1, x_2, \ldots, x_n$ mit $v_y \geqq h$. Wir könnten dann aus den m Linearformen (1) gewiß $h+1$ Linearformen herausgreifen, etwa

$$\begin{aligned} y_\alpha &= a_{\alpha 1} x_1 + \ldots + a_{\alpha n} x_n, \\ y_\beta &= a_{\beta 1} x_1 + \ldots + a_{\beta n} x_n, \\ &\vdots \\ y_\lambda &= a_{\lambda 1} x_1 + \ldots + a_{\lambda n} x_n, \end{aligned}$$

so daß für dieses neue System $v_y^* = h$ wird. Es sei h^* der Rang dieses Teilsystems. Es gibt nun zwei Möglichkeiten und beide ergeben einen Widerspruch.

1. Wenn $h^* < h$, dann ist unser Hilfssatz nach Voraussetzung schon bewiesen, also $v_y^* = h > h^*$ unmöglich, und

2. wenn $h^* = h$, so haben wir es mit dem schon erledigten Spezialfall $h = m - 1$ zu tun.

Damit ist der Hilfssatz bewiesen.

4. Wir können nun zum Beweis des Satzes 1 übergehen. Wir wollen zeigen, daß $v_y \leqq v_x$. Für $v_y = 0$ ist nichts zu beweisen, es sei also $v_y > 0$.

Man kann gewiß aus den Linearformen (1) ein Teilsystem von $m^* = v_y + 1$ Formen herausgreifen, so daß für dieselben $v_y^* = v_y = m^* - 1$ wird. Aus $v_y^* \leqq v_x$ würde also auch $v_y \leqq v_x$ folgen.

Wir können also von vornherein voraussetzen, daß

$$v_y = m - 1$$

ist, und müssen zeigen, daß $v_x \geqq m - 1$ *gilt.*

Nach Hilfssatz (2) ist $m - 1 = v_y \leqq h - 1$, also $h = m$, insbesondere auch $n \geqq m$.

Für $n = m$ *ist unser Satz richtig, und zwar ist* $v_x = m - 1$.

In der Tat, es ist $n = m = h$ und man kann dann (1) nach der Cramerschen Regel auflösen.

Aus den Cramerschen Formeln und den drei Tatsachen:

1. die $y_1, y_2, \ldots, y_m$ haben alternierende Vorzeichen ($v_y = m - 1$),

2. die Matrix $A = (a_{ik})$ ist minorendefinit,

3. es darf kein x_i verschwinden, denn sonst bekäme man ein neues System mit $m - 1$ Spalten, was $h = m$ widerspricht,

folgt, daß auch $x_1, x_2, \ldots, x_n$ alternierende Vorzeichen aufweisen, also $v_x = m - 1$ $(m = n)$.

Es sei nun $n > m$ *und der Satz* $v_x \geqq m - 1$ (*unter der Annahme* $v_y = m - 1$) *schon für kleinere Werte von* n *bewiesen.* Es gibt zwei Möglichkeiten:

1. $x_1, x_2, \ldots, x_n$ haben alternierende Vorzeichen, dann ist $v_x = n - 1 > m - 1 = v_y$, also der Satz richtig,

2. Die x_i haben nicht alternierende Vorzeichen und dann ist entweder ein $x_i = 0$, oder aber $x_i x_{i+1} > 0$. Im ersten Falle, also $x_i = 0$, streichen wir die i-te Spalte und das neue System genügt allen Bedingungen und hat weniger als n Spalten, also ist der Satz nach den vorangehenden richtig. Im Falle $x_i x_{i+1} > 0$ setzen wir

$$a_{k,i} x_i + a_{k,i+1} x_{i+1} = \varepsilon \cdot a'_k \qquad (k = 1, 2, \ldots, m)$$

mit $\varepsilon = \operatorname{sign} x_i$. Es ersetze ε die beiden Größen x_i und x_{i+1}. Das neue derart zusammengezogene System hat $n - 1$ Spalten, eine ebenfalls minorendefinite Matrix, es gilt $v_x^* = v_x$, so daß der Satz nach Voraussetzung richtig ist.

Damit ist Satz 1 bewiesen.

5. Der Beweis von Satz 2 wird durch vollständige Induktion in bezug auf n geführt.

Für $n = 1$ *ist der Satz richtig*, denn es gilt $v_x = 0$ und aus $v_y \leqq v_x$ folgt $v_y = 0$, woraus sich für $x_1 = 1$ ergibt, daß in

$$a_{11}, a_{21}, \ldots, a_{m1}$$

keine Zeichenwechsel vorkommen. Also ist A minorendefinit.

Es sei Satz 2 für $n - 1$ *schon bewiesen. Wir werden zeigen, daß er auch für* n *gilt.*

Wenn wir in den Linearformen

$$\begin{aligned} y_1 &= a_{11} x_1 + a_{12} x_2 + \ldots + a_{1n} x_n \\ y_2 &= a_{21} x_1 + a_{22} x_2 + \ldots + a_{2n} x_n \\ &\vdots \\ y_m &= a_{m1} x_1 + a_{m2} x_2 + \ldots + a_{mn} x_n \end{aligned} \tag{1}$$

irgendein x_i gleich Null setzen, was dem Streichen der entsprechenden Spalte gleichkommt, so entsteht ein neues System mit $n - 1$ Spalten, der Rang desselben ist genau gleich $n - 1$ und auch die Eigenschaft $v_y \leqq v_x$ bleibt erhalten. Nach unserer Voraussetzung sind also die n verschiedenen Matrizen dieser n Teilsysteme schon minorendefinit. Ich behaupte: *daraus folgt für die vollständige Matrix* A, *daß sie minorendefinit ist in bezug auf alle ihre Minoren inklusive bis zur Ordnung* $n - 2$.

In der Tat, es seien Q'_ν und Q''_ν zwei nichtverschwindende Minoren von A von der Ordnung $\nu \leqq n - 2$. Wir wollen zeigen, daß $Q'_\nu \cdot Q''_\nu > 0$ ist.

Falls A eine Spalte besitzt, welche weder Elemente von Q'_ν noch solche von Q''_ν enthält, dann stecken schon Q'_ν und Q''_ν in derselben minorendefiniten Matrix, die durch Streichung dieser Spalte entsteht, und wir sind fertig.

Wenn das nicht gilt, dann sei A' irgendeine solche Teilmatrix mit $n-1$ Spalten, welche Q'_ν enthält, und ebenso A'' eine Teilmatrix, welche Q''_ν enthält. Es haben nun A' und A'' gewiß $n-2$ gemeinsame Spalten. Wir greifen ν feste Spalten aus diesen heraus. In diesen gibt es gewiß einen nichtverschwindenden Minor ν-ter Ordnung, etwa Q^*_ν. Es stecken nun Q'_ν und Q^*_ν in A', also ist $Q'_\nu Q^*_\nu > 0$. Ebenso stecken Q''_ν und Q^*_ν in A'', also gilt $Q''_\nu Q^*_\nu > 0$. Mithin ist tatsächlich $Q'_\nu Q''_\nu > 0$, w. z. b. w.

6. *Wir wollen jetzt zeigen, daß alle nichtverschwindenden Minoren Q_{n-1} der Ordnung $n-1$ von A ebenfalls dasselbe Vorzeichen haben.*

Wir wissen schon, daß alle Q_{n-1} aus $n-1$ festen Spalten dasselbe Vorzeichen haben. Wir wollen nun zeigen, daß dieses Vorzeichen für alle $n = \binom{n}{n-1}$ Spaltenkombinationen dasselbe ist. Aus $h = n$ folgt, daß es einen nichtverschwindenden Q_n gibt. Dieser Minor Q_n hat also die Eigenschaft, in je $n-1$ Spalten einen von Null verschiedenen Q_{n-1} zu besitzen.

Jede Spaltenkombination hat also in Q_n einen nichtverschwindenden Repräsentanten Q_{n-1}, und es genügt, wenn wir unsere Behauptung für $m = n = h$ beweisen.

Zu diesem Ende seien A_{ik} die ersten Minoren der Matrix des Systems

$$\begin{aligned} y_1 &= a_{11} x_1 + \dots + a_{1n} x_n, \\ &\vdots \\ y_n &= a_{n1} x_1 + \dots + a_{nn} x_n. \end{aligned}$$

Es sei etwa

$$A_{ik} \cdot A_{rs} < 0 \qquad (k < s);$$

wir wollen zeigen, daß dies der Annahme $v_y \leqq v_x$ widerspricht ($k = s$ ist schon erledigt!).

Wir wissen, daß alle $A_{\alpha k}$ $(\alpha = 1, 2, \dots, n)$ dasselbe Vorzeichen, etwa $\varepsilon (= \pm 1)$ haben. Ebenso sei η das Vorzeichen der $A_{\alpha s}$, so daß $\varepsilon\eta = -1$.

Wir wollen das System

$$\begin{aligned} 1 &= a_{11} x_1 + \dots + a_{1n} x_n \\ -1 &= a_{21} x_1 + \dots + a_{2n} x_n \qquad D = |a_{ik}| \\ &\vdots \\ (-1)^{n-1} &= a_{n1} x_1 + \dots + a_{nn} x_n \end{aligned}$$

auflösen. Nach Cramer folgt

$$D x_k = (-1)^{k+1} \cdot \sum_{\alpha=1}^{n} A_{\alpha k} = (-1)^{k+1} \cdot \varepsilon \cdot K \qquad (K > 0)$$

und

$$D\,x_s = (-1)^{s+1}\cdot \sum_{\alpha=1}^{n} A_{\alpha s} = (-1)^{s+1}\cdot \eta \cdot K' \qquad (K' > 0),$$

also

$$\operatorname{sign}(x_k \cdot x_s) = -(-1)^{s-k},$$

mithin kann v_x nicht gleich $n-1$ sein, was aus $v_y = n-1$ und $v_y \leqq v_x$ folgen müßte. Damit ist die Behauptung bewiesen.

7. *Wir müssen endlich noch zeigen, daß auch alle nichtverschwindenden Q_n dasselbe Vorzeichen haben.*

Wir machen zunächst die Annahme $m = n+1$. Es gibt dann überhaupt nur $n+1$ Minoren Q_n n-ter Ordnung; wir wollen dieselben mit $A_1, A_2, \ldots, A_{n+1}$ bezeichnen. Aus $h = n$ folgt, daß die y_i bloß durch die Beziehung

$$y_1 A_1 - y_2 A_2 + \cdots + (-1)^n y_{n+1} A_{n+1} = 0$$

verbunden sind und nicht alle A_i verschwinden.

Wenn nicht alle A_i dasselbe Vorzeichen haben, so sei etwa

$$A_r > 0, \quad A_s < 0 \qquad (r < s).$$

Wenn wir $v_y = n$ ansetzen, also z. B. $y_i = (-1)^{i-1} |y_i|$, so folgt die Beziehung

$$|y_1| A_1 + \cdots + |y_r| A_r + \cdots + |y_s| A_s + \cdots + |y_{n+1}| A_{n+1} = 0$$

und dieser kann man tatsächlich mit sämtlich positiven $|y_i|$ genügen Man nehme etwa alle $|y_i|$ $(i \neq r, s)$ fest und positiv. Für sehr großes $|y_r|$ folgt für $|y_s|$ gewiß ein positiver Wert.

Die Möglichkeit $v_y = n$ widerspricht aber der Bedingung $v_y \leqq v_x$, wegen $v_x \leq n-1$. Wir haben also einen Widerspruch und die Behauptung ist bewiesen.

Wir betrachten jetzt den allgemeinen Fall $m > n$.

Es seien Q_n' und Q_n'' zwei von Null verschiedene Minoren. Es sei ν die Anzahl ihrer gemeinsamen Zeilen.

Es ist $0 \leqq \nu \leqq n-1$.

Für $\nu = n-1$ folgt aus dem schon bewiesenen Spezialfall $m = n+1$, daß tatsächlich $Q_n' Q_n'' > 0$.

Es sei diese Behauptung für $\nu + 1$ schon bewiesen. Wir wollen sie für ν beweisen.

Es seien $Z_1, Z_2, \ldots, Z_\nu$ die gemeinsamen ν Zeilen von Q_n' und Q_n'' ($\nu = 0$ nicht ausgeschlossen).

Es seien $Z_{\nu+1}'', Z_{\nu+2}'', \ldots, Z_n''$ die restlichen $n-\nu$ Zeilen von Q_n'' und es sei Z_0 irgendeine von $Z_1, Z_2, \ldots, Z_\nu$ verschiedene Zeile von Q_n'.

Wir betrachten die $n-\nu$ Determinanten, welche aus Q_n'' entstehen, wenn wir die Zeilen $Z_{\nu+1}'', Z_{\nu+2}'', \ldots, Z_n''$ nacheinander durch Z_0 ersetzen. *Ich behaupte, daß mindestens eine von diesen $n-\nu$ Determinanten nicht verschwindet.*

In der Tat, alle gleich Null gesetzt, ergeben für die Elemente von Z_0 offenbar $n-\nu$ lineare homogene Gleichungen. Aus $Q_n'' \neq 0$ folgt, daß diese linear unabhängig sind, also den Rang $n-\nu$ haben. Es sind aber gewiß $Z_1, Z_2, \ldots, Z_\nu$ Lösungen dieses Gleichungssystems und diese ν Lösungen sind als Zeilen von $Q_n''\ (\neq 0)$ linear unabhängig. Aber auch $Z_0, Z_1, \ldots, Z_\nu$ sind linear unabhängig [als Zeilen von $Q_n'(\neq 0)$]. Also ist Z_0 gewiß keine Lösung dieses Systems; denn ν ist die Maximalzahl linear unabhängiger Lösungen.

Also ist tatsächlich mindestens eine jener Determinanten von Null verschieden. Wir greifen eine solche heraus; diese ist bis auf das Vorzeichen ein Minor Q_n^* von A und von Null verschieden. Es hat aber Q_n' mit Q_n^* $\nu+1$ gemeinsame Zeilen (eben $Z_0, Z_1, \ldots, Z_\nu$), also ist $Q_n' Q_n^* > 0$.

Q_n'' und Q_n^* haben sogar $n-1$ gemeinsame Zeilen, also gilt auch $Q_n'' Q_n^* > 0$.

Es folgt also in der Tat $Q_n' Q_n'' > 0$, und Satz 2 ist vollständig bewiesen.

(Eingegangen am 4. Januar 1929.)

Reprinted from
Math. Z. 32 (1930), pp. 321–328

Zur Abzählung der reellen Wurzeln algebraischer Gleichungen.

Von

I. J. Schoenberg in Cambridge, Mass.

Das Problem der Abzählung der reellen Wurzeln einer reellen algebraischen Gleichung ist längst durch den Sturmschen Satz oder auch durch Hermites Methode der quadratischen Formen vollständig gelöst. Vor diesen Lösungen gab es hauptsächlich nur die Regel von Descartes, den Satz von Budan und Fourier und die von Jacobi auf ein endliches Intervall übertragenen Regel von Descartes, welche alle nur obere Schranken für die Wurzelanzahl geben. Aber auch nach der vollständigen Lösung des Problems hat anscheinend das Interesse der Mathematiker an den ungenauen Sätzen nicht nachgelassen. Es seien die Namen von Sylvester und Laguerre und in neuerer Zeit Runge, A. Hurwitz, Fejér, D. R. Curtiss, Fekete, Pólya, Szegö und Obreschkoff genannt.

In dieser Arbeit werden nicht weitere Sätze dieser Art aufgestellt, sondern ein von Felix Klein[1]) begonnener Vergleich der Leistungsfähigkeit dieser Sätze auf einem neuen Weg zu einem gewissen Abschluß gebracht. Felix Klein hat nachgewiesen, daß die von der Descartes-Jacobischen Regel gelieferte obere Schranke für die Wurzelanzahl für Gleichungen zweiten Grades nie die Schranke des Budan-Fourierschen Satzes übersteigt. Heinrich Weber, der in seinem Lehrbuch der Algebra (Bd. 1, 2. Auflage, S. 354—357) den geometrischen Beweis von Klein wiedergibt, wirft die Frage auf, ob derselbe Sachverhalt auch für Gleichungen beliebigen Grades gilt. In § 1 wird die Webersche Frage bejahend beantwortet. In § 2 wird in demselben Sinne die Überlegenheit der Descartes-Jacobischen Regel gegenüber einem für die Anwendungen äußerst bequemen Satz von Laguerre nachgewiesen. § 3 bringt die Hauptergebnisse dieser Arbeit; eine Bestimmung der allgemeinsten Polynomenfolge, welche im Sinne der Herren Pólya und Szegö der Descartesschen Regel genügen, zeigt die Überlegenheit der klassischen Regel von Descartes gegenüber einer ganzen Klasse von ähnlichen Sätzen, von denen gewisse Sätze von Runge und Obreschkoff typische Beispiele sind. Ferner gestatten diese

[1]) F. Klein, Geometrisches zur Abzählung der reellen Wurzeln algebraischer Gleichungen, Gesammelte Math. Abhandlungen 2, S. 198—208.

Ergebnisse und eine von Herrn Obreschkoff stammende Übertragung der Regel von Descartes auf komplexe Wurzeln alle hier besprochenen Sätze ebenfalls auf komplexe Wurzeln zu übertragen (§ 4).

Im Gegensatz zu Klein ist die hier benutzte Beweismethode analytisch. Da es hier auf Ungleichheitsbeziehungen zwischen den Anzahlen der Zeichenwechsel in verschiedenen linear voneinander abhängenden Zahlenfolgen ankommt, spielen naturgemäß die vom Verfasser schon früher untersuchten sogenannten variationsvermindernden Transformationen eine entscheidende Rolle.

§ 1.

Die Regel von Descartes-Jacobi und der Satz von Budan-Fourier.

1. Es sei

(1) $$f(x) \equiv a_0 x^n + a_1 x^{n-1} + \ldots + a_n = 0 \qquad (a_0 \neq 0)$$

eine algebraische Gleichung mit reellen numerischen Koeffizienten, und $N(\alpha, \beta)$ sei die Anzahl ihrer Nullstellen im offenen Intervall (α, β). Ferner sei $V(x)$ die Anzahl der Zeichenwechsel in der Folge

$$f(x), \quad f'(x), \quad f''(x), \quad \ldots, \quad f^{(n)}(x) \quad (-\infty < x < +\infty).$$

Nach Budan und Fourier gilt der folgende Satz:

I. *Es ist*

(2) $$N(\alpha, \beta) \leqq V(\alpha) - V(\beta)$$

und es ist, falls $f(\alpha) f(\beta) \neq 0$ *ist, die Differenz beider Seiten eine gerade Zahl*[2]).

Für $\alpha = 0$, $\beta \to +\infty$ folgt aus diesem Satz, als Spezialfall, die sogenannte Regel von Descartes[3])

(3) $$N(0, \infty) \leqq V(0) = v_a, \quad N(0, \infty) \equiv v_a \quad (\mathrm{mod}\ 2).$$

---

[2]) Ist $x = \beta$ eine k-fache Wurzel der Gleichung (1) ($k \geqq 0$), so ist

$$V(\alpha + 0) = V(\alpha), \quad V(\beta - 0) = V(\beta) + k + 2p \quad (p \geqq 0 \text{ und ganzzahlig}).$$

Aus Satz I folgt

$$N(\alpha, \beta) = \lim_{\varepsilon \to 0} N(\alpha + \varepsilon, \beta - \varepsilon) = \lim_{\varepsilon \to 0} [V(\alpha + \varepsilon) - V(\beta - \varepsilon) - 2q]$$
$$= V(\alpha) - V(\beta) - k - 2(p + q) \quad (q \geqq 0 \text{ und ganzzahlig}),$$

also

$$N(\alpha, \beta] = N(\alpha, \beta) + k = V(\alpha) - V(\beta) - 2r \quad (r \geqq 0 \text{ und ganzzahlig}).$$

Diese Verschärfung des Budan-Fourierschen Satzes stammt von A. Hurwitz, Math. Annalen 71 (1912), S. 584—591.

[3]) Hier und weiter unten wird folgende bequeme Bezeichnung benutzt: Ist $z_0, z_1, z_2, \ldots$ eine Folge reeller Zahlen, so sei v_z die Anzahl ihrer Zeichenwechsel.

Jacobi hat bemerkt, daß man mit der Descartesschen Regel ebenfalls die Aufgabe, $N(\alpha, \beta)$ abzuschätzen, lösen kann[4]). Er setzt $x = \frac{\alpha + \beta y}{1 + y}$. Dem Intervall $\alpha < x < \beta$ entspricht die positive Halbachse $0 < y < +\infty$, und es genügt die Anzahl der positiven Wurzeln der Gleichung

$$(4) \qquad (1+y)^n f\left(\frac{\alpha + \beta y}{1+y}\right) \equiv b_0 y^n + b_1 y^{n-1} + \dots + b_n = 0$$

zu ermitteln. Tatsächlich ist diese Anzahl ebenfalls gleich $N(\alpha, \beta)$ und die Regel von Descartes gibt die Ungleichung[3])

$$(5) \qquad N(\alpha, \beta) \leqq v_b, \quad N(\alpha, \beta) \equiv v_b \pmod 2.$$

Wir setzen die Descartessche Regel, also die Beziehungen (3) und (5), als bekannt voraus. Dann folgt sofort ein Beweis von Satz I, sowie die Beantwortung der in der Einleitung erwähnten Klein-Weberschen Frage aus folgendem

Hilfssatz 1. *Es ist immer*

$$(6) \qquad v_b \leqq V(\alpha) - V(\beta),$$

und es ist, falls $f(\alpha) f(\beta) \neq 0$ ist, die Differenz beider Seiten eine gerade Zahl.

Zum Beweis dieses Hilfssatzes setzen wir zunächst

$$(7) \qquad x = \alpha + (\beta - \alpha) z, \quad f(x) = f(\alpha + (\beta - \alpha) z) = f^*(z).$$

$N(\alpha, \beta)$ ist gleich der Anzahl $N^*(0,1)$ der Wurzeln von $f^*(z) = 0$ im Intervall $0 < z < 1$. Ferner folgt aus (4) und (7) sofort

$$V(\alpha) = V^*(0), \quad V(\beta) = V^*(1), \quad v_b = v_{b^*}, \quad f(\alpha) = f^*(0), \quad f(\beta) = f^*(1),$$

und deshalb genügt es, unseren Hilfssatz für $\alpha = 0$, $\beta = 1$ zu beweisen. Aus (1) folgt in unserer Bezeichnungsweise[3])

$$(8) \qquad V(0) = v_a, \quad V(1) = v_{b'},$$

wobei

$$(9) \qquad \begin{aligned} b_0' &= a_0, \\ b_1' &= \binom{n}{1} a_0 + a_1, \\ b_2' &= \binom{n}{2} a_0 + \binom{n-1}{1} a_1 + a_2, \\ &\vdots \\ b_{n-1}' &= \binom{n}{1} a_0 + \binom{n-1}{1} a_1 + \binom{n-2}{1} a_2 + \dots + a_{n-1}, \\ b_n' &= a_0 + a_1 + a_2 + \dots + a_{n-1} + a_n. \end{aligned}$$

---

[4]) C. G. J. Jacobi, Observatiunculae ad theoriam aequationum pertinentes, Crelles Journ. 13 (1835), S. 348.

Wir wollen nun das Polynom (4) ausrechnen. Aus

$$(1+y)^n f\left(\frac{y}{1+y}\right) = a_0 y^n + a_1 y^{n-1}(1+y) + a_2 y^{n-2}(1+y)^2 + \dots + a_n(1+y)^n$$
$$= b_0 y^n + b_1 y^{n-1} + \dots + b_n$$

folgt

$$(10)\qquad \begin{aligned} b_0 &= a_0 + a_1 + a_2 + \dots + a_n, \\ b_1 &= \quad a_1 + 2a_2 + \dots + \binom{n}{1} a_n, \\ b_2 &= \quad\quad a_2 + \dots + \binom{n}{2} a_n, \\ &\vdots \\ b_n &= \quad\quad\quad a_n. \end{aligned}$$

Aus einem sofort ersichtlichen Grund schreiben wir (9) und (10) als eine einzige lineare Transformation

$$(11)\qquad \begin{aligned} b'_0 &= a_0, \\ b'_1 &= \binom{n}{1} a_0 + a_1, \\ &\vdots \\ b'_n &= a_0 + a_1 + \dots + a_n, \\ b_0 &= a_0 + a_1 + \dots + a_n, \\ b_1 &= \quad a_1 + \dots + \binom{n}{1} a_n, \\ &\vdots \\ b_n &= \quad\quad a_n. \end{aligned}$$

Es sei $v_{b',b}$ die Zeichenwechselanzahl der Folge $b'_0, \dots, b'_n, b_0, \dots, b_n$. Die Transformation (11) ist *variationsvermindernd*, d. h. es gilt immer die Ungleichung

$$(12)\qquad v_{b',b} \leqq v_a.$$

Der Beweis dieser Behauptung wird später nachgeholt. Man bemerke ferner, daß

$$(13)\qquad v_{b'} + v_b \leqq v_{b',b},$$

und zwar gilt hier das Gleichheitszeichen, wenn $b'_n = b_0 \neq 0$, während die rechte Seite eventuell (um eine Einheit) überwiegt, falls $b'_n = b_0 = 0$. Aus (12) und (13) folgt aber

$$v_{b'} + v_b \leqq v_a$$

und das ist genau die Ungleichung (6) (für $\alpha = 0$, $\beta = 1$), wenn man (8) berücksichtigt. Wenn ferner $f(1) = b'_n = b_0 \neq 0$ und $f(0) = a_n \neq 0$, so ist wegen der besonderen Gestalt von (11) die Differenz $v_a - v_{b',b}$ gerade, also wegen $v_{b',b} = v_{b'} + v_b$, auch $v_a - (v_{b'} + v_b) = V(0) - V(1) - v_b$ eine gerade Zahl.

36*

2. Jetzt werde der Nachweis der variationsvermindernden Eigenschaft der Transformation (11) nachgeholt. Nach einem allgemeinen Satz[5]) genügt es, nachzuweisen, daß alle Minoren sämtlicher Ordnungen der Koeffizientenmatrix von (11) nichtnegativ sind. Matrizen mit dieser Eigenschaft hatte ich a. a. O. *total positiv* genannt.

Wir schicken zunächst einige allgemeine Bemerkungen über total positive Matrizen voraus. Es sei $C = \| c_{ik} \|$ $(i = 1, \ldots, m;\ k = 1, \ldots, n)$ eine total positive Matrix. Man kann an C einige einfache Veränderungen vornehmen, wir wollen sie *elementare Transformationen* nennen, welche C in eine ebenfalls total positive Matrix überführen. Elementare Transformationen sind offenbar: 1. Multiplikation einer Zeile (Spalte) von C mit einer positiven Zahl K. 2. Addition einer mit K (> 0) multiplizierten Zeile (Spalte) zu einer *unmittelbar benachbarten* Zeile (Spalte). 3. Einschiebung einer neuen Zeile (Spalte) die entweder aus lauter Nullen besteht, oder mit einer der unmittelbar benachbarten Zeilen (Spalten) identisch ist. 4. Streichung einer Zeile (Spalte). 5. Klappung der Matrix um eine der beiden Diagonalrichtungen. 6. Erweiterungen von C von der Art

$$\begin{Vmatrix} K & 0 \\ 0 & c_{ik} \end{Vmatrix} \quad \text{oder} \quad \begin{Vmatrix} c_{ik} & 0 \\ 0 & K \end{Vmatrix} \quad (K > 0).$$

Nach diesen Bemerkungen ist der Beweis der Ungleichung (12) in folgendem Hilfssatz enthalten.

Hilfssatz 2. *Die Matrix*

$$(14)\quad M_n = \begin{Vmatrix}
1 & \binom{n}{1} & \cdots & \binom{n}{2} & \binom{n}{1} & 1 & 1 & 0 & \cdots & 0 & 0 \\
0 & 1 & & \binom{n-1}{2} & \binom{n-1}{1} & 1 & 1 & 1 & \cdots & 0 & 0 \\
\vdots & \vdots & & \vdots & \vdots & \vdots & \vdots & \vdots & & \vdots & \vdots \\
0 & 0 & \cdots & 1 & 2 & 1 & 1 & \binom{n-2}{1} & \cdots & 0 & 0 \\
0 & 0 & \cdots & 0 & 1 & 1 & 1 & \binom{n-1}{1} & \cdots & 1 & 0 \\
0 & 0 & \cdots & 0 & 0 & 1 & 1 & \binom{n}{1} & \cdots & \binom{n}{1} & 1
\end{Vmatrix}$$

ist total positiv.

[5]) Siehe I. Schoenberg, Über variationsvermindernde lineare Transformationen, Math. Zeitschr. 32 (1930), S. 321—328. Hier wird nur der folgende Spezialfall des dort bewiesenen Satzes 1 angewendet: *Wenn alle Minoren sämtlicher Ordnungen der Koeffizientenmatrix* $C = \| c_{ik} \|$ *einer linearen Transformation* $y_i = \sum_{k=1}^{n} c_{ik} x_k$ $(i = 1, \ldots, m)$ *nichtnegativ sind, dann ist die Transformation variationsvermindernd, d. h. es ist immer* $v_y \leqq v_x$. In der demnächst erscheinenden Baseler Dissertation des Herrn Th. Motzkin findet man u. a. notwendige und hinreichende Bedingungen für variationsvermindernde Transformationen.

Für

$$M_0 = \| 1 \quad 1 \|,$$

stimmt die Behauptung. Bevor wir den Induktionsschluß ausführen, wollen wir zeigen, *daß man M_n durch elementare Transformationen* (*der 1. Art*) *in*

$$(15) \quad M'_n = \begin{Vmatrix} 1 & 1 & 1 \dots & 1 & 1 & 1 & 0 & \dots & 0 & 0 \\ 0 & 1 & 2 & \binom{n-1}{1} & \binom{n}{1} & \binom{n}{1} & 1 & \dots & 0 & 0 \\ 0 & 0 & 1 & \binom{n-1}{2} & \binom{n}{2} & \binom{n}{2} & \binom{n-1}{1} & \dots & 0 & 0 \\ \vdots & \vdots & \vdots & \vdots & \vdots & \vdots & \vdots & & \vdots & \vdots \\ 0 & 0 & 0 \dots & 1 & \binom{n}{1} & \binom{n}{1} & \binom{n-1}{1} & \dots & 1 & 0 \\ 0 & 0 & 0 \dots & 0 & 1 & 1 & 1 & \dots & 1 & 1 \end{Vmatrix}$$

überführen kann und umgekehrt. Mit $\binom{0}{0} = 1$ und $\binom{p}{q} = 0$ für $0 \leqq p < q$ ist in leicht verständlicher Schreibweise

$$(16) \qquad M_n = \left\| \binom{n-i}{n-k} \; \binom{i}{l} \right\| \qquad (i, k, l = 0, 1, \dots, n),$$

wobei i die $(i+1)$te Zeile, k und l die $(k+1)$te bezüglich $(n+1+l+1)$te Spalte angeben. Alle weiter unten auftretenden sinnlosen Symbole, wie etwa $\frac{1!}{(-1)!}$ sind durch die Null zu ersetzen. Nach (16) ist dann

$$M_n = \left\| \frac{(n-i)!}{(n-k)!\,(k-i)!} \quad \frac{i!}{l!\,(i-l)!} \right\|.$$

Man multipliziere nun die $(k+1)$te Spalte mit $(n-k)!\,k!$ und die $(n+1+l+1)$te Spalte mit $(n-l)!\,l!$ $(k, l = 0, 1, \dots, n)$. Die neue Matrix ist

$$\left\| \frac{(n-i)!\,k!}{(k-i)!} \quad \frac{i!\,(n-l)!}{(i-l)!} \right\|.$$

Dividiert man nun die $(i+1)$te Zeile durch $(n-i)!\,i!$ $(i = 0, \dots, n)$, so erhält man

$$\left\| \frac{k!}{i!\,(k-i)!} \quad \frac{(n-l)!}{(n-i)!\,(i-l)!} \right\| = \left\| \binom{k}{i} \; \binom{n-l}{n-i} \right\| = M'_n,$$

womit die Behauptung bewiesen ist.

Wir beweisen ferner: *M_n läßt sich durch elementare Transformationen in M_{n+1} überführen.* Es genügt natürlich von M'_n auszugehen[6]).

---

[6]) Übersichtlicher wird der Leser die nun folgenden Ausführungen an Hand der für $n = 2$, oder $n = 3$, ausgeschriebenen Matrizen verfolgen.

Man gehe von M'_n zur Matrix

$$(17)\qquad \begin{Vmatrix} 1 & 0 \\ 0 & M'_n \end{Vmatrix} = \begin{Vmatrix} 1 & 0 & 0 \\ 0 & \binom{k}{i} & \binom{n-l}{n-i} \end{Vmatrix}$$

über. Nun addiere man in der Matrix (17) die erste Spalte zur zweiten, ferner die neue zweite Spalte zur dritten usw., die neue $(n+1)$te Spalte zur $(n+2)$ten. Zuletzt erhält man wegen bekannter Additionseigenschaften der Binomialkoeffizienten die Matrix

$$(18)\qquad \begin{Vmatrix} 1 & 1 & 0 \\ 0 & \binom{k+1}{i+1} & \binom{n-l}{n-i} \end{Vmatrix}.$$

Man verdoppele nun in (18) die $(n+2)$te Spalte und erhält die Matrix

$$\begin{Vmatrix} 1 & 1 & 1 & 0 \\ 0 & \binom{k+1}{i+1} & \binom{n+1}{i+1} & \binom{n-l}{n-i} \end{Vmatrix} = \begin{Vmatrix} \binom{k'}{i'} & \binom{n+1-l'}{n+1-i'} \end{Vmatrix} = M'_{n+1}$$

$$(i', k', l' = 0, 1, \ldots, n+1).$$

Da nun M'_{n+1} wie oben in M_{n+1} transformiert werden kann, so ist die Behauptung bewiesen. Da wir durchgehend nur elementare Transformationen benutzt haben, so ist der Induktionsschluß vollständig und der Hilfssatz 2 bewiesen.

§ 2.

Ein Satz von Laguerre.

3. Nach Laguerre findet man eine obere Schranke für $N(\alpha, \beta)$ in folgender Weise. Man betrachte die $n+1$-Abschnitte des Ausdruckes

$$f(\beta) = f(\alpha) + \frac{\beta-\alpha}{1!} f'(\alpha) + \cdots + \frac{(\beta-\alpha)^n}{n!} f^{(n)}(\alpha),$$

also

$$(19)\qquad c_0 = f(\alpha),\quad c_1 = f(\alpha) + \frac{\beta-\alpha}{1!} f'(\alpha),\ \ldots,\ c_n = f(\beta).$$

Es gilt der folgende Satz

II[7]). *Es ist*

$$(20)\qquad N(\alpha, \beta) \leqq v_c,$$

und es ist, falls $f(\alpha) f(\beta) \neq 0$ ist, die Differenz beider Seiten eine gerade Zahl.

Ein Beweis dieses Satzes, sowie der Nachweis der Überlegenheit der Descartes-Jacobischen Regel (5) ist enthalten in folgendem

Hilfssatz 3. *Es ist immer*[8])

$$(21)\qquad v_b \leqq v_c,$$

[7]) E. Laguerre, Ouvres 1, S. 10. Auch sein Satz auf S. 77—78 ist im wesentlichen vom ersten nicht verschieden und fällt für $\alpha = 0$, $\beta = 1$ tatsächlich mit diesem zusammen.

und es ist, falls $f(\alpha) f(\beta) \neq 0$ ist, die Differenz beider Seiten eine gerade Zahl.

Zum Nachweis dieses Hilfssatzes bemerke man zunächst, daß (7) und (19)

$$v_b = v_{b^*}, \quad v_c = v_{c^*}, \quad f(\alpha) f(\beta) = f^*(0) f^*(1)$$

ergeben, so daß es wiederum genügt, den Fall $\alpha = 0$, $\beta = 1$ zu betrachten. Für $\alpha = 0$, $\beta = 1$ wird in umgekehrter Reihenfolge

$$\begin{aligned} c_0 &= a_0 + a_1 + a_2 + \dots + a_n, \\ c_1 &= \phantom{a_0 + {}} a_1 + a_2 + \dots + a_n, \\ c_2 &= \phantom{a_0 + a_1 + {}} a_2 + \dots + a_n, \\ &\vdots \\ c_n &= \phantom{a_0 + a_1 + a_2 + \dots + {}} a_n. \end{aligned} \tag{22}$$

Die Elimination der a_ν aus (10) und (22) ergibt sofort

$$\begin{aligned} b_0 &= c_0 \\ b_1 &= c_1 + c_2 + c_3 + \dots + c_n, \\ b_2 &= c_2 + 2c_3 + \dots + \binom{n-1}{1} c_n, \\ b_3 &= c_3 + \dots + \binom{n-1}{2} c_n, \\ &\vdots \\ b_n &= c_n. \end{aligned} \tag{23}$$

Diese Transformation ist aber variationsvermindernd [8]) und damit ist die Ungleichung (21) bewiesen. Falls ferner $f(0) = c_n \neq 0$, $f(1) = c_0 \neq 0$ ist, ist wegen (23) $v_c - v_b$ gerade, und also auch $v_c - N(0,1)$ eine gerade Zahl.

---

[8]) Tatsächlich ist ihre Koeffizientenmatrix total positiv, denn dieselbe ist eine einfache Erweiterung (elementare Transformation 6. Art) einer Teilmatrix der total positiven Matrix (15). Allerdings sei dazu folgendes bemerkt: Wenn C die Matrix einer variationsvermindernden Transformation $(y) = C(x)$ ist, so kann man an C die elementaren Transformationen 1, 2, 3 und 4 anwenden, ohne daß dabei C aufhört, Koeffizientenmatrix einer variationsvermindernden Transformation zu sein. Vgl. Pólya und Szegö, Aufgaben und Lehrsätze aus der Analysis, Berlin 1925, Bd. 2, 5. Abschnitt. Die Matrix von (23) läßt sich aber sehr einfach durch elementare Transformationen der 2. Art aus der Einheitsmatrix ableiten. Also ist (23) variationsvermindernd und der Umweg über den Satz der Fußnote [5]) ist vermieden. Diesen bequemen Beweis kann ich aber nicht auf die Transformation (11) anwenden. Um vielmehr die Matrix von (11) auf ähnliche Weise aus einfachen Matrizen abzuleiten (siehe Beweis des Hilfssatzes 2), konnte ich die Anwendung der elementaren Transformation 6 nicht vermeiden. Die Transformation 6 auf eine variationsvermindernde Matrix C angewandt, wird aber im allgemeinen keine variationsvermindernde Matrix ergeben.

4. Es sei hier noch auf eine andere Überlegenheit der Regel von Descartes-Jacobi gegenüber den Sätzen von Budan-Fourier und Laguerre hingewiesen. Laguerre (a. a. O., S. 81) schlug folgende Aufgabe vor:

Man bestimme eine obere und untere Schranke von $f(x)$ *in* (0,1), und er löste dieselbe auf folgende Weise.

Es ist gewiß λ z. B. eine obere Schranke, wenn die Gleichung

$$f^*(x) \equiv f(x) - \lambda' = 0$$

für jedes $\lambda' \geqq \lambda$ keine Wurzel zwischen 0 und 1 hat. Es ist

$$f^*(x) \equiv a_0 x^n + a_1 x^{n-1} + \dots + a_n - \lambda'$$

und, um den Laguerreschen Satz II anzuwenden, nach (22)

$$\begin{aligned} c_0^* &= a_0 + a_1 + \dots + a_n - \lambda' = c_0 - \lambda', \\ c_1^* &= \qquad a_1 + \dots + a_n - \lambda' = c_1 - \lambda', \\ &\vdots \\ c_n^* &= \qquad\qquad\qquad a_n - \lambda' = c_n - \lambda'. \end{aligned}$$

Es sind also

$$l = \operatorname*{Min}_{\nu} c_\nu, \quad L = \operatorname*{Max}_{\nu} c_\nu$$

brauchbare untere und obere Schranken, denn $v_{c^*} = 0$ für $\lambda' \leqq l$ oder $\lambda' \geqq L$. Laguerre weist a. a. O. darauf hin, daß der Satz von Budan-Fourier für diese Aufgabe im allgemeinen keine Lösung liefert, was sehr leicht einzusehen ist. *Die Regel von Descartes-Jacobi ist zur Lösung dieser Aufgabe ebensogut brauchbar wie der Satz von Laguerre und gibt sogar bessere Schranken als dieser.*

In der Tat, nach (10) ist

$$\begin{aligned} b_0^* &= a_0 + a_1 + a_2 + \dots + a_n - \lambda' = b_0 - \lambda', \\ b_1^* &= \qquad a_1 + 2a_2 + \dots + \binom{n}{1} a_n - \binom{n}{1}\lambda' = b_1 - \binom{n}{1}\lambda', \\ b_2^* &= \qquad\qquad a_2 + \dots + \binom{n}{2} a_n - \binom{n}{2}\lambda' = b_2 - \binom{n}{2}\lambda', \\ &\vdots \\ b_n^* &= \qquad\qquad\qquad a_n - \lambda' = b_n - \lambda', \end{aligned}$$

und man erhält aus demselben Grund wie oben für $f(x)$ in (0,1) die Schranken

$$j = \operatorname*{Min}_{\nu} \frac{b_\nu}{\binom{n}{\nu}}, \quad J = \operatorname*{Max}_{\nu} \frac{b_\nu}{\binom{n}{\nu}}.$$

Es ist immer $l \leqq j \leqq J \leqq L$, *also sind die Jacobischen Schranken enger als die Laguerreschen,* denn wegen (23) ist

$$b_0 = c_0, \frac{b_1}{\binom{n}{1}} = \frac{c_1 + c_2 + \dots + c_n}{n}, \quad \frac{b_2}{\binom{n}{2}} = \frac{c_2 + \binom{2}{1} c_3 + \dots + \binom{n-1}{1} c_n}{1 + \binom{2}{1} + \dots + \binom{n-1}{1}}, \dots, b_n = c_n.$$

Es sind also die Jacobischen Schranken gleich dem kleinsten und größten von gewissen Mittelwerten von Zahlen, von denen die kleinste und größte die Laguerreschen Schranken sind, und damit sind die Ungleichungen bewiesen.

§ 3.

Die Bestimmung der allgemeinsten der Descartesschen Regel genügenden Polynomenfolgen und die Vorzugsstellung der klassischen Regel von Descartes.

5. Es seien

(24) $$\varphi_0(x),\quad \varphi_1(x),\quad \ldots,\quad \varphi_n(x)$$

im Intervall $\alpha < x < \beta$ n-mal stetig differentierbare Funktionen. Die Herren Pólya und Szegö sagen, daß diese Funktionenfolge im offenen Intervall (α, β) der Descartesschen Regel genügt, wenn folgende Bedingung erfüllt ist: Wenn $c_0, c_1, \ldots, c_n$ irgendwelche reelle Zahlen bedeuten, die nicht sämtlich verschwinden, so soll die Anzahl $N(\alpha, \beta)$ der in (α, β) gelegenen Nullstellen (Vielfachheit inbegriffen) der linearen Kombination

(25) $$f(x) = c_0\,\varphi_0(x) + c_1\,\varphi_1(x) + \ldots + c_n\,\varphi_n(x)$$

die Anzahl v_c der Zeichenwechsel der Folge $c_0, c_1, \ldots, c_n$ nie übersteigen. Notwendige und hinreichende Bedingungen gibt das folgende wichtige Kriterium.

Kriterium von Pólya und Szegö. *Die Funktionenfolge* (24) *genügt dann und nur dann im Intervall* (α, β) *der Regel von Descartes, falls sie die folgenden Eigenschaften hat: Wenn* $i_0, i_1, \ldots, i_l$ *ganze Zahlen bezeichnen, wobei* $0 \leqq i_0 < i_1 < \ldots < i_l \leqq n$ *ist, dann sollen die Wronskischen Determinanten* $W[\varphi_{i_0}(x), \varphi_{i_1}(x), \ldots, \varphi_{i_l}(x)]$ *für* $\alpha < x < \beta$ *nicht verschwinden, und darüber hinaus sollen irgend zwei Wronskische Determinanten mit gleichviel Zeilen gleiches Vorzeichen besitzen*[9]).

Außer dem klassischen Beispiel der Folge $1, x, x^2, \ldots, x^n$ im Intervall $(0, \infty)$ gibt es noch eine beträchtliche Anzahl interessanter Funktionenfolgen, welche in gewissen Intervallen der Regel von Descartes genügen[10]). Ich beschränke mich auf Polynomenfolgen und zitiere die folgenden zwei Sätze von Runge und Obreschkoff.

---

[9]) Pólya und Szegö, a. a. O., Bd. 2. S. 52–53. Für die Anwendungen dieses Kriteriums ist die folgende Bemerkung manchmal nützlich: *Es genügt, die Determinantenbedingungen des Kriteriums nur für konsekutive Zahlen* $i_0 = i$, $i_1 = i+1$, $\ldots$, $i_l = i+l$ *nachzuweisen.* In der Tat, wenn die Determinanten $W[\varphi_i(x), \varphi_{i+1}(x), \ldots, \varphi_{i+l}(x)]$ $(0 \leqq i \leqq i+l \leqq n;\ \alpha < x < \beta)$ ein nur von l abhängendes Vorzeichen haben, so besitzen auch schon alle Determinanten $W[\varphi_{i_0}(x), \varphi_{i_1}(x), \ldots, \varphi_{i_l}(x)]$ dasselbe Vorzeichen. Dies folgt leicht durch Induktion in bezug auf l auf Grund des weiter unten angegebenen Hilfssatzes von Herrn Fekete.

[10]) Siehe Pólya und Szegö, a. a. O. S. 52.

III[11]). *Die Polynomenfolge*

(26) $1,\ x-\xi_1,\ (x-\xi_1)(x-\xi_2),\ \ldots,\ (x-\xi_1)(x-\xi_2)\ldots(x-\xi_n)\quad (\xi_\nu \text{ reell})$

genügt im Intervall $\mu < x < \infty$ ($\mu = \operatorname{Max} \xi_\nu$) *der Regel von Descartes.*

IV[12]). *Es sei* $p(x)$ *eine im endlichen oder unendlichen Intervall* $x_0 \leqq x \leqq x_1$ *positive Funktion, deren Momente* $\int_{x_0}^{x_1} p(x)\,x^r\,dx$ $(r = 0, 1, \ldots, 2n-1)$ *existieren und*

(27) $$P_0(x),\ P_1(x),\ \ldots,\ P_n(x)$$

seien Polynome, welche den Orthogonalitäts- und Normierungsbedingungen

(28) $$\int_{x_0}^{x_1} p(x)\,P_r(x)\,P_s(x)\,dx = 0 \quad (0 \leqq r < s \leqq n),\quad \lim_{x\to\infty} \frac{P_n(x)}{x^n} > 0$$

genügen. Es sei η_n *(bekanntlich* $< x_1$*) die größte Nullstelle von* $P_n(x)$. *Die Polynomenfolge* (27) *genügt im Intervall* (η_n, ∞) *der Regel von Descartes.*

Zur Aufklärung der Beziehung dieser und ähnlicher Sätze zur klassischen Regel von Descartes ist der folgende Satz von besonderem Interesse.

Hauptsatz. *Eine Polynomenfolge der Gestalt*

(29) $$\begin{aligned} \varphi_0(x) &= a_{00}, \\ \varphi_1(x) &= a_{10} + a_{11}x, \\ \varphi_2(x) &= a_{20} + a_{21}x + a_{22}x^2, \\ &\vdots \\ \varphi_n(x) &= a_{n0} + a_{n1}x + a_{n2}x^2 + \ldots + a_{nn}x^n \quad (a_{00} > 0,\ a_{11} > 0,\ \ldots,\ a_{nn} > 0) \end{aligned}$$

genügt dann und nur dann im Intervall $(0, \infty)$ *der Regel von Descartes, wenn die Koeffizientenmatrix*

(30) $$A = \| a_{ij} \| \quad (i, j = 0, 1, 2, \ldots, n;\ a_{ij} = 0 \text{ für } i < j)$$

total positiv ist.

Daß die Bedingung hinreicht, ist sehr leicht folgendermaßen einzusehen. Es sei

(31) $$f(x) = \sum_{i=0}^{n} c_i\,\varphi_i(x) = b_0 + b_1x + b_2x^2 + \ldots + b_nx^n,$$

---

11) Von Runge, siehe Pólya und Szegö, a. a. O. S. 43—44.

12) N. Obreschkoff, Über die Wurzeln algebraischer Gleichungen, Jahresbericht der Deutschen Math.-Vereinigung **33** (1924), S. 52—64. Siehe auch M. Marden, A rule of signs involving certain orthogonal polynomials, Annals of Math. (2) **33** (1931), S. 118—124.

also

$$(32)\qquad \begin{aligned} b_0 &= a_{00}c_0 + a_{10}c_1 + \dots + a_{n0}c_n, \\ b_1 &= \qquad\quad a_{11}c_1 + \dots + a_{n1}c_n, \\ &\vdots \qquad\qquad\qquad \ddots \qquad \vdots \\ b_n &= \qquad\qquad\qquad\qquad\; a_{nn}c_n. \end{aligned}$$

Wenn man die gewöhnliche Regel von Descartes auf (31) anwendet, so folgt die Ungleichung

$$(33)\qquad N(0, \infty) \leqq v_b.$$

Wenn ferner die Matrix (30) total positiv ist, so ist nach dem Satz der Fußnote [5]) die Transformation (32) variationsvermindernd, also

$$(34)\qquad v_b \leqq v_c.$$

Aus (33) und (34) folgt aber

$$(35)\qquad N(0, \infty) \leqq v_c,$$

und damit ist gezeigt, daß die Polynomenfolge (29) in $(0, \infty)$ der Regel von Descartes genügt[13]).

Wir schieben den etwas schwierigeren Beweis der zweiten Hälfte des Hauptsatzes auf, um zunächst eine interessante Folgerung hervorzuheben. Aus dem Hauptsatz und der Ungleichung (34) fließt sofort der folgende

Zusatz. *Die klassische Regel von Descartes gibt eine bessere obere Schranke für die Nullstellenanzahl (im Sinne* $\leqq$*), als irgendeine ähnliche, mit Hilfe von Polynomen der Gestalt* (29) *gebildete Regel.*

Wenn also $f(x)$ ein Polynom n-ten Grades ist, so ist zwar nach Herrn Obreschkoff die Anzahl v_c der Zeichenwechsel in der Folge der Fourierkoeffizienten

$$(36)\qquad c_i = \int_{x_0}^{x_1} p(x)\, f(x)\, P_i(x)\, dx \qquad (i = 0, 1, \dots, n)$$

eine obere Schranke von $N(\eta_n, \infty)$, aber die Anzahl $V(\eta_n)$ der Zeichenwechsel in der Folge der Taylorkoeffizienten

$$f(\eta_n), \quad f'(\eta_n), \quad \dots, \quad f^{(n)}(\eta_n)$$

ist ebenfalls eine solche obere Schranke und sogar eine bessere, denn nach (34) ist

$$V(\eta_n) \leqq v_c.$$

6. Der Beweis der zweiten Hälfte des Hauptsatzes stützt sich auf Hilfssatz 4. Zum Beweis desselben benutzen wir den folgenden

---

[13]) Diese erste Hälfte des Hauptsatzes beweist man leicht auch mit Hilfe des allgemeinen Kriteriums von Pólya und Szegö, doch ziehen wir den obigen Beweis vor, weil er gleichzeitig die an sich wichtige Ungleichung (34) ergibt (siehe den Zusatz).

Hilfssatz von Fekete[14]). *In der Matrix*

$$\left\| \begin{matrix} \gamma_{01} & \gamma_{02} & \cdots & \gamma_{0n}, \\ \gamma_{11} & \gamma_{12} & \cdots & \gamma_{1n}, \\ \gamma_{21} & \gamma_{22} & \cdots & \gamma_{2n}, \\ \vdots & \vdots & & \vdots \\ \gamma_{m1} & \gamma_{m2} & \cdots & \gamma_{mn} \end{matrix} \right\| \qquad (n > m+1)$$

seien diejenigen Minoren $(m+1)$*ter Ordnung, die man aus konsekutiven Spalten bilden kann, positiv. Ferner seien, wenn man die erste Zeile streicht, alle Minoren* m*ter Ordnung positiv. Dann sind auch alle Minoren* $(m+1)$*ter Ordnung positiv.*

Dieser bemerkenswerte Hilfssatz ist zur Untersuchung total positiver Matrizen sehr geeignet. Zum Beispiel folgt daraus leicht, durch wiederholte Induktionsschlüsse, der folgende Satz: *Wenn in einer Matrix C diejenigen Minoren (aller möglichen Ordnungen), die man aus konsekutiven Zeilen und Spalten bilden kann, positiv sind, so sind überhaupt sämtliche Minoren von C positiv, also insbesondere C total positiv.* Im Fall einer quadratischen Matrix, in der alle Elemente oberhalb der Hauptdiagonale verschwinden — wir wollen eine solche Matrix kurz eine *Dreiecksmatrix* nennen —, kann man auf Totalpositivität schon aus Bedingungen in viel geringerer Anzahl schließen. Diese vereinfachten hinreichenden Bedingungen werden im folgenden Hilfssatz aufgestellt.

Hilfssatz 4. *Es sei* $C = \|\gamma_{ij}\|$ $(i, j = 1, 2, \ldots, n)$ *eine Dreiecksmatrix (d. h.* $\gamma_{ij} = 0$ *für* $i < j$*), in der diejenigen Minoren aller Ordnungen, die man mit konsekutiven Zeilen und, von der ersten Spalte beginnend, mit konsekutiven Spalten bilden kann, positiv sind. Dann ist die Matrix C total positiv.*

Zur Abkürzung sei ein Minor $m$ter Ordnung durch das Symbol

$$(i_1, i_2, \ldots, i_m;\ j_1, j_2, \ldots, j_m)$$

bezeichnet, wobei i_μ und j_μ in wachsender Reihenfolge die Indizes der Zeilen und Spalten des Minors angeben. Die Voraussetzungen des Hilfssatzes lauten

$$(37)\quad (i, i+1, \ldots, i+m-1;\ 1, 2, \ldots, m) > 0 \text{ für } 1 \leqq i \leqq i+m-1 \leqq n.$$

Es ist sogar jeder Minor der Gestalt

$$(38)\quad (i_1, i_2, \ldots, i_m;\ 1, 2, \ldots, m) > 0 \text{ für } 1 \leqq i_1 < \cdots < i_m \leqq n \quad (m = 1, 2, \ldots, n).$$

[14]) M. Fekete, Rendiconti di Palermo **34** (1912), S. 92—93.

In der Tat, wenn (38) für $m-1$ (statt m) schon bewiesen ist, so folgt aus

$$(i_1, i_2, \ldots, i_{m-1};\quad 1, 2, \ldots, m-1) > 0$$

und (37), auch schon (38) auf Grund des Hilfssatzes von Herrn Fekete. Ferner können wir aus (38) auch auf die Ungleichungen

$$(39)\qquad (i_1, i_2, \ldots, i_m;\ p, p+1, \ldots, p+m-1) > 0$$
$$\text{für } p \leqq i_1 < i_2 < \ldots < i_m \leqq n \quad (p = 1, 2, \ldots, n),$$

schließen. In der Tat, aus (37) folgt $\gamma_{ii} > 0$ und, wegen $\gamma_{ij} = 0$ für $i < j$ und (38), ferner

$$0 < (1, 2, \ldots, p-1, i_1, \ldots, i_m;\ 1, 2, \ldots, p-1, p, \ldots, p+m-1)$$
$$= \gamma_{11}\gamma_{22}\cdots\gamma_{p-1,p-1}(i_1, i_2, \ldots, i_m;\ p, p+1, \ldots, p+m-1),$$

und damit ist (39) schon bewiesen. Aus (38) folgt insbesondere

$$\gamma_{ij} > 0 \text{ für } i \geqq j,$$

denn $0 < (1, 2, \ldots, j-1, i;\ 1, 2, \ldots, j-1, j) = \gamma_{11}\gamma_{22}\cdots\gamma_{j-1,j-1}\cdot\gamma_{ij}$.

Man bemerke nun, daß ein Minor Q_m ($m$ter Ordnung) von C gelegentlich auch dann verschwinden kann, wenn man seine positiven Elemente als unabhängige Variable betrachtet. In diesem Fall schreiben wir $Q_m \equiv 0$, sonst aber $Q_m \not\equiv 0$. Es ist offenbar dann und nur dann

$$(40)\qquad Q_m = (i_1, \ldots, i_m;\ j_1, \ldots, j_m) \not\equiv 0,$$

wenn alle Elemente seiner Hauptdiagonale (also auch unterhalb derselben) positiv sind, d. h. $i_\mu \geqq j_\mu$ für $\mu = 1, \ldots, m$. Die Zahl

$$b = j_m - j_1 + 1$$

wollen wir die *Breite* von Q nennen.

Wir behaupten, daß aus (40) *auch*

$$(41)\qquad Q_m > 0$$

folgt.

Die Behauptung sei für Minoren 1ter, 2ter, ... bis $(m-1)$ter Ordnung schon wahr. Ferner ist wegen (39) die Behauptung auch für Minoren $m$ter Ordnung der Breite m schon bewiesen. Es sei die Behauptung auch für Minoren $m$ter Ordnung und von den Breiten $m, m+1, \ldots, b-1$ schon wahr. Wir nennen den Ausdruck Q_m *zerfällbar*, wenn er die Gestalt

$$Q_m = \begin{vmatrix} Q_r & 0 \\ R & Q_{m-r} \end{vmatrix} = Q_r \cdot Q_{m-r} \qquad (1 \leqq r \leqq m-1)$$

hat. In diesem Fall sind Q_r und Q_{m-r} schon positiv und also auch Q_m. *Es sei also unser* Q_m *als unzerfällbar vorausgesetzt.*

Unter den Spaltenindizes von Q_m seien $j_1, j_2, \ldots, j_p$ konsekutive Zahlen, also $j_p = j_1 + p - 1$, dagegen $j_{p+1} - j_p > 1$. Es gibt ein solches p $(1 \leqq p \leqq m-1)$, weil die Breite $b > m$ ist. Aus unseren bisherigen Voraussetzungen und Bemerkungen schließen wir, daß

$$(42) \qquad \gamma_{i_r, j_r+1} > 0 \quad \text{für} \quad r = 1, \ldots, p,$$

$$(43) \qquad \gamma_{i_s, j_s} > 0 \quad \text{für} \quad s = p+1, \ldots, m.$$

In der Tat, wenn ein Element (42) verschwände, so wäre Q_m zerfällbar, und wenn ein Element (43) verschwände, so würde $Q_m \equiv 0$ folgen.

Man bezeichne mit Q^* die Matrix mit m Zeilen und $m+1$ Spalten die entsteht, wenn man zu Q_m noch die (j_p+1)te Spalte hinzufügt. Ferner seien Q'_m und Q''_m die Minoren von Q^*, die entstehen, wenn man die erste bzw. letzte Spalte von Q^* streicht. Die Elemente der Hauptdiagonale von Q'_m sind genau die positiven Elemente (42) und (43). Also ist $Q'_m \not\equiv 0$. Seine Breite aber ist $b-1$, also ist sogar

$$(44) \qquad Q'_m > 0.$$

Wegen der Dreiecksgestalt von C ist a fortiori $Q''_m \not\equiv 0$ und deshalb

$$(45) \qquad Q''_m > 0.$$

Es sei nun Q^{**} die Matrix der letzten $m-1$ Zeilen von Q^*. *Ich behaupte, daß sämtliche Minoren Q_{m-1} von Q^{**} positiv sind.* Für Minoren $(m-1)$ter Ordnung folgt aus $Q_{m-1} \not\equiv 0$ auch $Q_{m-1} > 0$. Deswegen und wegen der Dreiecksgestalt von C genügt es, nachzuweisen, daß $Q'_{m-1} \not\equiv 0$, wo Q'_{m-1} der mit den letzten $m-1$ Spalten von Q^{**} gebildete Minor von Q^{**} bedeutet. Es ist aber gewiß $Q'_{m-1} \not\equiv 0$, denn die Elemente seiner Hauptdiagonale sind die positiven Elemente (42) und (43), vom ersten Element γ_{i_1, j_1+1} abgesehen. Auf die Matrix Q^* können wir auf Grund des letzten Ergebnisses und der Ungleichungen (44) und (45) den Hilfssatz von Herrn Fekete anwenden und schließen, daß alle Minoren $m$ter Ordnung von Q^* positiv sind, also insbesondere auch $Q_m > 0$.

7. Jetzt können wir den zweiten Teil des Hauptsatzes in wenigen Worten beweisen. Man setze also voraus, daß die Polynomenfolge (29) in $(0, \infty)$ der Descartesschen Regel genügt und betrachte für irgendein positives x die Wronskische Matrix

$$(46) \qquad W(x) = \left\| \begin{matrix} \varphi_0(x) & 0 & 0 & \ldots & 0, \\ \varphi_1(x) & \varphi_1'(x) & 0 & \ldots & 0, \\ \varphi_2(x) & \varphi_2'(x) & \varphi_2''(x) & \ldots & 0, \\ \vdots & \vdots & \vdots & & \vdots \\ \varphi_n(x) & \varphi_n'(x) & \varphi_n''(x) & \ldots & \varphi_n^{(n)}(x) \end{matrix} \right\|.$$

Nach dem Kriterium von Pólya und Szegö und den Ungleichungen

$$W[\varphi_0(x), \quad \varphi_1(x), \ldots, \varphi_m(x)] = 1!\, 2! \ldots m!\; a_{00} a_{11} \ldots a_{mm} > 0$$

folgt, daß die Dreiecksmatrix $W(x)$ den Voraussetzungen des Hilfssatzes 4 genügt. Also ist $W(x)$ total positiv. Nach dem Grenzübergang $x \to 0$ ist aus Stetigkeitsgründen auch $W(0)$ und also auch die Matrix (30) total positiv.

8. Zur weiteren Klärung der Verhältnisse dient der folgende merkwürdige Satz

V. *Für irgendeine Polynomenfolge* (29) *gibt es eine (genügend große) Zahl* ξ, *so daß diese Polynomenfolge im Intervall* (ξ, ∞) *der Regel von Descartes genügt.*

Nach dem Kriterium von Pólya und Szegö und dessen Vereinfachung in Fußnote [9]) genügt es, eine Zahl ξ derart zu bestimmen, daß für $x > \xi$ alle Minoren der Wronskischen Matrix (46) mit konsekutiven Zeilen und mit der ersten Spalte beginnenden konsekutiven Spalten positiv sind. Diese Minoren sind gewisse Polynome $\Phi(x)$, deren höchste Glieder identisch sind mit den *entsprechenden* Minoren der folgenden Matrix der höchsten Glieder der Elemente der Matrix (46):

$$(47) \qquad \begin{Vmatrix} a_{00} & 0 & 0 & \dots & 0 \\ a_{11}x & 1!\,a_{11} & 0 & \dots & 0 \\ a_{22}x^2 & 2\,a_{22}x & 2!\,a_{22} & \dots & 0 \\ \vdots & \vdots & \vdots & & \vdots \\ a_{nn}x^n, & n\,a_{nn}x^{n-1}, & n(n-1)\,a_{nn}x^{n-2}, & \dots & n!\,a_{nn} \end{Vmatrix}$$

Wegen $a_{00} > 0,\ a_{11} > 0, \dots, a_{nn} > 0$ genügt die Folge $a_{00}, a_{11}x, \dots, a_{nn}x^n$ für $x > 0$ der Regel von Descartes, und deshalb sind nach dem Kriterium von Pólya und Szegö die höchsten Glieder jener Polynome $\Phi(x)$ für $x > 0$ positiv. Es gibt also gewiß ein ξ so, daß für $x > \xi$ alle $\Phi(x) > 0$ sind. Der kleinstmögliche Wert der Zahl ξ in Satz V ist offenbar gleich der Größten unter den reellen Wurzeln aller Gleichungen $\Phi(x) = 0$ [15]).

Satz V zeigt deutlich, daß das Interesse der Sätze III und IV nur in der genauen Bestimmung des kleinsten Wertes von ξ liegt, welcher

---

[15]) Man kann den Sachverhalt auch folgendermaßen fassen. Falls in (29) $a_{ij} = 0$ für $i > j$, so genügt die Folge (29) der Regel von Descartes in $(0, \infty)$. Bei beliebigen Werten der a_{ij} für $i > j$ können die „störenden" Glieder $a_{ij}x^j$ $(i > j)$ das Bestehen dieser Regel höchstens auf einer endlichen Strecke verhindern. Dieselbe Tatsache bleibt bestehen, wenn als störende Glieder auch noch etwa Glieder der Gestalt $\Sigma A_i \log(x + \alpha_i) + \Sigma B_i \cos\beta_i x + \Sigma C_i \sin\gamma_i x$ hinzukommen, welche es ebenfalls nicht verhindern können, daß die Minoren $\Phi(x)$ für große x positiv sind.

in beiden diesen Sätzen gleich der größten Wurzel der Gleichung $\varphi_n(x) = 0$ ist [16]).

9. Es ist vielleicht nicht uninteressant, zu zeigen, wie man mit den hier benutzten Hilfsmitteln die Sätze III und IV sehr einfach direkt beweisen kann. Zum Beweis des Satzes IV betrachte man die Wronskische Matrix (46) für $\varphi_i(x) = P_i(x)$. Nach den Ausführungen der vorhergehenden Nummer ist ein Minor der Gestalt

$$Q(x) = \begin{vmatrix} P_r(x) & P'_r(x) & \dots & P_r^{(s)}(x) \\ P_{r+1}(x) & P'_{r+1}(x) & \dots & P_{r+1}^{(s)}(x) \\ \vdots & \vdots & & \vdots \\ P_{r+s}(x) & P_{r+s}(x) & \dots & P_{r+s}^{(s)}(x) \end{vmatrix} \qquad (0 < r < r+s \leqq n)$$

für genügend großes x gewiß positiv. Der Satz IV ist bewiesen, wenn wir zeigen, daß der Minor $Q(x)$ sogar für $x > \eta_n$ schon positiv ist [$\eta_1 < \eta_2 < \dots < \eta_n$ sind die Wurzeln der Gleichung $P_n(x) = 0$]. Zu diesem Zweck betrachte man das Polynom

$$\Phi(x) = c_r P_r(x) + c_{r+1} P_{r+1}(x) + \dots + c_{r+s} P_{r+s}(x). \tag{48}$$

---

[16]) In diesem Zusammenhang ist die folgende, von Herrn Pólya angeregte Frage von Interesse: *Die Polynome $\varphi_2(x)$, $\varphi_3(x)$, ..., $\varphi_n(x)$ der Folge* (29) *sollen lauter reelle und einfache Nullstellen haben, und außerdem sollen die Nullstellen von $\varphi_i(x)$ $(1 \leqq i \leqq n-1)$ diejenigen von $\varphi_{i+1}(x)$ trennen. Kann man in diesem Fall die Zahl ξ des Satzes V bis auf die größte Wurzel η_n der Gleichung $\varphi_n(x) = 0$ herunterdrücken?* Das man diese Frage im allgemeinen verneinend beantworten muß, zeigt, für $n = 4$, das Beispiel der Folge

$$1,\ \left(x+\frac{1}{2}\right),\ (x+1)\left(x+\frac{1}{1000}\right),\ (x+10\,000)\left(x+\frac{1}{2}\right)\left(x+\frac{1}{1001}\right),$$
$$(x+100\,000)(x+1000)\left(x+\frac{1}{3}\right)\left(x+\frac{1}{1003}\right).$$

Hier ist $\eta_4 = -\frac{1}{1003} < 0$, während, wegen des Hauptsatzes, Satz V nur mit einem positiven ξ gelten kann. In der Tat ist in diesem Fall die Matrix (30) nicht total positiv, denn eine einfache Rechnung zeigt, daß der Minor

$$\begin{vmatrix} a_{20} & a_{21} & a_{22} \\ a_{30} & a_{31} & a_{32} \\ a_{40} & a_{41} & a_{42} \end{vmatrix} < 0$$

ist.

Noch unbeantwortet ist die folgende, ebenfalls von Herrn Pólya stammende Fragestellung: *Das Polynom n ten Grades $f(x)$ habe lauter reelle einfache Wurzeln $\eta_1 < \eta_2 < \dots < \eta_n$. Genügt die Folge*

$$f(x),\ f'(x),\ f''(x),\ \dots,\ f^{(n)}(x)$$

der Descartesschen Regel schon für $x > \eta_n$?

Wir behaupten: *Das Polynom $\Phi(x)$ hat im Intervall $\eta_1 \leqq x \leqq \eta_n$ mindestens r Nullstellen.* Angenommen, es hat in $[\eta_1, \eta_n]$ weniger als r Nullstellen, etwa $\alpha_1, \alpha_2, \ldots, \alpha_m$ $(m < r)$, dann ist etwa

$$(49) \quad \Phi(x) = \prod_{\nu=1}^{m} (x - \alpha_\nu) \cdot \Psi(x) \quad \text{mit} \quad \Psi(x) > 0 \ \text{für} \ \eta_1 \leqq x \leqq \eta_n.$$

Wegen (28) gelten die Gleichungen

$$\int_{x_0}^{x_1} p(x)\, \Phi(x)\, P_i(x)\, dx = 0 \quad (i = 0, \ldots, r-1)$$

und eine passende lineare Kombination derselben ergibt wegen (49)

$$(50) \quad \int_{x_0}^{x_1} p(x)\, \Phi(x) \prod_1^m (x - \alpha_\nu)\, dx = \int_{x_0}^{x_1} p(x)\, \Psi(x) \prod_1^m (x - \alpha_\nu)^2\, dx = 0.$$

Der Grad des Polynoms

$$G(x) = \Psi(x) \prod_1^m (x - \alpha_\nu)^2$$

übertrifft gewiß nicht die Zahl $2n - 1$, und die bekannte Formel der mechanischen Quadratur ergibt

$$\int_{x_0}^{x_1} p(x)\, \Psi(x) \prod_1^m (x - \alpha_\nu)^2\, dx = \prod_{i=1}^{n} \sigma_i G(\eta_i) > 0;$$

denn hier sind $\sigma_i > 0$, $G(\eta_i) \geqq 0$, und mindestens eine dieser letzten Größen ist positiv, wodurch wir einen Widerspruch zur Gleichung (50) erreicht haben.

Wir schließen nun sofort, daß $Q(x) > 0$ für $x > \eta_n$ ist, denn wäre $Q(\bar{x}) = 0$ für ein $\bar{x} > \eta_n$, so gäbe es ein Polynom $\Phi(x)$ $(\neq 0)$ der Form (48), welches in $x = \bar{x}$ eine $(s+1)$-fache Nullstelle hat, die zusammen mit den schon nachgewiesenen r Nullstellen im Intervall $[\eta_1, \eta_2]$, insgesamt $r + s + 1$ Nullstellen von $\Phi(x)$ ergibt; dies ist aber für ein Polynom vom Grad $\leqq r + s$ unmöglich.

Eine ähnliche, aber einfachere Schlußweise beweist auch Satz II; denn das mit den Polynomen (26) gebildete Polynom $\Phi(x)$ hat offenbar mindestens r Nullstellen im Intervall $\operatorname{Min} \xi_\nu \leqq x \leqq \operatorname{Max} \xi_\nu$ (und zwar die Nullstellen $x = \xi_1, \xi_2, \ldots, \xi_r$), und der Beweis des Satzes III wird sofort wie oben zu Ende geführt.

§ 4.

Einige Anwendungen zur Abzählung komplexer Wurzeln.

10. Man verdankt Herrn Obreschkoff (a. a. O. S. 61) den folgenden wichtigen Satz

VI. *Die Folge* 1, x, x^2, ..., x^n *genügt der Regel von Descartes nicht nur innerhalb des Intervalls* (0, ∞), *sondern sogar innerhalb des Winkels der komplexen Ebene*

(51) $$-\frac{\pi}{n} < \arg x < \frac{\pi}{n}.$$

Damit ist folgendes gemeint: Für eine reelle Gleichung (1) ist die rechte Seite $v_a = V(0)$ der Ungleichung (3) sogar eine obere Schranke für die Anzahl der Wurzeln innerhalb des Winkels (51).

Aus diesem Satz und der Identität (4) folgt sofort Satz

VII. *Die Schranke* v_b *der Descartes-Jacobischen Regel* (5) *ist sogar eine obere Schranke für die Anzahl der Wurzeln von* (1) *innerhalb der Kreislinse*

(52) $$-\frac{\pi}{n} < \arg\frac{x-\alpha}{\beta-x} < \frac{\pi}{n},$$

von deren Rand man die Strecke $\overline{\alpha\beta}$ *unter dem Winkel* $\frac{n-1}{n}\pi$ *sieht.*

Aus diesem Satz und den Ungleichungen (6) und (21) folgt sofort Satz

VIII. *Die Sätze* I *und* II *von Budan-Fourier und Laguerre geben obere Schranken* $V(\alpha) - V(\beta)$ *bzw.* v_c *für die Anzahl der Wurzeln von* (1) *sogar innerhalb der Kreislinse* (52).

Wenn die Polynomenfolge (29) in (0, ∞) der Descartesschen Regel genügt, so zeigt Satz VI und die Ungleichung (34), daß diese Polynomenfolge der Descartesschen Regel sogar innerhalb des Winkels (51) genügt. Diese Bemerkung läßt sich auf die Sätze III und IV anwenden und ergibt also insbesondere den folgenden Satz

IX. *Die Polynomenfolge* (27) *des Satzes* IV *genügt der Regel von Descartes sogar innerhalb des Winkels*

$$-\frac{\pi}{n} < \arg(x - \eta_n) < \frac{\pi}{n}.$$

(Eingegangen am 1. Juni 1933.)

Reprinted from
Math Z. **38** (1934), pp. 546–564

ON THE GENERATING FUNCTIONS OF TOTALLY POSITIVE SEQUENCES I[1]

By

Michael Aissen, I. J. Schoenberg and A. M. Whitney
in Philadelphia, Pa., U.S.A.

1. Introduction and statement of results.

A real matrix, finite or infinite, is called totally positive if and only if all its minors, of all orders = 1, 2, ..., are non-negative. An infinite sequence

$$a_0, a_1, a_2, \ldots, \qquad (a_0 = 1),$$

is said to be totally positive[2] if and only if the 2-way infinite matrix

$$(1) \qquad A = \| a_{i-k} \| = \begin{Vmatrix} a_0 & 0 & 0 & \ldots \\ a_1 & a_0 & 0 & \ldots \\ a_2 & a_1 & a_0 & \ldots \\ \cdot & \cdot & \cdot & \\ \cdot & \cdot & \cdot & \\ \cdot & \cdot & \cdot & \end{Vmatrix} \qquad \begin{matrix} (i, k = 1, 2, 3, \ldots) \\ (a_n = 0 \text{ if } n < 0), \end{matrix}$$

is totally positive, i. e. has only non-negative minors. We are concerned with the problem of describing the nature of the generating functions

$$(2) \qquad f(z) = a_0 + a_1 z + a_2 z^2 + \ldots$$

of totally positive sequences. The problem will be partially solved in the present note and its solution will be completed in the immediately following note by Albert Edrei.

We begin by stating the following

Lemma 1: *Let the formal power series* $\sum_0^\infty b_n z^n$ $(b_0 = 1)$ *generate a totally positive sequence* $\{b_n\}$, *and let*

---

1. The subject of this note is part of a Project sponsored by the Office of Naval Research at the University of Pennsylvania.

2. The related concept of an n-times positive sequence was first introduced by M. Fekete; see M. Fekete and G. Pólya, "Ueber ein Problem von Laguerre", *Rendiconti di Palermo*, 34 (1912), 1—32, especially pages 3—7.

(3) $$\frac{1}{\sum_0^\infty b_n(-z)^n} = \sum_0^\infty c_n z^n .$$

Then $\{c_n\}$ is also a totally positive sequence.

This elementary lemma, to be used extensively in our proofs below, is already mentioned here because it helps in obtaining the results one might expect in our investigation.

We now pass to six successive statements, the last of which will suggest our main problem: [3]

a). The function

(4) $$1 + \alpha z \qquad (\alpha \geqq 0)$$

generates a totally positive sequence. Indeed, the matrix (1) corresponding to the sequence $a_0 = 1$, $a_1 = \alpha$, $a_2 = a_3 = \ldots = 0$ is evidently totally positive.

b). If each of two formal power series generates a totally positive sequence, then so does their product. This becomes evident if we observe the following:

(i) If to the power series $\Sigma\, a_n z^n$ and $\Sigma\, b_n z^n$ correspond the matrices A and B respectively (according to (1)), then to their product

$$(\Sigma\, a_n z^n)(\Sigma\, b_n z^n) = \Sigma c_n z^n$$

corresponds the matrix C such that $C = AB$.

(ii) If the matrices A and B are totally positive, then so is their product $C = AB$.

c). If $\{a_n^{(k)}\}$ is a sequence of sequences, each of which is totally positive, for each fixed k, and if

$$\lim_{k\to\infty} a_n^{(k)} = a_n ,$$

then the limit sequence $\{a_n\}$ is also totally positive. This property is evident from the definition of total positivity of a sequence.

d). The function

(5) $$e^{\gamma z} \qquad (\gamma \geqq 0)$$

3. These remarks may be found in I. J. Schoenberg, Some analytical aspects of the problem of smoothing, *Courant Anniversary Volume*, New York 1948, pp. 351—370, especially §2.

generates a totally positive sequence. Indeed, by a) and b), we know that

$$\left(1 + \frac{\gamma z}{n}\right)^n$$

generates a totally positive sequence. Letting the integer $n \to \infty$ we obtain d) in view of c).

e). The function

$$\frac{1}{1 - \beta z} \qquad (\beta \geqq 0) \tag{6}$$

generates a totally positive sequence. This simple fact is easily proved from the definition of a totally positive sequence. It also follows from our Lemma 1: $1 + \beta z$ generates a totally positive sequence by a), hence so does (6) which is a special case of the left side of relation (3).

f). The function

$$f(z) = e^{\gamma z} \frac{\prod_1^\infty (1 + \alpha_\nu z)}{\prod_1^\infty (1 - \beta_\nu z)} \tag{7}$$

where $\gamma \geqq 0$, $\beta_\nu \geqq 0$, $\alpha_\nu \geqq 0$, and $\Sigma(\alpha_\nu + \beta_\nu)$ converges, generates a totally positive sequence.

This follows by applying all previous statements in the following order: a), d), e), b) and c).

The functions $f(z)$ of the form (7) generate a wide class of totally positive sequences. Do they generate all totally positive sequences? Our main result is the following

Theorem 1: If $a_0 = 1, a_1, a_2, \ldots$ is a totally positive sequence, then its generating function

$$f(z) = a_0 + a_1 z + a_2 z^2 + \ldots$$

is regular in a neighborhood of the origin and may be extended into the whole plane as a meromorphic function of the special form

$$f(z) = e^{g(z)} \frac{\prod_1^\infty (1 + \alpha_\nu z)}{\prod_1^\infty (1 - \beta_\nu z)} \tag{8}$$

$$(\alpha_\nu \geqq 0, \ \beta_\nu \geqq 0, \ \Sigma(\alpha_\nu + \beta_\nu) \text{ convergent}),$$

where $g(z)$ is an entire function ($g(0)=0$) with the property that the exponential factor

$$e^{g(z)}$$

will by itself generate a totally positive sequence.[4]

In the note immediately following this one, it will be shown by Albert Edrei[5] that $g(z)$ necessarily reduces to a linear function $g(z)=\gamma z$ ($\gamma \geqq 0$), thereby proving the full converse of our result f).

2. A proof of Lemma 1.

A proof will readily follow from the following

Lemma 2. Let $B=\|b_{ik}\|$ be a non-singular matrix of order n which is totally positive, and let

$$B^{-1}=\|(-1)^{i+k}c_{ik}\|,$$

then the matrix $C=\|c_{ik}\|$ is totally positive.

In order to prove Lemma 2 we first apply a classical theorem of Sylvester concerning the minors of the determinant which is adjoint to a given determinant.[6] Rephrasing this theorem in our present notation, we readily obtain the following statement: Let $|B_1|$ and $|C_1|$ be corresponding minors of B and C, respectively, (i. e. formed from the similar rows and columns of both matrices), and let $|B'_1|$ be that minor of B which is symmetric to $|B_1|$ with respect to the main diagonal of B. Then

$$|C_1|=\frac{K(|B'_1|)}{|B|},$$

where $K(|B'_1|)$ denotes the minor in B which is complementary to $|B'_1|$.

4. The problem solved by Theorem 1 is of the type discussed by J. Hadamard in his classical dissertation, "Essai sur l'étude des fonctions données par leur développement de Taylor", *Journal de Mathématiques*, 4th series, 8 (1892). Some of Hadamard's basic ideas are used in our proof below, adapted to our particular situation which gives the precise information concerning the nature of the zeros and poles of $f(z)$ implied by (8).

5. Albert Edrei, On the generating functions of totally positive sequences, II, this Journal, p. 104. See also the note by Michael Aissen, Albert Edrei, I. J. Schoenberg and Anne Whitney, On the generating functions of totally positive sequences, *Proceedings of the National Academy of Sciences*, 37 (1951), pp. 303—307, where the main results of these notes are stated and several applications are discussed.

6. See M. Bôcher, Introduction to Higher Algebra, Chapter II, §11.

Since B is assumed to be totally positive, we have $K(|B'_1|) \geqq 0$. $|B| > 0$, and the last relation shows that $|C_1| \geqq 0$, hence C is indeed totally positive.

Returning to Lemma 1, we notice that the identity (3), or

$$\left(\sum_0^\infty b_n z^n\right)\left(\sum_0^\infty c_n (-z)^n\right) = 1$$

is equivalent to the matrix relation

$$\left\|\begin{matrix} b_0 & 0 & 0 & \ldots \\ b_1 & b_0 & 0 & \ldots \\ b_2 & b_1 & b_0 & \ldots \\ \cdot & \cdot & \cdot & \\ \cdot & \cdot & \cdot & \\ \cdot & \cdot & \cdot & \end{matrix}\right\| \cdot \left\|\begin{matrix} c_0 & 0 & 0 & \ldots \\ -c_1 & c_0 & 0 & \ldots \\ c_2 & -c_1 & c_0 & \ldots \\ \cdot & \cdot & \cdot & \\ \cdot & \cdot & \cdot & \\ \cdot & \cdot & \cdot & \end{matrix}\right\| = \left\|\begin{matrix} 1 & 0 & 0 & 0 & \ldots \\ 0 & 1 & 0 & 0 & \ldots \\ 0 & 0 & 1 & 0 & \ldots \\ \cdot & \cdot & \cdot & \cdot & \\ \cdot & \cdot & \cdot & \cdot & \\ \cdot & \cdot & \cdot & \cdot & \end{matrix}\right\|$$

which in particular implies the following relation between non-singular matrices of order n:

$$\left\|\begin{matrix} b_0 & 0 & 0 & \ldots & 0 \\ b_1 & b_0 & 0 & \ldots & 0 \\ b_2 & b_1 & b_0 & \ldots & 0 \\ \cdot & \cdot & \cdot & & \cdot \\ \cdot & \cdot & \cdot & & \cdot \\ \cdot & \cdot & \cdot & & \cdot \\ b_n & b_{n-1} & b_{n-2} & \ldots & b_0 \end{matrix}\right\| \cdot \left\|\begin{matrix} c_0 & 0 & & \ldots & 0 \\ -c_1 & c_0 & & \ldots & 0 \\ c_2 & -c_1 & c_0 & \ldots & \cdot \\ \cdot & \cdot & \cdot & & \cdot \\ \cdot & \cdot & \cdot & & \cdot \\ \cdot & \cdot & \cdot & & \cdot \\ (-1)^n c_n & (-1)^{n-1} c_{n-1} & \ldots & & c_0 \end{matrix}\right\| = \left\|\begin{matrix} 1 & 0 & 0 & \ldots & 0 \\ 0 & 1 & 0 & \ldots & 0 \\ 0 & 0 & 1 & \ldots & 0 \\ \cdot & \cdot & \cdot & & \cdot \\ \cdot & \cdot & \cdot & & \cdot \\ \cdot & \cdot & \cdot & & \cdot \\ 0 & 0 & 0 & \ldots & 1 \end{matrix}\right\|.$$

Let us now assume the sequence $\{b_n\}$ to be totally positive. Then the first matrix B_n on the left-hand side of our last relation is totally positive. By Lemma 2, we see that

$$C_n = \left\|\begin{matrix} c_0 & 0 & \ldots & 0 \\ c_1 & c_0 & \ldots & 0 \\ \cdot & & & \\ \cdot & & & \\ \cdot & & & \\ c_n & c_{n-1} & \ldots & c_0 \end{matrix}\right\|$$

is totally positive. Since this is true for every value of n, we conclude that $\{c_n\}$ is totally positive.

3. A proof of the existence of $f(z)$.

Let

$$a_0 = 1 , \; a_1 , \; a_2 , \; \ldots \tag{9}$$

be a totally positive sequence. To show that its generating series

$$f(z) = a_0 + a_1 z + a_2 z^2 + \ldots \tag{10}$$

converges in a neighborhood of the origin, we distinguish two cases.

α) The sequence (9) contains zero elements, and let a_{m+1} be the first vanishing element, hence

$$a_\nu > 0 \quad (\nu = 0, 1, \ldots, m), \quad a_{m+1} = 0 .$$

Since all minors of (1) are non-negative, we have in particular

$$\begin{vmatrix} a_{m+1} & a_m \\ a_{m+p} & a_{m+p-1} \end{vmatrix} \geqq 0 \qquad \text{or } a_m a_{m+p} \leqq 0 .$$

From $a_m > 0$ and $a_{m+p} \geqq 0$, we conclude that

$$a_{m+1} = a_{m+2} = \ldots = a_{m+p} = \ldots = 0 .$$

Thus (10) reduces to a polynomial.

β) All elements of (9) are positive. We now have that

$$\begin{vmatrix} a_n & a_{n-1} \\ a_{n+1} & a_n \end{vmatrix} \geqq 0 \qquad (n = 0, 1, \ldots),$$

whence

$$\frac{a_1}{a_0} \geqq \frac{a_2}{a_1} \geqq \ldots \geqq \frac{a_{n+1}}{a_n} \geqq \ldots \tag{11}$$

and therefore

$$\lim_{n \to \infty} \frac{a_{n+1}}{a_n} = \beta_1 \geqq 0 . \tag{12}$$

This proves the convergence of (10) in the circle

$$|z| < R = \frac{1}{\beta_1} \qquad (\leqq + \infty). \tag{13}$$

4. Determination of the poles of $f(z)$.

We distinguish two cases depending on whether $f(z)$ is entire or not. If $f(z)$ reduces to a polynomial (case α above), or if $\beta_1 = 0$ in case β, we do nothing further at this point. The process now following applies only to the case when the relation (12) furnishes a $\beta_1 > 0$. Thus, assuming

(14) $$\lim_{n\to\infty} \frac{a_{n+1}}{a_n} = \beta_1 > 0\,,$$

we now multiply both sides of (10) by $1-\beta_1 z$ obtaining

(15) $$(1-\beta_1 z)\, f(z) = a'_0 + a'_1 z + a'_2 z^2 + \dots\,,$$

where

(16) $$a'_n = a_n - \beta_1 a_{n-1} \qquad \text{for all integral } n \ (a_\nu = a'_\nu = 0 \text{ if } \nu < 0)$$

and in particular

(17) $$a'_0 = a_0 = 1\,, \quad a'_1 = a_1 - \beta_1\,.$$

We now claim that the new sequence $a'_0 = 1,\ a'_1,\ a'_2, \dots$ *is totally positive.* Indeed, we know that the square matrix

$$\begin{Vmatrix} a_0 & a_{-1} & \dots & a_{-p} \\ a_1 & a_0 & \dots & a_{-p+1} \\ \cdot & \cdot & & \cdot \\ \cdot & \cdot & & \cdot \\ \cdot & \cdot & & \cdot \\ a_{p-1} & a_{p-2} & \dots & a_{-1} \\ a_{m+p} & a_{m+p-1} & \dots & a_m \end{Vmatrix} \qquad (m \geqq 0,\ a_\nu = 0 \text{ if } \nu < 0)$$

is totally positive as a submatrix of the matrix A defined by (1). Dividing the last row by the positive element a_m we obtain the likewise totally positive matrix

$$\begin{Vmatrix} a_0 & a_{-1} & \dots & a_{-p} \\ a_1 & a_0 & \dots & a_{-p+1} \\ \cdot & \cdot & & \cdot \\ \cdot & \cdot & & \cdot \\ \cdot & \cdot & & \cdot \\ a_{p-1} & a_{p-2} & \dots & a_{-1} \\ \dfrac{a_{m+p}}{a_m} & \dfrac{a_{m+p-1}}{a_m} & \dots & 1 \end{Vmatrix}.$$

Letting $m \to \infty$ we obtain, in view of (14), the totally positive matrix

(18) $$\begin{Vmatrix} a_0 & a_{-1} & \dots & a_{-p} \\ a_1 & a_0 & \dots & a_{-p+1} \\ \cdot & \cdot & & \cdot \\ \cdot & \cdot & & \cdot \\ \cdot & \cdot & & \cdot \\ a_{p-1} & a_{p-2} & \dots & a_{-1} \\ \beta_1^p & \beta_1^{p-1} & \dots & 1 \end{Vmatrix}.$$

If we now multiply the second column by β_1 and subtract it from the first, likewise multiply the third column by β_1 and subtract it from the second and continue to the last column, we obtain, by (16), the matrix

$$(19) \qquad \begin{Vmatrix} a'_0 & a'_{-1} & \dots & a'_{-p+1} & a_{-p} \\ a'_1 & a'_0 & \dots & a'_{-p+2} & a_{-p+1} \\ \cdot & \cdot & & \cdot & \cdot \\ \cdot & \cdot & & \cdot & \cdot \\ \cdot & \cdot & & \cdot & \cdot \\ a'_{p-1} & a'_{p-2} & \dots & a'_0 & a_{-1} \\ 0 & 0 & \dots & 0 & 1 \end{Vmatrix}.$$

If we reflect both matrices (18) and (19) about their secondary diagonals (an operation which does not alter the totally positive character of a matrix) we see that the process used in passing from (18) to (19) is precisely the reduction process which was investigated by Anne Whitney,[(7)] in her note immediately preceding the present one, and found by her to leave invariant the totally positive property of a matrix. Thus, by Anne Whitney's Theorem 2, we know that

$$\begin{Vmatrix} a'_0 & \dots & a'_{-p+1} \\ \cdot & & \cdot \\ \cdot & & \cdot \\ \cdot & & \cdot \\ a'_{p-1} & \dots & a'_0 \end{Vmatrix}$$

is totally positive. Since p is arbitrary, our assertion follows.

To the new generating function (15) we now apply the same process just applied to the function (10). If (15) is a polynomial, we do nothing; if $a'_n > 0$ for all $n = 0, 1, \dots$, we consider

$$\lim_{n\to\infty} \frac{a'_{n+1}}{a'_n} = \beta_2 \geqq 0 .$$

If $\beta_2 = 0$, then (15) is entire and we stop the process. If $\beta_2 > 0$, we repeat the process obtaining

$$(1 - \beta_2 z)(1 - \beta_1 z)\, f(z) = a''_0 + a''_1 z + a''_2 z^2 + \dots$$

which again generates a totally positive sequence.

7. A. M. Whitney, A reduction theorem for totally positive matrices, this Journal pp. 88—92.

Before we continue, we wish to show that in any case we have the inequality

(20) $$\beta_1 \geqq \beta_2 .$$

Indeed, since

$$0 \leqq \begin{vmatrix} a_n & a_{n-1} \\ a_{n+1} & a_n \end{vmatrix} = \begin{vmatrix} a_n - \beta_1 a_{n-1} & a_{n-1} \\ a_{n+1} - \beta_1 a_n & a_n \end{vmatrix} = \begin{vmatrix} a'_n & a_{n-1} \\ a'_{n+1} & a_n \end{vmatrix},$$

we find that

$$\begin{vmatrix} 1 & 1 \\ \dfrac{a'_{n+1}}{a'_n} & \dfrac{a_n}{a_{n-1}} \end{vmatrix} \geqq 0$$

and letting $n \to \infty$ we obtain

$$\begin{vmatrix} 1 & 1 \\ \beta_2 & \beta_1 \end{vmatrix} = \beta_1 - \beta_2 \geqq 0 .$$

Let us assume now that the process continues indefinitely and that at the k-th stage (for every k) we get a series

(21) $$\prod_{\nu=1}^{k} (1 - \beta_\nu z)\, f(z) = a_0^{(k)} + a_1^{(k)} z + a_2^{(k)} z^2 + \dots$$

having a finite radius of convergence $R_k = \dfrac{1}{\beta_{k+1}}$. We wish to show that the series

(22) $$\sum_1^\infty \beta_\nu \text{ converges.}$$

Indeed, from (17) we get

$$a'_1 = a_1 - \beta_1 , \quad a''_1 = a'_1 - \beta_2 = a_1 - \beta_1 - \beta_2 , \dots$$

and generally

$$a_1^{(k)} = a_1 - \beta_1 - \beta_2 - \dots - \beta_k ,$$

a relation which also follows from (21) by comparing the coefficients of z on both sides of that equation. Since $a_1^{(k)} > 0$, for all k, we find $\Sigma_1^k \beta_\nu < a_1$ and therefore

(23) $$\sum_1^\infty \beta_\nu \leqq a_1$$

which implies (22). It follows in particular that $\beta_\nu \to 0$, hence

$$R_k = \frac{1}{\beta_{k+1}} \to +\infty .$$

We therefore have that

(24) $$R \leqq R_1 \leqq \ldots \leqq R_k \to +\infty.$$

Let us now return to (21) and let $k \to \infty$. From the relation

$$a_n^{(k+1)} = a_n^{(k)} - \beta_{k+1}\, a_{n-1}^{(k)}$$

we see that the sequence $\{a_n^{(k)}\}$ $(k = 0, 1, \ldots)$ is a decreasing sequence of positive elements, for every fixed $n \geqq 1$. Setting

$$\lim_{k \to \infty} a_n^{(k)} = b_n \qquad (n = 0, 1, \ldots),$$

we may let $k \to \infty$ in (21) and obtain, in view of (22) and (24), the function

(25) $$\prod_{\nu=1}^{\infty} (1 - \beta_\nu z) \,.\, f(z) = b_0 + b_1 z + b_2 z^2 + \ldots \qquad (b_0 = 1),$$

which is entire and which again generates a totally positive sequence.

The final result (25) will cover all cases, i.e. whether the process was finite or infinite, except that the product would be finite (or perhaps even missing altogether) in the former case.

5. Determination of the zeros of $f(z)$.

We form the reciprocal of (25), with z replaced by $-z$, and expand in powers of z:

(26) $$\frac{1}{\Pi (1 + \beta_\nu z) \,.\, f(-z)} = c_0 + c_1 z + c_2 z^2 + \ldots \qquad (c_0 = 1).$$

This expansion also has totally positive coefficients $\{c_n\}$ by Lemma 1. We may now repeat the above process and find a finite or infinite sequence

$$\alpha_1 \geqq \alpha_2 \geqq \alpha_3 \geqq \ldots \quad \text{with } \Sigma\, \alpha_\nu \leqq c_1,$$

and such that

(27) $$\frac{\Pi (1 - \alpha_\nu z)}{\Pi (1 + \beta_\nu z) \,.\, f(-z)} = d_0 + d_1 z + d_2 z^2 + \ldots \qquad (d_0 = 1)$$

is entire and generates a totally positive sequence $\{d_n\}$. This last relation shows that $f(z)$ is a meromorphic function. We now claim that the entire function (27) has no zeros. Indeed, since the denominator of the left-hand side of (27) is entire, by (25), zeros of (27) could only be due to zeros of its numerator $\Pi(1-\alpha_\nu z)$. However, these zeros being positive, can not be zeros of the entire function (27) having an expansion $\Sigma d_n z^n$ with non-negative coefficients d_n. We may therefore write the function (27) in the form

$$\frac{\Pi(1-\alpha_\nu z)}{\Pi(1+\beta_\nu z)\,.\,f(-z)} = e^{-g(-z)} \tag{28}$$

for some appropriate entire function $g(z)$. Solving for $f(-z)$ and replacing z by $-z$, we obtain the desired relation

$$f(z) = e^{g(z)}\,.\,\frac{\Pi(1+\alpha_\nu z)}{\Pi(1-\beta_\nu z)}\,.$$

The last statement of Theorem 1 now follows by a renewed application of Lemma 1. Indeed, since

$$e^{-g(-z)} = \Sigma\, d_n z^n$$

generates a totally positive sequence, we conclude that also

$$\frac{1}{\Sigma\, d_n(-z)^n} = e^{g(z)}$$

will generate a totally positive sequence. This concludes a proof of Theorem 1.

University of Pennsylvania

(Received January 30, 1952)

Reprinted from
J. d'Analyse Math. **2** (1952), pp. 93–103

MATHEMATICA VOL. 8 (31), 1, 1966, pp. 61–82

Dedicated to T. POPOVICIU
on his 60th *birthday*

ON VARIATION DIMINISHING SPLINE APPROXIMATION METHODS

by

MARTIN MARSDEN and I. J. SCHOENBERG
University of Wisconsin, Madison, Wisconsin

CONTENTS

Introduction

Let the function $f(x)$ be defined in $[0, 1]$ and let us approximate it by a piece-wise linear function $S_l(x)$ obtained as follows: l being a natural number, we divide $[0, 1]$ into l equal parts and denote by $S_l(x)$ the continuous function which is linear in each subinterval $((i-1)/l, i/l)$ while interpolating $f(x)$ at its end points. Denoting by $N_j(x)$ the broken linear functions such that

$$N_j\left(\frac{i}{l}\right) = \delta_{ij}, \quad (i, j = 0, \ldots, l),$$

we may represent the approximation in the form

$$S_l(x) = \sum_{j=0}^{l} f\left(\frac{j}{l}\right) N_j(x). \tag{1}$$

We shall refer to this construction as the method of *linear interpolation*. If $\omega(\delta)$ denotes the modulus of continuity of $f(x)$, it is readily seen that

$$| S_l(x) - f(x) | \leqslant \omega\left(\frac{1}{l}\right) \text{ in } [0, 1]. \tag{2}$$

We wish to compare $S_l(x)$ with the Bernstein polynomial

$$B_l(x) = \sum_{j=0}^{l} f\left(\frac{j}{l}\right) \binom{l}{j} x^j (1 - x)^{l-j} \tag{3}$$

for which T. POPOVICIU ([8], [6]) established the error estimate

$$| B_l(x) - f(x) | \leqslant \frac{3}{2} \omega\left(\frac{1}{\sqrt{l}}\right). \tag{4}$$

Notice the following common properties of the approximations (1) and (3):

1. *They both reproduce linear functions.* This means that if $f(x)$ is linear, then $f(x) = S_l(x) = B_l(x)$.

2. *They are both variation diminishing.* This means that if $v(F)$ denotes the number of changes of sign in $[0, 1]$ of a function $F(x)$, then

$$v(S_l) \leqslant v(f), \quad v(B_l) \leqslant v(f). \tag{5}$$

The first inequality follows from the definition of $v(F)$ as

$$v(F) = \sup v(F(x_1), \quad F(x_2), \ldots, F(x_k)),$$

the supremum being formed subject to $0 \leqslant x_1 < \ldots < x_k \leqslant 1$ and for all finite k. Indeed

$$v(S_l) \leqslant v\left(f(0), \quad f\left(\frac{1}{l}\right), \ldots, f(1)\right) \leqslant v(f).$$

The second inequality (5) (first pointed out in [7]; see also [9]) is an easy corollary of Descartes' rule of signs.

Using both properties 1 and 2 we conclude that if $F(x) = S_l(x)$ or $B_l(x)$, then

$$v(F(x) - A - Bx) \leqslant v(f(x) - A - Bx)$$

for arbitrary reals A and B. It is known (see [7], [9]) that this inequality is the source of the shape-preserving properties of the Bernstein polynomials which were first discussed by T. POPOVICIU.

As far as the order of approximation is concerned, a comparaison of (2) and (4) reveals the evident superiority of linear interpolation. A serious drawback of $S_l(x)$ are, of course, its corners.

In a recent paper [10] the second author has shown how to construct for a function $f(x)$, defined in [0, 1], approximations of the following kind: l being a prescribed natural number, there are l different approximations

(6) $$S_{m,n}(x\,;\,f) = \sum_{j=0}^{l} f(\xi_j)\, N_j(x)$$

for $n = 1, 2, \ldots, l$, while m is determined by $l = m + n - 2$. The nodes ξ_j and the fundamental functions $N_j(x)$ depend on l and n, but not on $f(x)$. Their dependence on l and n is not explicitly indicated in order to simplify our notation. Furthermore, the approximations (6) have the following properties:

A. $S_{m,n}(x\,;\,f) \in C^{m-2}[0, 1]$ *and* $S_{m,n}(x\,;\,f)$ *is a polynomial of degree* $\leqslant m - 1$ *in each of the intervals* $((i-1)/n,\ i/n)$, $(i = 1, \ldots, n)$.

B. $S_{m,n}(x\,;\,f) = f(x)$ *in* $[0, 1]$ *whenever* $f(x)$ *is a linear function, while for any* $f(x)$ *the inequality*

$$v(S_{m,n}(x\,;\,f)) \leqslant v(f)$$

holds. In other words: the approximations (6) *are variation diminishing.*

C. *If* $n = l$ *hence* $m = 2$, *then*

$$S_{m,n}(x\,;\,f) = S_l(x).$$

If $n = 1$, *hence* $l = m - 1$ *then*

$$S_{m,n}(x\,;\,f) = B_l(x).$$

In his book, [2, page 116] P. J. DAVIS stresses the desirable shape preserving properties of the Bernstein polynomials, but also their slow rate of convergence to $f(x)$ as $l \to \infty$. In this paper we show that *for low values of* m the rate of convergence of $S_{m,n}(x\,;\,f)$ to $f(x)$ as $l \to \infty$, is much improved. By „low values of m"we mean that m is fixed, or bounded or at least such that

$$\lim_{l\to\infty} \frac{m}{n} = 0.$$

The approximations $S_{m,n}(x, f)$ are so-called *spline functions.* To make this paper readable we must reproduce here some of the results of the papers [10, 11, 5] with references for their proofs. This is done in §§ 1,2 and 3 below.

1. Spline functions and their variation diminishing property

Let

(1.1) $$u_0 < u_1 < \ldots < u_k,$$

while m is an integer $\geqslant 2$. Let $\mathcal{S} = \mathcal{S}_m(u_0, \ldots, u_k)$ denote the class of functions $S(x)$ with the following properties:

(i) $S(x) \in C^{m-2}(-\infty, \infty)$,

(ii) $S(x)$ is a polynomial of degree $\leqslant m-1$ in each of the intervals (u_{i-1}, u_i) $(i = 1, \ldots, k)$,

(iii) $S(x) = 0$ if $x \leqslant u_0$, or if $x \geqslant u_k$.

These functions are called spline functions for reasons previously explained (see e.g. [10, § 3]). The points (1.1) are called the *knots* of the spline functions. In order to avoid a trivial situation, namely that all $S(x) \equiv 0$, we must assume that $k \geqslant m$. For reasons which will become clear later we assume that

(1.2) $$k \geqslant 2m - 1.$$

For $m = 2$ we see that $S(x)$ is a continuous broken linear function with vertices at the points (1.1) and such that (iii) holds. These functions can be conveniently represented as linear combinations of $k-1$ „roof-functions", a method already used in writing the representation (1). The roof-functions generalize in a natural way as follows:

Writing

$$x_+ = \begin{cases} x \text{ if } x \geqslant 0 \\ 0 \text{ if } x < 0, \end{cases}$$

we consider the function of two variables

$$M(x\,;\,t) = m\,(t - x)_+^{m-1}.$$

In terms of $M(x\,;\,t)$ we define the sequence of functions

(1.3) $$M_j(x) = M(x\,;\,u_j, u_{j+1}, \ldots, u_{j+m}), \; (j = 0, 1, \ldots, k - m = l),$$

where the right side denotes the divided difference of order m of $M(x\,;\,t)$ with respect to the variable t and based on the $m+1$ points indicated in (1.3). These remarkable functions are called B-splines. They are surely familiar to many readers from the following fundamental property: If $F(x) \in C^m$, then the divided difference of $F(x)$ admits the representation

$$F(u_j, u_{j+1}, \ldots, u_{j+m}) = \frac{1}{m!}\int_{u_j}^{u_{j+m}} M_j(x)\, F^{(m)}(x)\, dx.$$

It is readily seen that $M_j(x) \in \mathcal{S}$. Moreover

(1.4) $$M_j(x) > 0 \text{ if } u_j < x < u_{j+m}, \quad M_j(x) = 0 \text{ if } x \leqslant u_j, \text{ or } x \geqslant u_{j+m}.$$

The functions $M_j(x)$ are called B-splines, or basis-splines for the following reason: *Every $S(x) \in \mathcal{S}$ may be uniquely represented in the form*

(1.5) $$S(x) = \sum_{j=0}^{l} c_j M_j(x), \quad (l = k - m),$$

(see [3, § 4]).

A further noteworthy property is the following:

The relation (1.5) *implies the inequality*

(1.6) $$v(S) \leqslant v(c_0, c_1, \ldots, c_l),$$

where the left side denotes the number of changes of sign of $S(x)$ for all real x (see [11]).

2. An identity and some corollaries

This identity states that (see [5])

(2.1) $$m(z - x)^{m-1} = \sum_{j=0}^{l} (u_{j+m} - u_j)(z - u_{j+1})(z - u_{j+2}) \ldots (z - u_{j+m-1})\, M_j(x)$$

for all real or complex z and all real x restricted to the interval

(2.2) $$I = \{u_{m-1} \leqslant x \leqslant u_{k-m+1}\}.$$

Observe that I is a proper interval in view of our assumption (1.2). By equating coefficients of like powers of z on both sides of (2.1) we obtain identities in x valid in the interval I. Equating the coefficients of z^{m-1} we obtain the following:

Defining the functions $N_j(x)$ by

$$(2.3) \qquad N_j(x) = \frac{1}{m}(u_{j+m} - u_j)\, M_j(x) \quad (j = 0, 1, \ldots, l)$$

we find that

$$(2.4) \qquad \sum_0^l N_j(x) = 1 \quad \text{if} \quad x \in I.$$

Moreover, a comparison of the coefficients of z^{m-2} yields the relation

$$(2.5) \qquad \sum_0^l \xi_j N_j(x) = x \quad \text{if} \quad x \in I,$$

where we have set

$$(2.6) \qquad \xi_j = \frac{1}{m-1}(u_{j+1} + u_{j+2} + \cdots + u_{j+m-1}) \quad (j = 0, 1, \ldots, l).$$

The relations (2.4), (2.5) and (2.6) were first described by T.N.E. GREVILLE in his supplement to SCHOENBERG'S paper [10].

From (2.6) and (1.1) it is clear that

$$(2.7) \qquad u_0 < \xi_0 < \xi_1 < \cdots < \xi_l < u_k.$$

The properties (2.4), (2.5) and (2.7) suggest the following approximation method.

If $f(x)$ is defined in the interval $[u_0, u_k]$ we construct the spline function

$$(2.8) \qquad S(x) = S(x\,; f) = \sum_0^l f(\xi_j)\, N_j(x)$$

which we regard as an approximation to $f(x)$.

The points ξ_j are called the *nodes* of the method because the values $f(\xi_j)$ are used in the construction of the approximation.

From (2.4) and (2.5) we conclude the following:

(2.9) *If $f(x)$ is linear, then $S(x, f) = f(x)$ in the interval* (2.2).

Moreover, from the property (1.6), in view of (2.3) and (2.7), we may argue as follows:

$$v(S) \leqslant v\left(\frac{1}{m}(u_{j+m} - u_j)\, f(\xi_j)\right) = v(f(\xi_j)) \leqslant v(f).$$

Thus the new approximation (2.8) is seen to possess the fundamental properties of the approximations (1) and (3). The only difference yet remaining is the fact that these properties of the approximation (2.8) do not hold in the interval $[u_0, u_k]$, but in its proper subinterval $I = [u_{m-1}, u_{k-m+1}]$.

3. The final approximation method

The difference just mentioned is easily overcome by the following device: *We coalesce the first m knots* (1.1) *and also the last m knots.*

In symbols: Instead of the inequalities (1.1) we shall now assume that

$$u_0 = u_1 = \ldots = u_{m-1} < u_m < \ldots < u_{k-m} < u_{k-m+1} = \ldots = u_k \tag{3.1}$$

We retain the old notation u_j and will also use the new notation described by

$$\begin{aligned} x_0 &= u_0 = \ldots = u_{m-1} \\ x_1 &= u_m \\ &\ \ \vdots \\ x_{n-1} &= u_{k-m} \\ x_n &= u_{k-m+1} = \ldots = u_k \end{aligned} \tag{3.2}$$

where $n = k - 2m + 2$. Since $l = k - m$ we also have

$$l = m + n - 2. \tag{3.3}$$

Our entire previous discussion, results and formulae, remain valid provided that the symbol $\mathcal{S}$ now denotes the class of functions $S(x)$, defined in $[x_0, x_n]$ such that

$$S(x) \in C^{m-2}[x_0, x_n] \tag{3.4}$$

(3.5) $S(x)$ is a polynomial of degree $\leqslant m - 1$ in each of the intervals (x_{i-1}, x_i) $(i = 1, \ldots, n)$.

Thus $S(x)$ is an arbitrary spline function of degree $m-1$ defined in $[x_0, x_n]$ and having the knots $x_1, \ldots, x_{n-1}$.

The definition (1.3) of the B-splines $M_j(x)$ retains its validity now in terms of divided differences with *coalescent* points. Again

$$S(x) = \sum_0^l c_j M_j(x) \quad (l = k - m = m + n - 2)$$

furnishes a unique representation of every $S(x)$ of the class $\mathcal{S}$ defined by (3.4), (3.5).

Also the identity (2.1) remains valid and its several consequences. However: *From* (2.2) *and* (3.1) *we see that now*

$$I = [x_0, x_n] \tag{3.6}$$

and the identities (2.4) *and* (2.5) *are seen to hold in the entire interval* $[x_0, x_n]$.

From (2.6) and (3.1) we see that the inequalities (2.7) are now to be replaced by

$$x_0 = \xi_0 < \xi_1 < \ldots < \xi_{l-1} < \xi_l = x_n. \tag{3.7}$$

The approximation (2.8) remains unchanged, namely

$$S(x) = S(x; f) = \sum_0^l f(\xi_j) \; N_j(x), \quad (x_0 \leqslant x \leqslant x_n) \tag{3.8}$$

and now enjoys all the properties previously described, in the full interval (3.6).

We specialize our approximation method still more: *For the remainder of this note we shall discuss the approximation* (3.8) *for the following special choice of the points* (3.2):

$$x_0 = 0, \quad x_1 = \frac{1}{n} \quad , x_2 = \frac{2}{n}, \ldots, \quad x_n = 1. \tag{3.9}$$

For the explicit expressions of the fundamental functions $N_j(x)$ in this case (particulary if $m \leqslant n$) we refer to [10].

Given the natural number l, we select $m \geqslant 2$ an $n \geqslant 1$ such that $l = m + n - 2$ and construct the approximation (3.8) corresponding to the points (3.9). We denote the approximating spline function by

$$S_{m,n}(x) = S_{m,n}(x\,;\, f) = \sum_0^l f(\xi_j) \; N_j(x), \quad (0 \leqslant x \leqslant 1). \tag{3.10}$$

Noteworthy special cases: 1. If we select $m = 2$, hence $n = l$, then

$$S_{2,l}\,(x) \tag{3.11}$$

is identical with the broken linear function (1). 2. If we select $n = 1$, hence $m = l + 1$ then

(3.12) $$S_{l+1,1}(x)$$

is identical with the Bernstein polynomial (3). (For details see [10].

4. The approximation of $f(x) = x^2$

It was pointed out in [9] that a variation diminishing approximation method such as (3.10) can not possibly reproduce exactly the function $f(x) = x^2$. Also its performance in this case is crucial for its quality and precise information for this special case is the main point of our discussion. Such information may be readily obtained from the dentity (2.1).

The approximating spline function being, by (3.10),

(4.1) $$S_{m,n}(x\,;\ t^2) = \sum_0^l (\xi_j)^2 N_j(x)$$

we define its error function by

(4.2) $$E(x) = E_{m,n}(x) = \sum_0^l (\xi_j)^2 N_j(x) - x^2.$$

We appeal once more to the identity (2.1), this time comparing the coefficients of z^{m-3} on both sides. Assuming that

(4.3) $$m \geqslant 3,$$

as we obviously must, we readily find that

(4.4) $$x^2 = \sum_0^l \xi_j^{(2)} N_j(x),\ 0 \leqslant x \leqslant 1,$$

where

4.5) $$\xi_j^{(2)} - \frac{1}{\binom{m-1}{2}} \sum_{j+1 \leqq r < s \leqq j+m-1} u_r\, u_s$$

From (4.2) and (4.4) we find that

(4.6) $$E_{m,n}(x) = \sum_0^l \lambda_j N_j(x)$$

where we set

$$\lambda_j = (\xi_j)^2 - \xi_j^{(2)}. \tag{4.7}$$

Finally, a simple algebric computation shows that (2.6), (4.5) and (4.7) imply the useful expression

$$\lambda_j = \frac{1}{(m-1)^2(m-2)} \sum_{j+1 \leqslant r < s \leqslant j+m-1} (u_s - u_r)^2. \tag{4.8}$$

Before we proceed we wish to describe the simple qualitative properties of the error function $E(x)$.

Lemma 1. *If* $m \geqslant 3$ *then* $E_{m,n}(x)$ *behaves as follows:*

$$E_{m,n}(0) = E_{m,n}(1) = 0, \ E_{m,n}(x) > 0 \ \text{if} \ 0 < x < 1, \tag{4.9}$$

$$E_{m,n}(x) = E_{m,n}(1-x), \ (0 \leqslant x \leqslant 1), \tag{4.10}$$

(4.11) $E_{m,n}(x)$ *is non-decreasing in* $\left[0, \frac{1}{2}\right]$, *non-increasing in* $\left[\frac{1}{2}, 1\right]$,

$$\max_x E_{m,n}(x) = E_{m,n}\left(\frac{1}{2}\right). \tag{4.12}$$

Proof: The vanishing of the error function at 0 and 1 is a general property of the approximation (3.10). If $0 < x < 1$ then $E_{m,n}(x) > 0$ for the following reasons: In (4.6) the λ_j $(j = 1, \ldots, l-1)$ are positive by (4.8) and (3.2). Moreover $N_j(x) \geqslant 0$ for all j and all x while for each x, such that $0 < x < 1$, we can find j, $1 \leqslant j \leqslant l-1$ such that $N_j(x) > 0$.

Concerning (4.10): (4.8), (3.2) and (3.9) show that λ_j $(j = 0, \ldots, l)$ is a symmetric sequence $(\lambda_j = \lambda_{l-j})$. This and the symmetry relation $N_j(x) = N_{l-j}(1-x)$ imply (4.10).

Again, by (4.8) λ_j increases in the first half of the range $j = 0, \ldots, l$, and decreases in the second. It follows that for every real A

$$v(\lambda_j - A) \leqslant 2.$$

In view of the relation

$$E_{m,n}(x) - A = \sum_0^l (\lambda_j - A) N_j(x),$$

we conclude from the variation diminishing property (1.6) of spline functions that also

$$v(E_{m,n}(x) - A) \leqslant 2.$$

The real A being arbitrary, (4.11) follows. Finally, (4.9) and (4.11) evidently imply (4.12).

For the extreme cases (3.11) and (3.12), the situation is as follows:
1. If $m = 2$, $n = l$, then Lemma 1 does not apply. However, the graph of $E_{2,l}(x)$ is evidently composed of n congruent parabolic arcs and a simple computation shows that

$$\max_x E_{2,l}(x) = \frac{1}{4n^2}. \tag{4.13}$$

2. If $n = 1$, $m = l + 1$, we may apply Lemma 1. However, familiar results on Bernstein polynomials show that

$$E_{l+1,1}(x) = \frac{x(1-x)}{l}. \tag{4.14}$$

Our general result is as follows.

THEOREM 1. *If*

$$l \to \infty \text{ and } \frac{m}{n} \to t \quad (0 \leqslant t \leqslant +\infty), \tag{4.15}$$

then

$$\lim nE_{m,n}\left(\frac{1}{2}\right) = \frac{1}{12}t \quad \text{if} \quad 0 \leqslant t \leqslant 1, \tag{4.16}$$

and

$$\lim (m-1)E_{m,n}\left(\frac{1}{2}\right) = \frac{1}{4} - \frac{1}{6t} \quad \text{if} \quad 1 \leqslant t \leqslant +\infty. \tag{4.17}$$

Our proof (§ 7) requires a certain amount of preparation.

5. On the nodes ξ_j and coefficients λ_j

Beyond the statement (3.7) and the symmetry and monotonicity of the sequence $\{\lambda_j\}$ we now require precise quantitative information which is readily available. To begin with, we observe that (3.2) and (3.9) imply the relations

$$\begin{gathered} u_0 = u_1 = \cdots = u_{m-1} = 0 \\ u_m = \frac{1}{n}, \; u_{m+1} = \frac{2}{n}, \ldots, u_l = \frac{n-1}{n} \\ u_{l+1} = u_{l+2} = \cdots = u_{l+m} = 1. \end{gathered} \tag{5.1}$$

There are essentially two distinct cases depending on whether $3 \leqslant m \leqslant n + 2$, or $m > n + 2$. In the first case the sequence of numbers

$$u_{j+1}, u_{j+2}, \ldots, u_{j+m-1}$$

(out of which ξ_j and λ_j are computed by (2.6) and (4.8)), forms an arithmetic progression of step $\frac{1}{n}$ for a certain range of values of j, while in the second case this never occurs.

Elementary computations which we omit lead to the following

Lemmma 2. *The beginning terms of the sequences ξ_j and λ_j are described by the formulae*

$$\xi_j = \frac{j(j+1)}{2n(m-1)}, \tag{5.2}$$

$$\lambda_j = \frac{1}{12n^2(m-1)^2(m-2)} j(j+1)\{j^2 - j + 2(m-j-1)(2j+1)\}. \tag{5.3}$$

Case 1. *If $m \leqslant n + 2$ then* (5.2) *and* (5.3) *are valid for $j = 0, 1, \ldots, m - 2$. The remaining terms are described by*

$$\left.\begin{aligned} &(5.4) \qquad \xi_j = \frac{1}{2} + \frac{1}{n}\left(j - \frac{l}{2}\right) \\ &(5.5) \qquad \lambda_j = \frac{m}{12n^2} \end{aligned}\right\} \quad (j = m-2, \ldots, n).$$

Case 2. *If $m > n + 2$ then* (5.2) *and* (5.3) *are valid for $j = 0, 1, \ldots, n - 1$. For the remaining terms we have*

$$\left.\begin{aligned} &(5.6) \quad \xi_j = \frac{1}{2} + \frac{1}{m+1}\left(j - \frac{l}{2}\right) \\ &(5.7) \quad \lambda_j = \frac{1}{(m-1)(m-2)}\left\{\frac{m-1}{4} - \frac{1}{6}\left(n - \frac{1}{n}\right) - \right. \\ &\qquad\qquad \left. - \frac{1}{m-1}\left(j - \frac{l}{2}\right)^2\right\} \end{aligned}\right\} \quad (j = n-1, \ldots, m-1).$$

Both sequences may be completed for $j = 0, \ldots, l$ by using their symmetry properties

$$\xi_j = 1 - \xi_{l-j}, \quad \lambda_j = \lambda_{l-j} \quad (j = 0, \ldots, l). \tag{5.8}$$

6. A convergence theorem

As a preparation of the proof of Theorem 1 we shall now discuss the behavior of $S_{m,n}(x; f)$ as $l = m + n - 2 \to \infty$. This will easily follow from Lemma 2. Indeed Lemma 2 yields the following:

(6.1) $$\max \lambda_j = \frac{m}{12n^2} \quad \text{if} \quad m \leqslant n + 2$$

(6.2) $$\max \lambda_j = \frac{1}{(m-1)(m-2)} \left\{ \frac{m-1}{4} - \frac{1}{6}\left(n - \frac{1}{n}\right) - \frac{1-(-1)^l}{8(m-1)} \right\} \text{if } m > n + 2.$$

In particular (6.2) evidently implies that

(6.3) $$\max \lambda_j \leqslant \frac{1}{4(m-2)} \quad \text{if } m > n + 2.$$

Using (6.1), respectively (6.3), together with (4.6) and (2.4) we obtain

Lemma 3. *If*

(6.4) $$3 \leqslant m \leqslant n + 2 \textit{ then } 0 \leqslant E_{m,n}(x) \leqslant \frac{m}{12n^2}, \quad (0 \leqslant x \leqslant 1).$$

If

(6.5) $$m > n + 2 \textit{ then } 0 \leqslant E_{m,n}(x) \leqslant \frac{1}{4(m-2)}, \quad (0 \leqslant x \leqslant 1).$$

As a corollary we obtain the following theorem.

THEOREM 2. *If* $f(x) \in C[0, 1]$, *then*

(6.6) $$\lim_{l \to \infty} S_{m,n}(x) = f(x) \textit{ uniformly in } x,$$

irrespective of how m *and* n *behave subject to the relation* $l = m + n - 2$.

Proof: We may disregard the case $m = 2$ of linear interpolation when the uniform convergence is evident. Let us assume that $l = m + n - 2 \to \infty$. If $3 \leqslant m \leqslant n + 2$ (Case 1) then $n \to \infty$ and (6.4) shows that

$$0 \leqslant E_{m,n}(x) = S_{m,n}(x; t^2) - x^2 \leqslant \frac{m}{12n^2} < \frac{1}{6n} \to 0,$$

hence $E_{m,n}(x) \to 0$ uniformly in x.

If $m > n + 2$ (Case 2), then $m \to \infty$ and (6.5) shows that

$$0 \leqslant E_{m,n}(x) \leqslant \frac{1}{4(m-2)} \to 0,$$

hence again $E_{m,n}(x) \to 0$ uniformly in x.

Since our approximation method is *positive* and reproduces linear functions, its uniform convergence for $f(x) = x^2$ establishes Theorem 2 as a corollary of the well known Bohman-Korovkin theorem ([1], [4]).

7. Proof of Theorem 1

BOHMAN has pointed out that an approximation method of the form (3.10) which converges uniformly to $f(x)$ for continuous functions, and ours does by Theorem 2, necessarily satisfies the following condition: *For a fixed x and $\delta > 0$*

$$\lim_{l \to \infty} \sum_{|\xi_j - x| \leqq \delta} N_j(x) = 1. \tag{7.1}$$

This we shall frequently apply for $x = \frac{1}{2}$ so that

$$\lim_{l \to \infty} \sum_{\left|\xi_j - \frac{1}{2}\right| < \delta} N_j\left(\frac{1}{2}\right) = 1. \tag{7.2}$$

The way in which we use this remark is as follows: From (4.6) we derive that

$$E_{m,n}\left(\frac{1}{2}\right) \geqslant \sum_{\left|\xi_j - \frac{1}{2}\right| < \delta} \lambda_j N_j\left(\frac{1}{2}\right)$$

and therefore

$$E_{m,n}\left(\frac{1}{2}\right) \geqslant \min_{\left|\xi_j - \frac{1}{2}\right| < \delta} \lambda_j \cdot \sum_{\left|\xi_j - \frac{1}{2}\right| < \delta} N_j\left(\frac{1}{2}\right). \tag{7.3}$$

This inequality will be used repeatedly below to obtain lower estimates for $E_{m,n}\left(\frac{1}{2}\right)$.

To begin with, we assume (4.15) and that

$$0 \leqslant t < 1. \tag{7.4}$$

From (4.15) we conclude that $m \leqslant n + 2$ for sufficiently large l. For such values of l we conclude from (6.4) (Lemma 3) that

$$n\, E_{m,n}\left(\frac{1}{2}\right) \leqslant \frac{1}{12}\frac{m}{n}$$

and letting $l \to \infty$ we obtain that

$$\overline{\lim_{l\to\infty}}\; n\, E_{m,n}\left(\frac{1}{2}\right) \leqslant \frac{1}{12}t. \tag{7.5}$$

In order to obtain the opposite relation which is to establish (4.16), we observe that the arithmetic progression (5.4) covers an interval with its midpoint at $x = \frac{1}{2}$ and having a length $(n - m + 2)/n$ which converges to the limit $1 - t$. This limit being positive by (7.4) we select a

$$\delta < \frac{1 - t}{2}$$

and may argue as follows in view of (5.5) and (7.3):

$$n\, E_{m,n}\left(\frac{1}{2}\right) \geqslant \frac{m}{12n} \sum_{\left|\xi_j - \frac{1}{2}\right| < \delta} N_j\left(\frac{1}{2}\right).$$

Letting $l \to \infty$ we conclude from (4.5) and (7.3) that

$$\underline{\lim}_{l\to\infty}\; n\, E_{m,n}\left(\frac{1}{2}\right) \geqslant \frac{1}{12}t.$$

This and (7.5) establish (4.15) under the assumption (7.2).

Let us now assume that

$$1 < t \leqslant +\infty. \tag{7.6}$$

From (6.2), (4.6) and (2.4) we conclude that

$$(m - 1)\, E_{m,n}\left(\frac{1}{2}\right) \leqslant \frac{1}{m - 2}\left\{\frac{m - 1}{4} - \frac{1}{6}\left(n - \frac{1}{n}\right)\right\},$$

for indeed now we have $m > n + 2$ for sufficiently large l.

Letting $l \to \infty$ we obtain

$$\overline{\lim_{l\to\infty}}\; (m - 1)\, E_{m,n}\left(\frac{1}{2}\right) \leqslant \frac{1}{4} - \frac{1}{6t}. \tag{7.7}$$

On the other hand we notice from (5.6) that the arithmetic progression (5.6) covers an interval, centered at 1/2, of length $(m - n)/(m - 1)$ which converges to the positive limit $1 - t^{-1}$. We select a δ such that

$$0 < \delta < \frac{1}{2} - \frac{1}{2t}. \tag{7.8}$$

In order to use the inequality (7.3) we need the quantity

$$\min \lambda_j$$

for the values of j satisfying $\left|\xi_j - \frac{1}{2}\right| \leqslant \delta$. The λ_j being described by the concave quadratic (5.7) it is clear that we can proceed as follows: Let j be a continuous variable and let $\lambda(j)$ denote the righthand side of (5.7). We determine j' from the equation

$$\frac{1}{m-1}\left(j' - \frac{l}{2}\right) = \delta$$

obtaining

$$j' - \frac{l}{2} = (m-1)\,\delta$$

and substituting into $\lambda(j)$ we evidently have

$$\min_{\left|\xi_j - \frac{1}{2}\right| < \delta} \lambda_j \geqslant \lambda(j').$$

Now (7.3) gives

$$(m-1)\, E_{m,n}\left(\frac{1}{2}\right) \geqslant \frac{1}{(m-2)}\left\{\frac{m-1}{4} - \frac{1}{6}\left(n - \frac{1}{n}\right) - \right.$$
$$\left. - \frac{1}{(m-1)}(m-1)^2\,\delta^2\right\} \cdot \sum N_j\left(\frac{1}{2}\right)$$

whence

$$\lim_{l \to \infty} (m-1)\, E_{m,n}\left(\frac{1}{2}\right) \geqslant \frac{1}{4} - \frac{1}{6t} - \delta^2.$$

Letting $\delta \to 0$ we obtain a result which together with (7.7) establishes (4.17) under the assumption (7.6).

We finally turn to the last case when

$$t = 1. \tag{7.9}$$

Depending on whether $m \leqslant n+2$ or $m > n+2$ we obtain from (6.1,) respectively (6.2), that

(7.10) $$\overline{\lim_{l \leftarrow \infty}}\; n\, E_{m,n}\left(\frac{1}{2}\right) \leqslant \frac{1}{12}.$$

Let us now derive the opposite tape of inequality. Since $m/n \to 1$ we might have either $m \leqslant n+2$ or $m > n+2$ for arbitratily large l. Moreover, in either case the arithmetic progressions (5.4) in the first case, and (5.6) in the second, cover intervals of lengths which shrink to zero as $l \to \infty$. For this reason we must now use the information described by (5.2) and (5.3). As these formulae apply to both cases and because the statements (4.16) and (4.17) are equivalent for $t = 1$, a distinction of the two cases becomes irrelevant.

Our procedure is much as before: We select δ, $0 < \delta < \frac{1}{2}$ and determine the real $j'(>0)$ such that

(7.11) $$\frac{j'(j'+1)}{2n(m-2)} = \frac{1}{2} - \delta.$$

We substitute j' so obtained into

$$\lambda(j) = \frac{1}{12n^2(m-1)^2(m-2)}\, j(j+1)\,\{j^2 - j + 2(m-j-1)(2j+1)\}$$

observing that

(7.12) $$\min_{\left|\xi_j - \frac{1}{2}\right| < \delta} \lambda_j \geqslant \lambda(j').$$

As $l \to \infty$, (7.11) implies that $j' \sim n\sqrt{1-2\delta}.$ A short computation shows that

$$\lim_{l \to \infty} n\lambda(j') = \frac{1}{12}(1-2\delta)^{\frac{3}{2}}\{4 - 3\sqrt{1-2\delta}\}.$$

Now (7.2), (7.3) and (7.12) imply

$$\underline{\lim}\; n\, E_{m,n}\left(\frac{1}{2}\right) \geqslant \frac{1}{12}(1-2\delta)^{\frac{3}{2}}(4 - 3\sqrt{1-2\delta}).$$

Letting $\delta \to 0$ we obtain the desired conclusion.

8. A comparison of the several approximations

Let l be prescribed and let us consider the l approximations

$$S_{m,n}(x;f)=\sum_{0}^{l} f(\xi_j)\,N_j(x), \quad (l=m+n-2), \tag{8.1}$$

each of which is constructed by means of $l+1$ approximate values of the function. *How are we to choose the parameters m and n, such that $m+n-2=l$, in order to obtain a good approximation?*

Theorem 1 of § 4 will help in answering this question. Indeed, it seems reasonable to judge our approximations by their performance for $f(x)=x^2$ which is approximated with the exact maximal error

$$E_{m,n}=\max_x E_{m,n}(x)=E_{m,n}\left(\frac{1}{2}\right)$$

discussed in Theorem 1.

We first note the following consequence of Theorem 1:

Corollary 1. *If*

$$l=m+n-2\to\infty \text{ and } \frac{m}{n}\to t \ (0\leqslant t\leqslant\infty), \tag{8.2}$$

then

$$\lim l\,E_{m,n}=\varphi(t) \tag{8.3}$$

where

$$\varphi(t)=\begin{cases}\dfrac{1}{12}t(t+1) & \text{if } 0\leqslant t\leqslant 1,\\[2mm] \dfrac{1}{\ }+\dfrac{1}{12t}-\dfrac{1}{6t^2} & \text{if } 1\leqslant t\leqslant+\infty.\end{cases} \tag{8.4}$$

Proof: (8.2) implies

$$\frac{l}{n}\to 1+t,\ \frac{l}{m}\to 1+t^{-1}.$$

$0\leqslant t\leqslant 1$. then (4.16) gives

$$l\,E_{m,n}=\frac{l}{n}\,n\,E_{m,n}\to(1+t)\frac{1}{12}t=\varphi(t)$$

If $1 \leqslant t \leqslant +\infty$ then (4.17) implies

$$l\,E_{m,n} = \frac{l}{m}\,m\,E_{m,n} \to (1+t^{-1})\left(\frac{1}{4}-\frac{1}{6t}\right) = \varphi(t).$$

One would expect the function $\varphi(t)$ to be monotone. This, however, is *not* the case and we easily find the following situation: $\varphi(t) \in C[0, \infty]$, *$\varphi(t)$ increases in $[0, 4]$ and decreases in $[4, \infty]$ while*

(8.5) $$\varphi(2) = \frac{1}{4} = \varphi(+\infty),\ \max_t \varphi(t) = \varphi(4) = \frac{1}{4}+\frac{1}{96}\cdot$$

Furthermore, since $\varphi(t) = 0$ if and only if $t = 0$ we obtain

Corollary 2. *Assuming* (8.2) *then*

(8.6) $$E_{m,n} = o\left(\frac{1}{t}\right) \text{ as } l \to \infty$$

holds if and only if $t = 0$ or equivalently

(8.7) $$l \to \infty \text{ and } \frac{m}{n} \to 0.$$

These results justify the following

Conclusions: 1. *The most efficient sequences of approximations* (3.10) *satisfying* (8.2) *are those for which m has a fixed value or at least such that* (8.7) *holds.*

2. *The sequences of approximations satisfying* (8.2) *with*

(8.8) $$2 < t < \infty$$

furnish approximations which are even slightly worse than the Bernstein polynomials, at least for sufficiently large l.

Indeed $\varphi(\infty) = 1/4$ while (8.8) implies that $\varphi(t) > 1/4$ by (8.5).

In numerical applications we do not use sequences but rather single approximations. For such purposes we recommend the choice of low values of m, such as $m = 4$, and n as large as feasible. From (6.4) we see that

(8.9) $$E_{m,n} \leqslant \frac{m}{12n^2}$$

which is only slightly worse than (4.13) or

(8.10) $$E_{2,n} = \frac{1}{4n^2},$$

if $m = 4$, say. The variation diminishing property easily implies the inequality

$$E_{2,n} \leqslant E_{m,n}.$$

This disadvantage in the quality of the approximation seems more than compensated by the fact that

$$S_{m,n}(x\,;f) \in C^{m-2}[0,1].$$

Further light will be shed on this question of comparison by our next and last section.

9. Error estimates

Let $f(x)$ be a function with the modulus of continuity $\omega(\delta)$ and let us estimate the error term $R(x)$ defined by

$$f(x) = S_{m,n}(x) + R(x), \qquad (0 \leqslant x \leqslant 1).$$

From (2.4) and (3.10) we obtain

$$(9.1) \qquad |S(x) - f(x)| \leqslant \sum_{0}^{l} |f(\xi_j) - f(x)| N_j(x).$$

Using the inequality $\omega(\lambda\delta) \leqslant (\lambda + 1)\, \omega(\delta)$, with $\lambda = |\xi_j - x|\delta^{-1}$, we get

$$|f(\xi_j) - f(x)| \leqslant \omega(|\xi_j - x|) \leqslant (|\xi_j - x|\delta^{-1} + 1)\, \omega(\delta)$$

whence, in view of (9.1) and (2.4),

$$(9.2) \qquad |S(x) - f(x)| \leqslant \omega(\delta) \left\{ \frac{1}{\delta} \sum_{0}^{l} |\xi_j - x| N_j(x) + 1 \right\}.$$

On the other hand, by (2.4) and (2.5) we have

$$\left(\sum |\xi_j - x| N_j(x)\right)^2 = \left(\sum |\xi_j - x| \sqrt{N_j} \cdot \sqrt{N_j}\right)^2 \leqslant \sum (\xi_j - x)^2 N_j(x) \sum \cdot N_j(x) =$$
$$= \sum (\xi_j - x)^2 N_j(x) = \sum (\xi_j^2 - 2x\xi_j + x^2) N_j(x) = \sum (\xi_j)^2 N_j(x) - x^2 =$$
$$= S_{m,n}(x\,; t^2) - x^2 = E_{m,n}(x).$$

Substituting this estimate into (9.2) we obtain

$$(9.3) \qquad |S_{m,n}(x) - f(x)| \leqslant \omega(\delta) \left\{ \frac{1}{\delta} \sqrt{E_{m,n}(x)} + 1 \right\}, \quad (0 \leqslant x \leqslant 1).$$

From this and previous results concerning $E_{m,n}(x)$ we can readily derive

THEOREM 3. *Let $f(x)$ be defined in $[0, 1]$ with the modulus of continuity $\omega(\delta)$ and let $S_{m,n}(x)$ be the spline function* (3.10) *of degree $m-1$ and knots $1/n, 2/n, \ldots, (n-1)/n$. We distinguish two cases:*

If $2 \leqslant m \leqslant n+2$ then

$$|S_{m,n}(x) - f(x)| \leqslant \left(1 + \sqrt{\frac{m}{12}}\right) \omega\left(\frac{1}{n}\right). \tag{9.4}$$

If $m > n+2$ then

$$|S_{m,n}(x) - f(x)| \leqslant \left\{1 + \sqrt{\frac{1}{m-2}\left(\frac{m-1}{4} - \frac{1}{6}\left(n - \frac{1}{n}\right)\right)}\right\} \omega\left(\frac{1}{\sqrt{m-1}}\right). \tag{9.5}$$

Proof: If $m = 2$ then (9.4) is implied by (2) which is even slightly better. If $3 \leqslant m \leqslant n+2$ then by (6.4)

$$E_{m,n}(x) \leqslant \frac{m}{12n^2}.$$

By substituting this estimate into (9.3) and choosing $\delta = 1/n$ we immediately obtain (9.4).

In the second case we may use the relation (6.2). From (4.6) we obtain

$$E_{m,n}(x) \leqslant \frac{1}{(m-1)(m-2)}\left\{\frac{m-1}{4} - \frac{1}{6}\left(n - \frac{1}{n}\right)\right\}.$$

Substituting this into (9.3) with $\delta = 1/\sqrt{m-1}$ we obtain (9.5).

Final remarks: 1. The estimates (9.4) and (9.5) corroborate our conclusions of § 8 concerning the advantage of choosing a large n as against a large m. For $m = 4$ (9.4) becomes

$$|S_{4,n}(x) - f(x)| \leqslant \left(1 + \sqrt{\frac{1}{3}}\right) \cdot \omega\left(\frac{1}{l-2}\right). \tag{9.6}$$

This estimate is slightly worse than the estimate (2) for linear interpolation. However, this seems to be more than balanced by the fact that $S_{4,n}(x)$ is an approximation having two continuous derivatives.

For $n = 1$, $l = m - 1$, (9.5) gives

$$|B_l(x) - f(x)| \leqslant \left(1 + \frac{1}{2}\sqrt{\frac{l}{l-1}}\right) \omega\left(\frac{1}{\sqrt{l}}\right)$$

which for large l is nearly as good as POPOVICIU's estimate (4). Our Theorem 3 and its derivation was suggested by POPOVICIU's way of deriving (4).

2. As a way of improving on the estimates (2), (4), (9.4) and (9.5) we offer the following suggestion: Observe that the left-hand sides of all these inequalities remain unchanged if we replace $f(x)$ by $f(x) - Cx$. Denoting the modulus of continuity of $f(x)$ by $\omega(\delta; f(x))$ we define a modified modulus by

$$\omega^*(\delta) = \inf_C \omega(\delta; f(x) - Cx).$$

Evidently $\omega^*(\delta) \leqslant \omega(\delta) = \omega(\delta; f(x))$ while it is clear that all the estimates mentioned remain valid if we replace $\omega(\delta)$ by $\omega^*(\delta)$. The new estimates so obtained have the obvious advantage that their right-hand sides vanish whenever $f(x)$ is a linear function.

Sponsored by the Mathematics Research Center, Unites States Army, Madison, Wisconsin under Contract No.: DA–31–124–ARO–D–462.

REFERENCES

[1] Bohman H., *On approximation of continuous and of analytic functions*, Arkiv för mat. **2**, 43–56 (1952).
[2] Davis P. J., *Interpolation and approximation*, New York, 1963.
[3] Curry H. B. and Schoenberg I. J., *On Polya frequence functions* IV: *The fundamental spline functions and their limits*, J. d'Analyse Math. **XVII**, 71–107 (1966).
[4] Korovkin P. P., *Linear operators and approximation theory*, translated from the 1959 Russian Edition, Delhi 1960.
[5] Marsden M., *An identity relating spline functions and polynomials*, to appear.
[6] Natanson I. P., *Konstruktive Funktionentheorie*, translation from Russian, Berlin, 1955.
[7] Polya G. and Schoenberg I. J., *Remarks on the de la Vallée Poussin means and convex conformal maps of the circle*, Pacific J. of Math, **8**, 295–334 (1958).
[8] Popoviciu T., *Sur l'approximation des fonctions convexes d'ordre supérieur*, Mathematica, **10**, 49–54 (1935).
[9] Schoenberg I. J., *On variation diminishing approximation methods*, Proceedings of MRC Symposium „On numerical approximation", Madison, Wisconsin, 249–274 (1958).
[10] Schoenberg I. J., *On spline functions*, with a supplement by T.N.E. Greville, MRC Technical Summary Report 625, July 1966; to appear in the Proceeding of the Symposium on „Inequalities" held August 1965 at the Wright-Patterson Air Force Base, Ohio.
[11] Schoenberg I. J., *On the variation diminishing properties of spline functions*, to appear.

Received 6. VII. 1966.

Total Positivity and Variation Diminishing Transformations

I. J. Schoenberg has made outstanding mathematical contributions in several areas of analysis. In particular, he fathered the subject of total positivity and continuously nurtured it with highly original and versatile research for more than five decades. I will review in this essay the concept of total positivity and the associated variation diminishing property emphasizing its wide scope in applications and its symbiotic mathematical relationships. I will also comment on recent developments (e.g., multivariate generalizations, connections to combinatorics) of this topic. In a companion essay I briefly discuss the important class of translation total positive kernels, the Pólya frequency functions.

Total positivity is a property of considerable power that plays an important role in various domains of mechanics, mathematics, statistics and operations research. Totally positive functions figure prominently in problems involving convexity and moment spaces (e.g., Karlin and Studden (1966)), in approximation theory (e.g., de Boor et al. (1982), Pinkus (1985)), in characterizations of the eigenstructure of certain integral and differential operators and particularly in the study of vibrating coupled mechanical systems (e.g., Gantmacher and Krein (1950)). For applications in the subject of statistical mechanics, see Spitzer (1974), Liggett (1976). In statistics, the variation diminishing property serves for assessments of parametric statistical decision procedures, (Karlin (1956, 1957), Brown et al. (1981)), in the analysis of diffusion-type-stochastic processes (Karlin and McGregor (1959)), and in terms of stochastic ordering relationships between monotone processes (Harris (1977), Karlin and Rinott (1980), Van Doorn (1981)). Total positivity is useful in describing optimal policies for inventory and production processes (e.g., Arrow et al. (1958)) and in evaluating the reliability of coherent systems and in characterizations of aging distributions of components and systems (e.g., Esary, Proschan and Walkup (1967), Barlow and Proschan (1975)).

A real function $K(x,y)$ of two variables ranging over linearly ordered sets X and Y is said to be *Totally Positive of order r* (TP_r) if for all $x_1 < \ldots < x_m$ and $y_1 < y_2 < \ldots < y_m$, $x_i \in X$, $y_j \in Y$ and $1 \le qm \le r$ we have the inequalities

$$\det\left[K(x_i, y_j)\right] \ge 0.$$

When strict inequality holds we say K is *strictly totally positive* of order r (STP_r). The composition operation (matrix multiplication) $M(x,y) = \int K(x,z)L(z,y)d\sigma(z),\ \ d\sigma(z) \ge 0$, preserves total positivity. For an abundant display of TP kernels connected with differential equation systems in statistics, stochastic processes, approximation theory, see e.g., Karlin (1968).

An important feature of totally positive kernels is their variation-diminishing property (abbreviated VDP). Let $f(t)$ be defined on I, where I is an ordered set. Let

$$S^-(f) = \sup S^-[f(t_1),\ f(t_2), \cdots, f(t_k)]$$

where the supremum is extended over all sets $t_1 < t_2 < \cdots < t_k (t_i \in I)$, k is arbitrary but finite, and $S^-(x_1, x_2, \cdots, x_k)$ is the number of sign changes of the indicated sequence, zero terms being discarded. A more refined method of counting sign changes is given by

$$S^+(f) = \sup S^+[f(t_1),\ f(t_2), \cdots, f(t_k)]$$

where $S^+(x_1,\ x_2, \cdots, x_k)$ denotes the maximum number of sign changes of the indicated sequence, the zero terms being permitted to take on arbitrary signs. Let f be bounded and Borel-measurable on Y, and consider the transformation (we assume all necessary smoothness so that all integrals appearing are well defined)

$$v(x) = (Tu)(x) = \int_Y K(x,y)u(y)d\sigma(y) \qquad d\sigma(y) \geq 0.$$

(Variation Diminishing Property) *If K is TP_r, then*

$$S^-(v) = S^-(Tu) \leq S^-(u) \qquad \textit{provided} \;\; S^-(u) \leq r-1.$$

Moreover, if $S^-(u) = S^-(v) \leq r-1$, then the functions u and v exhibit the same arrangement of signs going from left to right.

Under the stronger condition that K is STP_r and $u \not\equiv 0$ a.e. σ, we have

$$S^+(Tu) \leq S^-(u) \qquad \textit{provided} \;\; S^-(u) \leq r-1.$$

The conclusion of this theorem for a finite matrix transformation was the seminal work of Schoenberg (1930) [**4***]. Extensions were later made by Motzkin, Gantmacher, Krein and Karlin among others (see, Karlin (1968) for the history on this). The important theory of cyclic variation diminishing transformations was initiated in Mairhuber, Schoenberg and Williamson (1959) [**63**].

Schoenberg (1951, 1953) [**43***], [**48***] characterized the functions f entailing the VDP for the corresponding convolution transform

$$Tu = \int_{-\infty}^{\infty} f(x-y)u(y)dy.$$

Schoenberg presented many applications of the VDP in analysis. The refinement on the Descartes rule of signs and on locating zeros of generating functions, Schoenberg (1934, 1955) [**15***], [**53***] are particularly enticing. Chapters 6 and 10 of Karlin (1968) present various applications of the variation-diminishing property to the theory of summability, to problems of interpolation, and to studies on oscillation relationships of solutions for differential equations.

The analysis of non-linear integral transformations of the Hammerstein type where the Fréchet differential operator kernel (at the origin) is appropriately totally positive was also forthcoming, e.g., Rabinowitz (1971), Rabinowitz and Crandall (1970). They provide complete bifurcation determinations exploiting the special oscillation properties of eigenfunctions associated with operators induced by totally positive kernels.

Recent formulations of multivariate total positivity center on the following. A function $f(\mathbf{x})$ defined on $X = X_1 \otimes X_2 \otimes \ldots \otimes X_n$ is said to be *multivariate totally positive of order* 2 (MTP_2) if it satisfies $f(\mathbf{x} \vee \mathbf{y})f(\mathbf{x} \wedge \mathbf{y}) \geq f(\mathbf{x})f(\mathbf{y})$, where the lattice operations $\vee$ and $\wedge$ refer to the usual ordering on X induced by the order relations on X_i and each X_i is totally ordered. A random vector $Z = (Z_1, Z_2, \ldots, Z_n)$ of n real components is MTP_2 if its density is MTP_2. Examples include independent random variables, absolute values of a Gaussian vector variable whose covariance matrix Σ satisfies $-D\Sigma^{-1}D$ with non-negative off-diagonal elements for some diagonal matrix D, characteristic roots of random Wishart matrices, multivariate logistic, Gamma and F distributions, see Karlin and Rinott (1981).

Multivariate total positivity and generalized convexity play an important role with respect to measures of association among vector random variables, e.g., Karlin and Rinott (1980), in orderings among symmetric sampling plans from finite populations (Karlin (1974)), comparisons of probability measures in Euclidean spaces (Eaton (1982)), statistical applications including multivariate ranking and selection procedures (e.g., Rinott (1978)), and power functions assessments (e.g., Perlman and Olkin (1981)). The theory also applies in establishing correlation inequalities in statistical mechanics models including the FKG inequalities, e.g., Fortuin et al. (1973), Kemperman (1977), Liggett (1976), and in the theory of approximations in several variables, e.g., Dahmen and Micchelli (1981).

Total positivity is intimately connected with diffusion stochastic processes. It is proved in Karlin and McGregor (1959) that if $p(t,x,y)$ $(P(t,x,E)=\int_E p(t,x,y)dy$ is the probability that a particle initially at x is found in E at time t) is the transition density of a Markov process on the line with continuous paths then for each t, the kernel $K_t(x,y)=p(t,x,y)$ is TP. The determinant $\det[P(t,x_i,E_j)]$, $x_1<\cdots<x_n$, $E_1<\cdots<E_n$ ($E_i<E_j$ means that the set E_i lies to the left of E_j) has the following interpretation. Consider n labeled particles initially at positions $x_1, x_2, \ldots, x_n$ that execute the indicated process simultaneously and independently. The probability is $\det[P(t,\ x_i,\ E_j)]$ that particle i initially at x_i is in E_i at time t, $i=1,\ldots,n$ and no two of the particles are coincident in the intervening time epoch.

These determinant formulas (which we have called noncoincidence probabilities and also compound probabilities) are pertinent to a broad spectrum of classical combinatorial problems as exemplified below.

(i) Weak Ballot Problem. Let $A_1,\ldots,A_r$ be r candidates who have received $a_1,\ldots,a_r$ votes, respectively, $a_1\geq a_2\geq\ldots\geq a_r$. Assume the aggregate votes are randomly ordered. The probability that in the counting A_i leads (has at least as many votes as) A_j, for all pairs, $i\leq j$, is given by the formula

$$\det\left[\frac{a_i!}{(a_i+i-j)!}\right].$$

This result can be derived as a noncoincidence probability applying the compound transition probabilities to independent Poisson processes, Barton and Mallows (1965).

(ii) There are $n!\det\left[\frac{1}{\lambda_j+(i-j)!}\right]$ random Young Tableaux (see Stanley (1986)) of shape $\lambda_1\geq\lambda_2\geq\ldots\geq\lambda_r$, $\lambda_1+\cdots+\lambda_r=n$. This is again a compound probability.

(iii) Descent sets of a given structure for random permutations also possess determinantal representations, see Stanley (1986, Chapter 2).

(iv) The determinant of binomial coefficients $\binom{b_i}{a_j}$, $a_1<\cdots<a_r$, $b_1<\cdots<b_r$ has recently been interpreted as a count on certain classes of nonintersecting paths (Gessel and Veinnot (1985)). This is a version of the compound probability formula with transitions involving variable times.

The above examples are typical of a vast array of total positivity determinant expressions connecting to combinatorial identities. Obviously, the subject of total positivity and variation diminishing transformations in one and several variables is important, vibrant and forthcoming with mathematical discovery.

References

Arrow, K.J., S. Karlin, and H.E. Scarf (1958). *Studies in the Mathematical Theory of Inventory and Production*, Stanford University Press, Stanford, California.

Barlow, R.E. and F. Proschan (1975). *Statistical Theory of Reliability and Life Testing*, Holt, Rinehart & Winston, New York.

Barton, D.E. and C.L. Mallows (1965). "Some aspects of random sequences," *Ann. Math. Statist.* **37**, 236–260.

de Boor, C., S. Friedland, and A. Pinkus (1982). "Inverses of infinite sign regular matrices," *Trans. Amer. Math. Soc.* **274**, 59–68.

Brown, L.J., I.M. Johnstone, and B.K. MacGibbon (1981). "Variation diminishing transformations: A direct approach to total positivity and its statistical applications," *J. Amer. Statist. Assoc.* **76**, 824–832.

Dahmen, W. and C. Micchelli (1981). "On limits of multivariate B-splines," *J. Analyse Math.* **39**, 256–278.

Fortuin, C.M., S. Ginibre, and P.W. Kasteleyn (1971). "Correlation inequalities on some partially ordered sets," *Comm. Math. Phys.* **22**, 89–103.

Gantmacher, F.R. and M.G. Krein (1950). *Oscillatory Matrices and Kernels and Small Vibrations of Mechanical Systems*, 2nd ed., Moscow (Russian).

Gessel, I. and G. Viennot (1985). "Binomial determinants, paths, and hook length formulae," *Adv. in Math.* **58**, 300–321.

Harris, T.E. (1977). "A correlation inequality for Markov processes in partially ordered state space," *Ann. Probab.* **5**, 451-454.

Karlin, S. (1956). "Decision theory for Pólya distributions, case of two actions, I," *Proc. Third Berkeley Symposium on Probability and Statistics*, Vol. 1, University of California Press, Berkeley, California, pp. 115–29.

Karlin, S. (1957). "Pólya-Type distributions, II," *Ann. Math. Statist.* **28**, 281–308.

Karlin, S. (1964). "Total positivity, absorption probabilities and applications," *Trans. Amer. Math. Soc.* **111**, 33–107.

Karlin, S. (1968). *Total Positivity*, Volume I, Stanford University Press, Stanford, California.

Karlin, S. and Y. Rinott (1980). "Classes of orderings of measures and related correlation inequalities: I. Multivariate totally positive distributions," *J. Multivariate Anal.* **10**, 467–497.

Karlin, S. and Y. Rinott (1981). "Total positivity properties of absolute value multinormal variables with applications to confidence intervals estimates and related probabilistic inequalities," *Ann. Statist.* **9**, 1035–1049.

Karlin, S. and J.L. McGregor (1959). "Coincidence probabilities," *Pacific J. Math.* **9**, 1141–64.

Karlin, S. and W. Studden (1966). *Tchebycheff Systems: With Applications in Analysis and Statistics*, Interscience Publishers, New York.

Kemperman, J.H.B. (1977). "On the FKG-inequality for measures on a partially ordered space," *Indag. Math.* **39**, 313–331.

Liggett, T.M. (1976). "Coupling the simple exclusion process," *Ann. Probab.* **4**, 339–356.

Mairhuber, J.C., I.J. Schoenberg, and R.E. Williamson (1959). "On variation diminishing transformations of the circle," *Rend. Circ. Mat. Palermo (2)* **8**, 241–70. [**63**]

Perlman, M.D. and I. Olkin (1980). "Unbiasedness of invariant tests for manova and other multivariate problems," *Ann. Statist.* **8**, 1326–1341.

Pinkus, A. (1985). *n-Widths in Approximation Theory*, Springer-Verlag.

Rabinowitz, P. (1971). "Some global results for nonlinear eigenvalue problems," *J. Funct. Anal.* **7**, 487–513.

Rabinowitz, P. and M. Crandall, (1970). "Nonlinear Sturm Liouville eigenvalue problems," *J. Math. Mech.* **18**, 1083–1102.

Rinott, Y. (1978). "On two stage selection procedures and related probability inequalities," *Comm. in Stat* **8**, 799–811

Schoenberg, I.J. (1930). "Über variationsvermindernde lineare Transformationen," *Math. Z.* **32**, 321–28. [**4***]

Schoenberg, I.J. (1934). "Zur Abzählung der reellen Wurzeln algebraischer Gleichungen," *Math. Z.* **38**, 546–64. [**15***]

Schoenberg, I.J. (1951). "On Pólya frequency functions, I: The totally positive functions and their Laplace transforms," *J. Analyse Math.* **1**, 331–74. [**43***]

Schoenberg, I.J. (1953). "On smoothing operations and their generating functions," *Bull. Amer. Math. Soc.* **59**, 199–230. [**48***]

Schoenberg, I.J. (1955). "On the zeros of the generating functions of multiply positive sequences and functions," *Ann. of Math.* **62** (1955), 447–471. [**53***]

Schoenberg, I.J. and A. Whitney (1953). "On Pólya frequency functions, III: The positivity of translation determinants with an application to the interpolation problem by spline curves," *Trans. Amer. Math. Soc.* **74**, 246–59. [**47***]

Spitzer, F. (1974). "Introduction aux processus de Markov à parametre dans Z_ν." *Lecture Notes in Math.* **390**, Springer-Verlag, 115–189.

Stanley, R.P. (1986). *Enumerative Combinatorics*, Wadsworth Co., Belmont, California.

Van Doorn, E. (1981). *Stochastic Monotonicity and Queuing Applications of Birth-Death Processes*, Springer-Verlag Lecture Notes in Statistics.

Samuel Karlin
Stanford University

POLYA FREQUENCY FUNCTIONS

ON PÓLYA FREQUENCY FUNCTIONS

I. THE TOTALLY POSITIVE FUNCTIONS AND THEIR LAPLACE TRANSFORMS[1]

By

I. J. Schoenberg

in Philadelphia, Pa., U.S.A.

To the memory of Dolli Schoenberg née Landau.

Introduction

1. We denote by T_1 the class of entire functions which are limits, uniform in every finite domain, of real polynomials with only real non-positive zeros. Likewise we denote by T_2 the wider class of entire functions obtained if in the previous definition we only require that the approximating polynomials be real and have only real zeros. From the classical work of Laguerre and Pólya[2] we know that $\Phi(s) \in T_1$ if and only if $\Phi(s)$ admits a representation of the form

$$\Phi(s) = Ce^{\gamma s} s^m \prod_{\nu=1}^{\infty} (1 + \delta_\nu s), \tag{1}$$

$$(C \text{ real}, \ \gamma \geqq 0, \ \delta_\nu \geqq 0, \ \Sigma\delta_\nu < \infty),$$

and also that the elements $\Psi(s)$ of the class T_2 are characterized by the representation

$$\Psi(s) = Ce^{-\gamma s^2 + \delta s} s^m \prod_{\nu=1}^{\infty} (1 + \delta_\nu s) e^{-\delta_\nu s}, \tag{2}$$

$$(C \text{ real}, \ \gamma \geqq 0, \ \delta, \ \delta_\nu \text{ real}, \ \Sigma\delta_\nu^2 < \infty).$$

Let now $\Psi(s) \in T_2$, with $\Psi(0) > 0$, and such that $\Psi(s)$ does not reduce to the form $Ce^{\delta s}$, hence

$$\Psi(s) = Ce^{-\gamma s^2 + \delta s} \prod_{\nu=1}^{\infty} (1 + \delta_\nu s) e^{-\delta_\nu s}, \tag{3}$$

$$(C > 0, \ \gamma \geqq 0, \ \delta, \ \delta_\nu \text{ real}, \ 0 < \gamma + \Sigma\delta_\nu^2 < \infty),$$

---

1. The present work was partly sponsored by the Office of Naval Research at the University of Pennsylvania. A brief account of the main results appeared in 1947; see [15] in the bibliography at the end of the paper.

2. See [8] and [11].

and let us expand its reciprocal in powers of s

(4) $$\frac{1}{\Psi(s)} = \mu_0 - \frac{\mu_1}{1!}s + \frac{\mu_2}{2!}s^2 - \dots .$$

In 1915 Pólya[3] proved that the coefficients μ_n enjoy the following property: The determinants

(5) $$\begin{vmatrix} \mu_0 & \mu_1 & \dots & \mu_n \\ \mu_1 & \mu_2 & \dots & \mu_{n+1} \\ \cdot & & & \\ \cdot & & & \\ \cdot & & & \\ \mu_n & \mu_{n+1} & \dots & \mu_{2n} \end{vmatrix} \quad \text{are positive, for } n = 0, 1, \dots .$$

If $\Phi(s) \in T_1$, $\Phi(0) > 0$, not of the form $Ce^{\delta s}$, hence

(6) $$\Phi(s) = Ce^{\gamma s} \prod_{\nu=1}^{\infty} (1 + \delta_\nu s),$$
$$(C > 0,\ \gamma \geqq 0,\ \delta_\nu \geqq 0,\ 0 < \gamma + \Sigma\delta_\nu < \infty),$$

and if we expand its reciprocal in powers of s

(7) $$\frac{1}{\Phi(s)} = \mu_0 - \frac{\mu_1}{1!}s + \frac{\mu_2}{2!}s^2 - \dots,$$

then Pólya showed that also the determinants

(8) $$\begin{vmatrix} \mu_0 & \mu_1 & \dots & \mu_n \\ \mu_1 & \mu_2 & \dots & \mu_{n+1} \\ \cdot & & & \\ \cdot & & & \\ \cdot & & & \\ \mu_n & \mu_{n+1} & \dots & \mu_{2n} \end{vmatrix}, \quad \begin{vmatrix} \mu_1 & \mu_2 & \dots & \mu_{n+1} \\ \mu_2 & \mu_3 & \dots & \mu_{n+1} \\ \cdot & & & \\ \cdot & & & \\ \cdot & & & \\ \mu_{n+1} & \mu_{n+2} & \dots & \mu_{2n+1} \end{vmatrix} \quad \text{are positive, for } n = 0, 1, \dots .$$

In 1920 H. Hamburger[4] applied his solution of the Hamburger moment problem to conclude from (5) that the function (4) admits a representation as a Laplace-Stieltjes integral

(9) $$\frac{1}{\Psi(s)} = \int_{-\infty}^{\infty} e^{-xs}\, d\psi(x), \qquad (\psi(x) \text{ non-decreasing}),$$

in the maximal strip $\alpha < Rs < \beta$, containing the origin, in which the

3. See [12], theorems IV and V on pages 235—236.
4. See [5].

function $1/\Psi(s)$ is regular. He likewise concluded from (8) that the function (7) admits a representation

$$(10)\qquad \frac{1}{\Phi(s)} = \int_0^\infty e^{-xs}\, d\varphi(x), \qquad (\varphi(x) \text{ non-decreasing}),$$

in the maximal half-plane $Rs > \alpha$ of regularity of $1/\Phi(s)$.

2. The question as to the nature of the determining functions $\psi(x)$ and $\varphi(x)$ in the formulae (9) and (10) has so far remained unanswered.[5] An answer to this question is the main result of the present paper. It needs for its formulation the following definition: A *frequency function* $\Lambda(x)$, i.e. a non-negative measurable function satisfying the inequalities

$$0 < \int_{-\infty}^{\infty} \Lambda(x)\, dx < \infty,$$

is said to be a *Pólya frequency function* if and only if it satisfies the following condition[6]:

For every two sets of increasing numbers

$$x_1 < x_2 < \ldots < x_n, \quad t_1 < t_2 < \ldots < t_n, \qquad (n = 1, 2, \ldots),$$

we have the inequality

$$(11)\qquad \det \| \Lambda(x_i - t_k) \|_{1,n} \geqq 0.$$

An answer to the representation problem formulated above is now contained in the following two theorems.

I. *If $\Psi(s)$ is defined by (3) then its reciprocal may be represented in the form*

$$(12)\qquad \frac{1}{\Psi(s)} = \int_{-\infty}^{\infty} e^{-xs}\, \Lambda(x)\, dx,$$

where $\Lambda(x)$ is a Pólya frequency function, a representation which is valid in the maximal vertical strip containing the origin in which the left-hand side of (12) is regular.

5. We may interpret as a step in this direction some aspects of Wintner's paper [22]: For the special case when $\Psi(s)$, defined by (3), is even and $\gamma = 0$, Wintner's Theorem 1 gives certain necessary (but not sufficient) properties of the function $\psi(x)$ of (9).

6. The reasons for the name "Pólya frequency function" are given in [17].

Conversely, if $\Lambda(x)$ is a Pólya frequency function, then the integral (12) converges in a vertical strip containing the origin inside and represents there the reciprocal of a function $\Psi(s)$ of the form (3).[7]

II. If $\Phi(s)$ is defined by (6) then its reciprocal may be represented in the form

$$\frac{1}{\Phi(s)} = \int_0^\infty e^{-xs}\Lambda(x)\,dx, \tag{13}$$

within the maximal half-plane $Rs>\alpha$ of regularity of the left-hand side, where $\Lambda(x)$ is a Pólya frequency function such that $\Lambda(x)=0$ if $x<0$.

Conversely, if $\Lambda(x)$ is a Pólya frequency function such that $\Lambda(x)=0$ if $x<0$, then the integral (13) converges in a half-plane $Rs>\alpha$, $(\alpha<0)$, and represents there the reciprocal of a function $\Phi(s)$ of the form (6).

The totally positive functions mentioned in the title differ from Pólya frequency functions only by trivial exponential factors and give rise to Laplace integral representations of reciprocals $1/\Psi(s)$ in vertical strips not necessarily containing the origin.

Part I. The totally positive functions and their Laplace transforms

§ 1. Definition and elementary properties of totally positive functions.

3. Definition 1. A real-valued function $\Lambda(x)$, defined for all real x, is said to be totally positive if and only if it satisfies the following three conditions:

(α) $\Lambda(x)$ is measurable.

(β) If $x_1<x_2<\dots<x_n$, $t_1<t_2<\dots<t_n$, then

7. During the last few years Hirschman and Widder have investigated ways of inverting the convolution transformation (2.2) below, for the Pólya frequency functions $\Lambda(x)$ arising from (3) and (12), for $\gamma=0$; see [6] and [7]. Widder has recently announced an extension of this theory which includes all Pólya frequency functions; see [21].

(1.1) $$\det \| \Lambda(x_i - t_j) \|_{1,n} \geqq 0 \,.$$ [8]

This inequality should hold for all $n = 1, 2, 3, \ldots$.

(γ) $\Lambda(x) \neq 0$ for at least two distinct values of x.

For $n=1$, the inequality (1.1) is equivalent to $\Lambda(x) \geqq 0$ for all real x, justifying the name of total positivity for our set of conditions. A trivial example of a totally positive function is

$$\Lambda(x) = e^x,$$

because all determinants (1.1) vanish if $n > 1$.

The following remarks are obvious: 1. If $\Lambda(x)$ is totally positive then so is $\Lambda(-x)$ and generally $\Lambda(ax+b)$ $(a \neq 0)$. 2. If $\Lambda(x)$ is totally positive, then so is $\Lambda(x) \,.\, \exp(ax+b)$.

We shall now discuss two significant examples of totally positive functions: 1. The function

(1.2) $$\lambda_0(x) = \begin{cases} 1 & \text{if } x \geqq 0, \\ 0 & \text{if } x < 0, \end{cases}$$

is totally positive.

To see this we notice that the determinant $D_n = \det \| \lambda_0(x_i - t_j) \|$, of the form (1.1), has the following three properties: 1. Its elements are $=0$ or 1. 2. The elements in every row are non-increasing. 3. The elements in every column are non-decreasing. A moment's reflection will show these properties to imply that $D_n = 0$, unless D_n is of the form

$$D_n = \begin{vmatrix} 1 & 0 & 0 & \ldots & 0 \\ 1 & 1 & 0 & \ldots & 0 \\ 1 & 1 & 1 & \ldots & 0 \\ \cdot & & & & \\ \cdot & & & & \\ \cdot & & & & \\ 1 & 1 & 1 & \ldots & 1 \end{vmatrix} = 1 \,.$$

2. The function

(1.3) $$\Lambda(x) = e^{-x^2}$$

is totally positive.

---

8. See [16] for the necessary and sufficient conditions on the values of x_i, t_i $(i=1, \ldots, n)$ and the function $\Lambda(x)$ in order that the determinant (1.1) be actually positive.

Indeed

$$\det \| e^{-(x_i-t_j)^2} \| = \det \| e^{-x_i^2} . e^{-t_j^2} . e^{2x_it_j} \| = e^{-\Sigma x_i^2 - \Sigma t_j^2} . \det \| e^{2x_it_j} \| > 0 ,$$

the positivity of the last determinant being known. (9)

The following question arises: are the three conditions (α), (β), (γ), of Definition 1, independent? That they are, is shown by the following statement: *Of the three conditions* (α), (β), (γ), *neither two imply the third.*

Indeed, it is clear that (α) and (β) do not imply (γ), for $\Lambda(x) = 0$ satisfies (α), (β), but not (γ). Likewise (α) and (γ) do not imply (β). That (β) and (γ) do not imply (α) is seen as follows: Let $\varphi(x)$ be a discontinuous solution of Cauchy's functional equation

$$\varphi(x+y) = \varphi(x) + \varphi(y)$$

and let

$$\Lambda(x) = e^{\varphi(x)} .$$

It is known that $\varphi(x)$ and therefore $\Lambda(x)$ is non-measurable. However, (β) is satisfied because

$$\det \| e^{\varphi(x_i-t_j)} \| = \det \| e^{\varphi(x_i)+\varphi(-t_j)} \| = 0 \text{ if } n>1 ,$$

while (γ) is clearly verified.

4. We shall now derive a few properties of totally positive functions which follow readily from their definition. Some of the properties discussed in this section are shared by wider classes of functions described by the following

Definition 2. *Let k be a positive integer $\geqq 2$. A function $\Lambda(x)$ is called k-times positive (We shall write $\Lambda(x) \in P_k$ or say that $\Lambda(x)$ is an element of P_k) if and only if it satisfies the conditions* (α) *and* (γ) *of Definition* 1, *and also* (β) *for $n=1, n=2, \ldots, n=k$.*

Evidently $P_2 \supseteqq P_3 \supseteqq \ldots \supseteqq P_k \supseteqq \ldots$, the class of totally positive functions being the intersection of this sequence of classes. We shall be especially concerned with the class P_2 of functions which are *twice-positive*. This class may be completely described in terms of convex functions, the term "convex" being defined for our purpose as follows:

9. See [13], Problem 76, page 49.

Definition 3. A real function $\psi(x)$ is said to be convex (We shall write $\psi(x) \in \Gamma$) if and only if it is convex (and continuous) in the ordinary sense in an open interval $\alpha<x<\beta$ $(-\infty \leqq \alpha<\beta \leqq +\infty)$, $\psi(x)=+\infty$ for $x<\alpha$ and for $x>\beta$ (if such values of x exist), and if α or β are finite we also require that

$$\psi(\alpha+0) \leqq \psi(\alpha) \leqq +\infty, \quad \psi(\beta-0) \leqq \psi(\beta) \leqq +\infty .$$

Thus the three functions 1) x^2, 2) $\psi(x)=1/x$ if $x>0$ and $\psi(x)=+\infty$ if $x \leqq 0$, 3) $\psi(x)=x$ if $|x|<1$, $\psi(x)=+\infty$ if $|x| \geqq 1$, are examples of functions convex in our sense. The peculiar Definition 3 is justified by the following

Lemma 1. If $\psi(x)$ is convex, then

$$\Lambda(x)=e^{-\psi(x)} \tag{1.4}$$

is twice-positive. Conversely, if $\Lambda(x)$ is twice-positive, then

$$\psi(x)=-\log \Lambda(x) \tag{1.5}$$

is convex.

Proof: 1. Let us assume that $\psi(x) \in \Gamma$ and show that $\Lambda(x) \in P_2$. Clearly $\Lambda(x)$ satisfies the conditions (α) and (γ) (because in Def. 3 we assume that $\alpha<\beta$). Since $\Lambda(x) \geqq 0$, there remains to show that if

$$x_1<x_2 \text{ and } t_1<t_2, \tag{1.6}$$

then

$$\Lambda(x_1-t_1)\, \Lambda(x_2-t_2) \geqq \Lambda(x_1-t_2)\, \Lambda(x_2-t_1). \tag{1.7}$$

If we set

$$a=x_1-t_2, \quad b=x_2-t_1, \quad c=x_1-t_1, \quad d=x_2-t_2, \tag{1.8}$$

we see that (1.6) and (1.8) imply the relations

$$a<c<b, \quad a<d<b, \quad a+b=c+d. \tag{1.9}$$

Conversely, if a, b, c, d are four numbers satisfying (1.9) then the relations (1.8) allow to determine x_1, x_2, t_1, t_2 (up to an additive constant) satisfying (1.6). Rephrasing the conditions (1.6), (1.7) in terms of a, b, c, d, we see that we have to prove that the relations (1.9) always imply the inequality

$$\Lambda(a)\,\Lambda(b) \leqq \Lambda(c)\,\Lambda(d), \tag{1.10}$$

or that they imply the inequality

$$\phi(a) + \phi(b) \geqq \phi(c) + \phi(d). \tag{1.11}$$

If $\phi(a) = +\infty$, or $\phi(b) = +\infty$, there is nothing to prove. If $\phi(a)$ and $\phi(b)$ are both finite, then a and b are within, or perhaps at the end-points of the interval of ordinary convexity of $\phi(x)$. But then we have

$$\phi(c) \leqq \frac{b-c}{b-a}\phi(a) + \frac{c-a}{b-a}\phi(b), \quad \phi(d) \leqq \frac{b-d}{b-a}\phi(a) + \frac{d-a}{b-a}\phi(b).$$

Adding these inequalities we obtain (1.11), in view of the last relation (1.9). This proves that $\Lambda(x) \in P_2$.

2. Let us now assume that $\Lambda(x) \in P_2$ and show that $\phi(x) \in \Gamma$. In any case we already know that

$$-\infty < \phi(x) \leqq +\infty, \text{ for all } x. \tag{1.12}$$

Our previous discussion shows that $\phi(x)$ enjoys the following property: If a, b, c, d, satisfy (1.9) then the inequality (1.11) holds. However, $\Lambda(x)$ is positive in at least two distinct points x_0, x_1, $(x_0 < x_1)$ by the condition (γ) of Definition 2 (for $k=2$). But then the values $\phi(x_0)$ and $\phi(x_1)$ are finite. Now (1.12) and (1.11) (for $a = x_0$, $b = x_1$) imply that

$$\phi(x) < +\infty \text{ if } x_0 < x < x_1.$$

Let (α, β) $(-\infty \leqq \alpha \leqq x_0 < x_1 \leqq \beta \leqq +\infty)$ be the maximal open interval, containing (x_0, x_1), in which $\phi(x)$ assumes only finite values, i.e. let $\phi(x)$ be finite-valued inside (α, β) but in no other interval properly containing (α, β). Then we must have

$$\phi(x) = +\infty \text{ if } x > \beta \text{ or if } x < \alpha.$$

For if there were a γ such that $\beta < \gamma$, $\phi(\gamma) < +\infty$, then let us choose a point δ in the interval (α, β), hence with $\phi(\delta) < +\infty$. Since $\phi(x)$ is finite-valued at $x = \delta$, and $x = \gamma$, it follows by our previous argument that $\phi(x)$ is finite-valued in (δ, γ), hence also in (α, γ), in contradiction with our construction of (α, β) as the maximal interval of finite-valuedness.

In the range (α, β), where $\phi(x)$ is finite, we have that

$$\alpha < a < b < \beta, \quad c = d = \frac{a+b}{2},$$

by (1.11), imply that

$$\phi(a) + \phi(b) \geqq 2\phi\left(\frac{a+b}{2}\right).$$

This shows that $\phi(x)$ is a convex function in the sense of Jensen in the interval $\alpha < x < \beta$. Being also measurable in this range, a theorem of Sierpinski[10] implies that $\phi(x)$ is continuous in (α, β) and therefore also convex in the ordinary sense. This also implies the existence of the limits $\phi(\alpha+0)$, $\phi(\beta-0) \leqq +\infty$. Finally we must also have

$$\phi(\alpha+0) \leqq \phi(\alpha) \leqq +\infty, \quad \phi(\beta-0) \leqq \phi(\beta) \leqq +\infty.$$

Indeed, let us prove the first set of inequalities. If $\phi(\alpha) = +\infty$ there is nothing to prove. If $\phi(\alpha) < \infty$, choose δ such that $\alpha < \delta < \beta$, and let x, x' satisfy $\alpha < x < \delta$, $x + x' = \alpha + \delta$. By (1.11) we have

$$\phi(x) + \phi(x') \leqq \phi(\alpha) + \phi(\delta).$$

Letting $x \to \alpha + 0$, hence $x' \to \delta - 0$, we find $\phi(x) \to \phi(\alpha + 0)$, $\phi(x') \to \phi(\delta - 0) = \phi(\delta)$. In the limit, our last inequality now becomes $\phi(\alpha+0) + \phi(\delta) \leqq \phi(\alpha) + \phi(\delta)$ or $\phi(\alpha+0) \leqq \phi(\alpha)$, which was to be established. This completes a proof that $\phi(x) \in \Gamma$.

5. We conclude from Lemma 1 that a function $\Lambda(x)$ which is twice-positive has at most two points of discontinuity, both of the first kind. A further important property of such functions is as follows:

Lemma 2. *Let $\Lambda(x)$ be twice-positive. Then $\Lambda(x)$ is either monotone (in the wide sense) for all real x, or else it tends to zero exponentially as $x \to \pm\infty$.*

Proof: If $\Lambda(x)$ is monotone, there is nothing to prove. If $\Lambda(x)$ is non-monotone, we may find two pairs of numbers a, b and a', b' such that

$$a < b, \quad a' < b', \quad \Lambda(a) > \Lambda(b), \quad \Lambda(a') < \Lambda(b').$$

We want to show that for certain positive constants C, c, C', c' we shall have

(1.13) $\qquad \Lambda(x) < Ce^{-cx} \qquad$ for sufficiently large positive values of x,

and

(1.14) $\qquad \Lambda(x) < C'e^{c'x} \qquad$ for sufficiently large negative values of x.

10. See [19].

In order to prove (1.13), we observe that if $\Lambda(b)=0$ then $\Lambda(x)=0$ for all $x>b$. This is clear by Lemma 1, for $\Lambda(x)$ is positive in a certain interval and zero outside it. But then (1.13) becomes trivial. Hence we may assume that

$$\Lambda(a)>\Lambda(b)>0 .$$

By Lemma 1, $\psi(x)=-\log\Lambda(x)$ is convex, hence $\psi(x)$ has a graph whose branch corresponding to $x>b$ will not fall below the chord joining the points $(a, \psi(a))$ and $(b, \psi(b))$. Since $\psi(a)<\psi(b)$, this chord of equation

$$y = cx + \gamma$$

has a positive slope c. Thus

$$\psi(x) = -\log\Lambda(x) \geqq cx + \gamma \quad \text{if } x>b,$$

whence

$$\Lambda(x) \leqq e^{-cx-\gamma} \quad \text{if } x>b .$$

This proves (1.13) and a similar argument using a', b' will establish (1.14).

Examples: Of our two examples of totally positive functions we see that (1.2) is monotone, while (1.3) is not, and does indeed vanish exponentially as $x\to\pm\infty$.

6. We wish now to single out a certain type of totally positive functions by the following

Definition 4. *A totally positive function $\Lambda(x)$ which is non-monotone will be called a Pólya frequency function.*

These functions may also be described as follows:

Lemma 3. *A totally positive function $\Lambda(x)$ is a Pólya frequency function, if and only if it is summable.*

Indeed, in any case we have

$$0 < \int_{-\infty}^{\infty}\Lambda(x)\,dx \leqq +\infty .$$

If $\Lambda(x)$ is monotone, the integral evidently diverges. If $\Lambda(x)$ is not monotone, then by Lemma 2, $\Lambda(x)\to 0$ exponentially, hence the integral converges.

Of our examples (1.2) and (1.3), $\exp(-x^2)$ is a Pólya frequency function, while $\lambda_0(x)$ is not; nor is $\exp(ax+b)$ a Pólya frequency function.

As a matter of fact we lose no generality by confining our study of totally positive functions to Pólya frequency functions in view of the following

Lemma 4. *If $\Lambda(x)$ is a totally positive function, not of the form e^{ax+b}, then, for an appropriate real γ, the function*

$$\Lambda_1(x) = e^{-\gamma x}\Lambda(x)$$

is a Pólya frequency function.

As an example we know that $\lambda_0(x)$ is not a Pólya frequency function, while $\lambda(x) = e^{-x}\lambda_0(x)$ is evidently one.

Proof: We know by Lemma 1 that $-\log\Lambda(x)$ is convex while our assumption implies that it is not a linear function. There exist, therefore, two linear functions of different slopes $-a$, $-b$, $(a<b)$, such that

$$-\log\Lambda(x) > -ax-p, \quad -\log\Lambda(x) > -bx-q, \quad \text{for all real } x.$$

Thus

$$\Lambda(x) < e^{ax+p}, \quad \Lambda(x) < e^{bx+q}.$$

For

$$\Lambda_1(x) = e^{-\gamma x}\Lambda(x)$$

we now have the inequalities

$$\Lambda_1(x) < e^{(a-\gamma)x+p}, \quad \Lambda_1(x) < e^{(b-\gamma)x+q},$$

which show that $\Lambda_1(x) \to 0$ exponentially as $x \to \pm\infty$, provided we select γ in the range $a<\gamma<b$ when $a-\gamma<0$, $b-\gamma>0$. Thus $\Lambda_1(x)$ is a Pólya frequency function for such values of γ.

7. Of particular importance is

Lemma 5. *If $\Lambda_1(x)$ and $\Lambda_2(x)$ are two Pólya frequency functions, then also their convolution*

$$\Lambda(x) = \int_{-\infty}^{\infty} \Lambda_1(t)\Lambda_2(x-t)\,dt \tag{1.15}$$

is a Pólya frequency function.

Proof: The integral evidently exists for all x and defines a continuous $\Lambda(x)$, since Λ_1, Λ_2 are summable and bounded. We also see from (1.15) that $\Lambda(x)$ is summable and that its integral over the real axis is positive. There remains to show that

$$(1.16) \qquad \det \| \Lambda(x_i - t_j) \|_{1,n} \geqq 0,$$

according to the condition (β) of Definition 1, Art. 3. This follows immediately from the identity :

$$(1.17) \quad \det \| \Lambda(x_i - y_j) \| = \frac{1}{n!} \int_{-\infty}^{\infty} \cdots \int \det \| \Lambda_1(x_i - t_j) \| . \det \| \Lambda_2(t_i - y_j) \| \, dt_1 \dots dt_n.$$

Indeed, granting this identity, we see that it implies (1.16) because the integrand in (1.17) is always non-negative, not each factor separately but surely the product of the two determinants, because Λ_1 and Λ_2 are totally positive.

In order to establish (1.17), we mention first that (1.15) may be written as

$$\Lambda(x_1 - y_1) = \int_{-\infty}^{\infty} \Lambda_1(x_1 - t_1) \Lambda_2(t_1 - y_1) \, dt_1,$$

which is identical with (1.17) for $n = 1$. Let us now evaluate the determinant (1.17). We find

$$\det \| \Lambda(x_i - y_j) \| = \det \| \int_{-\infty}^{\infty} \Lambda_1(x_i - t_i) \Lambda_2(t_i - y_j) \, dt_i \|$$

$$= \sum_{(\nu_1, \dots, \nu_n)} (-1)^I \prod_{i=1}^{n} \int_{-\infty}^{\infty} \Lambda_1(x_i - t_i) \Lambda_2(t_i - y_{\nu_i}) \, dt_i,$$

where $(\nu_1, \dots, \nu_n)$ is a permutation of $1, \dots, n$ having I inversions, and where we sum over all $n!$ permutations. Writing the product of n integrals as an n-fold integral we have

$$\det \| \Lambda(x_i - y_j) \| = \sum_{(\nu_i)} (-1)^I \int_{-\infty}^{\infty} \cdots \int \prod_i \Lambda_1(x_i - t_i) \prod_i \Lambda_2(t_i - y_{\nu_i}) \, dt_1 \dots dt_n$$

$$= \int_{-\infty}^{\infty} \cdots \int \prod_i \Lambda_1(x_i - t_i) \left\{ \sum_{(\nu_i)} (-1)^I \prod_i \Lambda_2(t_i - y_{\nu_i}) \right\} dt_1 \dots dt_n$$

$$= \int_{-\infty}^{\infty} \cdots \int \prod_i \Lambda_1(x_i - t_i) . \det \| \Lambda_2(t_i - y_j) \| \, dt_1 \dots dt_n.$$

If in this formula we replace the variables $t_1, \dots, t_n$ by $t_{\nu_1}, \dots, t_{\nu_n}$, we find by applying the inverse permutation

$$\begin{pmatrix} \nu_1, & \ldots, & \nu_n \\ 1, & \ldots, & n \end{pmatrix}$$

to the rows of the determinant, the relation

$$\det \| \Lambda(x_i - y_j) \| = \int_{-\infty}^{\infty} \cdots \int (-1)^I \prod_i \Lambda_1(x_i - t_{\nu_i}) \,.\, \det \| \Lambda_2(t_i - y_j) \| \, dt_1 \ldots dt_n$$

By adding all $n!$ formulae obtained for all possible permutations and dividing by $n!$ we evidently obtain the desired relation (1.17).

Examples: The previous Lemma offers an effective method of obtaining new Pólya frequency functions from any two old ones.

1. If

$$\Lambda_1(x) = \Lambda_2(x) = e^{-x} \lambda_0(x) = \begin{cases} e^{-x} & \text{if } x \geqq 0, \\ 0 & \text{if } x < 0, \end{cases}$$

we easily find by (1.15) that

$$\Lambda(x) = \begin{cases} xe^{-x} & \text{if } x \geqq 0, \\ 0 & \text{if } x < 0, \end{cases}$$

is also a Pólya frequency function.

Convoluting again this last function $n-1$ times with the original function $e^{-x}\lambda_0(x)$ we find that

$$\Lambda(x) = \begin{cases} x^n e^{-x} & \text{if } x \geqq 0, \\ 0 & \text{if } x < 0, \quad (n = 1, 2, \ldots), \end{cases}$$

are all Pólya frequency functions.

2. Letting $\Lambda_1(x) = e^{-x}\lambda_0(x)$, $\Lambda_2(x) = \Lambda_1(-x)$, we find by (1.15) the Pólya frequency function

$$\Lambda(x) = \frac{1}{2} e^{-|x|}.$$

§ 2. Smoothing properties of Pólya frequency functions.

8. Let

$$(2.1) \qquad y_i = \sum_{k=1}^{n} a_{ik} x_k, \qquad (i = 1, \ldots, m),$$

be a real linear transformation and let us denote by $v(x)$ the number of variations of sign in the sequence of the variables $x_1, x_2, \ldots, x_n$ and by $v(y)$ the corresponding number in the sequence $y_1, y_2, \ldots, y_m$. In 1930 it was shown by the author [11] that the transformation (2.1) enjoys the property that we always have the inequality

$$v(y) \leqq v(x),$$

provided the matrix $A = \| a_{ik} \|$ is such that all its minors of all orders are non-negative.

Let the real function $f(x)$ be defined in an interval I which might be open or closed, finite or infinite. The number $v(f)$ of variations of sign of $f(x)$ in the interval I may now be defined as follows: If $x_1 < x_2 < \ldots < x_n$ is an increasing sequence of points of I, we set

$$v(f) = \sup v(f(x_1), f(x_2), \ldots, f(x_n))$$

for all n and all choices of the sequence $\{x_\nu\}$. Evidently $v(f)$ is a natural integer provided it is finite.

A continuous analogue of the theorem stated in the previous paragraph is contained in the following

Lemma 6. *Let $\Lambda(x)$ be a Pólya frequency function and let $f(x)$ be R-integrable in every finite interval and such that the integral*

$$g(x) = \int_{-\infty}^{\infty} \Lambda(x-t) f(t)\, dt \tag{2.2}$$

converges for every real x. Then

$$v(g) \leqq v(f), \tag{2.3}$$

where $v(g)$ and $v(f)$ denote the numbers of variations in sign of the respective functions, for all real x.

Proof: The inequality (2.3) is obviously valid if $v(g) = 0$, or if $v(f) = \infty$. Without loss of generality we may therefore assume that

$$v(g) > 0, \quad v(f) < \infty.$$

Let $x_0 < x_1 < \ldots < x_m$ be increasing numbers such that the numbers

(2.4) $\qquad g(x_0), \; g(x_1), \; \ldots, \; g(x_m) \qquad$ alternate in sign,

hence $v(g(x_i)) = m$. We wish to prove that we must have

11. See [14]. See also [9], Chapter IV, and [18] for recent developments concerning variation-diminishing transformations.

(2.5) $$m \leqq v(f).$$

If this is proved, our lemma follows, for then

$$v(g) = \sup m \leqq v(f).$$

In order to establish (2.5) we observe that

$$\lim_{\substack{\alpha \to -\infty \\ \beta \to +\infty}} \int_{\alpha}^{\beta} \Lambda(x-t) f(t)\,dt = \int_{-\infty}^{\infty} \Lambda(x-t) f(t)\,dt = g(x),$$

for every real x. Clearly this convergence is uniform for the $m+1$ values $x = x_0, x_1, \ldots, x_m$. We may therefore select α and β such that

$$g_1(x) = \int_{\alpha}^{\beta} \Lambda(x-t) f(t)\,dt$$

will also alternate in sign over the $m+1$ points $x_0, \ldots, x_m$, hence

(2.6) $$g_1(x_0), \ldots, g_1(x_m) \qquad \text{alternate in sign.}$$

Let us now subdivide the range $\alpha \leqq t \leqq \beta$ into n equal parts by

$$\alpha = t_0 < t_1 < \ldots < t_n = \beta, \qquad h = t_{j+1} - t_j = (\beta - \alpha)/n,$$

obtaining the approximating sums

(2.7) $$\gamma_i = h \sum_{j=1}^{n} \Lambda(x_i - t_j) f(t_j), \qquad (i = 0, 1, \ldots, m).$$

Surely if n is sufficiently large, then all differences $g_1(x_i) - \gamma_i$ become as small as we wish. In view of (2.6), we can select n such that also the $m+1$ numbers γ_i will alternate in sign, hence

$$v(\gamma) = m.$$

However, since $\Lambda(x)$ is totally positive, we see that all minors of the matrix of the transformation (2.7) are non-negative. By our theorem mentioned at the beginning of this section we find that

$$m = v(\gamma) \leqq v(f(t_1), \ldots, f(t_n)) \leqq v(f)$$

and (2.5) is established.

A closely related result concerning polynomials is as follows.

Lemma 7. *Let $\Lambda(x)$ be a Pólya frequency function. If $f(x)$ is a real polynomial of exact degree n, then*

$$g(x) = \int_{-\infty}^{\infty} \Lambda(x-t) f(t)\,dt \tag{2.8}$$

is also a real polynomial of exact degree n and such that

$$Z(g) \leqq Z(f), \tag{2.9}$$

where $Z(g)$ and $Z(f)$ denote the number of real zeros of the corresponding polynomials counted with their multiplicities.

Proof: By Lemma 6 we know that $v(g) \leqq v(f)$, a result related to (2.9) but not equivalent to it. To prove (2.9) we must investigate the transformation (2.8) algebraically. By Lemma 2, all the moments

$$\mu_\nu = \int_{-\infty}^{\infty} \Lambda(t) t^\nu\,dt, \quad (\nu = 0, 1, 2, \ldots),$$

exist. But then we find that

$$g(x) = \int_{-\infty}^{\infty} \Lambda(t) f(x-t) dt = \int_{-\infty}^{\infty} \Lambda(t) \left\{ \sum_{\nu=0}^{n} \frac{(-1)^\nu t^\nu}{\nu!} f^{(\nu)}(x) \right\} dt,$$

or

$$g(x) = \mu_0 f(x) - \frac{\mu_1}{1!} f'(x) + \ldots \pm \frac{\mu_n}{n!} f^{(n)}(x). \tag{2.10}$$

Since $\mu_0 > 0$ we see that $g(x)$ is indeed a polynomial of exact degree n. Setting

$$f(x) = a_0 x^n + \binom{n}{1} a_1 x^{n-1} + \ldots + a_n, \quad (a_0 \neq 0),$$

$$g(x) = b_0 x^n + \binom{n}{1} b_1 x^{n-1} + \ldots + b_n,$$

we find the transformation (2.10) to be equivalent to the following linear non-singular transformation among their coefficients:

$$\begin{aligned} b_0 &= \mu_0 a_0 \\ b_1 &= -\mu_1 a_0 + \mu_0 a_1 \\ b_2 &= \mu_2 a_0 - 2\mu_1 a_1 + \mu_0 a_2 \\ &\cdot \\ &\cdot \\ &\cdot \\ b_n &= \pm \mu_n a_0 \mp \binom{n}{1} \mu_{n-1} a_1 \pm \ldots + \mu_0 a_n. \end{aligned} \tag{2.11}$$

Let now $Z(g)$ be the number of real zeros of $g(x)$. If these are not all simple, we now displace the real zeros of $g(x)$ (keeping a_0 and b_0 fixed) so as to make them all simple, and we do this so slightly that the number $Z(f)$, of real zeros of $f(x)$ connected to $g(x)$ by (2.11), has thereby *not increased*. This last condition will automatically be verified, provided the changes in the coefficients $b_1, \ldots, b_n$ are sufficiently small. Let $f^*(x)$, $g^*(x)$ denote the modified polynomials so obtained. By Lemma 6 we now conclude that

$$Z(g) = Z(g^*) = v(g^*) \leqq v(f^*) \leqq Z(f^*) \leqq Z(f).$$

9. We now discuss briefly the particular case when the *Pólya* frequency function $\Lambda(x)$ has the additional property that

$$\Lambda(x) = 0 \quad \text{if} \quad x < 0.$$

The convolution transformation (2.2) may now be written as

$$g(x) = \int_{-\infty}^{x} \Lambda(x-t) f(t)\, dt,$$

and again has the property (2.3) by Lemma 6.

Suppose, however, that $f(t)$ is defined only for $t \leqq 0$. Then

$$g(x) = \int_{-\infty}^{x} \Lambda(x-t) f(t)\, dt = \int_{-\infty}^{0} \Lambda(x-t) f(t)\, dt$$

defines a function $g(x)$ for all $x \leqq 0$. Again we have a "variation-diminishing" property described by the following

Lemma 8. *Let $\Lambda(x)$ be a Pólya frequency function such that $\Lambda(x)=0$ if $x<0$. Let $f(x)$, defined for $x \leqq 0$, be R-integrable in every finite interval and such that*

$$g(x) = \int_{-\infty}^{0} \Lambda(x-t) f(t)\, dt \tag{2.12}$$

converges for $x \leqq 0$. Then

$$v_-(g) \leqq v_-(f), \tag{2.13}$$

where $v_-(g)$ and $v_-(f)$ denote the numbers of variations of sign of the respective functions in the range $-\infty < x \leqq 0$.

The proof corresponds exactly to the one for Lemma 6. As before we may assume that $v_-(g) > 0$, $v_-(f) < \infty$. Let

(2.14) $g(x_0), \ldots, g(x_m)$ alternate in sign $(x_0 < x_1 < \ldots < x_m \leqq 0)$.

We can then find an equidistant set of points

$$t_0 < t_1 < \ldots < t_n = 0, \quad (t_{i+1} - t_i = h > 0),$$

such that

(2.15) $$\gamma_i = h \sum_{j=1}^{n} \Lambda(x_i - t_j) f(t_j), \quad (i = 0, \ldots, m),$$

are so close to the numbers (2.14) that they also alternate in sign. Again, and for the same reason, (2.15) implies that $m \leqq v_-(f)$ whence

$$v_-(g) = \sup m \leqq v_-(f).$$

Lemma 9. *Let $\Lambda(x)$ be a Pólya frequency function such that $\Lambda(x) = 0$ if $x < 0$. If $f(x)$ is a real polynomial of degree n, then*

(2.16) $$g(x) = \int_0^\infty \Lambda(t) f(x-t)\, dt$$

is also a polynomial of degree n and

(2.17) $$Z_-(g) \leqq Z_-(f),$$

where $Z_-(g)$, $Z_-(f)$ denote the numbers of zeros of the corresponding functions in the range $-\infty < x \leqq 0$.

The argument used in the proof of lemma 7 applies with slight changes. Again we can so alter $g(x)$, to $g^*(x)$, by as little as we wish, such that the zero of $g(x)$ in $-\infty < x \leqq 0$ should stay in this range and become simple, whence

$$Z_-(g) = v_-(g^*) \leqq v_-(f^*)$$

and such that

$$Z_-(f^*) \leqq Z_-(f).$$

Then, as before, (2.17) follows on observing that $v_-(f^*) \leqq Z_-(f^*)$.

§ 3. The main theorems.

10. We may now prove the first theorem of our introduction which we re-state in the following only apparently more general form as follows.

Theorem 1.—1. *If $\Psi(s) \in T_2$, $\Psi(s) \not\equiv Ce^{\delta s}$, $\Psi(s) > 0$ in the interval $\alpha < s < \beta$, $(-\infty \leqq \alpha < \beta \leqq +\infty)$, then*

$$(3.1) \qquad \frac{1}{\Psi(s)} = \int_{-\infty}^{\infty} e^{-sx} \Lambda(x)\, dx, \qquad (\alpha < Rs < \beta),$$

where $\Lambda(x)$ is a totally positive function which is not an exponential e^{ax+b}.

2. *Conversely, if $\Lambda(x)$ is a totally positive function which is not an exponential e^{ax+b}, then the Laplace integral*

$$(3.2) \qquad \int_{-\infty}^{\infty} e^{-sx} \Lambda(x)\, dx$$

converges in a strip $\alpha < Rs < \beta$ and represents there a reciprocal $1/\Psi(s)$, where $\Psi(s) \in T_2$, $\Psi(s) \not\equiv Ce^{\delta s}$. The endpoints α and β, if finite, are zeros of the function $\Psi(s)$.

Proof: 1. In order to prove the direct part of the theorem we first assume that

$$(3.3) \qquad \alpha < 0 < \beta.$$

Now $\Psi(s)$ assumes the form (3) and therefore

$$(3.4) \qquad F(s) = \frac{1}{\Psi(s)} = C^{-1} e^{\gamma s^2 - \delta s} \prod_{\nu=1}^{\infty} \frac{e^{\delta_\nu s}}{1 + \delta_\nu s}$$

$$= \Phi_0(s) \prod_{\nu=1}^{\infty} \Phi_\nu(s), \quad (C > 0,\ 0 < \gamma + \Sigma \delta_\nu^2 < \infty).$$

We now wish to represent as *Laplace* integrals the factors $\Phi_0(s)$ and $\Phi_\nu(s)$ appearing in (3.4).

From the familiar formula

$$e^{\gamma s^2} = \frac{1}{2\sqrt{\pi\gamma}} \int_{-\infty}^{\infty} e^{-sx} e^{-x^2/4\gamma}\, dx, \qquad (\text{if } \gamma > 0),$$

we obtain that

$$(3.5) \qquad \Phi_0(s) = C^{-1} e^{\gamma s^2 - \delta s} = \frac{1}{2C\sqrt{\pi\gamma}} \int_{-\infty}^{\infty} e^{-sx} e^{-(x-\delta)^2/4\gamma}\, dx.$$

For the function

$$\lambda(x) = \begin{cases} e^{-x} & \text{if } x \geqq 0, \\ 0 & \text{if } x < 0, \end{cases}$$

we find the transform

$$\int_{-\infty}^{\infty} e^{-xs}\lambda(x)\,dx = \int_{0}^{\infty} e^{-x(s+1)}\,dx = \frac{1}{1+s}, \quad (Rs > -1),$$

whence, if $\delta \neq 0$,

$$\frac{1}{|\delta|}\int_{-\infty}^{\infty} e^{-xs}\lambda\left(\frac{x}{\delta}\right)dx = \frac{1}{1+\delta s}, \quad (-|\delta|^{-1} < Rs < |\delta|^{-1}),$$

and finally

$$\Phi_\nu(s) = \frac{e^{\delta_\nu s}}{1+\delta_\nu s} = \frac{1}{|\delta_\nu|}\int_{-\infty}^{\infty} e^{-xs}\lambda\left(\frac{x}{\delta_\nu}+1\right)dx, \tag{3.6}$$
$$(-|\delta_\nu|^{-1} < Rs < |\delta_\nu|^{-1}).$$

By remarks made in Art. 3 and Art. 6 we know that the determining function of the Laplace integrals (3.5) and (3.6) are Pólya frequency functions.

Let us now form the product

$$F_n(s) = \Phi_0(s)\prod_{\nu=1}^{n}\Phi_\nu(s). \tag{3.7}$$

By a fundamental property of the Laplace integral we obtain for $F_n(s)$ a representation

$$F_n(s) = \int_{-\infty}^{\infty} e^{-xs}\Lambda_n(x)\,dx, \tag{3.8}$$

where $\Lambda_n(x)$ is a Pólya frequency function (by Lemma 5) obtained by convoluting successively the $n+1$ Pólya frequency functions corresponding to the factors of $F_n(s)$. The integral (3.8) will converge in the maximal common strip of convergence of the factors (3.6), in particular on the imaginary axis.

If the product representation (3.4) has only a finite number of factors, n say, then there is nothing more to prove. If not, and assuming all $\delta_\nu \neq 0$, we wish to let $n \to \infty$ in our formula (3.8). Before we do that we invert (3.8) by the complex inversion formula to

$$\Lambda_n(x) = \frac{1}{2\pi i}\int_{-i\infty}^{i\infty} e^{xs}F_n(s)\,ds = \frac{1}{2\pi}\int_{-\infty}^{\infty} e^{itx}F_n(it)\,dt, \quad (n \geq 2). \tag{3.9}$$

The reasons why this formula is valid for all real x, if $n \geqq 2$, are as follows: $\Lambda_n(x)$ is continuous and summable and certainly of bounded variation in every finite interval (by Lemma 1). Also $F_n(s)$ is summable on the imaginary axis, for (3.4) implies

$$(3.10)\quad |F_n(it)| \leqq C^{-1} e^{-\gamma t^2} \cdot \frac{1}{|1+i\delta_1 t| \cdot |1+i\delta_2 t|}, \quad (-\infty<t<\infty,\ n\geqq 2),$$

where the right-hand side is summable even if $\gamma = 0$.

In (3.9) we may now let $n \to \infty$, for $F_n(s) \to F(s)$ in every point s which is not a pole of $F(s)$, in particular on the imaginary axis. Since, by (3.10), $|F_n(it)|$ is dominated by a summable function independent of n, we conclude by Lebesgue's convergence theorem that

$$(3.11)\qquad \lim_{n\to\infty} \Lambda_n(x) = \frac{1}{2\pi} \int_{-\infty}^{\infty} e^{ixt} F(it)\, dt = \Lambda(x),$$

where the limit $\Lambda(t)$ is again continuous as seen from its integral representation. From (3.11) we obtain

$$\int_{-X}^{X} \Lambda(x)\, dx = \frac{1}{2\pi} \int_{-\infty}^{\infty} F(it) \left(\int_{-X}^{X} e^{ixt}\, dx \right) dt = \frac{1}{\pi} \int_{-\infty}^{\infty} \frac{\sin Xt}{t} F(it)\, dt,$$

and letting $X \to \infty$ we obtain by Dirichlet's formula the result that

$$\int_{-\infty}^{\infty} \Lambda(x)\, dx = F(0) = C^{-1}.$$

Thus $\Lambda(x)$ being continuous, a limit of totally positive functions with a non-vanishing and finite integral, must be a Pólya frequency function (Lemma 3).

Again, for obvious reasons, we may invert the Fourier integral (3.11) to

$$F(it) = \int_{-\infty}^{\infty} e^{-ixt} \Lambda(x)\, dx.$$

On replacing in this integral it by s we find a Laplace integral which converges in a strip containing the origin inside (by Lemma 2). We may therefore conclude from the analyticity of both sides that

$$(3.12)\qquad F(s) = \int_{-\infty}^{\infty} e^{-sx} \Lambda(x)\, dx$$

in a certain strip about the imaginary axis.

Let $\alpha < Rs < \beta$ be the maximal strip of convergence of (3.12). That α and β, if finite, are neccessarily poles of $F(s)$ follows from a general property of Laplace integrals with a non-negative determining function $\Lambda(x)$ to the effect that α and β, if finite, must be singularities of the function represented by the integral.

Let us now drop the restriction (3.3) concerning the strip (α, β) of regularity of $F(s)$. Select a μ such that $\alpha<\mu<\beta$ and let

$$\Psi(s) = \Psi(s'+\mu) = \Psi_1(s').$$

Now $s'=\alpha-\mu$ and $s'=\beta-\mu$ are consecutive zeros of $\Psi_1(s')$ which is evidently a function of the class T_2 while $\alpha-\mu<0<\beta-\mu$. By the case just treated we have that

$$\frac{1}{\Psi_1(s')} = \frac{1}{\Psi(s'+\mu)} = \int_{-\infty}^{\infty} e^{-s'x}\Lambda_1(x)\,dx, \quad (\alpha-\mu<Rs'<\beta-\mu),$$

where $\Lambda_1(x)$ is a Pólya frequency function. Returning to the variable $s=s'+\mu$, we find that

$$\frac{1}{\Psi(s)} = \int_{-\infty}^{\infty} e^{-sx} e^{\mu x}\Lambda_1(x)\,dx, \quad (\alpha<Rs<\beta),$$

where $e^{\mu x}\Lambda_1(x) = \Lambda(x)$ is evidently totally positive.

2. In order to prove the converse part of Theorem 1 let us assume first that the given totally positive function $\Lambda(x)$ is a Pólya frequency function. Denoting its transform by $F(s)$ we know that the integral

$$F(s) = \int_{-\infty}^{\infty} e^{-sx}\Lambda(x)\,dx \tag{3.13}$$

converges in a strip $\alpha<Rs<\beta$ containing the origin. Let us consider the power series expansion

$$F(s) = \sum_{\nu=0}^{\infty} \frac{(-1)^{\nu}\mu_{\nu}}{\nu!}s^{\nu} \tag{3.14}$$

where

$$\mu_{\nu} = \int_{-\infty}^{\infty} x^{\nu}\Lambda(x)\,dx, \quad (F(0)=\mu_0<0), \tag{3.15}$$

and also the expansion of the reciprocal of $F(s)$:

(3.16) $$\Psi(s) = \frac{1}{F(s)} = \sum_{\nu=0}^{\infty} \frac{\varkappa_\nu}{\nu!} s^\nu .$$

We now return to the transformation (2.8) from the polynomial $f(x)$ to the polynomial $g(x)$. By (2.10) we know that this transformation may also be written as

(3.17) $$g(x) = F(D) f(x),$$

and that it has the property that

(3.18) $$Z(g) \leq Z(f).$$

The transformation (3.17) may now be inverted to

(3.19) $$f(x) = \frac{1}{F(D)} g(x) = \Psi(D) g(x) = \sum_{0}^{\infty} \frac{\varkappa_\nu}{\nu!} g^{(\nu)}(x)$$

and has, by (3.18) the property of transforming every real polynomial $g(x)$ into a polynomial $f(x)$ having no fewer real zeros than $g(x)$. If in particular we select $g(x) = x^n$ we find for $f(x)$ the polynomial

(3.20) $$P_n(x) = \Psi(D) x^n = \varkappa_0 x^n + \binom{n}{1} \varkappa_1 x^{n-1} + \ldots + \varkappa_n, \quad (n = 1, 2, \ldots).$$

Now (3.18) implies that

$$Z(P_n) \geqq Z(x^n) = n, \quad \text{or} \quad Z(P_n) = n,$$

showing that, for every n, the polynomial $P_n(x)$ has only real zeros. This implies that the function $\Psi(s)$, defined by the expansion (3.16), is an entire function of the class T_2 by virtue of a theorem of Pólya and Schur[12] to the effect that the reality of all the roots of the sequence of polynomials (3.20) is a necessary and sufficient condition that (3.16) should be the expansion of a function $\Psi(s)$ of the class T_2.

There remains to show that $\Psi(s)$ is not of the form $Ce^{\delta s}$. To this end we consider the quadratic form

$$\int_{-\infty}^{\infty} (\xi_0 + x\xi_1)^2 \Lambda(x)\, dx = \mu_0 \xi_0^2 + 2\mu_1 \xi_0 \xi_1 + \mu_2 \xi_1^2$$

which must be positive definite, hence necessarily

(3.21) $$\begin{vmatrix} \mu_0 & \mu_1 \\ \mu_1 & \mu_2 \end{vmatrix} > 0 .$$

12. See [10], Theorem IV on page 110.

However, assuming that we have $\Psi(s) = Ce^{\delta s}$ we find

$$F(s) = \frac{1}{\Psi(s)} = C^{-1} e^{-\delta s} = C^{-1}\left(1 - \frac{\delta}{1!}s + \frac{\delta^2}{2!}s^2 - \dots\right),$$

whence

$$\mu_0 = C^{-1}, \quad \mu_1 = C^{-1}\delta, \quad \mu_2 = C^{-1}\delta^2$$

and therefore

$$\begin{vmatrix} \mu_0 & \mu_1 \\ \mu_1 & \mu_2 \end{vmatrix} = C^{-2}\begin{vmatrix} 1 & \delta \\ \delta & \delta^2 \end{vmatrix} = 0,$$

in contradiction to the necessary inequality (3.21).

Let us now assume that the given function $\Lambda(x)$ is an arbitrary totally positive function, not of the form e^{ax+b}. By Lemma 4, we can find a real number μ such that

$$\Lambda_1(x) = e^{-\mu x}\Lambda(x)$$

is a Pólya frequency function. By the previous case already settled we know that

$$F_1(s') = \int_{-\infty}^{\infty} e^{-s'x} e^{-\mu x}\Lambda(x)\,dx$$

converges in a strip $\alpha' < Rs' < \beta'$ containing the origin and that

$$F_1(s') = \frac{1}{\Psi_1(s')}, \quad \text{where } \Psi_1(s') \in T_2, \quad \Psi_1(s') \not\equiv C'e^{\delta' s'}.$$

In terms of the variable $s = s' + \mu$ we conclude that

$$\int_{-\infty}^{\infty} e^{-sx}\Lambda(x)\,dx = \frac{1}{\Psi_1(s-\mu)} = \frac{1}{\Psi(s)}$$

converges in the strip $\alpha' + \mu < Rs < \beta' + \mu$, where $\Psi(s) \in T_2$ and evidently not of the form $Ce^{\delta s}$. This completes a proof of our theorem.

11. For functions $\Phi(s)$ of the narrower class T_1, of the form (1), we have the following

Theorem 2. 1. *Let* $\Phi(s) \in T_1$, $\Phi(s) \not\equiv Ce^{\gamma s}$, $\Phi(s) > 0$ *if* $s > 0$. *Then*

$$\frac{1}{\Phi(s)} = \int_{-\infty}^{\infty} e^{-sx}\Lambda(x)\,dx = \int_{0}^{\infty} e^{-sx}\Lambda(x)\,dx, \quad (Rs > 0), \tag{3.22}$$

where $\Lambda(x)$ is a totally positive function vanishing for $x<0$, the integral converging up to the first pole of $1/\Phi(s)$.

2. Conversely, let $\Lambda(x)$ be a totally positive function such that $\Lambda(x)=0$ for $x<0$, and such that the integral

$$\int_0^\infty e^{-sx}\Lambda(x)\,dx$$

converges for $Rs>0$. Then this integral represents in the half-plane $Rs>0$ the reciprocal of a function $\Phi(s)\in T_1$, not of the form $Ce^{\gamma s}$.

Proof: 1. Let us first assume that $\Phi(0)>0$ and let us apply to $\Phi(s)$ the direct part of Theorem 1 for the strip (α,β) where $\alpha<0$, $\beta=+\infty$. We are thus assured of the bilateral integral representation (3.22) valid for $Rs>\alpha$, $\Lambda(x)$ being totally positive. There remains to show that $\Lambda(x)=0$ for $x<0$. This is seen as follows: In our present case we have that

$$F(s)=\frac{1}{\Phi(s)}=C^{-1}e^{-\gamma s}\prod_{\nu=1}^{\infty}\frac{1}{1+\delta_\nu s},\qquad (\gamma\geqq 0,\ 0<\Sigma\delta_\nu<\infty,\ \delta_\nu\geqq 0).$$

For each factor we have a representation

$$\frac{1}{1+\delta_\nu s}=\frac{1}{\delta_\nu}\int_0^\infty e^{-sx}\lambda\left(\frac{x}{\delta_\nu}\right)dx,\qquad (\text{if } \delta_\nu>0),$$

as an ordinary Laplace integral. However, if $\Lambda_1(x)$, $\Lambda_2(x)$ are Pólya frequency functions both vanishing for $x<0$, then also their convolution

$$\Lambda(x)=\int_{-\infty}^{\infty}\Lambda_1(x-t)\,\Lambda_2(t)\,dt$$

enjoys this property, for if $x<0$ then at least one of the arguments $x-t$ and t (having a negative sum x) must be negative, showing that the integrand vanishes for all t, if $x<0$.

As before we now obtain the representation

$$\prod_1^\infty\frac{1}{1+\delta_\nu s}=\int_0^\infty e^{-sx}\Lambda_0(x)\,dx,$$

where $\Lambda_0(x)$ is totally positive, vanishing for $x<0$. Replacing x by $x-\gamma$

we get that

$$F(s) = C^{-1} e^{-\gamma s} \prod_{1}^{\infty} \frac{1}{1+\delta_\nu s} = C^{-1} \int_0^\infty e^{-sx} \Lambda_0(x-\gamma)\, dx,$$

where

$$\Lambda(x) = C^{-1} \Lambda_0(x-\gamma)$$

is a Pólya frequency function, vanishing for $x<0$ because $\gamma \geqq 0$.

If $\Phi(0) = 0$, let $\mu>0$ and set $s = \mu + s'$. Now $\Phi(s) = \Phi(\mu + s')$ is a function of s' which satisfies the assumptions of the previous case, hence

$$\frac{1}{\Phi(s'+\mu)} = \int_0^\infty e^{-s'x} \Lambda_0(x)\, dx, \quad (Rs' > -\mu).$$

In terms of the old variable s this means that

$$\frac{1}{\Phi(s)} = \int_0^\infty e^{-sx} e^{\mu x} \Lambda_0(x)\, dx, \quad (Rs > 0),$$

where $e^{\mu x} \Lambda_0(x)$ is a totally positive function vanishing for $x<0$. This completes a proof of the direct part of the theorem.

2. To prove the converse, we assume first that the given $\Lambda(x)$ vanishing for $x<0$, is a Pólya frequency function. This being a special case of the corresponding situation in Art. 10, we consider again integral (3.13), the expansions (3.14), (3.16), the corresponding polynomial transformation (3.17) and its inverse (3.19). The argument there presented shows that the polynomials (3.20) have only real zeros. However, in our case the transformation (2.8) reduces to the form (2.16). By Lemma 9 it has the property (2.17). The inverse transformation (3.19) now has the property that

$$Z_{-}(f) \geqq Z_{-}(g).$$

Since $Z_{-}(x^n) = n$, we now find that the polynomials (3.20) have only real and non-positive zeros. We may now use a second theorem of Pólya and Schur to the effect that the reality and non-positivity of all the roots of the sequence of the polynomials (3.20) implies that (3.16) is the expansion of an entire function of the class T_1.

If we only assume now that

$$\int_0^\infty e^{-sx} \Lambda(x)\, dx \qquad \text{converges for } Rs>0,$$

then, if $\mu > 0$, and setting $s = s' + \mu$, we see that

$$\int_0^\infty e^{-s'x} e^{-\mu x} \Lambda(x)\, dx$$

will converge for $Rs' > -\mu$, showing that $e^{-\mu x}\Lambda(x)$ is a Pólya frequency function. By the previous case

$$\int_0^\infty e^{-s'x} e^{-\mu x} \Lambda(x)\, dx = \frac{1}{\Phi_1(s')}, \quad (Rs' > -\mu),$$

where $\Phi_1(s') \in T_1$. Returning to $s = s' + \mu$, we find that

$$\int_0^\infty e^{-sx} \Lambda(x)\, dx = \frac{1}{\Phi_1(s-\mu)},$$

where $\Phi(s) = \Phi_1(s-\mu)$ is in the class T_1. This completes the proof of the theorem.

12. Theorem 1 allows to draw a few conclusions concerning the behavior and order of continuity of totally positive functions. We know already, by Lemma 1, that a totally positive function is positive and continuous in a certain open interval, vanishing outside this interval. We may now state the following

Corollary 1. *The maximal interval of positivity of a totally positive function must be infinite.*

Proof: Without loss of generality we may assume $\Lambda(x)$ to be a Pólya frequency function and let

$$F(s) = e^{\gamma s^2 - \delta s} \prod_{\nu=1}^{\infty} \frac{e^{\delta_\nu s}}{1+\delta_\nu s} = \int_{-\infty}^{\infty} e^{-sx} \Lambda(x)\, dx \tag{3.23}$$

be its Laplace transform. Suppose that $\Lambda(x) = 0$ outside the finite interval (a, b). This would imply that

$$F(s) = \int_a^b e^{-sx} \Lambda(x)\, dx,$$

showing that $F(s)$ is entire. But then in (3.23) we must have $\delta_\nu = 0$ for all ν, in which case $\Lambda(x)$ must be the normal frequency function in view of the formula (3.5). But then again $\Lambda(x)$ does not vanish outside (a, b).

The continuity properties of $\Lambda(x)$ are described by the following

Corollary 2. Let (3.23) be the transform of a Pólya frequency function $\Lambda(x)$.

1. $\Lambda(x)$ is of the continuity class C^∞ if and only if
   (α) either $\gamma > 0$,
   (β) or else $\gamma = 0$ and infinitely many δ_ν are $\neq 0$.
2. If $\gamma = 0$ and $\delta_1, \delta_2, \ldots, \delta_n$ are $\neq 0$, $\delta_\nu = 0$ for $\nu > n$ $(n \geq 1)$, then $\Lambda(x)$ is in C^{n-2} and in no higher class. In particular $\Lambda(x)$ is discontinuous if and only if $n = 1$, when $\Lambda(x)$ is essentially equivalent to the function $\lambda(x)$ of (3.6).

Proof: 1. Let us assume that assumptions (α) or (β) hold (in the latter case all $\delta_\nu \neq 0$). We invert (3.23) to

$$\Lambda(x) = \frac{1}{2\pi i} \int_{-i\infty}^{i\infty} e^{sx} F(s)\, ds. \tag{3.24}$$

If $\gamma > 0$ we have on the imaginary axis $s = it$

$$|F(s)| < e^{-\gamma t^2}.$$

If $\gamma = 0$, and all $\delta_\nu \neq 0$, we have

$$|F(s)| < \frac{1}{|1 + i\delta_1 t| \ldots |1 + i\delta_n t|}$$

for every n. But then the Fourier integral (3.24) may be differentiated under the integral sign any number of times leading to the absolutely convergent integral

$$\Lambda^{(k)}(x) = \frac{1}{2\pi i} \int_{-i\infty}^{i\infty} e^{sx} s^k F(s)\, ds. \tag{3.25}$$

These remarks prove the direct part of the first statement in Corollary 2. The converse part will be implied by the following discussion.

2. Assume $\gamma = 0$, $\delta_1, \ldots, \delta_n \neq 0$, $\delta_\nu = 0$ for $\nu > n$. In this case it is sufficient to discuss the representation of

$$F(s) = \prod_{\nu=1}^{n} \frac{1}{1 + \delta_\nu s} = \int_{-\infty}^{\infty} e^{-sx} \Lambda(x)\, dx \tag{3.26}$$ [13]

13. See [6], pp. 139—141 for a more detailed discussion of the Pólya frequency functions $\Lambda(x)$ of our formula (3.26).

and we are to show that $\Lambda(x) \in C^{n-2}$ but not in C^{n-1}. Assuming $n \geqq 2$, we may again invert (3.26) to (3.24). Since $F(s)$ vanishes at ∞ to the n-th order we see that again (3.25) is valid for $k = 0, 1, \ldots, n-2$ showing that $\Lambda(x) \in C^{n-2}$. On the other hand we may write

$$F(s) = \frac{A}{s^n} + \frac{\varphi(s)}{s^{n+1}},$$

where $A = 1/\Pi\delta_\nu$, and $\varphi(s)$ is regular at infinity. Substituting this expression into

$$\Lambda(x) = \frac{1}{2\pi i} \int_{\sigma-i\infty}^{\sigma+i\infty} e^{sx} F(s)\, ds,$$

where $\sigma > 0$ and sufficiently small, we find that

$$\Lambda(x) = A \frac{1}{2\pi i} \int_{\sigma-i\infty}^{\sigma+i\infty} \frac{e^{sx}}{s^n} ds + \frac{1}{2\pi i} \int_{\sigma-i\infty}^{\sigma+i\infty} \frac{e^{sx}\varphi(s)}{s^{n+1}} ds.$$

Now the second integral represents a function of the class C^{n-1}, while the first integral is known to equal

$$A \frac{x^{n-1}}{(n-1)!} \quad \text{if } x \geqq 0, \quad \text{and } 0 \quad \text{if } x < 0.$$

Thus the first integral is in C^{n-2} but not in C^{n-1}. The obvious discontinuity in the $(n-1)$st derivative of the first integral implies the relation

$$\Lambda^{(n-1)}(+0) - \Lambda^{(n-1)}(-0) = A = 1/\prod_1^n \delta_\nu.$$

We finally mention without going into details that an explicit expression of $\Lambda(x)$ is readily obtained from (3.26) by decomposing $F(s)$ into partial fractions and expressing each fraction as a Laplace integral.

Part II. Examples and applications

§ 4. Series expansions of totally positive functions.

13. We return for the moment to the twice positive functions of §1 and prove concerning their Laplace transforms the following

Lemma 10. *Let $\Lambda(x)$ be twice-positive (Def. 2 for $k=2$) and not an exponential function $\exp(ax+b)$. Then the limits*

$$(4.1) \qquad \alpha = \lim_{x \to +\infty} \frac{\log \Lambda(x)}{x}, \quad \beta = \lim_{x \to -\infty} \frac{\log \Lambda(x)}{x},$$

always exist, finite or infinite, where it is understood that

$$(4.2) \qquad \begin{aligned} &\alpha = -\infty, \text{ if } \Lambda(x) = 0 \text{ for large positive } x, \\ &\beta = +\infty, \text{ if } \Lambda(x) = 0 \text{ for large negative } x. \end{aligned}$$

These limits satisfy the condition

$$(4.3) \qquad -\infty \leqq \alpha < \beta \leqq +\infty,$$

and the integral

$$(4.4) \qquad F(s) = \int_{-\infty}^{\infty} e^{-xs} \Lambda(x)\, dx$$

has the exact strip of convergence

$$(4.5) \qquad \alpha < Rs < \beta.$$

Finally

$$(4.6) \qquad \lim_{x \to +\infty} e^{-xs} \Lambda(x) = 0, \text{ if } s > \alpha,$$

$$(4.7) \qquad \lim_{x \to -\infty} e^{-xs} \Lambda(x) = 0, \text{ if } s < \beta.$$

Proof: Since a shift of the variable does not affect our problem we may assume that $\Lambda(0) > 0$. The concavity of $\log \Lambda(x)$ implies that

$$\varphi(x) = \frac{\log \Lambda(x) - \log \Lambda(0)}{x}, \qquad (x \neq 0),$$

is a non-increasing function of x. This remark already implies the existence of the limits (4.1) and also the inequalities (4.3), since $\log \Lambda(x)$ is not linear, by assumption.

Let s be real and $s > \alpha$, and select α' such that $s > \alpha' > \alpha$. For $x > 0$ we have that

$$\log(e^{-xs} \Lambda(x)) = \left(-s + \frac{\log \Lambda(x)}{x}\right) x < (-s + \alpha')x,$$

provided x is sufficiently large. Thus

$$e^{-xs} \Lambda(x) < e^{-(s-\alpha')x} \text{ if } x > A.$$

This proves (4.6) and a similar argument establishes (4.7). These exponential estimates of the integrand of (4.4) actually prove its convergence in the strip (4.5).

Let us finally show that (4.4) can not converge outside (4.5). Assume α finite and let $s<\alpha_1<\alpha$. Then, if $x>0$, we find

$$\log(e^{-xs}\Lambda(x)) = \left(-s+\frac{\log\Lambda(x)}{x}\right)x > (-s+\alpha_1)x$$

for $x>C$, hence

$$e^{-xs}\Lambda(x) > e^{(\alpha_1-s)x} \quad \text{if } x>C,$$

showing that (4.4) diverges. A similar argument proves the divergence of (4.4) at its lower limit if $s>\beta$.

14. Let now

$$F(s) = \frac{1}{\Psi(s)} \tag{4.8}$$

be the reciprocal of a given function $\Psi(s)\in T_2$ which is positive between its consecutive zeros α, β $(\alpha<\beta)$. By Theorem 1 of § 3 we have a representation

$$F(s) = \int_{-\infty}^{\infty} e^{-xs}\Lambda(x)\,dx, \quad (\alpha<Rs<\beta), \tag{4.9}$$

where $\Lambda(x)$ is totally positive. By the complex inversion formula

$$\Lambda(x) = \frac{1}{2\pi i}\int_{\sigma-i\infty}^{\sigma+i\infty} e^{xs}F(s)\,ds, \quad (-\infty<x<\infty,\ \alpha<\sigma<\beta). \tag{4.10}$$

Let us now assume that

$$\ldots<\alpha_n<\alpha_{n-1}<\ldots<\alpha_2<\alpha_1<\alpha_0=\alpha \tag{4.11}$$

are all the zeros of $\Psi(s)$ to the left of α, arranged in decreasing order. Let $\varepsilon_n=\pm 1$ be such that

$$\varepsilon_n\Psi(s)>0 \quad \text{if } \alpha_n<s<\alpha_{n-1}.$$

Again by Theorem 1 we have a representation

$$\varepsilon_n F(s) = \int_{-\infty}^{\infty} e^{-xs}\Lambda_n(x)\,dx, \quad (\alpha_n<Rs<\alpha_{n-1}), \tag{4.12}$$

where $\Lambda_n(x)$ is also totally positive, and also

$$(4.13)\qquad \varepsilon_n \Lambda_n(x) = \frac{1}{2\pi i}\int_{\sigma_n - i\infty}^{\sigma_n + i\infty} e^{xs} F(s)\, ds, \qquad (\alpha_n < \sigma_n < \alpha_{n-1}).$$

By Cauchy's theorem we now obtain the relation

$$\frac{1}{2\pi i}\int_{\sigma - i\infty}^{\sigma + i\infty} e^{xs} F(s)\, ds - \frac{1}{2\pi i}\int_{\sigma_n - i\infty}^{\sigma_n + i\infty} e^{xs} F(s)\, ds = \sum_{\nu=0}^{n-1} \operatorname*{Res}_{s=\alpha_\nu} (e^{xs} F(s))$$

which, in view of (4.10) and (4.13), may be written as

$$(4.14)\qquad \Lambda(x) = \sum_{\nu=0}^{n-1} \operatorname*{Res}_{s=\alpha_\nu} (e^{xs} F(s)) + \varepsilon_n \Lambda_n(x).$$

This relation connects with each other the various totally positive functions associated with the meromorphic function $F(s)$ in its various strips of regularity. The residues are easily obtained as follows: If $s = \alpha$ is a k-fold pole and

$$(4.15)\quad F(s) = \frac{1}{\Psi(s)} = \frac{a_k}{(s-\alpha)^k} + \dots + \frac{a_1}{s-\alpha} + a_0 + \dots, \qquad (a_k \neq 0),$$

in view of

$$e^{xs} = e^{x\alpha} \cdot e^{x(s-\alpha)} = e^{\alpha x}\left(1 + \frac{x}{1!}(s-\alpha) + \dots + \frac{x^{k-1}}{(k-1)!}(s-\alpha)^{k-1} + \dots\right),$$

we find

$$(4.16)\quad \operatorname*{Res}_{s=\alpha}(e^{xs}F(s)) = \left(a_1 + \frac{x}{1!}a_2 + \dots + \frac{x^{k-1}}{(k-1)!}a_k\right) e^{x\alpha} = P(x;\, \alpha)\, e^{x\alpha},$$

where $P(x;\, \alpha)$ is a polynomial of exact degree $k-1$. Now (4.14) may be explicitly written as

$$(4.17)\qquad \Lambda(x) = \sum_{\nu=0}^{n-1} P(x;\, \alpha_\nu)\, e^{x\alpha_\nu} + \varepsilon_n \Lambda_n(x),$$

showing that the function $\varepsilon_n \Lambda_n(x)$ differs from $\Lambda(x)$ by an exponential sum with polynomial coefficients.

Also (4.17) in conjunction with Lemma 10 will give precise information concerning the behavior of $\Lambda(x)$ as $x \to +\infty$. Indeed by (4.6) we know that

$$e^{-xs}\Lambda(x) \to 0 \quad \text{if } s > \alpha.$$

However, from (4.17) for $n=1$ we obtain

$$\Lambda(x) = P(x;\ \alpha)\,e^{x\alpha} + \varepsilon_1 \Lambda_1(x),$$

where the remainder $\varepsilon_1 \Lambda_1(x)$ is of lower order than the main term, for again by (4.6) we see that

$$\varepsilon_1 \Lambda_1(x)\,e^{-xs} \to 0 \quad \text{if} \quad s > \alpha_1,$$

hence in particular

$$\varepsilon_1 \Lambda_1(x)\,e^{-x\alpha} \to 0.$$

Thus (4.17) leads to the following[14]

Theorem 3. *If $\Psi(s)$ has infinitely many zeros (4.11) to the left of α, then $\Lambda(x)$ admits the asymptotic expansion*

$$\Lambda(x) \sim \sum_{\nu=0}^{\infty} P(x;\ \alpha_\nu)\,e^{\alpha_\nu x} \quad \text{as } x \to +\infty. \tag{4.18}$$

For indeed by (4.17) and (4.6) we see that for $x \to +\infty$

$$\Lambda(x) - \sum_{\nu=0}^{n-1} P(x;\ \alpha_\nu)\,e^{\alpha_\nu x} = \varepsilon_n \Lambda_n(x) = o(e^{xs})$$

if $s > \alpha_n$, and in particular for $s = \alpha_{n-1}$.

15. We now consider the special case when all the zeros (4.11) are simple. Obviously $\varepsilon_n = (-1)^n$, and

$$\operatorname*{Res}_{s=\alpha_\nu} \left(e^{xs} F(s)\right) = \frac{1}{\Psi'(\alpha_\nu)}\,e^{x\alpha_\nu}, \tag{4.19}$$

so that our relation (4.14) becomes

$$\Lambda(x) = \sum_{\nu=0}^{n-1} \frac{1}{\Psi'(\alpha_\nu)}\,e^{x\alpha_\nu} + (-1)^n \Lambda_n(x), \qquad (n = 1, 2, \ldots). \tag{4.20}$$

Since $\Lambda_n(x) \geqq 0$ for all x and all n, we conclude that

$$\left|\Lambda(x) - \sum_{\nu=0}^{n-1} \frac{1}{\Psi'(\alpha_\nu)}\,e^{x\alpha_\nu}\right| \leqq \left|\frac{1}{\Psi'(\alpha_n)}\right| e^{x\alpha_n}, \qquad (-\infty < x < \infty). \tag{4.21}$$

An immediate corollary is the following

Theorem 4. *If $\Psi(s)$ has infinitely many zeros (4.11) to the left of α and all these are simple zeros, then*

14. See [3], Chapter 14, § 1, for a discussion of Mellin's work on asymptotic expansions of determining functions having meromorphic transforms, a very special case of which is our present Theorem 3.

$$\Lambda(x) = \sum_{\nu=0}^{\infty} \frac{1}{\Psi'(\alpha_\nu)} e^{x\alpha_\nu} \tag{4.22}$$

in the range of convergence $x > \gamma$ of this Dirichlet series, if any.

For indeed, if the series on the right-hand side of (4.22) converges, or if we only know that its nth term tends to zero, then (4.21) clearly implies the relation (4.22).

Entirely similar results hold for the poles of $F(s)$ which are to the right of β, if any:

$$\beta = \beta_0 < \beta_1 < \ldots < \beta_{n-1} < \beta_n < \ldots. \tag{4.23}$$

Assuming infinitely many such poles we obtain the asymptotic series

$$\Lambda(x) \sim -\sum_{\nu=0}^{\infty} P(x;\, \beta_\nu) e^{x\beta_\nu} \text{ for } x \to -\infty. \tag{4.24}$$

Moreover, if all these poles (4.23) are simple we obtain the relation

$$\Lambda(x) = -\sum_{\nu=0}^{\infty} \frac{1}{\Psi'(\beta_\nu)} e^{x\beta_\nu} \tag{4.25}$$

throughout the real range of convergence of this Dirichlet series, if any. The minus sign appearing in these formulae is due to the clockwise integration while applying Cauchy's residue theorem.

§5. The Gamma function.

16. The function

$$\Psi(s) = \frac{1}{\Gamma(s)} = e^{Cs} s \prod_{1}^{\infty} \left(1 + \frac{s}{n}\right) e^{-s/n} \tag{5.1}$$

is evidently of the class T_2. Its reciprocal $\Gamma(s)$ has the poles $0, -1, -2, \ldots$, and therefore by Theorem 1 of § 3 we have a representation

$$\Gamma(s) = \int_{-\infty}^{\infty} e^{-sx} \Lambda(x)\, dx, \quad (Rs > 0),$$

where $\Lambda(x)$ is totally positive. From (5.1) or even better from Gauss' equivalent representation

$$\frac{1}{\Gamma(s)} = \lim_{n \to \infty} \frac{s(s+1)\ldots(s+n)}{n!} n^{-s}$$

we easily find that

$$\operatorname*{Res}_{s=-n} \Gamma(s) = \frac{(-1)^n}{n!}, \qquad (n = 0, 1, \ldots).$$

By Theorem 3 of § 4 we find for $\Lambda(x)$ the asymptotic series

$$\Lambda(x) \sim \sum_{n=0}^{\infty} \frac{(-1)^n}{n!} e^{-nx} \quad \text{as} \quad x \to +\infty.$$

However, since this series converges for all values of x we find by Theorem 4 the relation

$$\Lambda(x) = \sum_{0}^{\infty} \frac{(-1)^n}{n!} e^{-nx} = e^{-e^{-x}}.$$

This proves Euler's integral representation

$$\Gamma(s) = \int_{-\infty}^{\infty} e^{-xs} e^{-e^{-x}} dx, \qquad (Rs > 0). \tag{5.2}$$

We also have

$$(-1)^n \Gamma(s) = \int_{-\infty}^{\infty} e^{-xs} \Lambda_n(x)\, dx, \qquad (-n < Rs < -n+1),$$

where $\Lambda_n(x)$ is totally positive. By (4.20) we find the relation

$$\Lambda(x) = \sum_{\nu=0}^{n-1} \frac{(-1)^\nu}{\nu!} e^{-\nu x} + (-1)^n \Lambda_n(x)$$

whence

$$\Lambda_n(x) = (-1)^n \left\{ e^{-e^{-x}} - \sum_{\nu=0}^{n-1} \frac{(-1)^\nu}{\nu!} e^{-\nu x} \right\}$$

a function already found by Cauchy[15] and now seen to be totally positive.

17. If the function $\Psi(s) \in T_2$ is given, we have seen how we may find the Laplace integral representation of its reciprocal in a certain strip. Indeed, a first perfectly general method is implicitly contained in our proof of the direct part of Theorem 1, § 3. A second method using series expansions for $\Lambda(x)$ has just been applied to the case of the Gamma function and will again be used in § 6.

15. See [2], page 92.

We wish now to discuss briefly the inverse problem. Suppose that a function $\Lambda(x)$ is given which is totally positive. We then know that

(5.3) $$F(s) = \int_{-\infty}^{\infty} e^{-xs}\Lambda(x)\,dx, \qquad (\alpha < Rs < \beta),$$

is the reciprocal $1/\Psi(s)$ of a function $\Psi(s)$ of the class T_2. How could we derive the product representation (2) of $\Psi(s)$?

A method which always works, at least in principle, is as follows. If $\Lambda(x)$ is discontinuous then we know by Corollary 2, that $\Lambda(x)$ must essentially reduce to $\lambda(x)$. Assuming $\Lambda(x)$ continuous, by an integration by parts in (5.3) we obtain

(5.4) $$sF(s) = \int_{-\infty}^{\infty} e^{-xs}\Lambda'(x)\,dx, \qquad (\alpha < Rs < \beta).$$

From (5.3) and (5.4) we now obtain

(5.4′) $$(1+\delta s)F(s) = \int_{-\infty}^{\infty} e^{-xs}(1+\delta D)\Lambda(x)\,dx$$

or

(5.5) $$(1+\delta s)e^{-\delta s}F(s) = \int_{-\infty}^{\infty} e^{-xs}(1+\delta D)\Lambda(x-\delta)\,dx,$$

where D denotes the differentiation operator.

We start by determining the range of convergence of (5.3) by the formulae (4.1) of Lemma 10. Assuming $\alpha \neq 0$, we now apply the formula (5.5) with

$$\delta = -\frac{1}{\alpha}.$$

Since $s = \alpha$ is known to be a pole of $F(s)$, which is of the form

(5.6) $$F(s) = C^{-1} e^{\gamma s^2 - \delta s} s^{-m} \prod \frac{e^{\delta_\nu s}}{1+\delta_\nu s},$$

we see that

(5.7) $$\Lambda_1(x) = (1+\delta D)\Lambda(x-\delta)$$

must again be totally positive by Theorem 1 and the unicity theorem for Laplace integrals. If $\alpha = 0$, we apply (5.4), instead of (5.5), with a similar result.

We now repeat the same procedure on $\Lambda_1(x)$ and continue the operation indefinitely, if necessary, determining successively the poles of $F(s)$, in order, which are to the left of α and to the right of β. The expression (5.6) shows that the totally positive function $\Lambda_n(x)$ obtained after n steps will converge, as $n \to \infty$, either to the normal frequency function having the transform (3.5), provided $\gamma > 0$, or else to the distribution $\chi(x)$ of transform

$$C^{-1} e^{-\delta s} = \int_{-\infty}^{\infty} e^{-xs} d\chi(x),$$

hence $\chi(x) = 0$ for $x < \delta$, $\chi(x) = C^{-1}$ for $x > \delta$.

18. The method just described [16] is best illustrated by the concrete case of

$$\Lambda(x) = e^{-e^{-x}}. \tag{5.8}$$

That this function is totally positive may be immediately verified from the Definition 1 of § 1. For if $x_1 < x_2 < \ldots < x_n$ and $t_1 < t_2 < \ldots < t_n$ then

$$\det \| \Lambda(x_i - t_j) \| = \det \| e^{-e^{x_i - t_j}} \| = \det \| e^{e^{x_i}(-e^{-t_j})} \|,$$

the last determinant being positive, because both sequences e^{x_i} and $-e^{-t_j}$ are increasing.

The range of convergence (α, β) of (5.3) is now found from

$$\frac{\log \Lambda(x)}{x} = -\frac{1}{x} e^{-x} \to \begin{cases} \alpha = 0 & \text{if } x \to +\infty, \\ \beta = +\infty & \text{if } x \to -\infty, \end{cases}$$

to extend from $\alpha = 0$ to $\beta = \infty$.

By (5.4) we find that

$$s F(s) = \int_{-\infty}^{\infty} e^{-xs} \Lambda_0(x)\, dx,$$

where

$$\Lambda_0(x) = D\Lambda(x) = e^{-x} e^{-e^{-x}},$$

and whose range of convergence is given by

16. The fundamental idea of the method just described goes back to Hadamard's dissertation [4]. For an application of this idea to the solution of the related problem of totally positive sequences see the recent note [1].

$$\frac{\log \Lambda_0(x)}{x} = -1 - \frac{1}{x} e^{-x} \to \begin{cases} \alpha_1 = -1 & \text{if } x \to +\infty, \\ \beta_1 = +\infty & \text{if } x \to -\infty. \end{cases}$$

We now conclude that

$$s\,F(s) = C^{-1} e^{\gamma s^2 - \delta s} \prod \frac{e^{\delta_\nu s}}{1+\delta_\nu s}, \qquad (\delta_\nu > 0).$$

Here $C^{-1} = 1$, since

$$\int_{-\infty}^{\infty} \Lambda_0(x)\,dx = 1.$$

Moreover $\delta_1 = -1/\alpha_1 = 1$. We now apply (5.4′) with $\delta = \delta_1 = 1$ and find

$$\Lambda_1(x) = (1+D)\,\Lambda_0(x) = e^{-2x} e^{-e^{-x}}$$

for which $\alpha_2 = -2$, hence $\delta_2 = -1/\alpha_2 = 1/2$.

After n steps we obtain the totally positive function

$$(5.9) \quad \Lambda_n(x) = D(1+D)\left(1+\frac{1}{2}D\right)\cdots\left(1+\frac{1}{n}D\right) e^{-e^{-x}} = \frac{1}{n!} e^{-(n+1)x} e^{-e^{-x}}$$

and therefore

$$F(s)\,s \prod_{\nu=1}^{n} \left(1+\frac{s}{\nu}\right) e^{-s/\nu} = \int_{-\infty}^{\infty} e^{-sx} \lambda_n(x)\,dx,$$

where

$$(5.10) \qquad \lambda_n(s) = \Lambda_n\left(x - 1 - \frac{1}{2} - \cdots - \frac{1}{n}\right).$$

These results prove already that

$$F(s) = e^{\gamma s^2 - \delta s} s^{-1} \prod_{\nu=1}^{\infty} \frac{e^{s/\nu}}{1+\frac{s}{\nu}},$$

where the constants γ and δ may be determined as follows: Using the formulae

$$1 + \frac{1}{2} + \cdots + \frac{1}{n} = C + \log n + O\left(\frac{1}{n}\right), \quad \log n! = \left(n+\frac{1}{2}\right)\log n - n + O(1),$$

we find by (5.10) and (5.9), that

$$\log \lambda_n(x) = n\,(C - x + 1 - e^{C-x}) + \frac{1}{2}\log n + O(1).$$

Since

$$C - x + 1 - e^{C-x} < 0 \quad \text{if} \quad x \lessgtr C,$$

we conclude that $\lambda_n(x) \to 0$ as $n \to \infty$, provided $x \neq C$. We conclude that $\Lambda(x)$ has no normal component, or $\gamma = 0$, and that $\lambda_n(x)$ converges to the distribution concentrated at $x = C$, hence $\delta = C$.

§ 6. Periodic meromorphic functions.

19. Perhaps the simplest application of the expansion Theorem 4, of § 4, is to the function

$$F(s) = \frac{\pi}{\sin \pi s}.$$

Indeed, let

$$\frac{\pi}{\sin \pi s} = \int_{-\infty}^{\infty} e^{-sx} \Lambda(x)\, dx, \qquad (0 < Rs < 1).$$

For $\Psi(s) = \sin \pi s/\pi$ we find $\Psi'(n) = \cos \pi n = (-1)^n$. Thus the two series (4.22) and (4.23) become

$$1 - e^{-x} + e^{-2x} - e^{-3x} + \dots \text{ and } e^{x} - e^{2x} + e^{3x} - \dots$$

respectively. Since the first converges for $x > 0$ to $1/(1 + e^{-x})$, and the second converges for $x < 0$ to $e^x/(1 + e^x)$, we find that

$$\Lambda(x) = \frac{1}{1 + e^{-x}} \qquad \text{for all } x.$$

Let us solve the corresponding problem for

$$(6.1) \qquad F(s) = \left(\frac{\pi}{\sin \pi s}\right)^k = \int_{-\infty}^{\infty} e^{-sx} \Lambda(x)\, dx, \qquad (0 < Rs < 1,\ k \text{ integer}).$$

By (4.17) we find

$$(6.2) \qquad \Lambda(x) = P(x;\ 0) + (-1)^k \Lambda_1(x).$$

If

$$\left(\frac{\pi}{\sin \pi s}\right)^h = \frac{a_k}{s^k} + \dots + \frac{a_1}{s} + a_0 + \dots,$$

or

$$\left(\frac{\pi s}{\sin \pi s}\right)^k = a_k + a_{k-1} s + \dots + a_1 s^{k-1} + \dots,$$

then we know by (4.16) that

$$P(x;\ 0) = a_1 + \frac{x}{1!} a_2 + \dots + \frac{x^{k-1}}{(k-1)!} a_k.$$

On the other hand we have that

$$\Lambda_1(x) = \frac{(-1)^k}{2\pi i} \int_{-1/2-i\infty}^{-1/2+i\infty} e^{sx} F(s)\, ds,$$

and substituting $s = -1 + z$ in this integral, we obtain by the periodicity of $F(s)$ that

$$\Lambda_1(x) = \frac{e^{-x}}{2\pi i} \int_{1/2-i\infty}^{1/2+i\infty} e^{zx} F(z)\, dz = e^{-x} \Lambda(x).$$

Substituting into (6.2) we have

$$\Lambda(x) = P(x;\ 0) + (-1)^k e^{-x} \Lambda(x),$$

whence [17]

$$\Lambda(x) = \frac{P(x;\ 0)}{1+(-1)^{k+1} e^{-x}}.$$

20. The method just used to derive the representation (6.1) may be used to obtain the Laplace integral representation for any periodic function $F(s)$ of our type. Indeed, let $\Psi(s) = 1/F(s)$ be a function of the class T_2 which is also periodic of period ω $(\omega > 0)$. Let α and β be consecutive zeros of $\Psi(s)$, $\Psi(s) > 0$ if $\alpha < s < \beta$. Let also

$$\alpha_{n-1} < \alpha_{n-2} < \ldots < \alpha_1 < \alpha_0 = \alpha \tag{6.3}$$

be all distinct zeros of $\Psi(s)$ within the period $\alpha - \omega < s \leqq \alpha$. Evidently

$$\Lambda(x) = \sum_{\nu=0}^{n-1} P(x;\ \alpha_\nu)\, e^{x\alpha_\nu} + \Lambda_n(x)$$

by (4.17). By the change of variable $s = z - \omega$ we obtain

$$\Lambda_n(x) = \frac{1}{2\pi i} \int_{\sigma-\omega-i\infty}^{\sigma-\omega+i\infty} e^{xs} F(s)\, ds = \frac{e^{-\omega x}}{2\pi i} \int_{\sigma-i\infty}^{\sigma+i\infty} e^{xz} F(z)\, dz = e^{-\omega x} \Lambda(x), \tag{6.4}$$

and from the two relations we obtain

$$\Lambda(x) = \frac{1}{1-e^{-\omega x}} \sum_{\nu=0}^{n-1} P(x;\ \alpha_\nu)\, e^{x\alpha_\nu} \tag{6.5}$$

---

17. This explicit expression is due to Widder; see [20], Theorem 17c, p. 561.

for the determining function in the representation formula

$$F(s) = \int_{-\infty}^{\infty} e^{-xs} \Lambda(x)\, dx, \qquad (\alpha < Rs < \beta = \alpha_{n-1} + \omega).$$

If the poles (6.3) are all simple, then (6.5) becomes

$$\Lambda(x) = \frac{1}{1-e^{-\omega x}} \sum_{\nu=0}^{n-1} \frac{1}{\Psi'(\alpha_\nu)} e^{x\alpha_\nu}. \tag{6.6}$$

This result may be made more explicit. Indeed, let us assume that the simple zero within a period $(0, \omega)$, necessarily even in number, are $\gamma_1, \gamma_2, \ldots, \gamma_{2n}$. Then we must have up to a constant factor

$$\Psi(s) = \prod_{\nu=1}^{2n} \sin\frac{\pi}{\omega}(s-\gamma_\nu), \qquad (0<\gamma_1<\ldots<\gamma_{2n}<\omega).$$

By (6.4) and (6.6) we find

$$\frac{1}{\Psi(s)} = \int_{-\infty}^{\infty} e^{-sx} \Lambda(x)\, dx, \qquad (\gamma_{2n}-\omega < Rs < \gamma_1),$$

where

$$\Lambda(x) = \frac{1}{e^{\omega x}-1} \sum_{\nu=1}^{2n} c_\nu e^{x\gamma_\nu},$$

with

$$\frac{1}{c_\nu} = \frac{\pi}{\omega} \prod_{\substack{\mu=1\\ \mu\neq\nu}}^{2n} \sin\frac{\pi}{\omega}(\gamma_\nu-\gamma_\mu).$$

As a particular case we have the formula

$$\frac{\sin 2\gamma}{\sin(\gamma+s)\sin(\gamma-s)} = \int_{-\infty}^{\infty} e^{-xs} \frac{\sinh\left(\frac{\pi}{2}-\gamma\right)x}{\sinh\frac{\pi}{2}x}\, dx, \quad \left(-\gamma<Rs<\gamma,\ 0<\gamma<\frac{\pi}{2}\right).$$

§7. Examples of representations of reciprocals of functions of the class T_1.

21. We have already seen such trivial examples like

$$\frac{n!}{s^{n+1}} = \int_0^{\infty} e^{-xs} x^n\, dx.$$

We shall now discuss two more interesting ones. Let

$$\Phi(s) = s\prod_{\nu=1}^{\infty}\left(1+\frac{s}{\nu^2}\right) = -\frac{1}{\pi}\sqrt{-s}\,\sin\pi\sqrt{-s},$$

whose reciprocal may be written as

$$\frac{1}{\Phi(s)} = \int_0^{\infty} e^{-sx}\Lambda(x)\,dx, \tag{7.1}$$

where $\Lambda(x)$ is a totally positive function vanishing if $x<0$. Let us derive the asymptotic series for $\Lambda(x)$. A simple computation shows that

$$\Phi'(-\nu^2) = \begin{cases} (-1)^\nu/2 & \text{if } \nu>0, \\ 1 & \text{if } \nu=0; \end{cases}$$

hence by Theorem 3

$$\Lambda(x) \sim 1 + \sum_{\nu=1}^{\infty} 2(-1)^\nu e^{-x\nu^2} \quad \text{for} \quad x \to +\infty.$$

This series being convergent for $x>0$, Theorem 4 shows that

$$\Lambda(x) = \begin{cases} \sum_{-\infty}^{\infty}(-1)^\nu e^{-x\nu^2} & \text{if } x>0 \\ 0 & \text{if } x \leqq 0. \end{cases} \tag{7.2}$$

Multiplying both sides of (7.1) by s and integrating by parts the right-hand side we find the representation

$$\prod_{\nu=1}^{\infty}\left(1+\frac{s}{\nu^2}\right)^{-1} = \int_0^{\infty} e^{-sx}\Lambda'(x)\,dx,$$

where

$$\Lambda'(x) = \begin{cases} \sum_{-\infty}^{\infty}(-1)^{\nu+1}\nu^2 e^{-x\nu^2} & \text{if } x>0, \\ 0 & \text{if } x \leqq 0, \end{cases}$$

is a Pólya frequency function.

As a second and last example let

$$\Phi(s) = \prod_{\nu=0}^{\infty}\left(1+\frac{s}{a^\nu}\right), \qquad (a>1).$$

Letting

$$A = \prod_{\nu=1}^{\infty}(1-a^{-\nu})^{-1},$$

we readily find by logarithmic differentiation of $\Phi(s)$ that

$$\frac{1}{\Phi'(-a^n)} = \frac{A a^n}{(1-a)(1-a^2)\dots(1-a^n)}.$$

By the arguments used in the previous example we find the representation

$$\prod_{\nu=0}^{\infty}\left(1+\frac{s}{a^\nu}\right)^{-1} = \int_0^\infty e^{-sx}\Lambda(x)\,dx, \qquad (Rs>-1),$$

where

$$\Lambda(x) = A\,.\sum_{\nu=0}^{\infty}\frac{(-1)^\nu a^\nu}{(a-1)(a^2-1)\dots(a^\nu-1)}e^{-a^\nu x}, \qquad (x>0), \tag{7.3}$$

is a Pólya frequency function if extended by $\Lambda(x)=0$ for $x<0$. Being infinitely often differentiable, the functions (7.2) and (7.3) must vanish at the origin together with all their derivatives.

BIBLIOGRAPHY

1. M. Aissen, A. Edrei, I. J. Schoenberg and A. Whitney, *On the generating functions of totally positive sequences*, Proceedings of the National Academy of Sciences, **37** (1951), 303—307.

2. A. L. Cauchy, Exercices de Mathématiques, IIe année, Paris, 1827.

3. G. Doetsch, *Theorie und Anwendung der Laplace-Transformation*, Berlin, 1937.

4. J. Hadamard, *Essai sur l'étude des fonctions données par leur développement de Taylor*, Journal de Mathématiques, 4e série, **8** (1892).

5. H. Hamburger, *Bemerkungen zu einer Fragestellung des Herrn Pólya*, Mathematishe Zeitschrift, **7** (1920), 302—322.

6. I. I. Hirschman and D. V. Widder, *The inversion of a general class of convolution transforms*, Transactions of the American Mathematical Society, **66** (1949), 135—201.

7. I. I. Hirschman and D. V. Widder, *Generalized inversion formulas for convolution transforms*, Duke Mathematical Journal, **15** (1948), 659—696.

8. E. Laguerre, *Sur les fonctions du genre zéro et du genre un*, Oeuvres de Laguerre I, Paris, 1898, 174—177.

9. Th. Motzkin, *Beitraege zur Theorie der linearen Ungleichungen*, Jerusalem 1936.

10. G. Pólya und I. Schur, *Ueber zwei Arten von Faktorenfolgen in der Theorie der algebraischen Gleichungen*, Journal für Mathematik, **144** (1914), 89—113.

11. G. Pólya, *Ueber Annaeherung durch Polynome mit lauter reelen Wurzeln*, Rendiconti di Palermo, **36** (1913), 1—17.

12. G. Pólya, *Algebraische Untersuchungen über ganze Funktionen vom Geschlechte Null und Eins*, Journal für Mathematik, **145** (1915), 224—249.

13. G. Pólya und G. Szegö, *Aufgaben und Lehrsaetze aus der Analysis*, II, Berlin, 1925.

14. I. J. Schoenberg, *Ueber variationsvermindernde lineare Transformationen*, Mathematische Zeitschrift, **32** (1930), 321—328.

15. I. J. Schoenberg, *On totally positive functions, Laplace integrals and entire functions of the Laguerre-Polya-Schur type*, Proceedings of the National Academy of Sciences, **33** (1947), 11—17.

16. I. J. Schoenberg et Anne Whitney, *Sur la positivité des déterminants de translation des fonctions de fréquence de Polya avec une application au problème d'interpolation par des fonctions "spline"*, Comptes Rendus des séances de l'Académie des Sciences, (1949).

17. I. J. Schoenberg, *On Polya frequency functions II: Variation-diminishing integral operators of the convolution type*, Acta Scientiarum Mathematicarum, Szeged, **12** (1950), 97—106.

18. I. J. Schoenberg and Anne Whitney, *A theorem on polygons in* n *dimensions with applications to variation-diminishing transformations*, Compositio Mathematica, 9 (1951), 141—160.

19. W. Sierpinski, *Sus les fonctions convexes mesurables*, Fundamenta Mathematicae, **1** (1920), 125—129.

20. D. V. Widder, *The Laplace transform*, Princeton, 1946.

21. D. V. Widder, *Weierstrass transforms of positive functions*, Proceedings of the National Academy of Sciences, **37** (1951), 315—317.

22. A. Wintner, *On a class of Fourier transforms*, American Journal of Mathematics, **58** (1936), 45—90.

(Received June 12, 1951)

Reprinted from
J. d'Analyse Math. **1** (1951), pp. 331–374

On Pólya frequency functions. II: Variation-diminishing integral operators of the convolution type.

By I. J. SCHOENBERG in Philadelphia, Pa.

1. Introduction and statement of results[1]).

A real matrix $A=\|a_{ik}\|$ $(i=1,\ldots,m;\ k=1,\ldots,n)$ is said to be *totally positive* if all its minors, of any order, are non-negative. In 1930[2]) the author showed that if A is totally positive, then the linear transformation

$$(1)\qquad y_i=\sum_{k=1}^{n} a_{ik}x_k \qquad (i=1,\ldots,m)$$

is variation-diminishing in the sense that if $v(x_k)$ denotes the number of variations of sign in the sequence $\{x_k\}$ and $v(y_i)$ the corresponding number in the sequence $\{y_i\}$, then we always have the inequality $v(y_i)\leqq v(x_k)$. In the same paper of 1930 the author showed that (1) is certainly variation-diminishing if the matrix A does not possess two minors of equal orders and of opposite signs; also the converse holds to a certain extent: If (1) is variation-diminishing, then A cannot have two minors of equal orders and of opposite signs, *provided the rank of A is* $=n$. The necessary and sufficient conditions in order that (1) be variation-diminishing were found in 1933 by TH. MOTZKIN[3]). Since they will be used in this paper we state them here as follows: Let r be the rank of A; then A should not have two minors of equal orders and of opposite signs if their common order is $<r$, while if their common order is $=r$ then again they should never be of opposite signs if they belong to the same combination of r columns of A.

A function $\Lambda(x)$, $-\infty<x<\infty$, is called a Pólya frequency function (abbreviated P. f. f.) if it satisfies the following three characteristic conditions:

1. $\Lambda(x)$ is measurable.

---

1) A résumé of the results of this paper has appeared under the same title in the *Proceedings of the National Academy of Sciences*, **34** (1948), pp. 164–169.

2) I. J. SCHOENBERG, Über variationsvermindernde lineare Transformationen, *Math. Zeitschrift*, **32** (1930), pp. 321–328.

3) TH. MOTZKIN, *Beiträge zur Theorie der linearen Ungleichungen*, Doctoral dissertation, Basel, 1933 (Jerusalem, 1936), 69 pp., especially Chap. IV.

E 7

2. If $x_1 < x_2 < \ldots < x_n$ and $t_1 < t_2 < \ldots < t_n$ then $\det \|\Lambda(x_i - t_j)\|_{1,n} \geqq 0$, $n = 1, 2, 3, \ldots$. For $n = 1$ this means that $\Lambda(x) \geqq 0$.

3. Finally

$$0 < \int_{-\infty}^{\infty} \Lambda(x)\, dx < +\infty.$$

In our previous paper on Pólya frequency functions[4]) it was shown that a P. f. f. $\Lambda(x)$ has a bilateral Laplace transform of the form

$$\text{(2)} \qquad \int_{-\infty}^{\infty} e^{-xs} \Lambda(x)\, dx = \frac{1}{\Psi(s)} \qquad (-c < Rs < c \text{ for some } c > 0),$$

where $\Psi(s)$ is an entire function of the type II of Laguerre, Pólya and Schur

$$\text{(3)} \quad \Psi(s) = c e^{-\gamma s^2 - \delta s} \prod_{\nu=1}^{\infty} (1 + \delta_\nu s)\, e^{-\delta_\nu s} \quad \left(c > 0,\ \gamma \geqq 0,\ 0 < \gamma + \sum_{1}^{\infty} \delta_\nu^2 < \infty\right),$$

and that conversely, the reciprocal of a function $\Psi(s)$ of this type allows of a representation (2) where $\Lambda(x)$ is a P. f. f.

All Pólya frequency functions $\Lambda(x)$ are everywhere continuous with the single exception of the so-called truncated exponential

$$\Lambda(x) = \begin{cases} e^{-x} & \text{if } x \geqq 0, \\ 0 & \text{if } x < 0, \end{cases}$$

and all functions arising from it by a change of scale and origin. Further notable examples of P. f. f. are $\Lambda(x) = e^{-x^2}$, $e^{-|x|}$, $e^{x - e^x}$, $1/\cosh x$.

Let $f(x)$ be a real function defined for all real x. The number $v(f)$, of variations of sign of $f(x)$ in the range $(-\infty, \infty)$ is defined as follows: If S: $x_1 < x_2 < \ldots < x_n$ is an arbitrary finite increasing sequence of reals, then

$$v(f) = \sup_{S} v(f(x_i)) \qquad (0 \leqq v(f) \leqq \infty).$$

Let now $L(t)$ be a given real function of bounded variation in the range $-\infty < t < \infty$ which we normalize by the conditions that $L(-\infty) = 0$, $2L(t) = L(t+0) + L(t-0)$; we also rule out the trivial case when $L(t) \equiv 0$. Let us consider the integral transformation

$$\text{(4)} \qquad g(x) = \int_{-\infty}^{\infty} f(x - t)\, dL(t),$$

where $f(x)$ is an arbitrary continuous and bounded function. We say that (4)

4) I. J. Schoenberg, On totally positive functions, Laplace integrals and entire functions of the Laguerre—Pólya—Schur type, *Proceedings of the National Academy of Sciences*, **33** (1947), pp. 11—17. A detailed paper will appear under the title "On Pólya frequency functions. I: Totally positive functions and their Laplace transforms" probably in the *Transactions of the American Mathematical Society.*

is *variation-diminishing* if (4) always implies the inequality

$$v(g) \leqq v(f). \tag{5}$$

Our main result is the following

Theorem 1. *The transformation* (4) *is variation-diminishing if and only if* $L(t)$ *is either, up to the sign, a cumulative Pólya frequency function*

$$L(t) = \varepsilon \int_{-\infty}^{t} \Lambda(u)\, du, \tag{6}$$

where $\varepsilon = \pm 1$ *and* $\Lambda(x)$ *is a Pólya frequency function, or else* $L(t)$ *is a step-function with only one jump.*

On combining this theorem with (2) and (3) we may restate our result without a distinction of two cases as

Theorem 2. *The transformation* (4) *is variation-diminishing if and only if* $L(t)$ *has a bilateral Laplace-Stieltjes transform of the form*

$$\int_{-\infty}^{\infty} e^{-st} dL(t) = C' e^{\gamma s^2 + \delta s} \prod_{\nu=1}^{\infty} \frac{e^{\delta_\nu s}}{1+\delta_\nu s} \quad (-c < Rs < c, \text{ for some } c > 0), \tag{7}$$

where $C' \gtrless 0$, $\gamma \geqq 0$, δ, δ_ν *real*, $\sum_{1}^{\infty} \delta_\nu^2 < \infty$.

It should be noticed that the trivial case when $\gamma = \delta_\nu = 0$, which was excluded in (3), corresponds to a step-function $L(t)$ with only one jump, in which case our transformation (4) becomes

$$g(x) = C' f(x+\delta) \tag{8}$$

and is evidently variation-diminishing.

2. **Proof of the direct part of Theorem 1.**

If (6) holds, the (4) becomes, assuming $\varepsilon = 1$,

$$g(x) = \int_{-\infty}^{\infty} \Lambda(x-t) f(t)\, dt \tag{9}$$

and we are to show that (9) is variation-diminishing within the class of bounded continuous function $f(x)$. This we shall now do even for the wider class of *measurable* and *bounded* functions $f(x)$. Indeed if $v(f) = \infty$ then (5) is trivially verified. Hence we may assume $v(f) < \infty$, $v(f) = m$ say.

Let

$$F(t) = \int_{0}^{t} f(t)\, dt \qquad (-\infty < t < \infty); \tag{10}$$

then (9) becomes

$$g(x) = \int_{-\infty}^{\infty} \Lambda(x-t)\, dF(t). \tag{11}$$

If $x_0 < x_1 < \ldots < x_n$ we are to show that

$$v(g(x_i)) \leqq m, \tag{12}$$

and without losing generality we may assume that $v(g(x_i)) = n$, that is

$$g(x_0), g(x_1), \ldots, g(x_n) \quad \text{alternate in sign.} \tag{13}$$

Clearly (11) converges uniformly in the $n+1$ points x_i; hence we can find A such that the quantities

$$g_i = \int_{-A}^{A} \Lambda(x_i - t)\, dF(t) \qquad (i = 0, 1, \ldots, n), \tag{14}$$

also alternate in sign, i. e. $v(g_i) = n$.

But (14) are ordinary Stieltjes integrals. We may therefore divide $(-A, A)$ into N equal parts by $t_0 = -A < t_1 < \ldots < t_N = A$ such that

$$\bar{g}_i = \sum_{\nu=1}^{N} \Lambda(x_i - t_\nu)\,(F(t_\nu) - F(t_{\nu-1})) \tag{15}$$

are all so close to the respective integrals (14) that we also have

$$v(\bar{g}_i) = n. \tag{16}$$

Since by assumption $v(f) = m$ we see, by (10), that the number of variations of sign in the sequence $F(t_\nu) - F(t_{\nu-1})$ $(\nu = 1, \ldots, N)$ is $\leqq m$. But (15) is a variation-diminishing linear transformation in view of the total positivity of the matrix $\|\Lambda(x_i - t_\nu)\|$. By (16) we conclude that $n \leqq m$, hence (12) is established.

3. Proof of the converse part of Theorem I in case when $L(t)$ is continuously differentiable.

Writing $\lambda(t) = L'(t)$ which we assume continuous and clearly also in the Lebesgue class $L(-\infty, \infty)$, we now assume the transformation

$$g(x) = \int_{-\infty}^{\infty} \lambda(x-t) f(t)\, dt \tag{17}$$

to be variation-diminishing for bounded continuous $f(t)$ and we are to conclude that $\lambda(x)$ is, up to the sign, a Pólya frequency function. By obvious continuity arguments the variation-diminishing property of (17) immediately generalizes to the case when $f(t)$ is a step-function with a finite number of jumps.

Let

(18) $$x_1 < x_2 < \ldots < x_n, \quad t_1 < t_2 < \ldots < t_n;$$

we want to show first that

(19) $$y_i' = \sum_{\nu=1}^{n} \lambda(x_i - t_\nu)\, y_\nu \qquad (i = 1, \ldots, n)$$

is a variation-diminishing linear transformation. To this end, given the y_ν, we define a step-function $f(t)$ as follows: $f(t) = \frac{1}{2h} y_\nu$ if $t_\nu - h \leqq t \leqq t_\nu + h$, $f(t) = 0$ outside the n intervals $[t_\nu - h, t_\nu + h]$ which are assumed not to overlap.

For the values of $g(x_i)$ as given by (17) we now find that

(20) $$g(x_i) = \sum_{\nu=1}^{n} \left(\frac{1}{2h} \int_{t_\nu - h}^{t_\nu + h} \lambda(x_i - t)\, dt \right) y_\nu \qquad (i = 1, \ldots, n).$$

Since $v(f) = v(y_\nu)$, while $v(g(x_i)) \leqq v(g)$, we have

(21) $$v(g(x_i)) \leqq v(y_\nu),$$

showing that (20) is variation-diminishing. Letting $h \to 0$, the continuity of $\lambda(x)$ implies $\lim_{h \to 0} g(x_i) = y_i' = \sum_{\nu=1}^{n} \lambda(x_i - t_\nu)\, y_\nu$, and, by (21), we have

$$v(y_i') \leqq \varliminf_{h \to 0} v(g(x_i)) \leqq v(y_\nu).$$

Thus (19) is indeed variation-diminishing.

This result implies in particular that $\lambda(x)$ never changes sign; for indeed, if it did, we could arrange to have $\lambda(x_1 - t_1)$ and $\lambda(x_2 - t_1)$ of opposite signs and this would contradict the inequality $v(y_i') \leqq v(y_\nu)$ for the values $y_1 = 1$, $y_2 = \ldots = y_n = 0$, if substituted into (19). Thus without loss of generality we may assume to have that

(22) $$\lambda(x) \geqq 0 \quad \text{for all } x, \text{ in particular} \quad \lambda(0) > 0,$$

since the last condition always obtains after a suitable shift of origin.

Let again $\{x_i\}$ *and* $\{t_j\}$ *satisfy* (18). *We wish to prove now that*

(23) $$D_n = \det \| \lambda(x_i - t_j) \|_{1,n} \geqq 0.$$

This is the main point of the proof. We shall establish it first for $n = 2$:

(24) *If* $x_1 < x_2$, $t_1 < t_2$, *then* $D_2 = \begin{vmatrix} \lambda(x_1 - t_1) & \lambda(x_1 - t_2) \\ \lambda(x_2 - t_1) & \lambda(x_2 - t_2) \end{vmatrix} \geqq 0.$

Indeed, suppose we had

(25) $$D_2 < 0.$$

Then let $x_3 = t_3 = \tau > \max(x_2, t_2)$ and consider

$$(26)\qquad D_3 = \begin{vmatrix} \lambda(x_1-t_1) & \lambda(x_1-t_2) & \lambda(x_1-\tau) \\ \lambda(x_2-t_1) & \lambda(x_2-t_2) & \lambda(x_2-\tau) \\ \lambda(\tau-t_1) & \lambda(\tau-t_2) & \lambda(0) \end{vmatrix}.$$

Since $\lambda(x) \geqq 0$ and the integral

$$\int_0^\infty (\lambda(x_1-\tau)+\lambda(x_2-\tau)+\lambda(\tau-t_1)+\lambda(\tau-t_2)\,d\tau$$

converges, we can certainly choose $\tau > \max(x_2, t_2)$ such that each of the four quantities under the integral sign are as small as we please. But then (25), (26) and $\lambda(0) > 0$ clearly imply that

$$D_3 = D_2\,\lambda(0) + \text{ (an arbitrarily small quantity) } < 0$$

for some appropriate value of τ. On the other hand, at least one of the four elements of D_2 is positive, $\lambda(x_1-t_2)$ say, and this implies that

$$(27)\qquad \begin{vmatrix} \lambda(x_1-t_2) & \lambda(x_1-\tau) \\ \lambda(\tau-t_2) & \lambda(0) \end{vmatrix} = \lambda(x_1-t_2)\,\lambda(0) + \text{(small quantity)} > 0.$$

But now we have a contradiction with the properties of variation-diminishing transformation stated in the first paragraph of our introduction: (26) is the determinant of a non-singular variation-diminishing transformation; as such it cannot have two minors, such as (25) and (27), which are of equal orders less than its rank 3, and of opposite signs.

We may now turn to a proof of the general inequality (23). Firstly we recall[5]) that the property (24) means that $\lambda(x)$ is logarithmically concave; this fact and the summability of $\lambda(x)$ imply that

$$(28)\qquad \lim_{x\to\pm\infty} \lambda(x) = 0,$$

even exponentially. Secondly, let $\xi_0, \xi_1, \ldots$ be an infinite sequence of real numbers having the following properties:

1. The sequence $\{\xi_\nu\}$ is monotone increasing.
2. The sequence $\{\xi_\nu\}$ contains every element of the finite sequences $\{x_i\}$ and $\{t_j\}$ appearing in (23).
3. For sufficiently large ν, $\{\xi_\nu\}$ is made up of consecutive integers.

N and r being positive integers let us consider the following three determinants

$$(29)\qquad \begin{aligned} D_{N+1} &= \det\|\lambda(\xi_i-\xi_j)\| && (i,j=0,1,\ldots,N), \\ D_n^* &= \det\|\lambda(\xi_{ri}-\xi_{rj})\| && (i,j=0,1,\ldots,n-1), \\ D_{n+1}^* &= \det\|\lambda(\xi_{ri}-\xi_{rj})\| && (i,j=0,1,\ldots,n). \end{aligned}$$

From the property 3 of the sequence $\{\xi_\nu\}$, together with (28), it is clear that

---

[5]) See the paper mentioned in footnote [4]).

$\lambda(\xi_{ri}-\xi_{rj})\to 0$ as $r\to\infty$, provided that $i\neq j$. By (29) we have $D_n^*\to(\lambda(0))^n$ and $D_{n+1}^*\to(\lambda(0))^{n+1}$; thus we now see that

$$D_n^*>0,\ D_{n+1}^*>0, \tag{30}$$

provided r is chosen large enough. If we now choose $N>rn$, then clearly D_n^* and D_{n+1}^* are minors of D_{N+1}; by the property 2 we may moreover assume also D_n to be a minor of D_{N+1}, by further increasing N, if necessary. We may now prove (23) as follows: We know from the first result of this section that D_{N+1} is the determinant of a variation-diminishing transformation; by (30) the rank of D_{N+1} is $\geqq n+1$. By MOTZKIN's theorem of our introduction we conclude that the two minors D_n and D_n^* cannot have opposite signs, hence $D_n^*>0$ implies that $D_n\geqq 0$. But then $\lambda(x)$ is indeed a Pólya f. f.

4. General proof of the converse part of Theorem 1.

Let $f(t)$ be continuous and bounded and let $g(x)$ be defined by

$$g(x)=\int_{-\infty}^{\infty} f(x-t)\,dL(t). \tag{31}$$

We define two new functions by

$$\lambda(x)=\frac{1}{\sqrt{\pi}}\int_{-\infty}^{\infty} e^{-(x-t)^2}\,dL(t) \tag{32}$$

and

$$h(x)=\frac{1}{\sqrt{\pi}}\int_{-\infty}^{\infty} g(x-t)\,e^{-t^2}\,dt. \tag{33}$$

Now (31), (32) and (33) imply that

$$h(x)=\int_{-\infty}^{\infty} f(x-u)\,\lambda(u)\,du. \tag{34}$$

Indeed, if we substitute (31) into (33) we find

$$h(x)=\frac{1}{\sqrt{\pi}}\int_{-\infty}^{\infty} e^{-t^2}dt\int_{-\infty}^{\infty} f(x-t-u)\,dL(u)=\frac{1}{\sqrt{\pi}}\int_{-\infty}^{\infty} e^{-t^2}dt\int_{-\infty}^{\infty} f(x-u)\,d_uL(u-t).$$

Since $f(t)$ is continuous and bounded and $L(u)$ of bounded variation it is easy to see that we may integrate first under the differential sign d_u obtaining

$$h(x)=\int_{-\infty}^{\infty} f(x-u)\,d_u\frac{1}{\sqrt{\pi}}\int_{-\infty}^{\infty} L(u-t)\,e^{-t^2}dt=\int_{-\infty}^{\infty} f(x-u)\,\frac{d}{du}\left(\frac{1}{\sqrt{\pi}}\int_{-\infty}^{\infty} e^{-(u-t)^2}\,L(t)\,dt\right)du.$$

Now

$$\frac{d}{du}\left(\frac{1}{\sqrt{\pi}}\int_{-\infty}^{\infty} e^{-(u-t)^2} L(t)\,dt\right) = -\frac{1}{\sqrt{\pi}}\int_{-\infty}^{\infty} 2(u-t)\,e^{-(u-t)^2} L(t)\,dt =$$

$$= -\frac{1}{\sqrt{\pi}}\int_{-\infty}^{\infty} L(t)\,d_t\, e^{-(u-t)^2} = \frac{1}{\sqrt{\pi}}\int_{-\infty}^{\infty} e^{-(u-t)^2} dL(t) = \lambda(u)$$

and (34) is established.

By assumption we have $v(g) \leqq v(f)$; but also (33) is a variation-diminishing transformation by the direct part of Theorem 1, hence $v(h) \leqq v(g)$. It follows that $v(h) \leqq v(f)$ and that (34) is therefore a variation-diminishing transformation. Since $\lambda(x)$ is obviously continuous and summable we learn from the previous case already settled, that $\lambda(x)$ is a P. f. f. Therefore

$$(35) \qquad \int_{-\infty}^{\infty} e^{-sx}\lambda(x)\,dx = \frac{1}{\Psi(s)} = Ce^{\gamma s^2 + \delta s} \prod_{\nu=1}^{\infty} \frac{e^{\delta_\nu s}}{1+\delta_\nu s} \qquad (-a < Rs < a),$$

where $C \gtrless 0$, $\gamma \geqq 0$, $0 < \gamma + \Sigma\delta_\nu^2 < \infty$.

Let us consider now the characteristic function of $L(t)$:

$$(36) \qquad H(it) = \int_{-\infty}^{\infty} e^{-itx}\,dL(x) \qquad (-\infty < t < \infty).$$

The convolution relation (32) now implies the characteristic function relation

$$\frac{1}{\Psi(it)} = e^{-\frac{t^2}{4}} H(it) \qquad (-\infty < t < \infty),$$

or

$$(37) \qquad H(it) = Ce^{\left(\frac{1}{4}-\gamma\right)t^2 + i\delta t} \prod_{\nu=1}^{\infty} \frac{e^{i\delta_\nu t}}{1+i\delta_\nu t} \qquad (-\infty < t < \infty).$$

Passing to norms we have

$$|H(it)|^2 = C^2 e^{2\left(\frac{1}{4}-\gamma\right)t^2} \Big/ \prod_{\nu=1}^{\infty} (1+\delta_\nu^2 t^2),$$

and from this relation we can easily conclude that

$$(38) \qquad \frac{1}{4} - \gamma \leqq 0.$$

Indeed, let $\varepsilon > 0$ be given and choose n such that $\sum_{n+1}^{\infty} \delta_\nu^2 < \varepsilon$; since $|H(it)|^2$ is bounded, by (36), we have for t^2 sufficiently large the relations

$$C^2 e^{2\left(\frac{1}{4}-\gamma\right)t^2} = |H(it)|^2 \prod_{1}^{n} (1+\delta_\nu^2 t^2) \prod_{n+1}^{\infty} (1+\delta_\nu^2 t^2) < e^{\varepsilon t^2} e^{\varepsilon t^2} = e^{2\varepsilon t^2},$$

hence $\frac{1}{4} - \gamma < \varepsilon$ and thence (38), on letting $\varepsilon \to 0$. With $\gamma_1 = \gamma - \frac{1}{4} \geqq 0$

we may now rewrite (37) as

$$H(it) = C e^{-\gamma_1 t^2 + i\delta t} \prod_{\nu=1}^{\infty} \frac{e^{i\delta_\nu t}}{1 + i\delta_\nu t} \qquad (-\infty < t < \infty). \tag{39}$$

Two cases may now arise:

1. $\gamma_1 = 0$ and $\delta_1 = \delta_2 = \ldots = 0$. In this case (39) becomes

$$H(it) = \int_{-\infty}^{\infty} e^{-itx} dL(x) = C e^{i\delta t},$$

hence

$$L(x) = \begin{cases} 0 & \text{if} \quad x < -\delta \\ C & \text{if} \quad x > -\delta. \end{cases}$$

2. $\gamma_1 + \sum_1^\infty \delta_\nu^2 > 0$, then $\Psi_1(s) = C^{-1} e^{-\gamma_1 s^2 - \delta s} \prod_{\nu=1}^{\infty} (1 + \delta_\nu s) e^{-\delta_\nu s}$

is of the type of (3); hence its reciprocal may be written as

$$\frac{1}{\Psi_1(s)} = \int_{-\infty}^{\infty} e^{-sx} \varepsilon \Lambda(x)\, dx \qquad (-a < Rs < a),$$

where $\varepsilon = \pm 1$ and $\Lambda(x)$ is a P. f. f. From the last relation, (39) and (36) we now obtain the identity $\int_{-\infty}^{\infty} e^{-itx} dL(x) = \int_{-\infty}^{\infty} e^{-itx} \varepsilon \Lambda(x)\, dx \qquad (-\infty < t < \infty),$

from which we conclude that $L(x) = \varepsilon \int_{-\infty}^{x} \Lambda(u)\, du.$

This concludes the proof of our Theorem 1.

5. The connection of Theorem 1 with a theorem of Pólya.

In 1915 PÓLYA[6]) discovered the following

Theorem 3 (PÓLYA). *Let*

$$\sum_{\nu=0}^{\infty} \frac{(-1)^\nu}{\nu!} \mu_\nu s^\nu \qquad (\mu_0 \neq 0) \tag{40}$$

be a given formal power series with the following property (*S*)*: If* $f(x)$ *is an arbitrary real polynomial, then the number of real zeros of the polynomial*

$$g(x) = \sum_{\nu=0}^{\infty} \frac{(-1)^\nu}{\nu!} \mu_\nu f^{(\nu)}(x) \tag{41}$$

should never exceed the number of real zeros of the given polynomial $f(x)$. *A power series* (40) *enjoys the property* (*S*) *if and only if it is the Taylor*

[6]) G. PÓLYA, Algebraische Untersuchungen über ganze Funktionen vom Geschlechte Null und Eins, *Journal für die reine und angewandte Math.*, **145** (1915), pp. 224–249, especially p. 231.

expansion of the reciprocal $1/\Psi(s)$ *of a function of the type* (3), *where, however, the possibility that* $\gamma=\delta_1=\delta_2=\ldots=0$ *is not excluded.*

This theorem is the source of the author's work on "Pólya frequency functions" and also explains the terminology. We now wish to point out that Theorem 3, combined with the representation (2), (3), for the Laplace transforms of Pólya frequency functions, will imply the converse part of our Theorem 1 without much difficulty, provided we make some additional assumption concerning the existence of the moments of $dL(t)$. The following assumption is perhaps stronger than necessary but applies easily: We add to Theorem 1 the additional assumption that

$$f(s)=\int_{-\infty}^{\infty} e^{-st}\,dL(t) \qquad \text{converges in a strip } -a<Rs<a. \tag{42}$$

Indeed, (4) is now meaningful if $f(x)$ is a polynomial of degree n, say. Setting

$\mu_\nu=\int_{-\infty}^{\infty} t^\nu\,dL(t),$ we find that

$$g(x)=\int_{-\infty}^{\infty} f(x-t)\,dL(t)=\int_{-\infty}^{\infty}\sum\frac{(-t)^\nu}{\nu\,!}f^{(\nu)}(x)\,dx,$$

which may now be written in the form (41). The variation-diminishing property of (4), assumed for bounded functions, is readily shown to hold for polynomials. Once this done, we easily see that the transformation (41) also enjoys the property (S) required by Theorem 3. That $\mu_0\neq 0$ is seen as follows: Consider $\lambda(x)$ as defined by (32). By assumption $v(\lambda)=0$. However, $\mu_0=0$ would clearly imply $\int_{-\infty}^{\infty}\lambda(x)\,dx=0$ which contradicts $v(\lambda)=0$ and the fact that $\lambda(x)\not\equiv 0$ (because $L(t)\not\equiv 0$). By Theorem 3 we now learn that

$$\int_{-\infty}^{\infty} e^{-st}\,dL(t)=\sum_0^\infty\frac{(-1)^\nu}{\nu\,!}\,\mu_\nu s^\nu=\frac{1}{\Psi(s)}\,.$$

If $\Psi(s)$ reduces to $Ce^{-\delta s}$, then $L(t)$ reduces to a step-function with only one jump. If $\gamma+\Sigma\delta_\nu^2>0$, then we have an identity

$$\int_{-\infty}^{\infty} e^{-st}\,dL(t)=\int_{-\infty}^{\infty} e^{-st}\varepsilon\Lambda(x)\,dx \qquad (-a<Rs<a),$$

where $\varepsilon=\pm 1$ and $\Lambda(x)$ is a P. f. f. This concludes our proof.

(Received July 15, 1949.)

ON PÓLYA FREQUENCY FUNCTIONS. III. THE POSITIVITY OF TRANSLATION DETERMINANTS WITH AN APPLICATION TO THE INTERPOLATION PROBLEM BY SPLINE CURVES(1)

BY

I. J. SCHOENBERG AND ANNE WHITNEY

INTRODUCTION

1. A frequency function $\Lambda(x)$, i.e., a non-negative measurable function satisfying the inequalities

$$0 < \int_{-\infty}^{\infty} \Lambda(x)dx < \infty,$$

is called a Pólya frequency function provided(2) it satisfies the following condition: *For every two sets of increasing numbers*

$$(1) \qquad x_1 < x_2 < \cdots < x_n, \qquad y_1 < y_2 < \cdots < y_n, \qquad n = 1, 2, \cdots,$$

we have the inequality

$$(2) \qquad D \equiv \det \|\Lambda(x_i - y_j)\|_{1,n} \geqq 0.$$

For $n=1$, (2) amounts to $\Lambda(x) \geqq 0$; for $n=2$, (2) is equivalent to the convexity of $-\log \Lambda(x)$.

It was shown in [6] that the following two functions are Pólya frequency functions: 1. The normal frequency function whose Laplace transform is

$$(3) \qquad \frac{1}{2(\pi\gamma)^{1/2}} \int_{-\infty}^{\infty} e^{-sx} e^{-(x-\delta)^2/4\gamma} dx = e^{\gamma s^2 - \delta s}.$$

2. The "one-sided" exponential function

$$(4) \qquad \lambda(x) = \begin{cases} e^{-x} & \text{if} \quad x \geqq 0, \\ 0 & \text{if} \quad x < 0, \end{cases}$$

of transform

Presented to the Society, April 30, 1949 under the title *On the positivity of translation determinants for Pólya frequency functions*; received by the editors April 4, 1952.

(1) This work was performed on a National Bureau of Standards contract with the University of California at Los Angeles and was sponsored in part by the Office of Scientific Research, USAF. Miss Whitney's contribution to this paper was accepted by the Graduate School of the University of Pennsylvania in partial fulfilment of the requirement for the Ph.D. degree. For a brief summary see [7] in the list of references at the end of this paper.

(2) In this paper we use the term "provided" in the sense of "if and only if."

$$\int_{-\infty}^{\infty} e^{-sx}\lambda(x)dx = \frac{1}{1+s}, \qquad (\mathrm{R}s > -1).$$

If $\delta \gtrless 0$, we obtain the formula

$$(5) \qquad \frac{1}{|\delta|}\int_{-\infty}^{\infty} e^{-sx}\lambda\left(\frac{x}{\delta}+1\right)dx = \frac{e^{\delta s}}{1+\delta s}, \qquad (-|\delta|^{-1} < \mathrm{R}s < |\delta|^{-1}),$$

giving the transform of the one-sided exponential, of mean zero and variance δ^2, which is descending if $\delta > 0$ and ascending if $\delta < 0$.

It was also shown in [6] that the convolution of Pólya frequency functions again leads to such functions. Moreover, that the most general Pólya frequency function $\Lambda(x)$ is obtained, up to a multiplicative positive constant, by convoluting the normal function (3) with a finite or infinite sequence of one-sided exponentials (5) which are such that the sum $\sum \delta_\nu^2$ of their variances converges. In other terms: the Laplace transform of a Pólya frequency function $\Lambda(x)$ converges in a vertical strip containing the origin and has there the form

$$(6) \qquad \int_{-\infty}^{\infty} e^{-xs}\Lambda(x)dx = \frac{1}{\Psi(s)},$$

where $\Psi(s)$ is an entire function of the form

$$(7) \qquad \Psi(s) = Ce^{-\gamma s^2+\delta s}\prod_{\nu=1}^{\infty}(1+\delta_\nu s)e^{-\delta_\nu s}$$

$$(C > 0,\ \gamma \geqq 0,\ \delta,\ \delta_\nu \text{ real},\ 0 < \gamma + \sum \delta_\nu^2 < \infty).$$

We single out the special case when $\gamma = 0$ and $\sum |\delta_\nu| < \infty$. Except for a trivial exponential factor which can be absorbed into the integral by means of a shift in the variable x, our function Ψ is of genus zero. We may therefore assume $\Lambda(x)$ to have the transform

$$(8) \qquad \int_{-\infty}^{\infty} e^{-xs}\Lambda(x)dx = \prod_{\nu=1}^{\infty}\frac{1}{1+\delta_\nu s}, \qquad (\delta_\nu \text{ real},\ \sum|\delta_\nu| < \infty).$$

These two types of transforms, (8) and (6), (7), will be referred to below as Case 1 and Case 2, respectively.

The present paper is divided into two sections. In §1 we answer (Theorem 1 below) the following question: *Given a Pólya frequency function* $\Lambda(x)$ *and a set of* $2n$ *numbers* (1), *how can we decide when the determinant* D, *defined by* (2), *is actually positive*? As an application of our answer to this question we solve in §2 the general problem of interpolation by so-called spline curves which were introduced in 1946 by one of us [5] for the purpose of approximation of infinitely many equidistant data.

1. The positivity of translation determinants

2. The following theorem answers our question.

THEOREM 1. *Let $\Lambda(x)$ be a Pólya frequency function of transform* (8) (*Case* 1), *or* (6), (7) (*where* $\gamma>0$, *or* $\gamma=0$ *and* $\sum|\delta_\nu|=\infty$) (*Case* 2). *Let the numbers* (1) *be given and let*

$$D = \det \|\Lambda(x_i - y_j)\|_{1,n}. \tag{1.1}$$

Case 1[3]. *Let k be the number of positive δ_ν and h the number of negative δ_ν ($0\leq k\leq\infty$, $0\leq h\leq\infty$, $k+h>1$). Then $D>0$ if and only if we have the inequalities*

$$x_{\nu-k} < y_\nu < x_{\nu+h}, \qquad \text{for } \nu = 1, \cdots, n, \tag{1.2}$$

with the convention that

$$x_r = \begin{cases} -\infty & \text{if} \quad -\infty \leq r < 1, \\ +\infty & \text{if} \quad n < r \leq +\infty. \end{cases} \tag{1.3}$$

Case 2. *We always have $D>0$.*

EXAMPLES. 1. It is clear from (1.3) that some of the inequalities (1.2) are automatically satisfied. Thus, if $k\geq n$ and $h\geq n$, then again $D>0$, no matter what the numbers (1) are. This is true for all n if $k=h=0$. If $k>0$ and $h>0$ we obtain, for $n=1$, the fact that $\Lambda(x)>0$, for all x.

2. The function $\Lambda(x)=\exp(-|x|)$ of transform

$$\int_{-\infty}^{\infty} e^{-xs}e^{-|x|}dx = \frac{2}{1-s^2} \qquad (-1 < \mathrm{R}s < 1)$$

illustrates Case 1 with $k=h=1$. From (1.2) we learn that $D>0$, provided $x_{\nu-1}<y_\nu<x_{\nu+1}$ ($\nu=1, \cdots, n$), or

$$y_1 < x_2,\ x_1 < y_2 < x_3,\ x_2 < y_3 < x_4, \cdots, x_{n-2} < y_{n-1} < x_n,\ x_{n-1} < y_n.$$

In this particular case we mention as a curiosity that D may be evaluated explicitly by the following formula, valid for any set (1),

$$\det \|e^{-|x_i-y_j|}\|_{1,n} = e^{-\Sigma x_i - \Sigma y_i}\prod_{i=1}^{n}\left\{e^{2\min(x_i,y_i)} - e^{2\max(x_{i-1},y_{i-1})}\right\}_+,$$

where the subscript "+" indicates the truncated function

$$x_+ = \begin{cases} x & \text{if} \quad x \geq 0, \\ 0 & \text{if} \quad x < 0. \end{cases} \tag{1.4}$$

(3) The Case 1 of Theorem 1 is related, without inclusion either way, to a theorem of M. Krein and G. Finkelstein [4], concerning the Green's function of linear differential operators. Concerning this subject see the comprehensive treatise [1] by Gantmakher and Krein.

3. The function

$$\Lambda(x) = e^{-x}x_+^k, \tag{1.5}$$

of transform

$$\int_0^\infty e^{-xs}e^{-x}x^k dx = \frac{k!}{(1+s)^{k+1}}, \tag{1.6}$$

corresponds to Case 1 with $h=0$ and $k+1$ positive δ_ν, all unity. Cancellation of (positive) exponential factors from the rows and columns of D leads to the following:

COROLLARY 1. *Let k be a natural integer. If $x_1<x_2< \cdots <x_n$, $y_1<y_2 < \cdots <y_n$, then*

$$D = \det \|(x_i - y_j)_+^k\| > 0 \tag{1.7}$$

if and only if the inequalities

$$x_{\nu-k-1} < y_\nu < x_\nu, \qquad \nu = 1, \cdots, n, \tag{1.8}$$

hold[4].

This is our special result to be applied in §2.

A PROOF OF THEOREM 1, CASE 1

3. We begin by noticing that the simplest of all cases, namely $k+h=1$, is not covered by Theorem 1. Although almost trivial and unimportant by itself, it happens to be fundamental for what follows. So let us assume that $k=1$, $h=0$, in (8), hence $\Lambda(x)$ to be the one-sided exponential

$$\Lambda(x) = \frac{1}{\delta}\lambda\left(\frac{x}{\delta}\right), \qquad \delta > 0. \tag{1.9}$$

Now $\lambda(x)=e^{-x}\cdot x_+^0$, where $x_+^0=1$ if $x\geqq 0$, $=0$ if $x<0$. Thus $D=\det \|\Lambda(x_i-y_j)\|>0$ provided $D'=\det \|(x_i-y)_+^0\|>0$. A moment's reflection will show that $D'>0$, provided all elements of D' are $=1$, except those above its main diagonal which are $=0$. By the definition of x_+^0 these conditions amount to the inequalities

$$y_1 \leqq x_1 < y_2 \leqq x_2 < \cdots < y_{n-1} \leqq x_{n-1} < y_n \leqq x_n,$$

or

$$x_{\nu-1} < y_\nu \leqq x_\nu \qquad \text{for } \nu = 1, \cdots, n. \tag{1.10}$$

(4) The nonvanishing of determinants of the form (1.7) appears explicitly as an assumption in some of the theorems of Chapter 9 in S. Mandelbrojt's *Dirichlet series*, Rice Institute Pamphlet, vol. 31, No. 4, October, 1944.

We prove next a special case of Case 1 ($h=0$, $1<k<\infty$) which, for convenience, we state as:

LEMMA 1. *If* $\Lambda(x)$ *has the transform*

$$\int_{-\infty}^{\infty} e^{-xs}\Lambda(x)dx = \prod_{\nu=1}^{k} \frac{1}{1+\delta_\nu s}, \qquad \delta_\nu > 0 \textit{ for all } \nu;\ k > 1, \tag{1.11}$$

then

$$D = \det \|\Lambda(x_i - y_j)\| > 0 \tag{1.12}$$

if and only if

$$x_{\nu-k} < y_\nu < x_\nu, \qquad \nu = 1, \cdots, n. \tag{1.13}$$

Proof. Let $\Lambda_1(x)$ and $\Lambda_2(x)$ have the transforms

$$\prod_{\nu=1}^{k-1} \frac{1}{1+\delta_\nu s} \quad \text{and} \quad \frac{1}{1+\delta_k s}$$

respectively. Thus $\Lambda(x)$ is the convolution of $\Lambda_1(x)$ with $\Lambda_2(x)$. From [6, Lemma 5], we recall the identity

$$\det \|\Lambda(x_i - y_j)\| = \frac{1}{n!}\int_{-\infty}^{\infty} \cdots \int \det \|\Lambda_1(x_i - t_j)\| \cdot \det \|\Lambda_2(t_i - y_j)\| dt_1 \cdots dt_n,$$

which, in view of the symmetry of the integrand in the variables $t_1, \cdots, t_n$, we may write as

$$\begin{aligned} \det \|\Lambda(x_i - y_j)\| &= \int_{t_1<t_2<\cdots<t_n} \cdots \int \det \|\Lambda_1(x_i - t_j)\| \\ &\quad \cdot \det \|\Lambda_2(t_i - y_j)\| dt_1 \cdots dt_n \\ &= \int \cdots \int D_1 D_2 dt_1 \cdots dt_n. \end{aligned} \tag{1.14}$$

Notice that $D_1 \geq 0$ and $D_2 \geq 0$, for all (t), Λ_1 and Λ_2 being Pólya frequency functions. D_2 is continuous in (t) except possibly on the hyperplanes $t_i = y_j$. D_1 is continuous in (t) if $k>2$ and possibly discontinuous on $t_j = x_i$ if $k=1$. We conclude that the integral (1.14) is positive if and only if there exists a point (t), with $t_1 < \cdots < t_n$, in a neighborhood of which *both* factors D_1, D_2 are positive. By (1.10), this is the case for D_2 provided

$$t_{\nu-1} < y_\nu < t_\nu, \qquad \nu = 1, \cdots, n. \tag{1.15}$$

Using induction by k and assuming Lemma 1 to hold for $k-1$, rather than k, we see that $D_1 > 0$, at (t) and near it, provided

$$x_{\nu-k+1} < t_\nu < x_\nu, \qquad \nu = 1, \cdots, n. \tag{1.16}$$

Thus (1.14) is positive provided (1.15) and (1.16) have a common solution of increasing t's. Inspecting the bounds imposed on t_ν by (1.15) and (1.16), we find the necessary inequalities

$$\max(x_{\nu-k+1}, y_\nu) < \min(x_\nu, y_{\nu+1}), \qquad \nu = 1, \cdots, n. \tag{1.17}$$

These, however, are also sufficient, for then the open intervals of the t-axis

$$I_\nu\colon \max(x_{\nu-k+1}, y_\nu) < t < \min(x_\nu, y_{\nu+1}), \qquad \nu = 1, \cdots, n, \tag{1.18}$$

form a sequence of nonoverlapping and advancing intervals, for increasing ν, in view of the obvious inequalities $\min(x_\nu, y_{\nu+1}) \leqq \max(x_{\nu-k+2}, y_{\nu+1})$. Choose $t_\nu \in I_\nu$ and both (1.15), (1.16) will be satisfied. By (1), the inequalities (1.17) are equivalent to

$$x_{\nu-k+1} < y_{\nu+1}, \qquad y_\nu < x_\nu, \qquad \text{or} \qquad x_{\nu-k} < y_\nu, \qquad y_\nu < x_\nu,$$

which are identical with (1.13). As the same argument applies if $k=2$, the proof by induction is complete.

REMARK. If we replace in Lemma 1 the assumption $\delta_\nu > 0$, for all ν, by

$$\delta_\nu < 0, \qquad \nu = 1, \cdots, n, \tag{1.19}$$

we have a problem which reduces to the previous case as follows: Replacing s by $-s$ in (1.11) we obtain

$$\prod \frac{1}{1 + |\delta_\nu| s} = \int_{-\infty}^{\infty} e^{sx}\Lambda(x)dx = \int_{-\infty}^{\infty} e^{-sx}\Lambda(-x)dx.$$

This interchanges the roles of x_i and y_j, and (1.12) holds now provided

$$y_{\nu-k} < x_\nu < y_\nu, \qquad \nu = 1, \cdots, n. \tag{1.20}$$

4. We may now complete a proof of Case 1 by a brief discussion of a few subcases:

Proof of Case 1 for finite positive k and h. Let the transform (8) be rational, there being k positive and h negative δ_ν. Let Λ_1 and Λ_2 have the transforms

$$\prod_{\nu=1}^{k} \frac{1}{1 + \delta_\nu s}, \quad \delta_\nu > 0, \qquad \prod_{\nu=1}^{h} \frac{1}{1 + \delta_{k+\nu} s}, \quad \delta_{k+\nu} < 0,$$

respectively. Then $\Lambda = \Lambda_1 * \Lambda_2$ and (1.14) holds. By Lemma 1 and its counterpart (1.19), (1.20), a point (t) and its neighborhood will produce a positive integrand in (1.14) provided we have

$$x_{\nu-k} < t_\nu < x_\nu, \qquad y_{\nu-h} < t_\nu < y_\nu, \qquad \nu = 1, \cdots, n. \tag{1.21}$$

These require that

$$\max(x_{\nu-k}, y_{\nu-h}) < \min(x_\nu, y_\nu), \qquad \nu = 1, \cdots, n, \tag{1.22}$$

which again are also sufficient. Indeed, if (1.22) hold, the intervals

$$I'_\nu: \max(x_{\nu-k}, y_{\nu-h}) < t < \min(x_\nu, y_\nu) \qquad \nu = 1, \cdots, n,$$

which may well overlap, have advancing right-hand end points (by (1)), hence an increasing sequence $\{t_\nu\}$ with $t_\nu \in I'_\nu$ is assured. Finally (1.22) amount to $x_{\nu-k} < y_\nu$, $y_{\nu-h} < x_\nu$, which in turn are equivalent to (1.2).

Proof of Case 1 for $k = \infty$, $h = 0$. Let the transform of $\Lambda(x)$ be (8) with all $\delta_\nu > 0$. We are to show that $D = \det \|\Lambda(x_i - y_j)\| > 0$ if and only if

$$y_\nu < x_\nu, \qquad \nu = 1, \cdots, n. \tag{1.23}$$

The necessity of these inequalities is seen directly as follows: In the present case of positive δ_ν's, (8) is an ordinary Laplace integral (see [6, Art. 11]), i.e., $\Lambda(x) = 0$ if $x \leqq 0$. But then $x_\mu \leqq y_\mu$, for some μ, implies $x_i \leqq y_j$ for all (i, j) such that $i \leqq \mu$, $j \geqq \mu$. Hence $\Lambda(x_i - y_j) = 0$, for such (i, j), and D vanishes, for instance by Laplace's expansion theorem.

In order to prove the sufficiency of (1.23), we choose a number $k' > n$ and write $\Lambda = \Lambda_1 * \Lambda_2$, where Λ_1 corresponds to the first k' factors of the product (8) and Λ_2 to all remaining ones. In (1.14), $D_1 > 0$ at (t) provided $x_{\nu-k'} < t_\nu < x_\nu$, or

$$t_\nu < x_\nu, \qquad \nu = 1, \cdots, n, \tag{1.24}$$

since $k' > n$ implies that $x_{\nu-k'} = -\infty$. On the other hand we certainly have $D_2 = \det \|\Lambda_2(t_i - y_j)\| > 0$ if

$$y_1 < t_1 < y_2 < t_2 < \cdots < y_n < t_n. \tag{1.25}$$

Indeed $\Lambda_2(x) > 0$ if $x > 0$ and $\Lambda_2(x) = 0$ if $x < 0$, so that D_2 has positive elements on the main diagonal and only zero elements above it. Now if (1.23) hold, it is clear that increasing t_ν can be found satisfying (1.24) and (1.25), which completes the proof. The case when $k = 0$, $h = \infty$, may be reduced to the previous one and leads to the reverse of (1.23), namely

$$x_\nu < y_\nu, \qquad \nu = 1, \cdots, n. \tag{1.26}$$

Proof of Case 1 for $k = h = \infty$. Let again $\Lambda = \Lambda_1 * \Lambda_2$, where Λ_1 corresponds to the negative δ_ν and Λ_2 to the positive ones. In (1.14), $D_1 > 0$ provided $x_\nu < t_\nu$, by (1.26), and $D_2 > 0$ provided $y_\nu < t_\nu$, by (1.23). Thus $D > 0$ always, as was to be shown.

Proof of Case 1 for $h = \infty$, k *finite*. With the same decomposition as before the positivity of (1.14) requires $x_\nu < t_\nu$, by (1.26), and $t_{\nu-k} < y_\nu < t_\nu$, by (1.13). Increasing t_ν will satisfy both conditions provided $x_\nu < y_{\nu+k}$, or

$$x_{\nu-k} < y_\nu, \qquad \nu = 1, \cdots, n,$$

which agree with (1.2). This completes a proof of Case 1 of our theorem.

A PROOF OF THEOREM 1, CASE 2

5. This case requires some information on the behavior of Pólya frequency functions $\lambda(x)$ of small variance. We know that $\lambda(x)$ attains its maximal value at just one point(6) $x=\mu$ and that $\lambda(x)$ is nonincreasing for $x \geqq \mu$ and nondecreasing for $x \leqq \mu$.

LEMMA 2. *Let $\{\lambda_m(x)\}$ be an infinite sequence of Pólya frequency functions such that*

$$(1.27)\quad \begin{gathered} \int_{-\infty}^{\infty} \lambda_m(x)dx = 1, \qquad \int_{-\infty}^{\infty} x\lambda_m(x)dx = 0, \\ \sigma_m^2 = \int_{-\infty}^{\infty} x^2\lambda_m(x)dx \to 0 \qquad \textit{as } m \to \infty. \end{gathered}$$

If $x=\mu_m$ denotes the point where $\lambda_m(x)$ attains its maximal value, then

$$(1.28)\quad \lim \mu_m = 0,$$

$$(1.29)\quad \lim \lambda_m(\mu_m) = +\infty,$$

$$(1.30)\quad \lim \lambda_m(x) = 0, \textit{ uniform in } x, \textit{ for } |x| \geqq \eta, \qquad \textit{for every } \eta > 0.$$

Proof. Clearly, for a fixed $\eta>0$,

$$\sigma_m^2 = \int x^2\lambda_m dx \geqq \int_{|x|\geqq\eta} x^2\lambda_m dx \geqq \eta^2 \int_{|x|\geqq\eta} \lambda_m dx,$$

hence by the last relation (1.27):

$$(1.31)\quad \int_{|x|\geqq\eta} \lambda_m dx \to 0 \quad \text{as} \quad m \to \infty, \qquad \text{for every } \eta > 0.$$

By the first relation (1.27) we conclude that

$$\int_{-\eta}^{\eta} \lambda_m dx \to 1 \quad \text{as} \quad m \to \infty, \qquad \text{for every } \eta > 0.$$

But then clearly

$$(1.32)\quad \max_{-\eta\leqq x\leqq\eta} \lambda_m(x) \to \infty \quad \text{as} \quad m \to \infty, \qquad \text{for every } \eta > 0.$$

We may now prove (1.28): If $x=\xi_m$ is a point where $\lambda_m(x)$ attains its maximum in $[-\eta, \eta]$, then by (1.32)

(6) This follows easily from Theorem 10, page 154, of Hirschman and Widder in [3]. Their $\lambda(x) = G_m(x)$ are not the most general Pólya frequency functions, since $\gamma=0$ in their case. However, in Art. 6 below we are applying Lemma 2 to a case where γ happens to vanish.

$$(1.33)\qquad -\eta \leqq \xi_m \leqq \eta, \qquad \lim_{m\to\infty} \lambda_m(\xi_m) = +\infty.$$

If (1.28) does not hold, then we would have $\mu_m > 2\eta$, say, for arbitrarily large values of $m = m_\nu$. Since $\lambda_m(x)$ is nondecreasing in the range $[\xi_m, \mu_m]$ we conclude that

$$\int_{\xi_m}^{2\eta} \lambda_m(x)dx \geqq \int_{\xi_m}^{2\eta} dx\cdot\lambda_m(\xi_m) \geqq \eta\lambda_m(\xi_m),$$

hence, by (1.33), that

$$\int_{\xi_m}^{2\eta} \lambda_m(x)dx \to \infty \quad \text{as} \quad m = m_\nu \to \infty,$$

which is absurd. This proves (1.28), and (1.29) follows from (1.32).

To prove (1.30), we use (1.28) and choose $M = M(\eta)$ such that $|\mu_m| < \eta/2$ if $m > M$. But then, $\lambda_m(x)$ being nonincreasing in $[\mu_m, \eta]$, we have

$$\frac{\eta}{2}\lambda_m(\eta) \leqq \int_{\eta/2}^{\eta} \lambda_m(x)dx,$$

where the last integral converges to zero by (1.31). Thus $\lim \lambda_m(\eta) = 0$ and (1.30) is established since $\lambda_m(x)$ is nonincreasing for $x \geqq \eta$.

6. **Proof of Case 2 if** $\gamma = 0$, $\delta_\nu > 0$ $(\nu = 1, 2, \cdots)$. Let the transform be

$$(1.34)\qquad \int_{-\infty}^{\infty} e^{-xs}\Lambda(x)dx = \prod_{\nu=1}^{\infty} \frac{e^{\delta_\nu s}}{1+\delta_\nu s}, \qquad \delta_\nu > 0, \sum_1^\infty \delta_\nu = \infty.$$

We are to show that

$$(1.35)\qquad D = \det \|\Lambda(x_i - y_j)\| > 0$$

always holds. Let $\Lambda = \Lambda_1 * \Lambda_2$ where Λ_1 corresponds to the product of the first m factors of (1.34) and Λ_2 to the remainder. Setting

$$\phi_m = \sum_1^m \delta_\nu,$$

we have that

$$(1.36)\qquad \prod_{\nu=1}^{m} \frac{1}{1+\delta_\nu s} = \int_{-\infty}^{\infty} e^{-(x+\phi_m)s}\Lambda_1(x)dx = \int_{-\infty}^{\infty} e^{-sx}\Lambda_1(x-\phi_m)dx.$$

Now

$$\det \|\Lambda_1(x_i - t_j)\| = \det \|\Lambda_1(x_i + \phi_m - t_j - \phi_m)\|$$

and the transform (1.36) shows that $\Lambda_1(x - \phi_m)$ is a function to which Lemma 1 applies. Thus

(1.37) $$D_1 = \det \|\Lambda_1(x_i - t_j)\| > 0$$

provided $x_{\nu-m}+\phi_m<t_\nu<x_\nu+\phi_m$, $\nu=1, \cdots, n$. If $m \geqq n$, these conditions reduce to

(1.38) $$t_\nu < x_\nu + \phi_m, \qquad \nu = 1, \cdots, n.$$

Since $\phi_m = \sum_1^m \delta_\nu \to \infty$, we see that (1.38) are satisfied in an arbitrarily large cube of the t-space, if only m is sufficiently large.

The transform of $\Lambda_2(x)$ is

$$\int_{-\infty}^{\infty} e^{-xs}\Lambda_2(x)dx = \prod_{\nu=m+1}^{\infty} \frac{e^{\delta_\nu s}}{1+\delta_\nu s} = 1 + \frac{s^2}{2}\sum_{m+1}^{\infty} \delta_\nu^2 + \cdots$$

showing that $\lambda_m(x) = \Lambda_2(x)$ is a sequence of Pólya frequency functions satisfying the assumptions of Lemma 2, with

$$\sigma_m^2 = \sum_{\nu=m+1}^{\infty} \delta_\nu^2.$$

In order to prove (1.35), we turn again to (1.14). Choose a fixed η such that

$$0 < \eta < \frac{1}{2} \min_\nu (y_{\nu+1} - y_\nu)$$

and let $t_i = y_i + \mu_m$, $i = 1, \cdots, n$, where $\{\mu_m\}$ is the sequence of maximum points in Lemma 2. We claim that for these values

(1.39) $$D_2 = \det \|\Lambda_2(t_i - y_j)\| > 0 \qquad \text{if } m > M.$$

Indeed, if $i \neq j$ and m is sufficiently large, we have

$$\begin{aligned} |t_i - y_j| &= |y_i - y_j + \mu_m| \\ &\geqq |y_i - y_j| - |\mu_m| > |y_i - y_j| - \eta > 2\eta - \eta = \eta, \end{aligned}$$

so that

$$\lim_{m\to\infty} \Lambda_2(t_i - y_j) = 0$$

by (1.30). On the other hand

$$\Lambda_2(t_i - y_i) = \Lambda_2(\mu_m) \to \infty, \qquad \text{as } m \to \infty,$$

by (1.29). Thus $D_2 \to \infty$, as $m \to \infty$, which amply proves (1.39). Now the t_ν, as defined above, are bounded, in view of (1.28). Therefore the inequalities (1.38) are also satisfied by our t_i, if only m is sufficiently large. Now (1.37), (1.39), and (1.14) imply (1.35), which completes the proof.

The remaining possibilities of Case 2 are now easily taken care of. Let us assume the transform (1.34), $\sum |\delta_\nu| = \infty$, but allow δ_ν of either sign. To fix the ideas, let the negative δ_ν have a divergent sum, and let again $\Lambda=\Lambda_1 * \Lambda_2$, Λ_1 and Λ_2 corresponding to the negative and positive δ_ν respectively. Then, in (1.14), we have $D_1>0$ always. Not so for D_2; however, the following is now clear: To whichever of the previously discussed cases the question of

$$D_2 = \left\|\Lambda_2(t_i - y_j)\right\| > 0$$

may belong, given the y_ν, we can always find such increasing t_ν as to make $D_2>0$. As $D_1>0$, (1.14) shows that $D>0$.

Finally, to treat the last remaining case, let

$$\int_{-\infty}^{\infty} e^{-xs}\Lambda(x)dx = e^{\gamma s^2}\cdot F(s), \qquad \gamma > 0, \tag{1.40}$$

where $F(s)$ is of either of the forms (8) or (1.34). Let for the last time $\Lambda=\Lambda_1 * \Lambda_2$, the transforms of these factors being the two factors of (1.40), respectively. The reasoning of the last paragraph again applies. In (1.14) we have $D_1>0$ for any t_ν(6), while by all previous cases, increasing t_i may be found as to produce $D_2>0$. Thus $D>0$ and a proof of Theorem 1 is completed.

2. Interpolation by spline curves

7. The problem. Let

$$\xi_1 < \xi_2 < \cdots < \xi_n \tag{2.1}$$

be n given points and k a natural integer. Let $F(x)$ be defined in each of the intervals $(-\infty, \xi_1)$, (ξ_1, ξ_2), $\cdots$, $(\xi_n, +\infty)$ by a separate polynomial, of degree not exceeding k, in such a way that the composite function $F(x)$ be continuous, for all real x, together with its first $k-1$ derivatives. $F^{(k)}(x)$ is a step-function with possible discontinuities at the points (2.1). A function $F(x)$ of this kind is called a *spline function*(7) (or curve) of degree k, while the points (2.1) are referred to as its *knots*.

The truncated function (1.4) and its powers are well suited for the representation of spline functions. Indeed, it is easily seen that

$$F(x) = P_k(x) + \sum_{\nu=1}^{n} A_\nu(x - \xi_\nu)_+^k, \tag{2.2}$$

where $P_k(x)$ is an arbitrary polynomial of degree at most k and A_ν are arbitrary parameters, represents uniquely the most general spline curve of degree k having the knots (2.1). Thus $F(x)$ depends on $n+k+1$ arbitrary parameters.

---

(6) See [6, Art. 3, Example 2].

(7) See [5, p. 67], for the connection with the flexible rod, called a *spline*, used for fitting smooth curves to experimental data.

We may now formulate our problem as follows: *Let*

$$x_1 < x_2 < \cdots < x_{n+k+1} \tag{2.3}$$

be a given set of $n+k+1$ *abscissae; when can we interpolate in the points* (2.3) *an arbitrary given set of ordinates* $Y_1, Y_2, \cdots, Y_{n+k+1}$, *by means of our spline curve* (2.2)?

8. **Its solution.** As a first step we observe the following: If we choose in any way $k+1$ fixed abscissae $y_1, \cdots, y_{k+1}$ such that

$$y_1 < y_2 < \cdots < y_{k+1} < x_1, \qquad y_{k+1} < \xi_1, \tag{2.4}$$

then the polynomial $P_k(x)$, of (2.2), may be uniquely written as

$$P_k(x) = \sum_{\nu=1}^{k+1} c_\nu(x - y_\nu)^k,$$

for appropriate values of the c_ν. But then we also have

$$P_k(x) = \sum_{\nu=1}^{k+1} c_\nu(x - y_\nu)_+^k, \qquad \text{if } x \geqq y_{k+1}. \tag{2.5}$$

If we now write

$$y_{k+2} = \xi_1,\ y_{k+3} = \xi_2, \cdots, y_{n+k+1} = \xi_n, \tag{2.6}$$

and correspondingly define $c_{\nu+k+1} = A_\nu$ $(\nu = 1, \cdots, n)$, we see that the arbitrary spline curve (2.2) of knots (2.1) may also be written in the form

$$F(x) = \sum_{\nu=1}^{n+k+1} c_\nu(x - y_\nu)_+^k, \qquad \text{in the range } x \geqq y_{k+1}. \tag{2.7}$$

We conclude that the interpolation problem of the previous article has a solution if and only if the determinant of order $n+k+1$

$$\det \|(x_i - y_j)_+^k\| \neq 0.$$

By Corollary 1, §1, this is the case provided

$$x_{\nu-k-1} < y_\nu < x_\nu \quad (\nu = 1, 2, \cdots, n + k + 1). \tag{2.8}$$

In view of (2.4) and (2.3) we see that the first $k+1$ inequalities (2.8) are automatically satisfied. Returning by (2.6) to our old notation we find the remaining n inequalities (2.8) to be

$$\begin{aligned} x_1 &< \xi_1 < x_{k+2}, \\ x_2 &< \xi_2 < x_{k+3}, \\ &\cdots\cdots\cdots, \\ x_n &< \xi_n < x_{n+k+1}. \end{aligned} \tag{2.9}$$

We state the result as

THEOREM 2. *We can interpolate in* $n+k+1$ *given abscissae* (2.3) *arbitrarily given ordinates, by a spline curve of degree* k *and knots* (2.1), *if and only if the inequalities* (2.9) *are satisfied.*

REMARK. The inequalities (2.9) may be described in words: The first n interpolation points $x_1, \cdots, x_n$ precede the knots $\xi_1, \cdots, \xi_n$, respectively, which in turn precede the last n interpolation points $x_{k+2}, \cdots, x_{n+k+1}$, respectively. Notice that all these conditions are satisfied automatically in case the knots $\xi_1, \cdots, \xi_n$ are chosen among the $n+k-1$ interpolation points $x_2, x_3, \cdots, x_{n+k}$.

We rephrase this last result as follows:

THEOREM 3. *Let there be given a set of* $N+1$ *abscissae*

$$x_0 < x_1 < \cdots < x_N, \tag{2.10}$$

and $N+1$ *arbitrary corresponding ordinates* y_ν $(\nu=0, \cdots, N)$. *As is well known, we may interpolate them, uniquely, by a polynomial of degree at most* N. *This corresponds to the classical case* $n=0$ *of no knots whatever.*

Also we may choose any combination of n $(1 \leqq n \leqq N-1)$ *among the interior abscissae*

$$x_1, x_2, \cdots, x_{N-1}, \tag{2.11}$$

to serve as the knots $\xi_1, \cdots, \xi_n$ *of an interpolating spline curve* $y=F(x)$, *in which case again we may interpolate uniquely by a spline curve of degree*

$$k = N - n. \tag{2.12}$$

REMARKS. 1. In the extreme case of $n=N-1$, all points (2.11) are knots of $F(x)$, which, by (2.12), is a spline curve of degree $k=N-(N-1)=1$. Therefore $y=F(x)$ is the ordinary polygonal line obtained by linear interpolation between each pair of consecutive points. Theorem 3 furnishes a sequence of interpolation procedures bridging the gap between the ordinary polynomial interpolation and the linear polygonal interpolation, each additional knot lowering the degree of the spline curve by one unit.

2. To each of these procedures there corresponds an interpolation formula expressing the interpolating spline curve $F(x)$ in terms of the ordinates y_ν, with coefficients which are the "fundamental functions" of the procedure. In the case of $N=4$, $n=1$ and therefore $k=3$, with

$$x_0 = -2,\ x_1 = -1,\ x_2 = 0,\ x_3 = 1,\ x_4 = 2 \text{ and } \xi_1 = 0,$$

we obtain the formula

$$F(x) = y_0 l_0(x) + y_1 l_1(x) + y_2 l_2(x) + y_{-1} l_1(-x) + y_{-2} l_2(-x), \tag{2.13}$$

with the following expressions for the fundamental functions:

$$l_0(x) = (1 + x)\left(1 + \frac{x}{2}\right)\left(1 - \frac{3}{2} x\right) + \frac{3}{2} x_+^3,$$

$$l_1(x) = \frac{1}{3} x(x + 1)(x + 2) - x_+^3,$$

$$l_2(x) = -\frac{1}{24} x(x + 1)(x + 2) + \frac{1}{4} x_+^3.$$

In (2.13) we have changed our notation for y, in order to exhibit the symmetry of this formula.

3. We conclude by raising the following question. Let $f(x)$ be given as continuous in the range $[a, b]$ which we divide into N equal parts by the points (2.10). It is well known that the interpolating polynomial of degree N does not always converge to $F(x)$, as $N \to \infty$ (8). However, the polygonal interpolating spline curve ($k=1$) always does. Is there a sequence of our new procedures, with $k=N-n\to\infty$ as $N\to\infty$, producing a spline curve $F(x)$ of degree k, which will always converge to $f(x)$ in $[a, b]$? And how strongly may $k\to\infty$ with N, subject to this condition? Similar questions can be raised concerning the quadrature formulae obtained by integrating our interpolation formulae. Further problems will undoubtedly occur to the reader.

References

1. F. Gantmakher and M. Krein, *Oscillatory matrices and kernels and small vibrations of mechanical systems* (in Russian), 2d ed., Moscow, 1950.

2. H. Hahn, *Über das Interpolationsproblem*, Math. Zeit. vol. 1 (1918) pp. 115–142.

3. I. I. Hirschman and D. V. Widder, *The inversion of a general class of convolution transforms*, Trans. Amer. Math. Soc. vol. 66 (1949) pp. 135–201.

4. M. Krein and G. Finkelstein, *Sur les fonctions de Green complètement non-négatives des opérateurs differentiels ordinaires*, C. R. (Doklady) Acad. Sci. URSS. vol. 24 (1939) pp. 220–223.

5. I. J. Schoenberg, *Contributions to the problem of approximation of equidistant data by analytic functions*, Parts A and B, Quarterly of Applied Mathematics vol. 4 (1946) pp. 45–99, 112–141.

6. ———, *On Pólya frequency functions* I. *The totally positive functions and their Laplace transforms*, Journal d'Analyse Mathématique vol. 1 (1951) pp. 331–374.

7. I. J. Schoenberg and Anne Whitney, *Sur la positivité des déterminants de translations des fonctions de fréquence de Pólya avec une application au problème d'interpolation par les fonctions "spline,"* C. R. Acad. Sci. Paris vol. 228 (1949) pp. 1996–1998.

University of Pennsylvania,
Philadelphia, Pa.

(8) See [2] concerning the general problem mentioned and also for further references.

Reprinted from *Trans. Amer. Math. Soc.* **74** (1953), pp. 246–259

ON PÓLYA FREQUENCY FUNCTIONS IV: THE FUNDAMENTAL SPLINE FUNCTIONS AND THEIR LIMITS(*)

By

H. B. CURRY AND I. J. SCHOENBERG

CONTENTS

Introduction

The present paper was written in 1945 and completed by 1947 (see the abstract [3]) but for no good reason has so far not been published. It appears now in a somewhat revised and improved form.

The paper is divided into two parts. The spline functions for equidistantly spaced knots were introduced in [9]. In Part I we discuss such functions, of

* Sponsored by the Mathematics Research Center, United States Army, Madison, Wisconsin, under Contract No.: DA-11-022-ORD-2059.

degree $n-1$, for arbitrarily spaced knots. The possibility of such an extension was already implied by Curry in his review [2] of Schoenberg's paper [9].

The so-called fundamental spline functions are introduced and shown to be bell-shaped (Theorem 1). They are also shown to be the projections onto the x-axis of the volumes of appropriate n-dimensional simplices (Theorem 2). This geometric interpretation allows us to conclude, by means of Brunn's theorem, that the fundamental spline functions are logarithmically concave (Theorem 3). They are also shown to form a basis for all spline functions of degree $n-1$ and given knots (Theorem 4).

The present interest in spline functions is one of the reasons for the publication of this paper at the present time; in fact Theorem 4 is quoted and used in the paper [14] which is about to appear. An extension of the theory to the case of multiple knots (Section 3 and 4) has recently been added. This extension (Theorem 5), in particular Corollary 1 of Section 4, allows us to generalize to the case of multiple nodes the fundamental results of spline interpolation. This generalization of spline interpolation was stated in [14] and was applied to the construction of best quadrature formulae in the recent paper [15].

The fundamental spline functions $M_n(x)$ are frequency functions, which means that they are non-negative and their integral over the reals is unity. In Part II it is shown that the Pólya distribution functions are the only possible limits of a convergent sequence of distribution functions of the form

$$\int_{-\infty}^{x} M_n(t)dt, \quad \text{as } n \to \infty,$$

and conversely (Theorem 6). This was actually the way in which the Pólya frequency functions were originally discovered, although this approach is not mentioned in any of the previous papers of this series ([10], [11], [12]). Theorem 6 adds a fourth to the three previously known characterizations of Pólya frequency functions (See Section 5). Finally, it is shown that the fundamental spline functions themselves converge to the Pólya frequency function of the limit distribution (Theorem 7). The proof of Theorem 7 is shown to depend on a limit theorem (Theorem 8) concerning the class $\mathscr{P}_2$, of logarithmically concave frequency functions. A proof of Theorem 8, together with further extremum properties of the class $\mathscr{P}_2$, will be published elsewhere.

I. ON SPLINE FUNCTIONS

1. The fundamental spline functions $M_n(x)$. Let

$$\cdots < x_{-2} < x_{-1} < x_0 < x_1 < \cdots < x_\nu < \cdots \tag{1.1}$$

be a sequence of reals, such that $x_\nu \to \pm\infty$ as $\nu \to \pm\infty$, and let n be a natural number. By a spline function $S_n(x)$, of order n, or of degree $n-1$, having the knots (1.1), we mean a function of the class $C^{n-2}(-\infty,\infty)$, such that in each interval $(x_\nu, x_{\nu+1})$ it reduces to a polynomial of degree not exceeding $n-1$. $S_n(x)$ will also be referred to as an n-spline.

A remarkable example of an n-spline appears in connection with the integral representation of divided differences. We write

$$x_+^{n-1} = \begin{cases} x^{n-1} & \text{if } x \geqq 0, \\ 0 & \text{if } x < 0, \end{cases} \tag{1.2}$$

and define

$$M_n(x;y) = n(y-x)_+^{n-1}. \tag{1.3}$$

Finally, we consider the divided difference of order n of the function (1.3) with respect to the variable y and based on the values $y = x_0, x_1, \cdots, x_n$. Using Steffensen's notation, we denote this divided difference by $M_n(x; x_0, x_1, \cdots, x_n)$, which we will often abbreviate to $M_n(x)$ or even to $M(x)$. In terms of $\omega(x) = (x-x_0)\cdots(x-x_n)$, its explicit expression is

$$M_n(x; x_0, \cdots, x_n) = \sum_{\nu=0}^{n} \frac{n(x_\nu - x)_+^{n-1}}{\omega'(x_\nu)}. \tag{1.4}$$

Let us look at this function more closely. By (1.2) and (1.4) it is clear that $M_n(x)$ is an n-spline having the knots $x_0, \cdots, x_n$. Moreover, it evidently vanishes if $x > x_n$. However, it vanishes also if $x < x_0$, for then we may remove in (1.4) the subscript "+" and the sum then vanishes as a divided difference of order n of a polynomial of degree $n-1$.

By a fundamental theorem of Peano (see [4, Chap. III, Sec. 3.7]) applied to divided differences the following holds:

If $f(x) \in C^n$ then

$$f(x_0, x_1, \cdots, x_n) = \frac{1}{n!} \int_{x_0}^{x_n} M_n(x; x_0, \cdots, x_n) f^{(n)}(x) dx. \tag{1.5}$$

In particular, if $f(x) = x^n$ we obtain

$$\int_{-\infty}^{\infty} M_n(x; x_0, \cdots, x_n) dx = 1. \tag{1.6}$$

Further properties of $M(x) = M_n(x; x_0, \cdots, x_n)$ are as follows.(1)

Theorem 1. *The νth derivative $M^{(\nu)}(x)$ $(\nu = 0, \cdots, n-2)$ has exactly ν distinct simple zeros in the interval (x_0, x_n). In particular $M(x) > 0$ in (x_0, x_n).*

Proof. By (1.4)

$$M(x) = n(x_n - x)^{n-1} / \omega'(x_n) \text{ if } x_{n-1} < x < x_n ,$$

(1) From Lemma 6 (Section 6 below) we easily find (by expanding in powers of $s = ti$) for the mean and the standard deviation of the frequency function $M(x) = M_n(x; x_0, .., x_n)$ the values

$$\mu_1 = \int_{-\infty}^{\infty} xM(x)dx = \frac{1}{n+1} \sum_0^n x_\nu,$$

$$\sigma^2 = \int_{-\infty}^{\infty} (x - \mu_1)^2 M(x) dx = \frac{1}{(n+1)^2(n+2)} \sum_{i>j} (x_i - x_j)^2.$$

hence $M(x) > 0$ in this interval. We can therefore find three increasing values of x, the first and last being x_0 and x_n, such that the corresponding values of $M(x)$ have the signs $0, +, 0$. By the mean-value theorem we can find four increasing values of x, again x_0, x_n being the extreme ones, such that the corresponding values of $M'(x)$ have the signs 0, $+$, $-$, 0 (one change of sign). Continuing in like manner we finally obtain $n+1$ increasing values of x, such that the corresponding values of $M^{(n-2)}(x)$ have the signs 0, $+$, $-$, $+$, $-$, $\cdots$, 0 ($n-2$ changes of sign). On the other hand, differentiation of (1.4) gives

$$M^{(n-2)}(x) = (-1)^{n-2} \; n! \sum_{0}^{n} \frac{(x_\nu - x)_+}{\omega'(x_\nu)}$$

showing that the graph of this function is an ordinary (continuous) polygonal line having vertices at $x = x_0, \cdots, x_n$ and vanishing at the extreme points. As seen above, this graph must exhibit *at least* $n-2$ changes of sign. From this we easily conclude that the elements of the sequence $M^{(n-2)}(x_\nu)$, $(\nu = 1, \cdots, n-1)$ must be all different from zero and alternate in sign. Therefore *$M^{(n-2)}(x)$ has exactly $n-2$ simple zeros in (x_0, x_n)*. The theorem now follows easily: We have seen above that $M^{(\nu)}$ has at least ν distinct zeros in (x_0, x_n). If it had more, or if some of them were not simple, then Rolle's theorem would imply that $M^{(n-2)}(x)$ has more than $n-2$ zeros, which we have shown not to be the case.

2. A geometric interpretation of $M_n(x)$. According to Hermite and Gennochi (see [8, p.16]) we can write

$$f(x_0, \cdots, x_n) = \int_{\tau_n} \cdots \int f^{(n)}(x_0 t_0 + \cdots + x_n t_n)\, dt_1 \cdots dt_n \tag{2.1}$$

where $t_0 = 1 - t_1 - \cdots - t_n$ and the integration is performed over the simplex

$$\tau_n\colon t_1 \geqq 0, \cdots, t_n \geqq 0, \quad 1 - \sum t_\nu \geqq 0.$$

This leads to the following interpretation of $M_n(x)$ which is hard to forget:

Theorem 2. *The fundamental spline function $M_n(x; x_0, \cdots, x_n)$ is the linear density function obtained by projecting orthogonally onto the x-axis the volume of an n-dimensional simplex σ_n, of volume unity, so located that its $n+1$ vertices project orthogonally into the points $x_0, x_1, \cdots, x_n$ of the x-axis, respectively.*

Proof: Let R^n be the space of points $X = (x, x', \cdots, x^{(n-1)})$ which is Cartesian and referred to orthogonal axes obtained by adjoining to the x-axis $n-1$ additional axes for the variables $x', \cdots, x^{(n-1)}$. Denoting by $\Phi(x)$ $(x_0 \leqq x \leqq x_n)$ the linear density described in the theorem we obtain, by Cavalieri's principle, the relation

$$\int_{\sigma_n} \cdots \int f^{(n)}(x)\, dx\, dx' \cdots dx^{(n-1)} = \int_{x_0}^{x_n} \Phi(x) f^{(n)}(x)\, dx. \tag{2.2}$$

Let $X_0, X_1, \cdots, X_n$ be the vertices of σ_n having coordinates given by

$$X_\nu = (x_\nu, x'_\nu, \cdots, x_\nu^{(n-1)}), \quad (\nu = 0, \cdots, n).$$

In the n-fold integral (2.2) we change to the new variables of integration $t_1, \cdots, t_n$ defined by the relations

$$\begin{aligned} x &= x_0(1 - \textstyle\sum t_\nu) + x_1 t_1 + \cdots + x_n t_n \\ &\vdots \\ x^{(n-1)} &= x_0^{(n-1)}(1 - \textstyle\sum t_\nu) + x_1^{(n-1)} t_1 + \cdots + x_n^{(n-1)} t_n. \end{aligned} \tag{2.3}$$

The absolute value of the Jacobian is found to be

$$\left| \frac{\partial(x, \cdots, x^{(n-1)})}{\partial(t_1, \cdots, t_n)} \right| = \text{abs. val.} \begin{vmatrix} x_1 - x_0, x_2 - x_0, \cdots, x_n - x_0 \\ x'_1 - x'_0, \cdots, \qquad x'_n - x'_0 \\ \vdots \\ x_1^{(n-1)} - x_0^{(n-1)}, \cdots, x_n^{(n-1)} - x_0^{(n-1)} \end{vmatrix}$$

$$= n! \operatorname{vol} \sigma_n = n!,$$

In view of the first relation (2.3), our integral (2.2) is transformed into the integral on the right side of (2.1) multiplied by $n!$. Thus (2.2) yields the relation

$$\int_{x_0}^{x_n} \Phi(x) f^{(n)}(x)\,dx = f(x_0, x_1, \cdots, x_n) \cdot n!$$

which is valid for all $f \in C^n$. Using the relation (1.5) and the continuity of $\Phi(x)$, we conclude that $\Phi(x) = M_n(x)$, which proves our theorem.

The above discussion implies that $M_n(x) = M_n(x; x_0, \cdots, x_n)$ is equal to the $(n-1)$-dimensional volume of the intersection of the simplex σ_n with the hyperplane orthogonal to the x-axis at the point x $(x_0 \leqq x \leqq x_n)$. Since σ_n is a convex body we may now use the famous theorem of H. Brunn (see [1, p. 71]) according to which

$(M_n(x))^{1/(n-1)}$ *is a concave function of* x *in* $x_0 \leqq x \leqq x_n$.

Writing

$$g(x) = (M_n(x))^{1/(n-1)} \tag{2.4}$$

we conclude that

$$g(x) > 0, \; g''(x) \leq 0 \quad \text{if } x_0 < x < x_n. \tag{2.5}$$

For the function

$$h(x) = \log g(x) = \frac{1}{n-1} \log M_n(x) \tag{2.6}$$

we now obtain

$$h'(x) = g'(x)/g(x), \; h''(x) = (g''(x)g(x) - (g'(x))^2)/(g(x))^2 \leqq 0, \tag{2.7}$$

where the last inequality follows from (2.5). In view of (2.6) we have just established the following

Theorem 3. *The function*

$$\log M_n(x; x_0, \cdots, x_n) \tag{2.8}$$

is concave in the open interval (x_0, x_n).

It is, of course, desirable to give an analytic proof of Theorem 3 which does not use the geometry of convex domains and Brunn's theorem. This can be done by means of the variation diminishing properties of the class $\mathcal{P}_2$ of Section 8 and will be done in a future paper which will also contain a proof of Theorem 8 below.

3. Spline functions with multiple knots. It is easy and important to extend the above theory to the case when some of the knots (1.1), which so far were assumed distinct, become coalescent. The meaning of multiple knots, a priori devoid of sense, must be defined as follows.

Definition 1. *In* (1.1) *we assume that*

$$x_{-1} < x_0 = x_1 = \cdots = x_{r-1} < x_r, \tag{3.1}$$

while the spline function $S_n(x)$, *of Section* 1, *satisfies the condition*

$$S_n(x) \in C^{n-1-r}(x_{-1}, x_r). \tag{3.2}$$

We then say that $x = x_0$ *is an r-fold knot of* $S_n(x)$, *or a knot of multiplicity* r.

In particular, the case when $r = n$ *means that there are no continuity requirements whatever, at* $x = x_0$, *between the two polynomials defining* $S_n(x)$ *in the adjacent intervals* (x_{-1}, x_0) *and* (x_0, x_r). *The multiplicity* r *of a knot is therefore restricted by the inequality*

$$r \leqq n, \tag{3.3}$$

which we assume throughout this paper.

We now generalize our class of spline functions $S_n(x)$ by allowing in (1.1) any number of equality signs subject to the condition

$$x_\nu < x_{\nu+n} \text{ for all } \nu. \tag{3.4}$$

Most of our previous discussion remains valid: The formulae (1.4) and (1.6) still hold, but the right side of (1.4) must be replaced by the expression for divided differences for multiple arguments. As an illustration, we mention the extreme case when $x_0 = x_1 = \cdots = x_{n-1} < x_n = \xi$. From Newton's general interpolation formula we obtain

$$f(\xi)=f(x_0)+(\xi-x_0)f'(x_0)+\cdots+\frac{(\xi-x_0)^{n-1}}{(n-1)!}f^{(n-1)}(x_0)+(\xi-x_0)^n f(x_0,\cdots,x_0,\xi).$$

Equating its remainder with its familiar integral form we obtain

$$(\xi-x_0)^n f(x_0,\cdots,x_0,\xi)=\frac{1}{(n-1)!}\int_{x_0}^{\xi}(\xi-x)^{n-1}f^{(n)}(x)\,dx.$$

Comparing with (1.5) we obtain

$$\text{(3.5)}\qquad M_n(x;x_0,\cdots,x_0,\xi)=\begin{cases}\dfrac{n(\xi-x)^{n-1}}{(\xi-x_0)^n} & \text{in } (x_0,\xi)\\[2mm] 0 & \text{outside } (x_0,\xi).\end{cases}$$

$M(x)$ is discontinuous at $x = x_0$ and Theorem 2, which remains valid, shows that the entire base of the simplex σ_n projects into x_0 and the opposite vertex into ξ. Theorem 1 needs modification; it can be shown that its conclusions concerning the zeros of $M^{(\nu)}(x)$ remain valid as long as these derivatives are continuous.

4. The fundamental n-splines form a basis. We return for the moment to the knots (1.1) and the n-splines which they define. The fundamental n-splines having consecutive knots $x_\nu, x_{\nu+1}, \cdots, x_{\nu+n}$ will now be denoted by

$$\text{(4.1)}\qquad M_\nu(x) = M_n(x; x_\nu, x_{\nu+1}, \cdots, x_{\nu+n}).$$

Their interest is due to the following

Theorem 4. 1. *If* $0 < N < n$ *and if* $S(x)$ *is an n-spline having the knots* (1.1) *and such that*

$$S(x) = 0 \text{ everywhere outside the interval } (x_0, x_N), \tag{4.2}$$

then

$$S(x) = 0 \quad \text{for all } x. \tag{4.3}$$

2. *If* $N \geqq n$ *and* (4.2) *holds, then* $S(x)$ *can be uniquely represented in the form*

$$S(x) = \sum_0^{N-n} c_\nu M_\nu(x). \tag{4.4}$$

3. *An n-spline* $S(x)$ *vanishing if* $x < x_0$ *can be uniquely represented in the form*

$$S(x) = \sum_0^{\infty} c_\nu M_\nu(x). \tag{4.5}$$

4. *Every n-spline* $S(x)$ *can be uniquely represented in the form*

$$S(x) = \sum_{\nu=-\infty}^{\infty} c_\nu M_\nu(x), \tag{4.6}$$

where the c_ν *are constants, and conversely, any such series represents an n-spline.*

We shall not establish this theorem as it stands, but rather its generalization, Theorem 5 below, to the case of multiple knots. For this purpose we must now allow also equality relations between consecutive knots in (1.1), subject to the restriction (3.4).

We shall assume that the knots (1.1) are located at the distinct points

$$\cdots < y_{-1} < y_0 < y_1 < y_2 < \cdots \quad (y_n \to \pm\infty \text{ as } n \to \pm\infty), \tag{4.7}$$

where

$$y_i \text{ is a knot of multiplicity } \alpha_i, \ (\alpha_i \leqq n). \tag{4.8}$$

Our discussion requires a kind of "double book-keeping" where multiple knots (4.7) are also represented by the non-decreasing sequence $\{x_\nu\}$ with the correct multiplicities. To fix the ideas, we assume this representation such that

$$\cdots, x_0 = y_0, x_1 = x_2 = \cdots = x_{\alpha_1} = y_1, x_{\alpha_1+1} = y_2, \cdots.$$

Thus in terms of the original sequence (1.1) the knots are

$$\cdots < x_{-\alpha_0+1} = \cdots = x_0 < x_1 = \cdots = x_{\alpha_1} < x_{\alpha_1+1} = \cdots = x_{\alpha_1+\alpha_2} < \cdots. \tag{4.9}$$

Again, as in (4.1), we associate with the knots (4.9) the sequence of fundamental n-splines

$$M_j(x) = M_n(x; x_j, x_{j+1}, \cdots, x_{j+n}), \quad (-\infty < j < \infty). \tag{4.10}$$

The generalization of Theorem 4 is as follows.

Theorem 5. *Let $S(x)$ be an n-spline having the knots* (4.7), *of multiplicities as described by* (4.8).

1. *If*

$$s = \sum_1^N \alpha_\nu \leqq n \tag{4.11}$$

and

$$S(x) = 0 \textit{ everywhere outside the interval } (y_1, y_N), \tag{4.12}$$

then

$$S(x) = 0 \textit{ for all } x. \tag{4.13}$$

2. *If*

$$s = \sum_1^N \alpha_\nu \geqq n+1 \tag{4.14}$$

and (4.12) *holds, then $S(x)$ can be uniquely represented in the form*

$$S(x) = \sum_{1}^{s-n} c_j M_j(x). \tag{4.15}$$

3. *If* $S(x) = 0$ *whenever* $x < y_1$, *then* $S(x)$ *can be uniquely represented in the form*

$$S(x) = \sum_{1}^{\infty} c_j M_j(x). \tag{4.16}$$

4. *Every* $S(x)$ *can be uniquely represented in the form*

$$S(x) = \sum_{-\infty}^{\infty} c_j M_j(x). \tag{4.17}$$

Proof: The four parts of the theorem are logically related in the order in which they are stated.

1. Let us construct an $S(x)$ such that (4.12) holds. By the continuity requirement (3.2), which defines the meaning of the multiplicity α_1 of the knot y_1, we conclude that

$$S(x) = a_1^1(x - y_1)^{n-1} + a_2^1(x - y_1)^{n-2} + \cdots + a_{\alpha_1}^1(x - y_1)^{n-\alpha_1} \text{ in } y_1 < x < y_2.$$

Proceeding in like fashion from each (y_i, y_{i+1}) to the next, new terms are added, until we arrive at the expression

$$S(x) = \sum_{i=1}^{\alpha_1} a_i^1(x - y_1)^{n-i} + \cdots + \sum_{i=1}^{\alpha_N} a_i^N(x - y_N)^{n-i} \quad \text{if } x > y_N.$$

By (4.12) $S(x)$ is supposed to vanish in this range, whence the identity

$$\sum_{i=1}^{\alpha_1} a_i^1(x - y_1)^{n-i} + \cdots + \sum_{i=1}^{\alpha_N} a_i^N(x - y_N)^{n-i} = 0 \quad \text{for all } x. \tag{4.18}$$

Conversely, an identity (4.18) implies an $S(x)$ satisfying all conditions and defined by

$$S(x) = \sum_{i=1}^{\alpha_1} a_i^1(x - y_1)_+^{n-i} + \cdots + \sum_{i=1}^{\alpha_N} a_i^N(x - y_N)_+^{n-i}, \quad (-\infty < x < \infty). \tag{4.19}$$

Let us assume for the moment a little more than (4.11), namely

$$s = \sum_{1}^{N} \alpha_\nu = n \tag{4.20}$$

and let us show that the identity (4.18) implies the vanishing of all coefficients a_i^ν. In other words, we have to show that (4.20) implies the linear independence of the n polynomials

$$\begin{array}{c} \frac{(x-y_1)^{n-1}}{(n-1)!}, \frac{(x-y_1)^{n-2}}{(n-2)!}, \cdots, \frac{(x-y_1)^{n-\alpha_1}}{(n-\alpha_1)!}, \\ \vdots \\ \frac{(x-y_N)^{n-1}}{(n-1)!}, \frac{(x-y_N)^{n-2}}{(n-2)!}, \cdots, \frac{(x-y_N)^{n-\alpha_N}}{(n-\alpha_N)!}. \end{array} \tag{4.21}$$

To decide this point we expand them in increasing powers of x and find that the $n \times n$ determinant of their coefficients is (up to a positive numerical factor) equal to the determinant $\mathscr{D}$ having N groups of α_i rows each, the ith group being described by the formula

$$\mathscr{D} = \begin{vmatrix} \frac{y_i^{n-1}}{(n-1)!}, \frac{y_i^{n-2}}{(n-2)!}, \cdots, & 1 \\ \frac{y_i^{n-2}}{(n-2)!}, \frac{y_i^{n-3}}{(n-3)!}, \cdots, & 0 \\ \vdots & \\ \frac{y_i^{n-\alpha_i}}{(n-\alpha_i)!}, \frac{y_i^{n-\alpha_i-1}}{(n-\alpha_i-1)!}, \cdots, & 0 \end{vmatrix} \quad i = 1, 2, \cdots, N. \tag{4.22}$$

This is a generalization of the Vandermonde determinant, to which it reduces if all the α_i are equal to unity. That $\mathscr{D} \neq 0$ will now follow from the unicity of the solution

$$P(x) = A_0 \frac{x^{n-1}}{(n-1)!} + A_1 \frac{x^{n-2}}{(n-2)!} + \cdots + A_{n-1} \tag{4.23}$$

of Hermite's interpolation problem ([5, p. 432])

(4.24) $$P(y_i) = f(y_i),\ P'(y_i) = f'(y_i), \cdots, P^{(\alpha_i - 1)}(y_i) = f^{(\alpha_i - 1)}(y_i)$$

$$(i = 1, 2, \cdots, N;\ \alpha_i + \cdots + \alpha_N = n).$$

Indeed, the relation $\mathscr{D} = 0$ would imply the existence of a polynomial (4.23), $P(x) \not\equiv 0$, having an α_i-fold zero at $x = y_i$ $(i = 1, \cdots, N)$, hence a total of $\sum \alpha_t = n$ zeros whose number exceeds its degree. Our assertion is therefore established if (4.20) holds.

We now drop the assumption (4.20) and assume instead that (4.11) holds with the inequality sign. Given $S(x)$ satisfying (4.12), we then add beyond y_N additional fictitious simple knots $y'_{n+1}, y'_{n+2}, \cdots$ in sufficient number to produce the previous case of the equality (4.20). By the previous case we now know that $S(x)$ must vanish identically.

2. We need the following

Lemma 1. *If*

(4.25) $$\alpha_1 + \alpha_2 + \cdots + \alpha_N = n + 1$$

then the fundamental spline function

(4.26) $$M_1(x) = M_n(x; \overbrace{y_1, \cdots, y_1}^{\alpha_1}, \cdots, \overbrace{y_N, \cdots, y_N}^{\alpha_N})$$

is precisely of the continuity class $C^{n-\alpha_i-1}$ in the neighborhood of the point $x = y_i$ $(i = 1, \cdots, N)$. "Precisely" means that it belongs to no higher class.

Proof: The construction of $M_1(x)$ is a special case of our problem when we have equality in (4.14), hence (4.25). By (4.19) it appears that $M_1(x)$ must be of the form

(4.27) $$\begin{aligned} M_1(x) = {} & a_1^1(x - y_1)_+^{n-1} + \cdots + a_{\alpha_1}^1(x - y_1)_+^{n-\alpha_1} \\ & \vdots \\ & a_1^N(x - y_N)_+^{n-1} + \cdots + a_{\alpha_N}^N(x - y_N)_+^{n-\alpha_N}, \qquad (-\infty < x < \infty), \end{aligned}$$

which expression should vanish identically on removing everywhere the subscript "+". The resulting identity (4.18) now expresses the linear *dependence* of the functions (4.21), and we wish to show that (up to a constant factor) only one such relation exists. This much is clear: In (4.27), or (4.18) we must have

$$a^{\nu}_{\alpha_\nu} \neq 0 \text{ for } \nu = 1, \cdots, N, \tag{4.28}$$

for if one were to vanish, $a^{1}_{\alpha_1} = 0$ say, then we could diminish the multiplicity α_1 (of the knot y_1) by one unit and thereby fall back on the case (4.20) when we know that only trivial identities (4.18) are possible. Using (4.28) the same kind of reasoning shows that, up to factors, only one relation (4.18) exists.

Finally (4.27) and (4.28) show that

$$M_1^{(n-\alpha_\nu)}(y_\nu + 0) - M_1^{(n-\alpha_\nu)}(y_\nu - 0) = (n - \alpha_\nu)! a^{\nu}_{\alpha_\nu} \neq 0 \ (\nu = 1, \cdots, N)$$

which proves our lemma.

A proof of (4.15) under the assumptions (4.12) and (4.14), is now straightforward. Indeed, observe that by (4.12) and (4.8)

$$S(x) = (x - y_1)^{n-\alpha_1} \cdot \phi(x), \ (\phi \text{ is a polynomial}),$$

in a right-neighborhood of y_1. By Lemma 1 also

$$M_1(x) = c(x - y_1)^{n-\alpha_1} + \text{higher powers of } x - y_1, \ (c \neq 0),$$

in a right-neighborhood of y_1. We can therefore determine c_1 uniquely so that

$$S_1(x) = S(x) - c_1 M_1(x) = (x - y_1)^{n-\alpha_1+1} \phi_1(x), (y_1 < x < y_2).$$

However, again by Lemma 1,

$$M_2(x) = c'(x - y_1)^{n-\alpha_1+1} + \cdots \ (y_1 < x < y_2)$$

and a unique c_2 will produce

$$S_2(x) = S(x) - c_1 M_1(x) - c_2 M_2(x) = (x - y_1)^{n-\alpha_1+2} \phi_2(x), \quad (y_1 < x < y_2),$$

and so on.

Two cases are possible:

1. $\alpha_2 + \alpha_3 + \cdots + \alpha_N \leqq n$. The construction may continue until we obtain

(4.29) $S_t(x) = S(x) - c_1 M_1(x) - \cdots - c_t M_t(x) = (x - y_1)^{n-\alpha_1+t}\phi_t(x), \quad (y_1 < x < y_2),$

where $t = \alpha_1 + \alpha_2 + \cdots + \alpha_N - n \leqq \alpha_1$. However, $S_t(x)$ is a spline function which may be thought of, by (4.29), as having at $x = y_1$ a knot of multiplicity $\alpha_1 - t$. But then the sum of all multiplicities of knots of $S_t(x)$ is

$$(\alpha_1 - t) + \alpha_2 + \cdots + \alpha_N = n$$

and by Case 1 of Theorem 5 we conclude that $S_t(x) = 0$ for all x, while $t = s - n$.

2. $\alpha_2 + \alpha_3 + \cdots + \alpha_N > n$. In this case we may surely proceed until we reach

$$S_{\alpha_1-1}(x) = S(x) - c_1 M_1(x) - \cdots - c_{\alpha_1-1} M_{\alpha_1-1}(x) = (x - y_1)^{n-1}\phi(x),$$

in $y_1 < y < y_2$, where $\phi(x)$ is necessarily a constant. But then, with an appropriate unique c_{α_1}

$$S_{\alpha_1}(x) = S_{\alpha_1-1}(x) - c_{\alpha_1} M_{\alpha_1}(x) = 0 \quad \text{if } x < y_2.$$

From this point we continue with

$$M_{\alpha_1+1}(x) = M_n(x; \overbrace{y_2, \cdots, y_2}^{\alpha_2}, y_3, \cdots)$$

using right-neighborhoods of y_2. We can surely continue this successive subtraction process until we reach the spline function

$$S_{s-n}(x) = S(x) - \sum_1^{s-n} c_j M_j(x)$$

and we wish to show that it vanishes for all values of x.

In order to describe and justify the termination of this process, let the integers r, α and β be defined by the relations

$$\alpha_{r+1} + \cdots + \alpha_N < n + 1 \leqq \alpha_r + \alpha_{r+1} + \cdots + \alpha_N, \quad (2 \leqq r < N),$$

$$\alpha + \alpha_{r+1} + \cdots + \alpha_N = n + 1, \quad (0 < \alpha \leqq n).$$

$$\alpha + \beta = \alpha_r.$$

The functions $S(x), S_1(x), \cdots, S_{\alpha_1 - 1}(x)$ are near the point $x = y_1$ of the continuity classes $C^{n-\alpha_1-1}, C^{n-\alpha_1}, \cdots, C^{n-2}$, respectively. Likewise $S_{\alpha_1}(x), \cdots, S_{\alpha_1+\alpha_2-1}(x)$ vanish for $x < y_2$ and are near $x = y_2$ of the continuity classes $C^{n-\alpha_2-1}, C^{n-\alpha_2}, \cdots, C^{n-2}$ respectively. Finally

$$S_{\alpha_1+\cdots+\alpha_{r-1}}(x), \; S_{\alpha_1+\cdots+\alpha_{r-1}+1}(x), \cdots, S_{\alpha_1+\cdots+\alpha_{r-1}+\beta+1}(x)$$

vanish for $x < y_r$ and are near $x = y_r$ of the continuity classes

$$C^{n-\alpha_r-1}, C^{n-\alpha_r}, \cdots, C^{n-\alpha}, \text{ respectively.}$$

Observe now that by our definition of r, α and β we have

$$\alpha_1 + \cdots + \alpha_{r-1} + \beta + 1 = s - n.$$

Also that $S_{s-n}(x)$ is a spline function which may be thought of as having at $x = y_r$ a knot of multiplicity $\alpha - 1$. Thus the total sum of the multiplicities of its knots is $(\alpha - 1) + \alpha_{r+1} + \cdots + \alpha_N = n$. As $S_{s-n}(x)$ vanishes outside (y_r, y_N), we may apply the result of Case 1 and conclude that $S_{s-n}(x) = 0$ for all real x. This settles the Case 2 of Theorem 5.

3. The previous construction applies with the difference that it continues indefinitely resulting in (4.16). Clearly Case 2 is a special case of this when the c_j vanish for sufficiently large j.

4. To deal with this last case we need

Lemma 2. *The fundamental functions*

(4.30) $$M_{-n+1}(x), M_{-n+2}(x), \cdots, M_0(x)$$

are linearly independent in the interval $(x_0, x_1) = (y_0, y_1)$ *and therefore form in this interval a basis for* π_{n-1}.[2]

[2] π_{n-1} denotes the class of polynomials of degree not exceeding $n - 1$.

Proof: Let us assume that

$$S(x) \equiv \sum_{-n+1}^{0} c_j M_j(x) = 0 \text{ in } (y_0, y_1) \tag{4.31}$$

and let us show that this implies that

$$c_j = 0 \ (j = -n+1, \cdots, 0). \tag{4.32}$$

Consider the restriction of $S(x)$ to the interval $(y_0, +\infty)$. By (4.31) this restriction is a spline function vanishing everywhere outside the interval (x_1, x_n) provided that we define it as $=0$ if $x \leqq y_0$. It may therefore be thought of as an n-spline having at most n knots, namely $x_1, x_2, \cdots, x_n$; it must therefore vanish everywhere by Case 1 of Theorem 5. Thus (4.31) implies that $S(x) = 0$ in (y_0, ∞). Similarly it is shown that $S(x) = 0$ in the interval $(-\infty, y_1)$. Thus $S(x) = 0$ for all x and now (4.32) follows from the unicity in Case 2 of our theorem. This completes a proof of Lemma 2.

Let us finally consider an arbitrary n-spline $S(x)$. Let

$$S(x) = P(x) \text{ if } y_0 < x < y_1, \ (P \in \pi_{n-1}).$$

By Lemma 2 we can write

$$P(x) = \sum_{-n+1}^{0} c_j M_j(x)$$

which implies that

$$S^*(x) = S(x) - \sum_{-n+1}^{0} c_j M_j(x) \tag{4.33}$$

is an n-spline vanishing in the interval $(y_0, y_1) = (x_0, x_1)$. We may therefore write

$$S^*(x) = S_1(x) + S_0(x), \tag{4.34}$$

where S_1 and S_0 are n-splines vanishing in the intervals $(-\infty, x_1)$ and $(x_0, +\infty)$ respectively. By Case 3 of our theorem we may therefore write uniquely

$$S_1(x) = \sum_{1}^{\infty} c_j M_j(x), \quad S_0(x) = \sum_{-\infty}^{-n} c_j M_j(x).$$

Now (4.33) and (4.34) imply the desired representation (4.17). This completes a proof of Theorem 5.

In view of its relevance to spline interpolation we restate here the Case 2 of Theorem 5 as follows.

Corollary 1. *Let* $x_1, x_2, \cdots, x_s$, $(s > n)$, *be non-decreasing reals such that*

$$x_\nu < x_{\nu+n} \quad (1 \leqq \nu < \nu + n \leqq s).$$

This implies that at most n of the x_ν *may coalesce at a point. Then the fundamental spline functions*

$$M_j(x) = M_n(x; x_j, x_{j+1}, \cdots, x_{j+n}), \quad (j = 1, 2, \cdots, s-n), \tag{4.35}$$

are linearly independent. Every spline function $S(x)$, *of degree* $n-1$, *having the points* $x_1, \cdots, x_s$ *as knots, with multiplicities as they occur in this sequence, and such that*

$$S(x) = 0 \text{ everywhere outside } (x_1, x_s), \tag{4.36}$$

may be uniquely represented in the form

$$S(x) = \sum_{1}^{s-n} c_j M_j(x). \tag{4.37}$$

By means of Corollary 1 the proofs of the main results concerning spline interpolation as given in [14], extend easily to the case of interpolation with *multiple nodes*. By a *node* of an interpolation problem we mean a point where a functional value is given; a node is *multiple* if an appropriate number of consecutive derivatives of a function are also to be interpolated at that point.

Fig. 1 shows a sketch of the five fundamental quadratic spline functions for the case when $n = 3$, $s = 9$ and

$$x_1 = x_2 = 0, \; x_3 = 1, \; x_4 = 2, \; x_5 = x_6 = 3, \; x_7 = x_8 = x_9 = 4.$$

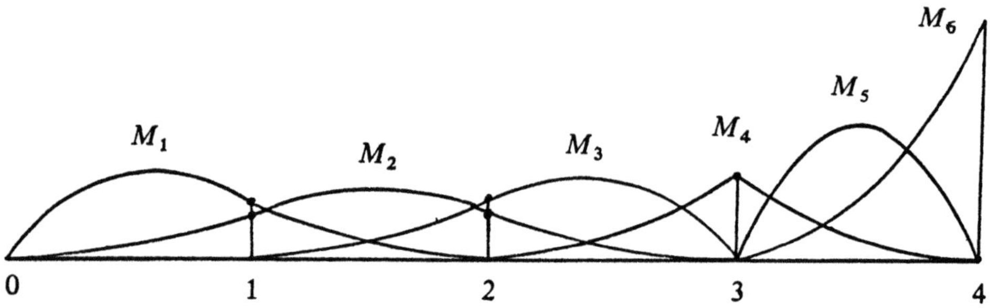

Figure 1

In terms of the expression (4.10) we may write

$$M_1(x) = M_3(x;0,0,1,2)$$

$$M_2(x) = M_3(x;0,1,2,3)$$

$$M_3(x) = M_3(x;1,2,3,3)$$

$$M_4(x) = M_3(x;2,3,3,4)$$

$$M_5(x) = M_3(x;3,3,4,4)$$

$$M_6(x) = M_3(x;3,4,4,4).$$

All arcs between consecutive integral values of x are parabolic and are easily found explicitly. Fig. 1 illustrates nicely Lemma 1. By Theorem 5 we know that

$$S(x) = \sum_1^6 c_j M_j(x)$$

represents the most general quadratic spline function in the interval $[0,4]$ satisfying the following conditions

1. It has the knots 1, 2, 3.
2. It belongs to the classes C^1, C^1, C^0 near the points $x = 1,2,3$, respectively.
3. $S(0) = 0$.

As illustrations of Cases 3 and 4 of Theorem 5 we mention the following identities. The knots being integral ($x_i = i$) and writing

$$M_j(x) = M_m(x; j, j+1, \cdots, j+m)$$

we find that

$$x_+^{m-1} = \sum_{j=0}^{\infty} (j+1)(j+2)\cdots(j+m-1)M_j(x),$$

$$x^{m-1} = \sum_{j=-\infty}^{\infty} (j+1)(j+2)\cdots(j+m-1)M_j(x).$$

II. THE LIMITS OF FUNDAMENTAL SPLINE FUNCTIONS

5. The problem and main results. Interesting results in Analysis are sometimes of the following nature: We are given a certain class $\{F\}$ of functions F and we are able to describe independently, in terms of structural properties, the *closure* of the class $\{F\}$ in terms of a suitable topology. A classical result of this nature is the following theorem:

Let

$$\left\{P_n(s) = \prod_{\nu=1}^{n} (1 + \alpha_{n,\nu}s)\right\} \tag{5.1}$$

be the class of real polynomials having only real zeros, so normalized that $P_n(0) - 1$. The closure of this class with respect to uniform convergence in every finite domain of the complex s-plane is identical with the Laguerre-Pólya class of entire functions described by

$$\Psi(s) = e^{-\gamma s^2 + \delta s} \prod_{1}^{\infty} (1 + \delta_\nu s) e^{-\delta_\nu s} \tag{5.2}$$

$$(\gamma \geqq 0,\ \delta,\ \delta_\nu \text{ real},\ \sum \delta_\nu^2 < \infty),$$

(see [6, Chap. III, Section 3] where also original references are given).

By a distribution function we mean a monotone function $F(x)$ such that $F(-\infty) = 0$, $F(+\infty) = 1$. $F_n(x)$ and $F(x)$ being distribution functions, we say that $F_n(x)$ converges to $F(x)$ and write $F_n(x) \to F(x)$, provided that the relation holds for all x at which $F(x)$ is continuous. Our problem, closely related with the class of functions (5.2), is as follows. We consider the class of spline distribution functions (see (1.6))

$$F_n(x) = \int_{-\infty}^{x} M_n(x; x_{0,n}, x_{1,n}, \cdots, x_{n,n})\, dx, \tag{5.3}$$

the integrand being a fundamental n-spline (Sections 1 and 4). Concerning its knots we make the widest possible assumption, namely

$$x_{0,n} \leqq x_{1,n} \leqq \cdots \leqq x_{n,n}, x_{0,n} < x_{n,n}. \tag{5.4}$$

We ask: *Assuming*

$$\lim_{n \to \infty} F_n(x) = F(x), \tag{5.5}$$

what is the nature of the limit distribution functions $F(x)$?

An answer to this question requires the class of Pólya distribution functions $F(x)$. These are distribution functions having a bilateral Laplace transform of the form

$$\int_{-\infty}^{\infty} e^{-sx} dF(x) = \frac{1}{\Psi(s)}, \tag{5.6}$$

where $\Psi(s)$ is described by (5.2). If $\Psi(s) = e^{\delta s}$, then $dF(x)$ has its entire unit mass located at $x = \delta$. If

$$\Psi(s) \neq e^{\delta s} \tag{5.7}$$

then

$$\int_{-\infty}^{\infty} e^{-sx} \Lambda(x) dx = \frac{1}{\Psi(s)}, \tag{5.8}$$

where $\Lambda(x)$ is a Pólya frequency function. The Pólya frequency functions may also be characterized by the structural property of being totally positive (see [10], [6]).

Theorem 6. *The limit $F(x)$ of a convergent sequence of spline distribution functions $F_n(x)$ is a Pólya distribution function. Conversely, every Pólya distribution function $F(x)$ is the limit of an appropriate sequence of spline distribution functions $F_n(x)$, of the form* (5.3), *as $n \to \infty$.*

The question arises as to when the limit relation (5.5) may be differentiated, so that we may infer that the fundamental spline functions M_n themselves converge. The answer will depend on the continuity of $\Lambda(x)$. Writing

$$\lambda(x) = \begin{matrix} e^{-x} & \text{if } x \geqq 0 \\ 0 & \text{if } x < 0 \end{matrix}$$

we know (see [10]) that the only discontinuous $\Lambda(x)$ is the one-sided exponential

$$\Lambda_0(x) = \frac{1}{|\delta_1|}\lambda\left(\frac{x-\delta+\delta_1}{\delta_1}\right) \quad (\delta_1 \neq 0) \tag{5.9}$$

having the transform

$$\int_{-\infty}^{\infty} e^{-sx}\Lambda_0(x)dx = 1/\{e^{\delta s}(1+\delta_1 s)e^{-\delta_1 s}\} \tag{5.10}$$

Theorem 7. *Let* (5.5) *and* (5.7) *hold, hence $dF(x) = \Lambda(x)dx$, where $\Lambda(x)$ is a Pólya frequency function. Then*

$$\lim_{n\to\infty} M_n(x; x_{0,n}, \cdots, x_{n,n}) = \Lambda(x), \tag{5.11}$$

uniformly for all x, provided that $\Lambda(x)$ is not of the form (5.9). *For $\Lambda(x) = \Lambda_0(x)$, again* (5.11) *holds uniformly for x outside an arbitrarily small neighborhood of the point $x = \delta - \delta_1$, where $\Lambda_0(x)$ is discontinuous.*

If $\Psi(s) = e^{\delta s}$ then

$$\lim_{n\to\infty} M_n(x; x_{0,n}, \cdots, x_{n,n}) = 0 \tag{5.12}$$

uniformly in x outside an arbitrarily small neighborhood of the point $x = \delta$.

Theorem 7 describes a new characterization of the Pólya frequency functions. It adds a fourth to the three previously known characterizations: 1. The special nature (5.8), (5.2), of their Laplace transforms. 2. Their totally positive character ([10]). 3. Their variation diminishing property ([11]).

6. A few auxiliary propositions.(3) Since Paul Levy's work of the 1920's, the characteristic functions are usually the most convenient tools in dealing with convergence problems for distribution functions. Here we also use some related but different integrals. Their use will be justified by two propositions. Below, $F_n(x)$ and $F(x)$ are arbitrary distribution functions.

Lemma 3. *If*

$$F_n(x) \to F(x) \tag{6.1}$$

then the functions

$$g_n(t) = \int_{-\infty}^{\infty} \left(1 + \frac{itx}{n}\right)^{-n} dF_n(x) \tag{6.2}$$

have the property

$$\lim_{n\to\infty} g_n(t) = \int_{-\infty}^{\infty} e^{-itx} dF(x) \text{ locally uniformly.} \tag{6.3}$$

Lemma 4. *Let $\{F_n(x)\}$ be a sequence of distribution functions which are equi-continuous at infinity. By this we mean that for every positive ε there corresponds an $A = A(\varepsilon)$ such that*

$$\int_{|x| \geq A} dF_n(x) < \varepsilon \text{ for all } n. \tag{6.4}$$

(3) In connection with Lemma 4 we acknowledge helpful conversations with Professor Joshua Chover.

Then the pointwise convergence of the sequence of functions (6.2), *for all real* t, *implies the convergence relation* (6.1) *and therefore the relation* (6.3) *holds by Lemma* 3.

Proof of Lemma 3: The assumption (6.1) implies that

$$(6.5)\qquad f_n(t) = \int_{-\infty}^{\infty} e^{-itx}dF_n(x) \to \int_{-\infty}^{\infty} e^{-itx}dF(x) \text{ locally uniformly.}$$

On the other hand

$$(6.6)\ |f_n(t) - g_n(t)| \leqq \int_{-A}^{A} \left|e^{-itx} - \left(1 + \frac{itx}{n}\right)^{-n}\right| dF_n(x) + 2(1-F_n(A)) + 2F_n(-A).$$

Let ε and T be positive and let us choose A such that A and $-A$ are continuity points of $F(x)$ and such that

$$(6.7)\qquad 2(1 - F(A)) < \varepsilon, \quad 2F(-A) < \varepsilon.$$

Furthermore, let $N = N(\varepsilon, TA)$ be such that

$$(6.8)\qquad \left|e^{-z} - \left(1 + \frac{z}{n}\right)^{-n}\right| < \varepsilon \text{ for } |z| \leqq TA, \text{ if } n > N.$$

By (6.6), (6.7), (6.8) and $F_n(\pm A) \to F(\pm A)$ we conclude that

$$|f_n(t) - g_n(t)| < \varepsilon + \varepsilon + \varepsilon = 3\varepsilon \text{ if } -T \leqq t \leqq T,$$

provided that n is sufficiently large. Now (6.5) implies (6.3) and Lemma 3 is established.

Proof of Lemma 4. The result will follow from the fact that a distribution function is uniquely defined by its characteristic function. Indeed, applying Helly's selection principle, we obtain an infinite sequence N of increasing natural numbers such that $F_n(x)$ converges pointwise for all real x, as $n \to \infty$, $n \in N$. Thus

$$(6.9)\qquad \lim F_n(x) = F(x), \text{ for all real } x, \text{ as } n \to \infty,\ n \in N.$$

$F(x)$ is not yet a distribution function as we may well have $F(\infty) - F(-\infty) < 1$. However, (6.9) and the equi-continuity assumption (6.4) easily imply that $F(-\infty) = 0$, $F(\infty) = 1$, by an argument which we may omit. Thus $F(x)$ is a distribution function and (6.1) holds if we restrict n to the sequence N. However, Lemma 3 is applicable and shows that

$$\lim g_n(t) = \int_{-\infty}^{\infty} e^{-itx} dF(x) \text{ as } n \to \infty,\ n \in N. \tag{6.10}$$

Since the entire sequence $\{g_n(t)\}$ converges pointwise, by assumption, it is clear that (6.10) holds for every t as $n \to \infty$ through all integers. The limit distribution $F(x)$ being *uniquely* defined by its characteristic function (6.10) we may have essentially only one such function, hence (6.1) holds. Finally, the relation (6.3) is implied by Lemma 3.

Lemma 5. *Let*

$$P_n(s) = \prod_{\nu=1}^{n} (1 + \alpha_{n,\nu} s)$$

be a sequence of real polynomials having only real zeros, $P_n(0) = 1$, *which converges uniformly on a segment*

$$-\rho \leqq t \leqq \rho \ (\rho > 0;\ s = it)$$

of the imaginary axis. Then the sequence $\{P_n(s)\}$ *converges locally uniformly throughout the s-plane to an entire function* $\Psi(s)$ *of the class described by* (5.2).

This result was apparently novel in 1947 but is no longer so now (see [6, Theorem 3.4 on page 47]). For related recent results see [7].

In this section we simplify our notation for the knots (5.4) by dropping the second subscript and denoting them by $x_0, x_1, \cdots, x_n$.

Lemma 6. *The relation*

$$(6.11)\quad 1\Big/\prod_{\nu=0}^{n}\left(1+\frac{x_\nu}{n+1}ti\right) = \int_{-\infty}^{\infty}\left(1+\frac{xti}{n+1}\right)^{-n-1} M_n(x;x_0,\cdots,x_n)\,dx$$

holds for all t which are not purely imaginary.

Proof: Let $f(x) = 1/(1-ax)$, where a is a constant. By induction we verify that

$$f(x_0, x_1, \cdots, x_n) = a^n\Big/\prod_0^n (1-ax_\nu)$$

while

$$f^{(n)}(x) = n!a^n(1-ax)^{-n-1}.$$

The relation (1.5) now implies

$$1\Big/\prod_0^n (1-ax_\nu) = \int_{-\infty}^{\infty} (1-ax)^{-n-1} M_n(x;x_0,\cdots,x_n)dx.$$

Setting $a = -it/(n+1)$ we obtain (6.11).

7. A proof of Theorem 6. Let $F_n(x)$ denote the spline distribution function (5.3), our assumption being that

$$F_n(x) \to F(x),$$

where $F(x)$ is a distribution function. By Lemma 3 we conclude that

$$\lim_{n\to\infty}\int_{-\infty}^{\infty}\left(1+\frac{itx}{n+1}\right)^{-n-1} M_n(x)dx = \int_{-\infty}^{\infty} e^{-itx}\,dF(x),$$

locally uniformly for all real t and by Lemma 6 we may rewrite this as

$$(7.1)\qquad \lim_{n\to\infty} 1\Big/\prod_{\nu=0}^{n}\left(1+\frac{x_{\nu,n}}{n+1}ti\right) = \int_{-\infty}^{\infty} e^{-itx}dF(x),$$

locally uniformly for all real t.

Let us select a positive ρ such that

$$\int_{-\infty}^{\infty} e^{-itx} dF(x) \neq 0 \qquad \text{if } -\rho \leqq t \leqq \rho .$$

In this range we may form the reciprocals of both sides of (7.1) and conclude that

$$\lim_{n\to\infty} \prod_{\nu=0}^{n} \left(1 + \frac{x_{\nu,n}}{n+1} ti\right) = 1 \Big/ \int_{-\infty}^{\infty} e^{-itx} dF(x), \tag{7.2}$$

uniformly in the interval $-\rho \leqq t \leqq \rho$. Now Lemma 5 becomes applicable and implies the relation

$$\lim_{n\to\infty} \prod_{\nu=0}^{n} \left(1 + \frac{x_{\nu,n}}{n+1} s\right) = \Psi(s), \tag{7.3}$$

locally uniformly in the complex s-plane, where $\Psi(s)$ is an entire function of the form (5.2). The zeros of $\Psi(s)$ being all real, we now realize that the limit (7.1) is different from zero for all real t and therefore that (7.2) holds for all real t. By (7.2) and (7.3) we obtain

$$\int_{-\infty}^{\infty} e^{-itx} dF(x) = \frac{1}{\Psi(it)} \qquad (-\infty < t < \infty)$$

and a comparison with (5.6) shows that $F(x)$ is a Pólya distribution function. This concludes a proof of the direct part of Theorem 6.

Conversely, let $F(x)$ be a Pólya distribution function having a transform defined by (5.6) and (5.2). The easy converse part of the Laguerre-Pólya theorem asserts the existence of reals $x_{0,n}, x_{1,n}, \cdots, x_{n,n}$ satisfying (5.4), defined for a sequence N of values of n, and such that (7.3) holds locally uniformly in the s-plane. With these reals $x_{\nu,n}$ as knots, we construct the spline distribution function $F_n(x)$, defined by (5.3), and wish to show that

$$F_n(x) \to F(x), \qquad (n \in N). \tag{7.4}$$

Let us choose a positive number a such that

$$\Psi(s) \neq 0 \qquad \text{if } -a < \operatorname{Re} s < a.$$

From (5.6), (7.3) and Lemma 6 we conclude that

$$\lim_{n\to\infty} \int_{-\infty}^{\infty} \left(1 + \frac{x}{n+1}s\right)^{-n-1} M_n(x)dx = \int_{-\infty}^{\infty} e^{-sx}dF(x), \tag{7.5}$$

where $M_n(x) = M_n(x; x_{0,n}, \cdots, x_{n,n})$, holds locally uniformly in the same strip $-a < \operatorname{Re} s < a$. In particular, (7.5) holds uniformly in a certain circular neighborhood of $s = 0$. The expansions in powers of s of these integrals are readily obtained in terms of the moments of the distributions and we find

$$\int_{-\infty}^{\infty} e^{-sx}dF(x) = \sum_{0}^{\infty} (-1)^{\nu} \frac{\mu_\nu}{\nu!} s^{\nu}$$

$$\int_{-\infty}^{\infty} \left(1 + \frac{xs}{n+1}\right)^{-n-1} M_n(x)dx = \sum_{0}^{\infty} (-1)^{\nu} \frac{(n+1)(n+2)\cdots(n+\nu)}{(n+1)^{\nu}} \frac{\mu_\nu^{(n)}}{\nu!} s^{\nu},$$

where μ_ν and $\mu_\nu^{(n)}$ are the moments of $F(x)$ and $F_n(x)$, respectively. These expansions and the uniform convergence relation (7.5) show that

$$\lim_{n\to\infty} \mu_\nu^{(n)} = \mu_\nu \qquad \text{for all } \nu, \qquad n \in N.$$

In particular for $\nu = 2$ we conclude that

$$\lim_{n\to\infty} \int_{-\infty}^{\infty} x^2 M_n(x)dx = \int_{-\infty}^{\infty} x^2 dF(x), \qquad (n \in N). \tag{7.6}$$

A proof of (7.4) now follows readily. Indeed, on the one hand (7.5) shows that

$$\lim_{\substack{n\to\infty \\ n\in N}} \int_{-\infty}^{\infty} \left(1+\frac{itx}{n+1}\right)^{-n-1} dF_n(x) = \int_{-\infty}^{\infty} e^{-itx}\, dF(x) \text{ locally uniformly.} \tag{7.7}$$

On the other hand (7.6) implies that the left side of the inequality

$$A^2 \int_{|x|\geqq A} dF_n(x) \leqq \int_{-\infty}^{\infty} x^2 M_n(x)dx$$

remains below a constant K for all n and all positive A. Thus

$$\int_{|x|\geqq A} dF_n(x) < KA^{-2} \text{ for all } A > 0,$$

which evidently implies the equi-continuity condition (6.4) required by Lemma 4. In view of this and (7.7), Lemma 4 implies our conclusion (7.4), completing our proof of Theorem 6.

8. On frequency functions which are logarithmically concave and proof of Theorem 7. It was pointed out that Pólya frequency functions are *totally positive.* A wider class of functions is formed by the so-called *multiply positive functions* of a certain order (see [13]). Here we are concerned with the widest such class, namely the class of functions which are *twice-positive.* The precise definition is as follows.

Definition 2. *We say that a frequency function $f(x)$, i.e. a non-negative finite-valued function whose integral over the reals is unity, belongs to the class $\mathscr{P}_2$ of twice-positive functions, provided that*

$$\begin{vmatrix} f(x_1-t_1) & f(x_1-t_2) \\ f(x_2-t_1) & f(x_2-t_2) \end{vmatrix} \geqq 0 \quad \text{if } x_1 < x_2, t_1 < t_2. \tag{8.1}$$

It was shown in [10, p. 337] that an equivalent definition(4) is as follows: A frequency function $f(x)$ is an element of the class $\mathscr{P}_2$ if and only if it is of the form

$$f(x) = e^{-\phi(x)} \qquad (-\infty < x < \infty), \tag{8.2}$$

the function $\phi(x)$ having the following properties

1. $-\infty < \phi(x) \leqq +\infty$
2. There is an interval (α, β) $(-\infty \leqq \alpha < \beta \leqq +\infty)$ such that

$$\phi(x) = \begin{cases} \text{a continuous convex function in } \alpha < x < \beta \\ +\infty \text{ outside } [\alpha, \beta] \end{cases} \tag{8.3}$$

while

$$\phi(\alpha + 0) \leqq \phi(\alpha) \leqq +\infty, \ \phi(\beta - 0) \leqq \phi(\beta) \leqq +\infty, \tag{8.4}$$

in case that α, respectively β, are finite.

Examples of elements of the class $\mathscr{P}_2$ are all Pólya frequency functions, e.g. e^{-x^2}, $e^{-|x|}$and the one-sided exponential (5.9). From the properties (8.2), (8.3) and (8.4), it is clear that if $f(x) \in \mathscr{P}_2$, then $f(x)$ is everywhere continuous with the possible exception of at most two points, α and (or) β, where $f(x)$ may have discontinuities of the first kind.

We denote by $\bar{\mathscr{P}}_2$ the class $\mathscr{P}_2$ enlarged by including also the Dirac functions $f(x) = \delta_\xi(x)$. These are the frequency functions corresponding to the distribution functions

$$F(x) = (x - \xi)_+^0. \tag{8.5}$$

Theorem 8. *Let $\{f_n(x)\}$ be a sequence of frequency functions,*

---

(4) This definition describes functions $f(x)$ which are appropriately called "logarithmically concave". However, this (descriptive) definition of the class $\mathscr{P}_2$ is not as searching, in some ways, as the first (structural) definition (8.1). For instance, the important fact that *the class $\mathscr{P}_2$ is closed with respect to the operation of convolution* is not at all apparent from the descriptive definition, while it is easily derived from the property (8.1) (see [10, the proof of Lemma 5, 341–342]).

$$(8.6) \qquad f_n(x) \in \mathscr{P}_2 \qquad (n = 1, 2, \cdots).$$

Let $F_n(x)$ *denote the corresponding distribution functions and let us assume that*

$$(8.7) \qquad F_n(x) \to F(x),$$

where $F(x)$ *is a distribution function.*

Then $dF(x) = f(x)dx$, *where*

$$(8.8) \qquad f(x) \in \overline{\mathscr{P}}_2.$$

Moreover the relation (8.7) *may be differentiated giving*

$$(8.9) \qquad \lim_{n\to\infty} f_n(x) = f(x),$$

which holds uniformly for all x, *provided that we remove arbitrarily small neighborhoods of the one or two possible discontinuity points of* $f(x)$.

A proof of this theorem will be given in a future paper devoted to a study of the class $\mathscr{P}_2$. Here we only observe that Theorem 8 implies Theorem 7 immediately. Indeed, the alternative definition (8.2), (8.3), (8.4) and Theorem 3 shows that

$$M_n(x; x_0, x_1, \cdots, x_n) \in \mathscr{P}_2.$$

9. Examples and Applications. By Theorems 6 and 7 we know that the relation

$$(7.3) \qquad \lim_{n\to\infty} \prod_{\nu=0}^{n} \left(1 + \frac{x_{\nu,n}}{n+1} s\right) = \Psi(s)$$

implies either the relation

$$(5.11) \qquad \lim_{n\to\infty} M_n(x; x_{0,n}, \cdots, x_{n,n}) = \Lambda(x)$$

or

$$(5.12) \qquad \lim_{n\to\infty} M_n(x; x_{0,n}, \cdots, x_{n,n}) = 0, \quad (|x - \xi| \geqq \eta),$$

depending on whether the limit frequency $\Lambda(x)$ is a Pólya frequency function or the Dirac function $\delta_\xi(x)$. Here we wish to illustrate this result by a few concrete examples.

1. Let $\{x_\nu\}$ be an increasing sequence, $x_n \to \xi$, and let us choose the knots

$$x_{\nu,n} = x_\nu \qquad (\nu = 0, 1, \cdots, n).$$

As it is easy to show that (7.3) holds, with $\Psi(s) = e^{\xi s}$, we conclude by Theorem 7 that

$$\lim_{n\to\infty} M_n(x; x_0, x_1, \cdots, x_n) = \delta_\xi(x).$$

2. Let

$$x_{0,n} = x_{1,n} = \cdots = x_{n-1,n} = 0, \; x_{n,n} = n.$$

Now (7.3) holds, with $\Psi(s) = 1 + s$, the corresponding Pólya frequency function being

$$\Lambda(x) = \lambda(x) = \begin{matrix} e^{-x} & \text{if } x \geqq 0, \\ 0 & \text{if } x < 0. \end{matrix}$$

By Theorem 7 we conclude that

$$\lim_{n\to\infty} M_n(x; \overbrace{0, \cdots, 0}^{n}, n) = \lambda(x)$$

uniformly in x if we remove a neighborhood of $x = 0$, a result which is also easily verified directly from the explicit expression of the fundamental spline function (3.5).

3. The easiest way to obtain in (7.3) the limit

$$\Psi(s) = e^{-s^2/4},$$

which corresponds to

$$\Lambda(x) = \frac{1}{\sqrt{\pi}} e^{-x^2},$$

is to set $n+1=2k$ and to write

$$\prod_{\nu=0}^{n}\left(1+\frac{x_{\nu,n}}{n+1}s\right)=\left(1-\frac{s}{2\sqrt{k}}\right)^k\left(1+\frac{s}{2\sqrt{k}}\right)^k=\left(1-\frac{s^2}{4k}\right)^k.$$

Indeed, Theorem 7 shows that

$$\lim_{k\to\infty} M_{2k-1}(x;\overbrace{-k^{1/2},\cdots,-k^{1/2}}^{k},\overbrace{k^{1/2},\cdots,k^{1/2}}^{k})=\frac{1}{\sqrt{\pi}}e^{-x^2}, \tag{9.1}$$

uniformly for all real x.

4. The normal frequency function appears as the limit of spline functions also under different circumstances. Let $n=2k$ and let us consider the frequency function with equidistant knots

$$M_{2k}(x) = M_{2k}(x;-k,-k+1,\cdots,k-1,k).$$

Setting $f(x)=e^{itx}$ in (1.5) we easily obtain its well-known characteristics function which turns out to be

$$\int_{-\infty}^{\infty} M_{2k}(x)e^{itx}dx=\left(\frac{\sin t/2}{t/2}\right)^{2k}=\left(1-\frac{t^2}{24}+\cdots\right)^{2k}.$$

Appropriate changes of variables now produce the relation

$$\sqrt{\frac{k}{3}}\int_{-\infty}^{\infty} M_{2k}\left(x\sqrt{\frac{k}{3}}\right)e^{itx}dx=\left(1-\frac{t^2}{4.2k}+\cdots\right)^{2k}. \tag{9.2}$$

Notice that the integrand in (9.2) is a spline function of degree $2k-1$ having $2k+1$ knots at the points x obtained from

$$x\sqrt{\frac{k}{3}}=\nu,\quad (\nu=-k,\cdots,k).$$

Since its total area is unity, we must have

$$\sqrt{\frac{k}{3}}M_{2k}\left(x\sqrt{\frac{k}{3}};-k,\cdots,k\right)=M_{2k}\left(x;-k\sqrt{\frac{3}{k}},\cdots,k\sqrt{\frac{3}{k}}\right).$$

On the other hand, the right hand side of (9.2) converges to $\exp(-t^2/4)$. From the fundamental limit theorem for characteristic functions and Theorem 7 (which assures us that we can work with the frequency functions) we conclude that

$$\lim_{k\to\infty} M_{2k}(x;-k(3/k)^{1/2},(-k+1)(3/k)^{1/2},\cdots,k(3/k)^{1/2})=\frac{1}{\sqrt{\pi}}e^{-x^2}. \tag{9.3}$$

We now apply our theory (in reverse order), to conclude that

$$\lim_{k\to\infty}\prod_{\nu=-k}^{k}\left(1+\frac{\nu\sqrt{3}\,s}{\sqrt{k}\,(2k+1)}\right)=e^{-s^2/4}. \tag{9.4}$$

However, the relation (9.4) is easily shown to hold locally uniformly, and then we conclude again that (9.3) holds, this time without the appeal to characteristic functions.

The relation (9.3) has been noticed before in different notations, e.g. by F. Tricomi [16].

5. Let us now consider the Pólya frequency function

$$\Lambda(x)=\frac{1}{\pi\cosh x}$$

having the transform $1/\Psi(x)$, where $\Psi(s)=\cos\frac{\pi s}{2}$. From

$$\cos\frac{\pi s}{2}=\prod_{1}^{\infty}\left(1-\frac{s^2}{(2\nu-1)^2}\right)$$

we conclude that

$$\lim_{k\to\infty}\prod_{\nu=-k+1}^{k}\left(1+\frac{s}{2\nu-1}\right)=\cos\frac{\pi s}{2}.$$

Theorem 7 shows that if we set $n=2k-1$ and select the $2k$ knots

$$\frac{2k}{2\nu - 1} \quad (\nu = -k+1, \cdots, k)$$

we obtain, uniformly for all x, the relation

(9.5)

$$\lim_{k\to\infty} M_{2k-1}\left(x; -\frac{2k}{1}, -\frac{2k}{3}, \cdots, -\frac{2k}{2k-1}, \frac{2k}{2k-1}, \frac{2k}{2k-3}, \cdots, \frac{2k}{1}\right) = \frac{1}{\pi \cosh x}.$$

It is instructive to compare the formulae (9.3) and (9.5). The knots in (9.3) are equidistant. In (9.5), however, their density is higher near the origin in order to produce the function (9.5) which decays so much more slowly than the normal function.

6. As a last example, we consider the Pólya frequency function

$$\Lambda(x) = e^{-x-e^{-x}}$$

having the transform $1/\Psi(s) = s\Gamma(s)$. By Gauss' formula

$$\begin{aligned} 1/(s\Gamma(s)) &= \lim_{k\to\infty} \frac{(s+1)(s+2)\cdots(s+k)}{k!} k^{-s} \\ &= \lim_{k\to\infty} \left(1+\frac{s}{1}\right)\left(1+\frac{s}{2}\right)\cdots\left(1+\frac{s}{k}\right)\left(1-\frac{s\log k}{k}\right)^k = \Psi(s). \end{aligned}$$

On comparing with (7.3) we see that we have $2k$ knots, hence $n = 2k-1$, and that

$$\lim_{k\to\infty} M_{2k-1}\left(x; \overbrace{-2\log k, \cdots, -2\log k}^{k}, \frac{2k}{k}, \frac{2k}{k-1}, \cdots, \frac{2k}{1}\right) = e^{-x-e^{-x}}.$$

Notice the k-fold negative knot which $\to -\infty$ very slowly in order to produce a limit which tends to zero very fast as $x \to -\infty$. We also notice that in the ever increasing interval

$$-2\log k \leqq x \leqq 2$$

the spline function is a polynomial of degree $2k-2$, of the form

$$(x + 2\log k)^{k-1} \cdot \pi_{k-1}(x),$$

converging uniformly to $\exp(-x - e^{-x})$. Here $\pi_{k-1}(x)$ denotes an appropriate polynomial of degree $k-1$.

REFERENCES

1. T. Bonnesen and W. Fenchel, Theorie der Konvexen Körper, Berlin, 1934.

2. H. B. Curry, Review of the paper [9], Math, Tables and other Aids to Comp. **2** (1947), 167–169 and 211–213.

3. H. B. Curry and I. J. Schoenberg, On Pólya frequency functions IV: The spline functions and their limits. *Bull. Amer. Math. Soc.*, Abstract 380t, **53** (1947), 1114.

4. P. J. Davis, Interpolation and approximation, New York, 1963.

5. Ch. Hermite, Sur la formule d'interpolation de Lagrange, Oeuvres, vol. 3, 432–443.

6. I. I. Hirschman and D. V. Widder, The convolution transform, Princeton, 1955.

7. J. Korevaar and Ch. Loewner, Approximation on an arc by polynomials with restricted zeros, *Indagationes Math.*, **26** (1964), 121–128.

8. N. E. Nörlund, Vorlesungen über Differenzenrechnung, Berlin, 1923.

9. I. J. Schoenberg, Contributions to the problem of approximation of equidistant data by analytic functions, *Quart. Appl. Math.* **4** (1964) Part A, 45–99, Part B, 112–141.

10. I. J. Schoenberg, On Pólya frequency functions I: The totally positive functions and their Laplace transforms, *J. d'Analyse Math.*, **1** (1951), 331–374.

11. I. J. Schoenberg, On Pólya frequency functions II: Variation diminishing integral operators of the convolution type, *Acta Sci. Math. Szeged*, **12** (1950), 97–106.

12. I. J. Schoenberg and A. Whitney, On Pólya frequency functions III: The positivity of translation determinants with an application to the interpolation problem by spline curves, *Trans. A.M.S.*, **74** (1953), 246–259.

13. I. J. Schoenberg, On the zeros of the generating functions of multiply positive sequences and functions, *Annals of Math.* **62** (1955), 447–471.

14. I. J. Schoenberg, On interpolation by spline functions and its minimal properties, On Approximation Theory, International Series of Numerical Mathematics-ISNM, vol. 5 (1964), 109–129, Basel, Switzerland.

15. J. J. Schoenberg, On monosplines of least deviation and best quadrature formulae, *J. SIAM Numer Analysis, Ser. B.* vol. **2** (1965), 144–170.

16. F. Tricomi, Über die Summe mehrerer zufälliger Veränderlichen mit konstanten Verteilungsgesetzen, *Jahresber. der dentschen math. Ver.*, **42** (1933), 174–179.

PENNSYLVANIA STATE UNIVERSITY
AND
MATHEMATICS RESEARCH CENTER, US ARMY,
UNIVERSITY OF WISCONSIN

(Received May 24, 1965)

Reprinted from *J. d'Analyse Math.* **17** (1966), pp. 71–107

Pólya frequency functions and sequences

The highly important totally positive kernels of the form $K(x,y) = f(x-y)$ where x, y traverse the real line (or the integers), Schoenberg called Pólya frequency (PF) functions (sequences). The two prime examples of PF functions are the normal function

$$f(u) = e^{-\gamma u^2} \tag{1}$$

where γ is a positive parameter, and the (truncated) exponential function

$$f(u) = \begin{cases} e^{-\lambda u} & u \geq 0 \\ 0 & u < 0 \end{cases} \quad \text{or} \quad f(v) = \begin{cases} e^{\lambda v} & v \leq 0 \\ 0 & v > 0 \end{cases} \tag{2}$$

where λ is a free positive parameter. In a remarkable series of papers, Schoenberg (see his review paper (1953) [**48***]) set the basis of the theory of Pólya frequency functions, and established the fundamental representation theorems. Earlier works of Pólya, Laguerre and Schur aided these developments; their concern was the approximation of functions by polynomials with only real zeros.

Schoenberg (1951) [**43***] established the representation formula that an integrable function f is PF iff its Laplace transform exists in an open strip containing the imaginary axis and has the form

$$\phi(s) = \frac{ce^{+\gamma s^2 + \delta s}}{\prod_{i=1}^{\infty}(1 + a_i s)e^{-a_i s}}$$

$$c > 0, \gamma \geq 0; \; a_i \text{ and } \delta \text{ real}; \; 0 < \gamma + \sum_{i=1}^{\infty} a_i^2 < \infty.$$

Thus f is a convolution of the examples (1) and (2) and their translates.

The corresponding result for the case of a one-sided PF summable sequence

$$\cdots, a_{-2}, a_{-1}, a_0, a_1, a_2, \cdots \qquad a_k = 0 \text{ for } k < 0$$

(a_{n-m} is TP in the variables n and m) affirms that $a(z) = \sum_{i=0}^{\infty} a_i z^i$ is a meromorphic function of the form

$$a(z) = cz^k e^{\gamma z} \frac{\prod_{i=0}^{\infty}(1 + \alpha_i z)}{\prod_{i=0}^{\infty}(1 - \beta_i z)} \tag{3}$$

where k is an integer, $c > 0$, $\gamma \geq 0$, $\alpha_i \geq 0$, $\beta_i \geq 0$ $(i = 0, 1, 2, \cdots)$, and $\sum_{i=0}^{\infty}(\alpha_i + \beta_i) < \infty$, Schoenberg (1953) [**48***], Edrei (1953a).

The representation formula for the two-sided PF sequence was mostly done by Schoenberg. The final step provided by Edrei (1953b), relies on the powerful machinery of the Nevanlinna fundamental theorem of meromorphic functions. Edrei's efforts with the theory of PF sequences seem to have led to an extensive development of deep function-theoretic analysis concerning deficiency values of meromorphic functions (Edrei and Fuchs (1962)).

Mairhuber, Schoenberg, and Williamson (1959) contributed a notable paper concerning PF functions on the circle (cyclic PF functions). A representation formula in this case is not known. An enticing example of a CPF function is the de la Vallée Poussin kernel (Pólya and Schoenberg (1958) [**58***]). For other geometric perspectives on this subject, see Schoenberg and Whitney (1951).

Hirschman and Widder (1955) investigated inversion and representation theorems for convolution operators T determined by a function f where the transformation T acts by the formula

$$v(x) = (Tu)(x) = \int_{-\infty}^{\infty} f(x-y)u(y)dy.$$

They found that provided f is a PF function inversion formulas involving polynomial differential operators are available. This class of convolution linear operators includes the classical Laplace, Stieltjes, and Weierstrass transforms.

Two other general developments relate to the theory of Pólya frequency functions. Hirschman (1961, 1962) has described a class of variation-diminishing transformations and representation formulas for certain kernels determined by Bessel functions and kernels based on ultra-spherical polynomials. His results can be interpreted as a nonhomogeneous extension of Schoenberg's theory generalizing the translation operation. See also Cholewinski (1965) and Haimo (1965).

The class of PF_r functions (so $f(x-y)$ is TP_r) is richer and more complex than the class of PF_∞ functions. The class $r = 2$ is especially important and has multifaceted applications to statistical theory (Karlin and Rubin (1956)), to reliability theory (Barlow and Proschan (1975)), in game theory, mathematical economics and stochastic inventory control (Arrow et al. (1958)). Every PF_2 function is of the form $f(x) = e^{-\psi(x)}$, where $\psi(x)$ is convex (so f is log concave). There exists no such simple characterization for PF_3 functions. PF_r functions possess certain smoothness properties; e.g., if f is strictly PF_r, then f is automatically of class C^{2r-2} (see Chapter 4 in Karlin (1968)).

The theory and applications of *multivariate* log concave functions has been a focus of intense recent activity. This class includes all multivariate normal densities, the central Wishart density, multivariate exponential, the uniform density for any convex set, multivariate B-splines and marginals and convolutions of these (e.g., Tong (1980), Karlin et al. (1986)).

The log concave (PF_2) endowment and related properties (e.g., unimodality, quasi-concavity) are basic along with their relations to (multivariate) Schur functions and totally positive kernels. Applications abound in statistical decision theory, statistical mechanics, stochastic processes, multivariate approximation theory and stochastic ordering inequalities (e.g., Brascamp and Lieb (1976); Rinott (1976); Simon (1979); Tong (1980)). These relate also to multivariate majorization theory and its applications (e.g., Fan and Lorentz (1954); Prekopa (1971); Rinott (1973); Marshall and Olkin (1979); Karlin and Rinott (1983)).

The connections (limit theorems) of basic spline densities and PF densities was noted early in Schoenberg (1946) [**31***], see also Curry and Schoenberg (1966) [**86***]. Multivariate versions are described in Dahmen and Micchelli (1981) and Karlin et al. (1986).

In concluding this essay, I hasten to underscore the plethora of log concave sequences that occur in contexts of combinatorial mathematics. These include: Stirling numbers of the first and second kind, Bell numbers, counts of inversions for random permutations of size n, multiset generalized binomial coefficients (Heine coefficients), ascent counts, among others. There is no doubt that the attributes attendant to Pólya frequency functions and sequences will aid with characterizations and interpretations of formulas and identities in enumerative mathematics.

References

Arrow, K.J., S. Karlin, and H.E. Scarf (1958). *Studies in the Mathematical Theory of Inventory and Production*, Stanford University Press, Stanford, California.

Barlow, R.E. and F. Proschan (1975). *Statistical Theory of Reliability and Life Testing*, Holt, Rinehart & Winston, New York.

Brascamp, H. and E.H. Lieb (1976). "On extensions of the Brunn-Minkowski and Prékopa Leindler theorems, including inequalities for log concave functions, and with an application to the diffusion equation," *J. Funct. Anal.* **22**, 366–389.

Cholewinski, F. (1965). "A Hankel convolution complex inversion theory," *Mem. Amer. Math. Soc.* **58**.

Curry, H.B. and I.J. Schoenberg (1966). "On Pólya frequency functions. IV. The fundamental spline functions and their limits," *J. Analyse Math.* **18**, 71–107.

Dahmen, W. and C. Micchelli (1981). "On limits of multivariate B-splines," *J. Analyse Math.* **39**, 256–278.

Edrei, A. (1953a). "On the generating function of a doubly infinite totally positive sequence," *Trans. Amer. Math. Soc.* **74**, 367–83.

Edrei, A. (1953b). "Proof of a conjecture of Schoenberg on the generating function of a totally positive sequence," *Canad. J. Math.* **5**, 86–94.

Edrei, A. and W. Fuchs (1962). "Bounds for the number of deficient values of certain classes of meromorphic functions," *Proc. London Math. Soc. (3)* **12**, 315–344.

Fan, K. and G.G. Lorentz (1954). "An integral inequality," *Amer. Math. Monthly* **61**, 626–631.

Haimo, Deborah Tepper (1965). "Integral equations associated with Hankel convolutions," *Trans. Amer. Math. Soc.* **116**, 330–75.

Hirschman, I.I. (1961). "Variation diminishing transformations and Sturm-Liouville systems," *Comment. Math. Helv.* **36**, 214–33.

Hirschman, I.I. (1961/1962). "Variation diminishing transformations and orthogonal polynomials," *J. Analyse Math.* **9**, 177–93.

Hirschman, I.I. and D.V. Widder (1955). *The Convolution Transform*, Princeton University Press, Princeton, New Jersey.

Karlin, S. (1968). *Total Positivity*, Volume I, Stanford University Press, Stanford, California.

Karlin, S. and H. Rubin (1956). "The theory of decision procedures for distributions with monotone likelihood ratio," *Ann. Math. Statist.* **27**, 272–99.

Karlin, S. and Y. Rinott (1983). "Comparison of measures, multivariate majorizations and applications to statistics," *Studies in Econometrics, Time Series and Multivariate Statistics* (S. Karlin, T. Amemiya, and L.A. Goodman, Eds.), Academic Press, New York.

Karlin, S., C.A. Micchelli, and Y. Rinott (1986). "Multivariate Splines: A Probabilistic Perspective," *J. Multivariate Anal.* **20**, 69–90.

Marshall and Olkin (1979). *Inequalities: Theory of Majorization and Its Applications*, Academic Press, New York.

Mairhuber, J.C., I.J. Schoenberg, and R.E. Williamson (1959). "On variation diminishing transformations of the circle," *Rend. Circ. Mat. Palermo (2)* **8**, 241–70. [**63**]

Pólya, G. and I.J. Schoenberg (1958). "Remarks on de la Vallée Poussin means and convex conformal maps of the circle," *Pacific J. Math.* **8**, 295–334. [**58***]

Prekopa, A. (1971). "On logarithmic concave measures and functions," *Acta Sci. Math. (Szeged)* **34**, 335–343.

Rinott, Y. (1973). "Multivariate majorization and rearrangement inequalities with some applications to probability and statistics," *Israel J. Math.* **15**, 60–77.

Rinott, Y. (1976). "On convexity of measures," *Ann. Probab.* **4**, 1020–1026.

Schoenberg, I.J. (1946). "Contributions to the problems of approximation of equidistant data by analytic functions," Part A in *Quant. Appl. Math.* **4**, 45–99. Part B in *ibid* **4**, 112–41. [**31***], [**32***]

Schoenberg, I.J. (1951). "On Pólya frequency functions, I: The totally positive functions and their Laplace transforms," *J. Analyse Math.* **1**, 331–74. [**43***]

Schoenberg, I.J. (1953). "On smoothing operations and their generating functions," *Bull. Amer. Math. Soc.* **59**, 199–230. [**48***]

Schoenberg, I.J. (1954). "On multiply positive sequences and functions," *Bull. Amer. Math. Soc.* **60**, 160. cf. [**52**]

Schoenberg, I.J. and A. Whitney (1951). "A theorem on polygons in n dimensions with applications to variation-diminishing and cyclic variation-diminishing linear transforms," *Compositio Math.* **9**, 141–60. [**42**]

Simon, B. (1979). *Functional Integration and Quantum Physics*, Academic Press, New York.

Tong, Y.L. (1980). *Probability Inequalities in Multivariate Distributions*, Academic Press, New York.

Samuel Karlin
Stanford University

THE VARIATIONAL APPROACH TO SPLINES

SPLINE INTERPOLATION AND BEST QUADRATURE FORMULAE

BY I. J. SCHOENBERG

Communicated by Felix Browder, October 9, 1963

1. **The spline interpolation formula.** A spline function $S(x)$, of degree $k(\geqq 0)$, having the knots

$$x_0 < x_1 < \cdots < x_n, \tag{1}$$

is by definition a function of the class C^{k-1} which reduces to a polynomial of degree not exceeding k in each of the $n+2$ intervals in which the points (1) divide the real axis. The function $S(x)$ is seen to depend linearly on $n+k+1$ parameters. In [5, Theorem 2, p. 258] are given the precise conditions under which we can interpolate uniquely by $S(x)$ arbitrarily given ordinates at $n+k+1$ points on the real axis.

For the remainder of this note we set $k=2m-1$ $(1\leqq m\leqq n)$ and single out from this family of spline functions the

CLASS Σ_m: *The class of spline functions $S(x)$ of degree $2m-1$, knots* (1), *and the additional property that $S(x)$ reduces to polynomials of degree not exceeding $m-1$ in each of the ranges $(-\infty, x_0)$ and $(x_n, +\infty)$.*

The restriction that $m\leqq n$ is essential, otherwise Σ_m reduces to π_{m-1} (here and below π_k denotes a generic polynomial of degree $\leqq k$, as well as their class). In a paper [1] soon to appear C. de Boor observes that [5, Theorem 2] implies the following interesting

THEOREM 1 (C. DE BOOR). *Given m $(1\leqq m\leqq n)$, the points* (1) *and also arbitrary reals y_i $(i=0,\cdots,n)$, then there is a unique $S(x)$ such that*

$$S(x_i) = y_i \qquad (i = 0, \cdots, n). \tag{2}$$

Let us now consider this interpolating spline function $S(x)$ in a given finite interval $a\leqq x\leqq b$ containing the points (1). Its particular interest is due to the following

THEOREM 2 (C. DE BOOR). *Let $f(x)\in C^{m-1}[a,b]$, having an absolutely continuous $f^{(m-1)}(x)$, and be such that*

$$f(x_i) = y_i \qquad (i = 0, \cdots, n).$$

If $S(x)$ denotes the interpolating spline function of Theorem 1 *then*

$$\int_a^b (f^{(m)}(x))^2 dx \geqq \int_a^b (S^{(m)}(x))^2 dx$$

with equality if and only if $f(x)=S(x)$.[1]

In other words: The interpolating $S(x)\in\Sigma_m$ minimizes the integral $\int_a^b(y^{(m)})^2dx$ among *all* interpolating functions and in this sense $S(x)$ is the best interpolating function for the order m. For $m=1$ $S(x)$ reduces to the linear interpolation between consecutive points while $S(x)=y_0$ if $x\leq x_0$ and $S(x)=y_n$ if $x\geq x_n$. For $m=2$ these results are due to Holladay [2]. The periodic analogue for $m=2$ was found by Walsh, Ahlberg and Nilson [8] and for an arbitrary m see [9]. The results of [1] and [9] were also found independently, but later (September 1963), by the author (see [7]). Holladay and de Boor consider only the case when $a=x_0$, $x_n=b$.

We may express the interpolating $S(x)$ of Theorem 1 in the form

$$S(x) = \sum_0^n y_\nu L_\nu(x),$$

where $L_\nu(x)$ are the fundamental functions, i.e. $L_i(x_j)=\delta_{ij}$. The function $f(x)$ being defined in $[a, b]$ we consider the interpolation formula (I. F.)

$$f(x) = \sum_0^n f(x_\nu)L_\nu(x) + R(x;f) \tag{3}$$

which we call the spline I. F. of order m $(m\leq n)$. Observe that $R(x;\pi_{m-1})=0$.

2. **Best quadrature formulae.** By integrating (3) over the range $[a, b]$ and setting

$$A_\nu = \int_a^b L_\nu(x)dx \tag{4}$$

we obtain the quadrature formula (Q. F.)

$$\int_a^b f(x)dx = \sum_0^n A_\nu f(x_\nu) + R_A(f) \tag{5}$$

which is evidently also exact if $f\in\pi_{m-1}$, i.e. $R_A(\pi_{m-1})=0$.

[1] As a matter of fact the interpolating spline $S(x)$ of Theorem 1 enjoys yet another fundamental minimal property to be found in [7, Theorem 1]. It may be stated as follows: *Let $f(x)$ be defined as in Theorem 2 and let $S(x)$ be the unique element of Σ_m interpolating $f(x)$ at the points* (1). *Then $S(x)$ is also uniquely characterized, up to an additive π_{m-1} by the condition:* (*) *To minimize $\int_a^b(S^{(m)}(x)-f^{(m)}(x))^2dx$ for $S(x)\in\Sigma_m$.* In [7] Theorems 1 and 2 are derived from the minimal property (*) of $S(x)$. The property (*) was discovered by Walsh, Ahlberg and Nilson [8] for the periodic case of cubic splines ($m=2$) and apparently now generalized by them to all m in the periodic case (see [9]).

Concerning quadrature formulae A. Sard [4] proposed in 1949 the following very reasonable scheme: Let $1 \leqq m \leqq n$ and let

$$\int_a^b f(x)dx = \sum_0^n B_\nu f(x_\nu) + R_B(f) \tag{6}$$

be a Q. F. which is exact if $f(x) \in \pi_{m-1}$. This requirement is equivalent to the relations

$$\frac{b^{r+1} - a^{r+1}}{r+1} = \sum_{\nu=0}^n B_\nu x_\nu^r \qquad (r = 0, \cdots, m-1), \tag{7}$$

which still leave $n-m+1$ free parameters. These are disposed of as follows: By Peano's theorem [3] we can write

$$R_B(f) = \frac{1}{(m-1)!}\int_a^b K(x) f^{(m)}(x)dx \tag{8}$$

and we now determine the B_ν by the requirement that

$$\int_a^b (K(x))^2 dx = \text{minimum}, \tag{9}$$

with the side-conditions (7). This method leads to a Q. F. (6) which Sard very appropriately calls the Best Q. F. for the given x_ν, given $[a, b]$ and given order m.

Sard computes the coefficients B_ν of (6), for equidistant x_ν at unit steps, $(a, b) = (x_0, x_n)$ and $m = 1, 2, 3, 4$ and $n = 1, \cdots, 6$ $(m \leqq n)$. Holladay [2] computes the coefficients A_ν of (5), for $m = 2$, equidistant x_ν, $(a, b) = (x_0, x_n)$ and $n = 2, 3, \cdots, 10$. On comparing their tables it is seen that Holladay's formulae are identical with Sard's for $m = 2$.[2]

That these are no mere numerical accidents is shown by

THEOREM 3. *The Q. F.* (5), *obtained by integrating the spline I. F.* (3), *is identical with Sard's Best Q. F.* (6).

The key to the proof is the following remark: The I. F. (3) is not only exact if $f \in \pi_{m-1}$ but also if $f \in \Sigma_m$, because of the unicity in Theorem 1. Therefore also *the Q. F.* (5) *is exact if* $f \in \Sigma_m$. Conversely, this property uniquely characterizes (5), for if $f(x) = L_\nu(x)$ and $R_A(L_\nu) = 0$, then (5) implies (4).

The identity of (5) with (6) will therefore follow as soon as we have established the following

[2] The author is here performing the meritorious service of decreasing the number of known Q. F. by showing that two known sets are identical.

LEMMA. *Sard's Q. F. (6) is exact if $f\in\Sigma_m$.*

PROOF. We assume the relations (7) and then by Peano's theorem

$$K(x) = \frac{1}{m}(b-x)_+^m - \frac{1}{m}(a-x)_+^m - \sum_0^n B_\nu(x_\nu - x)_+^{m-1} \qquad (-\infty < x < \infty),$$

where we use the function $y_+ = (y+|y|)/2$. We now solve the problem (9), (7) by Lagrange's multiplier method and consider the expression

$$E = \frac{1}{2}\int_a^b (K(x))^2 dx + \sum_{r=0}^{m-1} \lambda_r \left(\sum_{\nu=0}^n B_\nu x_\nu^r - (b^{r+1} - a^{r+1})(r+1)^{-1} \right).$$

Its partial derivatives $\partial E/\partial B_i$ furnish the $n+1$ relations

$$\int_a^b \left\{ -\frac{1}{m}(b-x)^m + \sum_0^n B_\nu (x_\nu - x)_+^{m-1} \right\} (x_i - x)_+^{m-1} dx \tag{10}$$
$$+ \sum_{r=0}^{m-1} \lambda_r x_i^r = 0 \qquad (i = 0, \cdots, n),$$

which are satisfied by the coefficients B_ν of (6) together with appropriate constants λ_r.

As a preliminary step towards the proof of the Lemma let us evaluate the integral appearing in (10) and show that (10) may also be written in the equivalent form

$$-\frac{1}{2m}(x_i - a)^{2m} + \sum_{\nu=0}^n B_\nu (x_i - x_\nu)_+^{2m-1} + \sum_{r=0}^{m-1} \lambda_r' x_i^r = 0 \tag{11}$$
$$(i = 0, \cdots, n),$$

where the λ_r' are proportional to the λ_r. To this end we observe that (7) are equivalent to the identity

$$\frac{1}{m}(a-x)^m - \frac{1}{m}(b-x)^m + \sum_0^n B_\nu (x_\nu - x)^{m-1} = 0.$$

Subtracting this vanishing expression within the curly brackets in (10) and using the identity $y_+^{m-1} - y^{m-1} = (-1)^m(-y)_+^{m-1}$, we may write (10) as

$$\int_a^b \left\{ -\frac{1}{m}(a-x)^m + (-1)^m \sum_0^n B_\nu (x - x_\nu)_+^{m-1} \right\} (x_i - x)_+^{m-1} dx \tag{10'}$$
$$+ \sum_{r=0}^{m-1} \lambda_r x_i^r = 0.$$

Now the integrals can be evaluated explicitly:

$$\int_a^b (a-x)^m (x_i - x)_+^{m-1} dx = (-1)^m \gamma_0 \cdot (x_i - a)^{2m},$$

$$\int_a^b (x-x_\nu)^{m-1} (x_i - x)_+^{m-1} dx = \gamma_1 \cdot (x_i - x_\nu)_+^{2m-1}$$

where

$$\gamma_0 = \int_0^1 t^m (i-t)^{m-1} dt, \qquad \gamma_1 = \int_0^1 t^{m-1}(1-t)^{m-1} dt.$$

Observing that $\gamma_1 = 2\gamma_0$ and substituting the values just found into (10′), we see that it reduces to (11) where $\lambda_r' = (-1)^m \lambda_r/(2\gamma_0)$.

A proof of the Lemma now follows easily. Indeed, the most general $S \in \Sigma_m$ has the form

(12) $S(x) = \sum_0^n C_i (x_i - x)_+^{2m-1} + P_{m-1}(x)$ (C_i constants, $P_{m-1} \in \pi_{m-1}$)

in which the condition that $S(x) \in \pi_{m-1}$ in $(-\infty, x)$ requires the relations

(13) $$\sum_{i=0}^n C_i x_i^r = 0 \qquad (r = 0, \cdots, m-1).$$

But then

$$R_B(S) = \sum C_i R_B((x_i - x)_+^{2m-1}) + R_B(P_{m-1})$$

or, since the last term vanishes,

$$R_B(S) = \sum_0^n C_i \frac{1}{2m} (x_i - a)^{2m} - \sum_{i=0}^n C_i \sum_{\nu=0}^n B_\nu (x_i - x_\nu)_+^{2m-1}.$$

But the right-hand side expression vanishes: Indeed, we obtain it by multiplying the left side of (11) by $-C_i$ and summing over i and then take into account the relations (13). Thus $R_B(S) = 0$ and the Lemma is established.

3. **Concluding remarks.** Let me point out again, as I already did in [6], that the connection between the problems of interpolation and of mechanical quadratures becomes closer and more illuminating if we interpolate not only by polynomials but by the wider class of splien functions. The Q. F. (6) was derived by a method entirely foreign

to interpolation and yet, as seen, it arises also from the I. F. (3) by integration. There are many other similar examples.

REFERENCES

1. C. de Boor, *Best approximation properties of spline functions of odd degree*, J. Math. Mech. **12** (1963), 747–749.

2. J. C. Holladay, *Smoothest curve approximation*, Math. Tables Aids Comput. **11** (1957), 233–243.

3. G. Peano, *Residuo in formulas de quadratura*, Mathesis **34** (1914), 1–10.

4. A. Sard, *Best approximate integration formulas; best approximation formulas*, Amer. J. Math. **71** (1949), 80–91.

5. I. J. Schoenberg and Anne Whitney, *On Polya frequency functions*. III. *The positivity of translation determinants with an application to the interpolation by spline curves*, Trans. Amer. Math. Soc. **74** (1953), 246–259.

6. I. J. Schoenberg, *Spline functions, convex curves and mechanical quadratures*, Bull. Amer. Math. Soc. **64** (1958), pp. 352–357.

7. ———, *On interpolation by spline functions and its minimal properties*, Proceedings of the Conference on Approximation, Oberwolfach, Germany, August 4–10, 1963. (to appear)

8. J. L. Walsh, J. H. Ahlberg and E. N. Nilson, *Best approximation properties of the spline fit*, J. Math. Mech. **11** (1962), 225–234.

9. ———, *Best approximation and convergence properties of higher-order spline fits*, Abstract 63t-103, Notices Amer. Math. Soc. **10** (1963), 202.

UNIVERSITY OF PENNSYLVANIA AND
INSTITUTE FOR ADVANCED STUDY

Reprinted from *Bull. Amer. Math. Soc.* **70** (1964), pp. 143–148

KONINKL. NEDERL. AKADEMIE VAN WETENSCHAPPEN – AMSTERDAM
Reprinted from Proceedings, Series A, 67, No. 2 and Indag. Math., 26, No. 2, 1964

MATHEMATICS

ON BEST APPROXIMATIONS OF LINEAR OPERATORS

BY

I. J. SCHOENBERG

(Communicated by Prof. N. G. DE BRUIJN at the meeting of November 30, 1963)

1. *Introduction*

We recently gave in the note [8] some applications of the so-called spline interpolation formula to real variable theory. The purpose of the present note is to point out the fundamental role of spline interpolation in the numerical analysis of functions of one real variable. The role which spline interpolation is here called upon to play is based on the ideas of A. SARD [4, 5] and is inherently due to the two familiar requirements for approximation formulae of numerical analysis: 1) That they be linear, 2) That they be exact for polynomials of a certain specified degree.

In the course of the paper it will also be shown (Corollary 1) that spline interpolation is not a new mathematical object but is in fact identical with Sard's best special interpolation formula. We therefore have the interesting situation wherein an object is being rediscovered from an entirely new point of view and with a new set of properties. Herewith we unify its split personality and hope that it will flourish in the glory of its many virtues.

Our discussion must commence by recalling the known but not yet "well known" definitions and properties of spline interpolation [1]).

2. *Spline interpolation*

Let Δ denote a set of $n+1$ reals x_ν, $x_0 < x_1 < \ldots < x_n$. By a spline function

$$S(x) = S_{2m}(x|\Delta), \qquad (m \geqslant 1),$$

of degree $2m-1$ (or equivalently of order $2m$) having the knots x_ν, we mean a function such that $S(x) \in C^{2m-2}(-\infty, \infty)$ with the property that [2]) $S(x) \in \pi_{2m-1}$ in each of the intervals $(-\infty, x_0), (x_0, x_1), \ldots, (x_n, \infty)$.

---

[1]) The results of Sections 2 and 3 were discovered in stages by different people: HOLLADAY [2] gave the Interpolation Theorem and the Second Minimal Property for the special case when $m = 2$ (cubic spline functions). The First Minimal Property was first found by WALSH–AHLBERG–NILSON [9] again for cubic splines and periodic data. The three theorems as stated here (except for the "extrapolation feature" which allows the interval $[x_0, x_n]$ to be properly contained in $I = [a, b]$) were first published by DE BOOR [1]. As stated here these theorems were also found independently by the author in [6].

[2]) Here and below we denote by π_k the class of polynomials of degree k while $\pi_k(x)$ denotes an unspecified element of this class.

Moreover, let

(1) $$m < n+1$$

and let us denote by $\sum = \sum_{2m}(\Delta)$ the subclass of those $S(x)$ satisfying the additional requirement

(2) $$S(x) \in \pi_{m-1} \text{ in } (-\infty, x_0) \text{ and also in } (x_n, \infty).$$

The first property of the family of spline functions $\sum_{2m}(\Delta)$ is the following

Interpolation theorem. *If $y_0, \ldots, y_n$ are reals then there is a unique $S(x) \in \sum_{2m}(\Delta)$ such that*

$$S(x_\nu) = y_\nu, \quad (\nu = 0, \ldots, n).$$

If $m=1$ then $S(x)$ is the continuous broken linear function which interpolates the y_ν linearly between consecutive x_ν while $S(x)=y_0$ if $x<x_0$ and $S(x)=y_n$ if $x>x_n$. In the other extreme case when $m=n+1$ it is easy to show that $\sum_{2m}(\Delta)=\pi_{m-1}=\pi_n$ so that the interpolating $S(x)$ is identical with the familiar interpolating polynomial $\pi_n(x)$. If we should choose $m>n+1$ then again $\sum_{2m}(\Delta)=\pi_{m-1}$ and from $m-1>n$ we conclude that the interpolating $S(x)$ is no longer unique. To avoid this trivial and undesirable case we shall assume (1) to hold.

Let $L_\nu(x)$ be the fundamental functions of our interpolation process, i.e.

(3) $$L_\nu(x) \in \sum_{2m}(\Delta), \quad L_\nu(x_i) = \begin{matrix} 1 \text{ if } i=\nu, \\ 0 \text{ if } i \neq \nu. \end{matrix}$$

We also choose a fixed finite interval $I=[a,b]$ containing our points x_ν and consider a function $f(x)$ defined in I. We interpolate to $f(x)$ in the points x_ν and may write the result in the form

(4) $$f(x) = \sum_0^n f(x_\nu)\, L_\nu(x) + R(x;f) = S(x) + R(x;f), \qquad (x \in I),$$

which will be referred to below as the *spline interpolation formula* (S.I.F.).

3. *The minimal properties of spline interpolation*

We assume that the function $f(x)$ to be interpolated has an mth derivative. More precisely we consider the function class
$F_m[a,b] = \{f(x) | f \in C^{m-1}[a,b],\ f^{(m-1)} \text{ absolutely continuous},\ f^{(m)} \in L^2(a,b)\}$.
If we assume that $f(x) \in F_m[a,b]$ then the interpolating spline function $S(x)$, defined by (4), enjoys the following properties:

First minimal property. *If $s(x) \in \sum_{2m}(\Delta)$ then*

(5) $$\int_I (s^{(m)}(x) - f^{(m)}(x))^2\, dx \geq \int_I (S^{(m)}(x) - f^{(m)}(x))^2\, dx,$$

with equality if and only if $s(x) = S(x) + \pi_{m-1}(x)$.

Second minimal property.

(6)
$$\int_I (f^{(m)}(x))^2\,dx > \int_I (S^{(m)}(x))^2\,dx,$$

with equality if and only if $f(x)=S(x)$ *in* $I=[a, b]$.

For $m=n+1$, when $\sum_{2m}(\Delta)=\pi_n$, both properties are trivially satisfied. For $m<n+1$ the first property shows that the construction of $S(x)$ is equivalent with the solution of a least squares problem. The second property shows that $S(x)$ is the solution of an important minimum problem.

4. *Linear approximations of linear operators*

Let $\mathscr{L}f$ denote a linear operator defined for $f(x)\in C^{m-1}[a, b]$. Following Sard, we shall from the outset assume that the operator has the form

(7)
$$\mathscr{L}f = \sum_{\nu=0}^{m-1} \int_I f^{(\nu)}(x)\,d\mu_\nu(x),$$

where $\mu_\nu(x)$ are functions of bounded variation. This class contains as special cases the operators

(8)
$$\int_I f(u)\,du,\ f(x),\ f'(x),\ \ldots,\ f^{(m-1)}(x), \qquad (x \text{ fixed, } x\in I),$$

and many others.

The problem of linear approximation to $\mathscr{L}f$ is as follows: Given $\Delta=\{x_0, \ldots, x_n\}$ we wish to determine the constants B_ν such that $\sum B_\nu f(x_\nu)$ should approximate $\mathscr{L}f$ in a sense to be made more precise. In fact we write

(9)
$$\mathscr{L}f = \sum_0^n B_\nu f(x_\nu)+R(f)$$

and impose on this formula the following familiar condition:

(10)
$$R(f)=0 \text{ if } f\in\pi_{m-1},$$

which is equivalent to the relations

(11)
$$\mathscr{L}x^r = \sum_{\nu=0}^n B_\nu x_\nu{}^r, \qquad (r=0, \ldots, m-1).$$

If $m=n+1$ then the B_ν are uniquely defined by (11) and there is no further problem. This is precisely the situation for most of the formulae of interpolation, approximate integration or differentiation known from the elements of numerical analysis. We shall from now on assume the more interesting assumption

(12)
$$m\leqslant n,$$

in which case the relations (11) still leave $n-m+1$ free parameters. Following SARD's general procedure [4, § 5] we shall dispose of these free parameters as follows: By Peano's theorem we may write for the remainder of (9)

$$R(f) = \frac{1}{(m-1)!} \int_I K(t)\, f^{(m)}(t)\, dt \tag{13}$$

where [3])

$$K(t) = R_x (x-t)_+^{m-1}. \tag{14}$$

We now dispose of the $n-m+1$ free parameters by requiring that

$$J = \int_I (K(t))^2\, dt = \text{minimum} \tag{15}$$

under the side-conditions (11). The resulting formula (9) will be referred to as the best approximation of $\mathscr{L}f$ in I, for the set of knots x_ν and the order of differentiation m.

SARD, and MEYERS and SARD, apply this procedure to the first three operators (8). In particular, for $\mathscr{L}f = f(x)$ they obtain the so-called *special formulae* (See [5, page 91] and also [3]).

The main result of the present paper is the following

Theorem 1. *The best approximation* (9) *to a linear operator* $\mathscr{L}f$ *of the form* (7) *is obtained by operating with* $\mathscr{L}$ *on the spline function*

$$S(x) = \sum_0^n f(x_\nu)\, L_\nu(x)$$

which interpolates $f(x)$ *at the points* x_ν, *i.e.*

$$\sum B_\nu f(x_\nu) = \mathscr{L} \sum f(x_\nu)\, L_\nu(x) = \sum f(x_\nu)\, \mathscr{L} L_\nu(x)$$

or

$$B_\nu = \mathscr{L} L_\nu(x), \qquad (\nu = 0, \ldots, n). \tag{16}$$

For the special case of the interpolation operator $\mathscr{L}f = f(x)$ (x fixed, $x \in I$) Theorem 1 yields the

Corollary 1. *Sard's best special interpolation formula* (9), *for* $\mathscr{L}f = f(x)$, *is identical with the spline interpolation formula* (4), *i.e.*

$$B_\nu = L_\nu(x), \qquad (\nu = 0, \ldots, n;\ a \leqslant x \leqslant b).$$

The identification of the S. I. F. with SARD's best special interpolation formula (B. S. I. F.) uncovers a new characteristic minimal property of spline interpolation, namely the property (15). It also makes available SARD's valuable tables giving the explicit expressions for the $L_\nu(x)$ for equidistant x_ν [5, pp. 94–100].

From the other point of view we now know that Sard's B. S. I. F. has the following new properties:

1. The B. S. I. function is uniquely defined by the property of being

[3]) We use here the truncated function $x_+ = x$ if $x \geqslant 0$ and $x_+ = 0$ if $x < 0$.

an interpolating spline function of degree $2m-1$, knots x_ν and subject to the condition (2) [4]).

2. It produces a function $S(x)$ enjoying the two minimal properties (5) and (6).

However, the special role of spline interpolation is best shown by Theorem 1. Indeed, Theorem 1 shows that best approximation formulae are effectively constructed as soon as we are in possession of the interpolating spline function. Thus for

$$\mathscr{L}f = \int_I f\,dx$$

the best quadrature formula (9) is obtained by integrating the S. I. F. (4) obtaining

$$B_\nu = \int_I L_\nu(x)\,dx, \tag{17}$$

(See [7] where this special result was first obtained). Likewise if $\mathscr{L}f = f'(x)$, then the best approximate differentiation formula (9) is obtained by setting

$$B_\nu = L_\nu'(x). \tag{18}$$

For $m=n+1$, a case excluded by (12) but for which Theorem 1 is still valid, we obtain Newton's procedure of performing $\mathscr{L}$ on the interpolating polynomial. Theorem 1 shows that Newton's idea of the fundamental role of interpolation generalizes to the cases when (12) holds, by means of spline interpolation.

5. *Proof of Theorem* 1

We begin by attacking the minimum problem (15), (14), (11) and showing that it has a solution which is unique. By (14) and (9) we find

$$K(t) = \mathscr{L}(x-t)_+^{m-1} - \sum_\nu B_\nu (x_\nu - t)_+^{m-1}.$$

Applying Lagrange's multiplier method to (15) and equating to zero the partial derivatives $\partial E/\partial B_i$ of

$$E - \tfrac{1}{2}J + \sum_r \lambda_r \left(\sum_\nu B_\nu x_\nu^r - \mathscr{L}x^r\right)$$

we obtain

$$\int_I \{-\mathscr{L}(x-t)_+^{m-1} + \sum B_\nu (x_\nu - t)_+^{m-1}\}\,(x_i - t)_+^{m-1}\,dt + \sum_r \lambda_r x_i^r = 0, \quad (i=0,\ldots,n). \tag{19}$$

These are $n+1$ linear equations which together with the m relations (11)

---

[4]) Concerning the case when $x_\nu = \nu (\nu = 0, \ldots, n)$ SARD [5, p. 93] observes that the fundamental functions of his B.S.I.F. are broken polynomials in x of degree $2m$–1 with transitions at $x = 0, 1, \ldots, n$ and having $2m$–2 continuous derivatives for all real x. However, he does not mention the vital property (2), nor the fact that such a description might characterize these functions uniquely.

are to determine the unknowns B_ν, λ_r. We also know that (19) are necessary conditions for a minimum in the following sense: If the B_ν are to minimize J then they must satisfy (19) and (11) together with appropriate λ_r.

Let us show that *the system of simultaneous equations* (19) *and* (11) *has a unique solution.* To see this we observe that the relations (11) imply the identity in t

$$-\mathscr{L}(x-t)^{m-1}+\sum_\nu B_\nu(x_\nu-t)^{m-1}=0.$$

On subtracting the vanishing left side of this relation within the curly brackets of (19) and keeping in mind the identiy

$$y_+^{m-1}-y^{m-1}=(-1)^m(-y)_+^{m-1}$$

we see that we may rewrite (19) as

$$(20)\quad (-1)^m\int_I\{-\mathscr{L}(t-x)_+^{m-1}+\sum_\nu B_\nu(t-x_\nu)_+^{m-1}\}(x_i-t)_+^{m-1}\,dt+\sum_r\lambda_r x_i^r=0.$$

Assuming that $x\in I$, a linear change of variables shows that

$$\int_I(t-x)_+^{m-1}(x_i-t)_+^{m-1}\,dt=\gamma(x_i-x)_+^{2m-1},$$

where

$$\gamma=\int_0^1 u^{m-1}(1-u)^{m-1}\,du,$$

and therefore also

$$\int_I(t-x_\nu)_+^{m-1}(x_i-t)_+^{m-1}\,dt=\gamma(x_i-x_\nu)_+^{2m-1}.$$

But then (20) may equivalently be written as

$$(21)\quad -\mathscr{L}(x_i-x)_+^{2m-1}+\sum_\nu B_\nu(x_i-x_\nu)_+^{2m-1}+\sum_r\lambda_r' x_i^r=0,\qquad (i=0,\ldots,n),$$

where $\lambda_r'=(-1)^m\gamma^{-1}\lambda_r$.

Arranging terms, let us rewrite (21) and (11) as the system

$$(22)\quad \begin{cases}\sum_\nu B_\nu(x_i-x_\nu)_+^{2m-1}+\sum_r\lambda_r' x_i^r=\mathscr{L}(x_i-x)_+^{2m-1}\\ \sum_\nu B_\nu x_\nu^r \qquad\qquad\qquad\qquad\quad =\mathscr{L}x^r,\end{cases}$$

and let us show that its determinant of order $m+n+1$

$$(23)\qquad D=\begin{vmatrix}(x_i-x_\nu)_+^{2m-1} & x_i^r\\ x_\nu^r & 0\end{vmatrix}\neq 0.$$

We do this by showing that the statement $D\neq 0$ is equivalent to the Interpolation Theorem of Section 2. Indeed, $S(x)\in\sum_{2m}(\Delta)$ if and only if $S(x)$ has the form

$$(24)\qquad S(x)=\sum_0^n C_\nu(x-x_\nu)_+^{2m-1}+\sum_0^{m-1}\gamma_r x^r,$$

where the C_ν are subject to the relations

$$\sum_\nu C_\nu x_\nu^r = 0, \quad (r=0, \ldots, m-1),$$

which originate from the requirement that $S(x) \in \pi_{m-1}$ for $x > x_n$ (See (2)). If we now require that $S(x_i)=0$ for all i, we obtain the system

$$(25) \quad \begin{cases} \sum_\nu C_\nu (x_i - x_\nu)_+^{2m-1} + \sum_r \gamma_r x_i^r = 0, & (i=0, \ldots, n), \\ \sum_\nu C_\nu x_\nu^r = 0, & (r=0, \ldots, m-1). \end{cases}$$

However, the Interpolation Theorem says that (25) imply that $C_\nu = 0$ and $\gamma_r = 0$ for all ν and r. This establishes (23).

We now turn to our S. I. F. (4) and operating on both sides with $\mathscr{L}$, as required by Theorem 1, we obtain the approximation formula

$$(26) \quad \mathscr{L}f = \sum_\nu B_\nu^* f(x_\nu) + R^*(f),$$

where

$$(27) \quad B_\nu^* = \mathscr{L}L_\nu(x), \quad (\nu = 0, \ldots, n).$$

We observe next that the formula (26) has the following property:

$$(28) \quad R^*(f) = 0 \text{ if } f \in \textstyle\sum_{2m}(\Delta).$$

Indeed, by (26)

$$R^*(L_\nu) = \mathscr{L}L_\nu - B_\nu^* = 0$$

in virtue of (27). We have shown: *The approximation formula* (26) *is exact for all spline functions of the class* $\sum_{2m}(\Delta)$.

Let us express this fact for the spline functions $S(x)$ which vanish if $x > x_n$. These are of the form

$$S(x) = \sum_0^n C_i'(x_i - x)_+^{2m-1},$$

where the C_i' satisfy

$$(29) \quad \sum_i C_i' x_i^r = 0, \quad (r=0, \ldots, m-1),$$

as required by (2). Now

$$(30) \quad R^*(S) = \sum_i C_i'\left(\mathscr{L}(x_i - x)_+^{2m-1} - \sum_\nu B_\nu^*(x_i - x_\nu)_+^{2m-1}\right)$$

and this is supposed to vanish as a consequence of the relations (29), which means that the linear form (30) in the variables C_i', is a linear combination of the forms (29). This means that constants μ_r exist such that

$$(31) \quad \mathscr{L}(x_i - x)_+^{2m-1} - \sum_\nu B_\nu^*(x_i - x_\nu)_+^{2m-1} = \sum_r \mu_r x_i^r, \quad (i=0, \ldots, n).$$

However, also $x^r \in \sum_{2m}(\Delta)$ and therefore also the relations

$$\mathscr{L}x^r - \sum_\nu B_\nu^* x_\nu^r = 0, \qquad (r = 0, \ldots, m-1), \tag{32}$$

hold. Now (31) and (32) are seen to form a system in the unknowns B_ν^*, μ_r which is identical with the system (22) which was shown to have a *unique* solution. Thus $B_\nu^* = B_\nu$ for $\nu = 0, \ldots, n$. This and (27) imply (16) and Theorem 1 is established.

6. *Numerical examples to Theorem* 1

(i) Integration. It was pointed out in [7] that the quadrature formulae computed by Holladay [2] for $m=2$ and equidistant x_ν, are identical with the formulae computed by Sard for $m=2$ (See [4, Table 1 on page 84] and also [5, Table 1 on page 53]; in Sard's notation the role of our m and n is interchanged). Since Holladay computed his formulae by integrating the interpolating spline function while Sard computed his by the minimum property (15), the identity of the resulting tables illustrates our Theorem 1.

In [5, pp. 94–100] Sard gives explicitly the fundamental functions $L_\nu(x)$ for $x_0 = 0, x_1 = 1, \ldots, x_n = n$ up to and including the case when $m = 3$ and $n = 5$. According to (17), by integration of these $L_\nu(x)$ we may obtain most of his Table 1 of best approximate integration formulas [5, page 53].

(ii) Differentiation. Sard computes explicitly [5, pp. 112–115]. the coefficients $B_\nu = B_\nu(x)$ of the best approximate differentiation formula

$$f'(x) = \sum_{\nu=0}^{4} B_\nu(x) f(\nu - 2) + R(x; f)$$

for $m = 3$ and the range $2 \leqslant x \leqslant 3$. The resulting fourth degree polynomials [5, p. 115] are readily seen to be the derivatives of the $L_\nu(x)$ for $m = 3$, $n = 4$ as given by Sard [5, p. 98] for the range (2, 3), as they should be according to our relation (18).

University of Pennsylvania
and
Institute for Advanced Studies

REFERENCES

1. Boor, C. de, Best approximation properties of spline functions of odd degree, J. of Math. and Mech., **12**, 747–749 (1963).
2. Holladay, J. C. Smoothest curve approximation, Math. Tables and other Aids to Comput., **11**, 233–243 (1957).
3. Meyers, L. F. and A. Sard, Best interpolation formulas, J. of Math. and Phys., **29**, 198–206 (1950).
4. Sard, A., Best approximate integration formulas; best approximation formulas, Amer. J. of Math., **71**, 80–91 (1949).
5. ———, Linear Approximations, Math. Surveys No. 9, Amer. Math. Society, Providence, R.I., 1963.

6. Schoenberg, I. J., On interpolation by spline functions and its minimal properties to appear in the Proceedings of the Conference on Approximation Theory, Oberwolfach, Germany, August, 1963.
7. ———, Spline interpolation and best quadrature formulae, to appear in the Bull. Amer. Math. Society.
8. ———, Spline interpolation and the higher derivatives, to appear in the Proceedings of the Nat. Academy of Sciences.
9. Walsh, J. L., J. H. Ahlberg and E. N. Nilson, Best approximation properties of the spline fit, J. of Math. and Mech., **11**, 225–234 (1962).

Reprint from ISNM, Vol. 5 (1964)
ON APPROXIMATION THEORY
BIRKHÄUSER VERLAG BASEL/STUTTGART

ON INTERPOLATION BY SPLINE FUNCTIONS AND ITS MINIMAL PROPERTIES [1)]

by I. J. Schoenberg

TABLE OF CONTENT

INTRODUCTION

Polynomials are wonderful even after they are cut into pieces, but the cutting must be done with care. One way of doing the cutting leads to the so-called spline functions. These were introduced by the writer in 1944 as a tool for the approximation of functions and were suggested by the work of T. N. E. Greville and other actuarial writers on the subject of osculatory interpolation (for the connections and references see [8], Part B). The other tool used was the law of the flow of heat in a homogeneous wire and with both tools combined extensive ballistic tables were approximated and thereby smoothed (see [8]). For recent large-scale physical applications of cubic spline functions see [6]. The spline functions used in [8] had equidistant knots. In 1945 H.B. Curry and the author wrote a paper on spline functions with arbitrarily spaced knots which for no good reason remained unpublished (see the Abstract [2]). For further work on spline functions and related topics see the list of references which claims no completeness.

I am writing to-day again on this subject because of the interesting work of J.C.Holladay [3] and J.L.Walsh, J.H.Ahlberg and E.N.Nilson [12], devoted to the special case of cubic spline functions, in which these writers introduced new and stimulating ideas. Using their approach to obtain an extension of their results to spline functions of an arbitrary degree would seem to be a formidable task [2)]. In these lines I wish to show how this can be done with ease by means of the approach to spline functions found by Curry and Schoenberg in 1945.

In Section 1 are assembled a few needed facts from [2], some of which are well known by now and were perhaps known before 1945 in view of the early und fundamental work of G. Peano on Remainders in numerical analysis (see e.g. J. Kuntzmann's book [5]). For this reason their elementary proofs are omitted. Concerning the main problems there are two distinct cases: 1. The finite case, 2. The periodic case. These are discussed in Part I (Sections 2 and 3) and Part II (Sections 5 and 6), respectively. Section 4 prepares our discussion of the periodic case with a few remarks on periodic spline functions.

Part III extends the theory to the case of interpolation problems in which not only ordinates are given but also a few successive derivatives (Section 7). Sections 8 and 9 use the connection with the Euler and Bernoulli polynomials to solve a few simple problems for periodic functions.

PART I: THE FINITE CASE

1. DEFINITION AND SOME PROPERTIES OF SPLINE FUNCTIONS

Let

$$\dots < x_{-2} < x_{-1} < x_0 < x_1 < \dots < x_\nu < \dots \tag{1.1}$$

be a given infinite sequence of reals such that $\lim x_\nu = \pm\infty$ if $\nu \longrightarrow \pm\infty$, and let k be a given natural integer. By a ***spline function*** $S_k(x)$ ***of order*** k, having the ***knots*** (1.1), we mean a function of the class C^{k-2} for all real x such that in each interval $(x_\nu, x_{\nu+1})$ it reduces to a polynomial of degree not exceeding $k-1$. These polynomials may well vary from one interval to the next but they must agree in value at their common endpoint (or knot) as well as their first $k-2$ derivatives. A spline function of order k will also be referred to as a k-spline.

This definition shows that a 1-spline $S_1(x)$ is a step function with possible discontinuities at the points (1.1). Likewise an $S_2(x)$ is a continuous broken linear function with possible corners at some or all of the points (1.1). Finally, an $S_3(x)$ has a graph composed of a sequence of parabolas which join at the knots continuously together with their slopes.

Observe that the operation of differentiation (or integration) transforms a spline into another spline of an order decreased (or increased) by one unit. Also that the k-splines of knots (1.1) form a class which is closed with respect to the formation of linear combinations of its elements with constant coefficients. Let us finally mention that the family of $S_k(x)$ contains the family π_{k-1} of polynomials of degree $\leq k-1$ of which it is a generalization.

If $S_k(x)$ vanishes outside the range $(x_\nu, x_{\nu+N})$ but not outside any subrange, then we say that our $S_k(x)$ has the *span* N. We now ask: What is the smallest span N of a k-spline? Thus if $k=2$ then the answer is evidently $N=2$. The general answer for k-splines is $N=k$ in which case the corresponding $S_k(x)$ is uniquely determined up to a nonvanishing constant factor. These are the so-called fundamental k-splines which may be constructed as follows:

We use the truncated powers

$$x_+^{k-1} = \begin{cases} x^{k-1} & (x \geq 0) \\ 0 & (x < 0) \end{cases}$$

and write

$$M_k(x\,;\,y) = k\,(y-x)_+^{k-1} \quad .$$

Let us now consider the kth order divided difference of the function $M_k(x\,;\,y)$, with respect to the variable y, and based on the points $y = x_0, x_1, \ldots, x_k$. Using Steffensen's notation for divided differences (see [11]) we obtain

$$\text{(1.2)} \qquad M_k(x\,;\,x_0, x_1, \ldots, x_k) = \sum_{\nu=0}^{k} \frac{k(x_\nu - x)_+^{k-1}}{\omega'(x_\nu)} \quad ,$$

where $\omega(x) = (x-x_0)\ldots(x-x_k)$. Since $(x_\nu - x)_+^{k-1}$ is clearly a k-spline having as only genuine knot the point $x = x_\nu$, we see that the function (1.2) is a k-spline having as actual knots only the points $x_0, \ldots, x_k$. Observe, moreover, that the function (1.2) visibly vanishes if $x > x_k$. However, it vanishes also if $x < x_0$ for the following reason: If $x < x_0$, we may omit the subscript "+" in the right-hand expression (1.2) and then the sum is seen to vanish as the kth order divided difference of a polynomial of degree $k-1$.

The functions

$$M_k(x; x_\nu, x_{\nu+1}, \ldots, x_{\nu+k}) \quad \text{for all integers } \nu,\ .$$

are called the fundamental k-splines for the set of knots (1.1) (see [7, pp. 70-71] for the explicit expressions and graphs of the fundamental k-splines for equidistant knots and $k = 1, 2, 3$ and 4).

Their fundamental role is brought out by the following

LEMMA 1. *Every k-spline having the knots* (1.1) *may be represented in the form*

$$S_k(x) = \sum_{\nu=-\infty}^{\infty} c_\nu M_k(x; x_\nu, \ldots, x_{\nu+k}) \tag{1.3}$$

where the c_ν *are appropriate uniquely determined constants. Conversely,* (1.3) *always represents a k-spline.*

Observe that the series (1.3) always reduces to a finite sum. Lemma 1 is familiar only if $k = 2$ when $M_2(x; x_\nu, x_{\nu+1}, x_{\nu+2})$ is a so-called "roof-function".

A further basic property of the fundamental splines is stated as

LEMMA 2. *If* $f(x) \in C^{k-1}[x_0, x_k]$ *while* $f^{(k-1)}(x)$ *is absolutely continuous then its divided difference of order k may be expressed as*

$$f(x_0, x_1, \ldots, x_k) = \frac{1}{k!} \int_{x_0}^{x_k} M_k(x; x_0, \ldots, x_k) f^{(k)}(x)\, dx\ . \tag{1.4}$$

This follows from Peano's general theorem concerning linear functionals which vanish for polynomials of degree at most $k-1$. Although not needed below, let us mention a few further properties: If we apply (1.4) to the function $f(x) = x^k$ we obtain

$$\int_{-\infty}^{\infty} M_k(x; x_0, \ldots, x_k)\, dx = 1\ .$$

Moreover, the function (1.2) is positive in the interval (x_0, x_k) where its graph has a very nicely balanced shape due to the following property: Its νth derivative has in (x_0, x_k) exactly ν distinct and simple zeros $(\nu = 1, \ldots, k-2)$.

2. THE INTERPOLATION PROBLEM BY $2m$-SPLINES

Let m be a given natural integer and let us consider the class of functions

$$F_m[a,b] = \{f(x) \mid f \in C^{m-1}[a,b],\ f^{(m-1)} \text{ absolutely continuous},\ f^{(m)} \in L^2(a,b)\} \tag{2.1}$$

in a given finite interval $[a, b]$. Let the integer $n \geq m$ be given and suppose that we are also given $n+1$ abscissae

$$x_0 < x_1 < \ldots < x_n \quad \text{in} \quad [a,b]\ , \qquad (n \geq m)\ . \tag{2.2}$$

We now chose arbitrary but fixed reals x_ν for $\nu > n$ and for $\nu < o$ such as to obtain a sequence of knots (1.1).

We consider now the family of splines $S_m(x)$ having the knots (1.1) and such that

(2.3) $$S_m(x) = o \quad if \quad x < x_o \quad or \quad x > x_n$$

and we propose to construct an m-spline which will solve the problem

(2.4) $$\int_a^b (S_m(x) - f^{(m)}(x))^2\, dx = minimum .$$

This problem is easily solved, at least in theory. Indeed, let us look at the fundamental m-splines for the knots (1.1) using for them the simplified notation

$$M_j(x) = M_m(x; x_j, x_{j+1}, \ldots, x_{j+m}) .$$

By Lemma 1 we see that we may represent the family of m-splines defined by the condition (2.3) by

$$S_m(x) = \sum_o^{n-m} c_j M_j(x) .$$

The problem (2.4) thus becomes

$$\int_a^b (\sum_o^{n-m} c_j M_j(x) - f^{(m)}(x))^2\, dx = minimum$$

and by partial differentiations we obtain the system of $n-m+1$ normal equations

(2.5) $$\int_a^b (\sum_o^{n-m} c_j M_j(x) - f^{(m)}(x))\, M_i(x)\, dx = o , \qquad (i = o, \ldots, n-m).$$

The determinant of this linear system in the unknowns c_j is the Gramian of the functions $M_j(x)$ $(j = o, \ldots, n-m)$. These being linearly independent in $[a, b]$ by Lemma 1, we conclude that the problem (2.4) has a unique solution which is characterized by the relations

(2.6) $$\int_a^b (S_m(x) - f^{(m)}(x))\, M_i(x)\, dx = o , \qquad (i = o, \ldots, n-m).$$

We now integrate the spline function $S_m(x)$ so determined m times, obtaining a $2m$-spline $S_{2m}(x)$ which still contains an arbitrary additive polynomial $P_{m-1}(x)$ of degree at most $m-1$. In any case, by (2.3), $S_{2m}(x)$ is seen to satisfy the following

(2.7) CONDITION: *$S_{2m}(x)$ reduces to a polynomial of degree at most $m-1$ in each of the intervals $(-\infty, x_o)$ and $(x_n, +\infty)$.*

Writing

$$\varphi(x) = f(x) - S_{2m}(x)$$

(2.6) becomes

9

$$\int_a^b \varphi^{(m)}(x)\ M_i(x)\ dx = o, \qquad (i = o,\ldots,n-m).$$

At this point we apply Lemma 2 from which we conclude the vanishing of the following divided differences

$$\varphi(x_i, x_{i+1}, \ldots, x_{i+m}) = o, \qquad (i = o,\ldots,n-m).$$

But this means that the additive polynomial $P_{m-1}(x)$ of $S_{2m}(x)$ may be so determined uniquely that

(2.8) $$S_{2m}(x_\nu) = f(x_\nu) \quad for \quad \nu = o,1,\ldots,n.$$

We have just shown the existence of a $2m$-spline $S_{2m}(x)$ subject to the condition (2.7) which interpolates our function at the points (2.2). This spline function is moreover uniquely determined because $S_{2m}^{(m)}(x) = S_m(x)$ was uniquely determined by the system (2.6).

This proves the following

THEOREM 1. *Let there be given in* $[a,b]$ *a function* $f(x)$ *in the class* $F_m[a,b]$ *defined by* (2.1) *and a set of* $n+1$ *points* (2.2) *with* $n \geq m$. *Then there is a uniquely determined* $2m$*-spline* $S_{2m}(x)$ *having as knots the points* (2.2) *and subject to the condition* (2.7) *which interpolates the function* $f(x)$ *at the points* (2.2). *This* $2m$*-spline is also uniquely characterized, up to an additive polynomial* $P_{m-1}(x)$, *among all* $2m$*-splines with knots* (2.2) *and subject to* (2.7) *by the condition that*

(2.9) $$\int_a^b (S_{2m}^{(m)}(x) - f^{(m)}(x))^2\ dx = minimum.$$

It is clear from the unicity of the solution of this interpolation problem that the interpolating $S_{2m}(x)$ is already uniquely defined by the ordinates $f(x_\nu)$ alone. We state this as

COROLLARY 1. *Given* $n+1$ *points in the plane*

(2.10) $$(x_\nu, y_\nu), \qquad (\nu = o,\ldots,n;\ x_o < x_1 < \ldots < x_n),$$

and an integer m, $1 \leq m \leq n$, *then there exists a unique* $2m$*-spline* $S_{2m}(x)$, *with the knots* x_ν *and satisfying the condition* (2.7), *such that*

(2.11) $$S_{2m}(x_\nu) = y_\nu \qquad (\nu = o,\ldots,n).$$

In Corollary 1 the condition (2.7) may also be replaced by the boundary conditions

(2.12) $$S_{2m}^{(i)}(x_o) = S_{2m}^{(i)}(x_n) = o, \qquad (i = m, m+1,\ldots,2m-2).$$

Let us mention a few examples: In the case when $m = 1$, then the boundary conditions are vacuous and the solution $S_2(x)$ $(x_o \leq x \leq x_n)$ is the broken linear function obtained by linear interpolation between consecutive points. If $m = 2 \leq n$, then $S_4(x)$

$(x_o \leq x \leq x_n)$ is piece-wise cubic while (2.12) reduce to $S_4''(x) = S_4''(x) = o$. If our data (2.10) can be interpolated by a polynomial $P_{m-1}(x)$, then this polynomial is our solution because it is also a $2m$-spline and satisfies all required conditions.

3. THE MINIMAL PROPERTY OF THE INTERPOLATING SPLINE FUNCTION

As discovered by Holladay for the cubic case $(m = 2)$, the interpolating spline of Corollary 1 may be directly characterized by a minimal property which we state as

THEOREM 2. *Let all assumptions of Corollary 1 be satisfied and let* $S_{2m}(x)$ *be the unique* $2m$*-spline which solves the interpolation problem* (2.11). *Let us choose an interval* $[a,b]$ *containing all* x_ν *and let* $f(x) \in F_m[a,b]$ *such that*

$$f(x_\nu) = y_\nu \qquad (\nu = o,\ldots,n). \tag{3.1}$$

Then

$$\int_a^b (f^{(m)}(x))^2\,dx \geq \int_a^b (S_{2m}^{(m)}(x))^2\,dx$$

with equality, only if $f(x) = S_{2m}(x)$.

In other words: The interpolating $2m$-spline minimizes the integral

$$\int_a^b (f^{(m)}(x))^2\,dx$$

among all interpolating functions in $F_m[a,b]$. Even for the special case $m = 2$ Theorem 2 is slightly more general than Holladay's theorem who assumes that $[a,b] = [x_o, x_n]$.

To prove Theorem 2 we adapt the argument given in [11] for the periodic case and $m = 2$. 3/

Let us show that (3.1) imply the relation

$$\int_a^b (f^{(m)})^3\,dx = \int_a^b (f^{(m)} - S_{2m}^{(m)})^2\,dx + \int_a^b (S_{2m}^{(m)})^2\,dx$$

from which Theorem 2 easily follows. To prove this we start from the identity

$$\int_a^b (f^{(m)})^2\,dx = \int_a^b (f^{(m)} - S^{(m)})^2\,dx + 2\int_a^b S^{(m)}(f^{(m)} - S^{(m)})\,dx + \int_a^b (S^{(m)})^2\,dx$$

and proceed to show that the middle term on the right side vanishes. Indeed, this term becomes by an integration by parts

$$\int_a^b S^{(m)}(f^{(m)} - S^{(m)})\,dx = \left[S^{(m)}(f^{(m-1)} - S^{(m-1)})\right]_a^b - \int_a^b S^{(m+1)}(f^{(m-1)} - S^{(m-1)})\,dx$$

and the "finite parts" on the right side vanish because of (2.3). For the same reason we may repeat the integrations by parts until we obtain

$$\pm\int_a^b S^{(2m-1)}(f' - S')\,dx = \pm\sum_1^n \int_{x_{\nu-1}}^{x_\nu} S^{(2m-1)}(f' - S')\,dx \;,$$

as the end-integrals

$$\int_a^{x_o} \quad \text{and} \quad \int_{x_n}^{b} S^{(2m-1)}(f' - S')\, dx$$

both vanish since $S_{2m}^{(2m-1)}(x) = o$ in these ranges. However, $S_{2m}^{(2m-1)}(x)$ is a constant in each of the intervals $(x_{\nu-1}, x_\nu)$ while $f(x) - S_{2m}(x)$ vanishes at their endpoints in view of (2.11) and (3.1). This completes our proof.

PART II: THE PERIODIC CASE

4. PERIODIC SPLINE FUNCTIONS

We now wish to describe those k-splines $S(x)$ which are periodic functions of period 1. For rough structural reasons it is clear that as a first necessary condition the set of knots (1.1) must itself have the period 1. For this reason and throughout Part II we shall make concerning the knots (1.1) the following additional assumptions: We assume that

$$o < x_1 < x_2 < \ldots < x_n \leq 1 \tag{4.1}$$

while all other knots are obtained from these n knots by translations of all integral sizes. The knots (1.1) forming an increasing sequence we must have, besides (4.1), the relations

$$x_{\nu+ln} = x_\nu + l \qquad (1 \leq \nu \leq n, \ -\infty < l < \infty) . \tag{4.2}$$

Geometrically all this means that the set of knots (1.1) will go over in itself if we translate it by one unit.

In order to avoid the trivial case when all periodic k-splines are merely constants, we shall assume that there are at least two knots for each period, hence that $n > 1$.

Let us now look at the fundamental k-splines of a periodic set of knots as just described. Keeping the order k fixed, let us simplify their notation by writing

$$M_k(x; x_\nu, \ldots, x_{\nu+k}) = M_\nu(x) .$$

It is geometrically clear that the functions $M_\nu(x)$ and $M_{\nu+n}(x)$ are identical except for a shift of one unit. Indeed, from (1.2) and (4.2) it is easily verified that

$$M_{\nu+ln}(x) = M_\nu(x - l) . \tag{4.3}$$

If now $\{c_\nu\}$ is a periodic sequence with the period n then the spline

$$S_k(x) = \sum_{-\infty}^{\infty} c_\nu M_\nu(x) \tag{4.4}$$

has the period 1 because by (4.3)

$$S(x+1) = \sum c_\nu M_\nu(x+1) = \sum c_\nu M_{\nu-n}(x) = \sum c_{\nu+n} M_\nu(x)$$

$$= \sum c_\nu M_\nu(x) = S(x) .$$

Conversely, if (4.4) represents a spline of period 1 then as above we obtain the identity

$$\sum c_\nu M_\nu(x) = \sum c_{\nu+n} M_\nu(x) .$$

The unicity of the representation (4.4) (Lemma 1) now implies that $c_\nu = c_{\nu+n}$, i.e. the periodicity of the sequence $\{c_\nu\}$ with period n. This proves

LEMMA 3. *The most general k-spline of period 1 is represented uniquely by the formula* (4.4), *where $\{c_\nu\}$ is an arbitrary periodic sequence of period n.*

A periodic $S_k(x)$ therefore depends on exactly n parameters. A base for this linear family is provided by the n periodic k-splines

(4.5) $$N_\nu(x) = \sum_{l=-\infty}^{\infty} M_\nu(x+l) , \qquad (\nu = 1,\dots,n) .$$

Indeed, if $S_k(x)$ is a periodic spline then the periodicity of $\{c_\nu\}$ gives

$$S_k(x) = \sum_{\nu=1}^{n} \sum_{l} c_{\nu+ln} M_{\nu+ln}(x) = \sum_{\nu,l} c_\nu M_\nu(x-l)$$

and finally

(4.6) $$S_k(x) = \sum_{1}^{n} c_\nu N_\nu(x) .$$

We have thus established the following

LEMMA 4. *The periodic k-splines of period 1 are uniquely represented by the formula* (4.6) *in which the periodic k-splines $N_\nu(x)$ are defined by* (4.5). *These n periodic k-splines $N_\nu(x)$ $(\nu = 1,\dots,n)$ are linearly independent in the interval* $[0,1]$.

We have excluded the case $n = 1$ for a reason already stated: *The only periodic k-splines, for $n = 1$, are constants.* Indeed, if $n = 1$, then (4.2) shows that $x_\nu = x_1 + (\nu - 1)$. If

$$S_k(x) = P_{k-1}(x) \text{ in } x_0 \le x \le x_1 ,$$

then the periodicity of $S_k(x)$ implies the relations

$$P_{k-1}^{(\nu)}(x_0) = P_{k-1}^{(\nu)}(x_0+1).$$

for $\nu = 0,1,\dots,k-2$ and therefore also for all values of ν . Thus the polynomial $P_{k-1}(x)$ has the period 1 and must reduce to a constant.

5. THE INTERPOLATION PROBLEM BY PERIODIC $2m$-SPLINES

Our discussion in this section and the next is very close to the discussion of the similar problems in Sections 2 and 3. However, the present periodic case is slightly more complicated in its details, although no new ideas are needed. Moreover, the important inequality $m \leq n$ (of Theorems 1, 2 and Corollary 1) between the order of the interpolating spline and the number of points to be interpolated, has no analogue in the present periodic case. To avoid ambiguities resulting from mere references to the previous case, we discuss the problems and their solutions in detail.

Let m be a given natural integer and let us consider the class of functions

$$(5.1) \qquad F_m(per.1) = \{ f(x) \mid f(x+1) = f(x),\ f \in C^{m-1},\ f^{(m-1)} \text{ abs. continuous}, \ f^{(m)} \in L^2(0,1) \}.$$

Given $f(x) \in F(per.1)$, we propose to construct a periodic m-spline $S_m(x)$, of period 1, which will solve the problem

$$(5.2) \qquad \int_0^1 (S_m(x) - f^{(m)}(x))^2\, dx = \text{minimum}.$$

In our notation of Section 4 $k = m$ and (4.6) shows that we are to determine the constants $c_1, \ldots, c_n$ such that

$$\int_0^1 \Big(\sum_1^n c_j N_j(x) - f^{(m)}(x) \Big)^2 dx = \text{minimum}.$$

By partial differentiation this problem leads to the system of n normal equations

$$(5.3) \qquad \int_0^1 \Big(\sum_1^n c_j N_j(x) - f^{(m)}(x) \Big) N_i(x)\, dx = 0, \qquad (i = 1, \ldots, n).$$

Their determinant is the Gramian of the $N_\nu(x)$. By Lemma 4 this Gramian is positive and our system (5.3) has a unique solution.

Let us denote by $\tilde{S}_m(x)$ the unique periodic m-spline just found and which is characterized by the equations

$$(5.4) \qquad \int_0^1 (\tilde{S}_m(x) - f^{(m)}(x)) N_i(x)\, dx = 0, \qquad (i = 1, \ldots, n).$$

If we write $\varphi(x) = \tilde{S}_m(x) - f^{(m)}(x)$ then (5.4) and (4.6) show that

$$\int_0^1 \varphi(x) S_m(x)\, dx = 0$$

for *all* periodic *m-splines* $S_m(x)$. Since $S_m(x) \equiv 1$ is such a spline we conclude that $\int_0^1 \varphi(x)\, dx = 0$ and hence

$$(5.5) \qquad \int_0^1 \tilde{S}_m(x)\, dx = \int_0^1 f^{(m)}(x)\, dx = 0,$$

because of the periodicity of $f^{(m-1)}(x)$.

At this point we wish to construct a periodic $2m$-spline $\tilde{S}_{2m}(x)$ such that

$$\tilde{S}_{2m}^{(m)}(x) = \tilde{S}_m(x) \tag{5.6}$$

and we will show that there is a unique solution up to an arbitrary additive constant. This is readily seen as follows: The integral

$$\int_0^x \tilde{S}_m(u)\,du$$

is an $(m+1)$ -spline of period 1 because of (5.5). Adding an appropriate constant (which is uniquely defined) we obtain an integral of $S_m(x)$ whose average value over $(0,1)$ vanishes and thereby in its turn admits a periodic integral. Repeating this operation m times we obtain a $2m$ -spline $\tilde{S}_{2m}(x)$ satisfying (5.6) and determined up to an additive constant because the last additive constant is unrestricted.

Now our relations (5.4) may equivalently be written as

$$\int_0^1 (\tilde{S}_{2m}^{(m)}(x) - f^{(m)}(x))\, N_i(x)\,dx = 0, \qquad (i = 1,\ldots,n). \tag{5.7}$$

Let us show that, because of (4.5), the relations (5.7) are equivalent to

$$\int_{-\infty}^{+\infty} (\tilde{S}_{2m}^{(m)}(x) - f^{(m)}(x))\, M_i(x)\,dx = 0, \qquad (i = 1,\ldots,n). \tag{5.8}$$

Indeed, setting

$$\varphi(x) = \tilde{S}_{2m}(x) - f(x), \tag{5.9}$$

we may write (5.7) as

$$\int_0^1 \varphi^{(m)}(x) \left(\sum_l M_i(x+l)\right) dx = 0,$$

where we need sum only over a finite range $l = -N,\ldots,N$, dropping those $M_i(x+l)$ which have no support in $(0,1)$. But then

$$\begin{aligned} 0 &= \sum_l \int_0^1 \varphi^{(m)}(x)\, M_i(x+l)\,dx = \sum_l \int_l^{l+1} \varphi^{(m)}(x-l)\, M_i(x)\,dx \\ &= \sum_l \int_l^{l+1} \varphi^{(m)}(x)\, M_i(x)\,dx = \int_{-N}^{N+1} \varphi^{(m)}(x)\, M_i(x)\,dx \end{aligned}$$

and the equivalence of the relations (5.8) with (5.7) is established.

We now appeal to Lemma 2 and may conclude from (5.8) that the divided differences

$$\varphi(x_i, x_{i+1}, \ldots, x_{i+m}) = 0 \tag{5.10}$$

for $i = 1, \ldots, n$ and by periodicity for *all* integral i. However, the relations (5.10) mean that there exists a polynomial $P_{m-1}(x)$ of degree $\leq m-1$ such that

$$\varphi(x_i) = \tilde{S}_{2m}(x_i) - f(x_i) = P_{m-1}(x_i) \qquad \text{for all } i\,. \tag{5.11}$$

Since the sequence $\varphi(x_i)$ is periodic and therefore bounded we conclude that $P_{m-1}(x) \equiv C$ (a constant) and now

$$S_{2m}(x) = \tilde{S}_{2m}(x) - C$$

is a periodic $2m$-spline interpolating $f(x)$ at all its knots x_i. This interpolating periodic $2m$-spline is moreover unique because we can retrace our steps to the unique solution $\tilde{S}_m(x)$ of the minimum problem (5.2). We have therefore proved the following

THEOREM 3. *Let m be a natural integer and let $f(x)$ be an element of the class of functions F_m(per. 1), defined by* (5.1). *Then there is a unique periodic 2m-spline $S_{2m}(x)$ such that*

$$S_{2m}(x_i) = f(x_i) \qquad \text{for all } i\,. \tag{5.12}$$

Among all periodic 2m-splines this interpolating $S_{2m}(x)$ is also characterized up to an additive constant as the solution of the problem

$$\int_0^1 (S_{2m}^{(m)}(x) - f^{(m)}(x))^2\, dx = \text{minimum}\,. \tag{5.13}$$

It is clear by (5.12), that the interpolating $S_{2m}(x)$ is already uniquely defined by the ordinates $f(x_i)$. We state this as

COROLLARY 2. *Let $\{x_i\}$ be a periodic set of knots, i.e. satisfying the conditions* (4.1) *and* (4.2) *and let $\{y_i\}$ be a given sequence of ordinates of period n ($y_i = y_{i+n}$). Given the natural integer m, there exists a unique periodic 2m-spline $S_{2m}(x)$ such that*

$$S_{2m}(x_i) = y_i \qquad \text{for all } i\,. \tag{5.14}$$

For $m = 1$ we obtain the trivial case of linear interpolation between consecutive points. For $m = 2$ Theorem 3 and Corollary 2 reduce to results on *cubic* splines (of order 4 in our notation) due to Walsh, Ahlberg and Nilson [12].

6. THE MINIMAL PROPERTY OF THE INTERPOLATING PERIODIC SPLINE

The periodic analogue of Theorem 2 is

THEOREM 4. *If the function $f(x) \in F_m$(per. 1) interpolates all points (x_i, y_i) of Corollary 2 and if $S_{2m}(x)$ is the interpolating periodic 2m-spline of that Corollary, then*

$$(6.1)\qquad \int_0^1 (f^{(m)}(x))^2\,dx \geq \int_0^1 (S_{2m}^{(m)}(x))^2\,dx$$

with equality, only if $f(x) = S_{2m}(x)$.

In other words: The interpolating periodic $S_{2m}(x)$ is characterized by the condition of minimizing the integral

$$\int_0^1 (f^{(m)}(x))^2\,dx$$

among all interpolating functions of period 1 having $m-1$ absolutely continuous derivatives. A proof of Theorem 4 follows from the relation

$$(6.2)\qquad \int_0^1 (f^{(m)})^2\,dx = \int_0^1 (f^{(m)} - S_{2m}^{(m)})^2\,dx + \int_0^1 (S_{2m}^{(m)})^2\,dx$$

which will be shown to hold for any interpolating $f(x)$ of the class described. To establish (6.2) we start from the identity

$$\int_0^1 (f^{(m)})^2\,dx = \int_0^1 (f^{(m)} - S^{(m)})^2\,dx + 2\int_0^1 S^{(m)}(f^{(m)} - S^{(m)})\,dx + \int_0^1 (S^{(m)})^2\,dx$$

and will show that the middle term on the right side vanishes. Indeed, this term becomes by an integration by parts over the period (x_0, x_n)

$$\int_{x_0}^{x_n} S^{(m)}(f^{(m)} - S^{(m)})\,dx = \Big[S^{(m)}(f^{(m-1)} - S^{(m-1)})\Big]_{x_0}^{x_n} - \int_{x_0}^{x_n} S^{(m+1)}(f^{(m-1)} - S^{(m-1)})\,dx\,.$$

The first term on the right side vanishes because of the periodicity of our functions. Repeated integrations by parts finally furnish

$$\pm \sum_{\nu=1}^{n} \int_{x_{\nu-1}}^{x_\nu} S^{(2m-1)}(f' - S')\,dx\,.$$

Now each term of the sum vanishes because $S^{(2m-1)}(x)$ is a constant in $(x_{\nu-1}, x_\nu)$ while $f-S$ vanishes at its endpoints.

PART III. EXTENSION TO COALESCENT KNOTS AND EXAMPLES

7. A FINITE PROBLEM WITH COALESCENT KNOTS

In Sections 2 and 3 the knots $x_0, \ldots, x_n$ were assumed to be *strictly* increasing but there is no reason why they should. Spline functions with coalescent knots appear naturally e.g. as the kernels in Peano's expression for the Remainders of quadrature formulae involving also the derivatives of the function at isolated points. If the knots of $S_k(x)$

satisfy $x_{-1} < x_o = x_1 = \dots = x_{r-1} < x_r$, i.e. we have r knots coalescing at x_o, then all this means is that near the point $x = x_o$ the function $S_k(x)$ is only required to be of the class C^{k-r-1} rather than C^{k-2} as usual. Thus for $r = k$ there is no continuity requirement whatever and therefore ***k is the largest number of knots which are allowed to coalesce***, as $r > k$ is meaningless.

It is seen that a $S_2(x)$ having all its knots double is a succession of linear functions with possible discontinuities at the knots. Thus, if $B_1(x) = x - \frac{1}{2}$ $(o \le x \le 1)$ is the first Bernoulli polynomial and if we denote its periodic extension by $\bar{B}_1(x)$, then $\bar{B}_1(x)$ is a periodic 2-spline having double knots at the integers. By $n-1$ integrations with the proper constants of integration (namely the Bernoulli numbers) we obtain the periodic Bernoulli function $\bar{B}_n(x)$. These integrations increase progressively their order of continuity, so that $\bar{B}_n(x)$ is a periodic $(n+1)$-spline with double knots at the integers.

There is not much to change in our discussion of spline functions (Sections 1 and 4) to include coalescent knots. The knots are again described by (1.1) except that we allow anywhere equality signs, but not too many consecutive ones, because we require that

$$x_\nu < x_{\nu+k} \qquad \text{for all } \nu.$$

For this reason the fundamental splines are still defined by the divided difference $M_k(x; x_o, \dots, x_k)$ except that the right-hand side expression (1.2) is no longer valid, as coalescence of knots must be taken into account (see [11]). Lemmas 1 and 2 remain valid as they stand.

After this long introduction we return to the finite interpolation problem of Sections 2 and 3, allowing coalescence of knots. To avoid a long and cumbersome statement we shall discuss only a single case described as follows: We assume $n = 2r-1$ and

$$x_\nu = o \ \text{ for } \nu = o, 1, \dots, r-1, \qquad x_\nu = 1 \ \text{ for } \nu = r, r+1, \dots, n. \tag{7.1}$$

We therefore have two knots, o and 1, each of multiplicity r. All other knots x_ν ($\nu > n$ and $\nu < o$), outside $[o, 1]$, are simple as before.

For the order m of the spline $S_m(x)$ we need the inequalities

$$r \le m \le 2r - 1$$

of which the first is due to our restriction on the order of coalescence while the second is the basic inequality $m \le n$ of Section 2. With these assumptions our former proofs go over without any change. Choosing $[a, b] = [o, 1]$ we state Corollary 1 and Theorem 2 for this situation as

COROLLARY 3. *Given natural integers m and r such that*

$$r \le m \le 2r - 1 \tag{7.2}$$

and two sets of reals $y_o^{(\nu)}$, $y_1^{(\nu)}$ $(\nu = o, 1, \dots, r-1)$, then there exists a unique polyno-

mial P(x) of degree $\leq 2m-1$ *satisfying the two sets of relations*

$$(7.3)\qquad P^{(\nu)}(0) = y_0^{(\nu)}, \quad P^{(\nu)}(1) = y_1^{(\nu)} \qquad (\nu = 0,\dots,r-1),$$

and

$$(7.4)\qquad P^{(s)}(0) = P^{(s)}(1) = 0 \quad \text{for } s = m, m+1, \dots, 2m-r-1.$$

This polynomial P(x) is the unique solution of the problem

$$(7.5)\qquad \int_0^1 (f^{(m)}(x))^2\,dx = \text{minimum}$$

among all functions f(x) having in $[0,1]$ *an absolutely continuous derivative* $f^{(m-1)}(x)$ *and satisfying the boundary conditions*

$$(7.6)\qquad f^{(\nu)}(0) = y_0^{(\nu)}, \quad f^{(\nu)}(1) = y_1^{(\nu)} \qquad (\nu = 0,\dots,r-1).$$

Observe that there is no mention of splines in this statement. Corollary 3 can be proved by precisely the method of Sections 2 and 3 without the use of splines and their properties. This makes a nice exercise in Integral Calculus which the writer may discuss in a journal with didactic interests.

The origin of the curious conditions (7.4) is seen from the requirement (2.3) on the solutions of the problem (2.4). As a result of (2.3) the final solution $S_{2m}(x)$ of the problem (2.9) must satisfy the condition (2.7). However, the knots $x_0 = 0$ and $x_n = 1$ being r-fold by (7.1), we require for $S_{2m}(x)$ (which is $= P(x)$ in $[0,1]$) at these points only $2m-r-1$ continuous derivatives. But then (2.7) implies (7.4).

Observe that $m \leq 2m-r$ by (7.2). If the problem is such that $m > 2m-r-1$, then the conditions (7.4) are vacuous and are to be disregarded. Thus $r = m = 2$ is such a case and the cubic solution $P(x)$ of the interpolation problem (7.3) is the solution of the minimum problem (7.5). However, if $r = 2$ and $m = 3$ then the solution of (7.5) is the quintic $P(x)$ uniquely defined by (7.3) and the additional conditions

$$P'''(0) = P'''(1) = 0.$$

8. TWO EXAMPLES OF PERIODIC CASES

We now discuss the periodic problem of Corollary 2 for $n = 2$ and the following two sets of knots:

$$(8.1)\qquad x_1 = \tfrac{1}{2}, \quad x_2 = 1,$$

and

$$(8.2)\qquad x_1 = x_2 = 1.$$

In either case we know by (4.6) that the family of periodic $2m$-splines depends on two parameters and we only need two such $2m$-splines which are linearly independent in $[0,1]$. In either case $f(x) \equiv 1$ is an acceptable function. Let us show that the second element of a base is given in terms of Euler polynomials in the first case and Bernoulli polynomials in the second.

Indeed, let (8.1) hold. Following Nörlund [7, p. 23] we define the Euler polynomial $E_\nu(x)$ as the polynomial satisfying the identity

$$\frac{1}{2}(f(x+1)+f(x)) = x^\nu$$

and it is shown there that it satisfies the relations

$$E'_\nu(x) = \nu E_{\nu-1}(x) \qquad \text{for } \nu = 1,2,\ldots \,. \tag{8.3}$$

Nörlund [7, p. 34] defines the extension $\bar{E}_\nu(x)$ satisfying

$$\bar{E}_\nu(x+1) = -\bar{E}_\nu(x) \quad \text{for all } x \tag{8.4}$$

and is such that

$$\bar{E}_\nu(x) = E_\nu(x) \quad \text{if } 0 \le x < 1\,.$$

Evidently (8.4) implies that $\bar{E}_\nu(x)$ has the period 2 and it is shown that this function is continuous for all x. But then (8.3) and (8.4) imply that the relation

$$\bar{E}'_\nu(x) = \nu \bar{E}_{\nu-1}(x) \tag{8.5}$$

holds for all x. Since $E_0(x)$ is evidently a 1-spline of period 2 with knots at all integers, it follows from (8.5) by successive integrations that $\bar{E}_\nu(x)$ is a periodic $(\nu+1)$-spline with integral knots. Therefore

$$\bar{E}_{2m-1}(2x)\,, \qquad (m \ge 1)\,,$$

is a $2m$-spline with knots described by (8.1).

Theorem 4 now furnishes

COROLLARY 4. *Within the family F_m (per. 1) of functions of period 1 satisfying*

$$f(0) = 1, \qquad f(\tfrac{1}{2}) = -1\,,$$

the integral

$$\int_0^1 (f^{(m)}(x))^2\,dx$$

is minimized by and only by the Euler function

$$X_m(x) = \frac{\bar{E}_{2m-1}(2x)}{E_{2m-1}(0)} \tag{8.6}$$

For $m = 1$ (8.6) reduces to the piece-wise linear interpolation of the ordinates $y_\nu = (-1)^\nu$. For $m = 2$ we obtain

$$X_2(x) = \begin{cases} (4x-1)(8x^2-4x-1) & \text{if } 0 \le x \le \frac{1}{2}, \\ -X_2(x-\frac{1}{2}) & \text{if } \frac{1}{2} \le x \le 1. \end{cases}$$

The case of (8.2) is even simpler because we have already observed in our Introduction to Section 7 that the Bernoulli function $\overline{B}_\nu(x)$ is a periodic $(\nu+1)$-spline for the integer knots counted double.

Again Theorem 4 implies

COROLLARY 5. *Among all functions* $f(x) \in C^{(m-1)}[0,1]$ $(m > 1)$, *having an absolutely continuous* $f^{(m-1)}(x)$ *and satisfying the boundary conditions*

$$f^{(\nu)}(0) = f^{(\nu)}(1), \qquad (\nu = 0,1,\dots,m-1),$$

and

$$f(0) = 0, \quad f'(0) = 1,$$

the integral

$$\int_0^1 (f^{(m)}(x))^2\,dx$$

is minimized by and only by the Bernoulli polynomial

$$Y_m(x) = \frac{\overline{B}_{2m-1}(x)}{B_{2m-1}(0)}. \tag{8.7}$$

Thus, for $m = 2$ we find the solution

$$Y_2(x) = x(x-1)(2x-1) \qquad (0 \le x \le 1).$$

We come now to an interesting question : *What happens to the solution* $S_{2m}(x)$ *of the interpolating problem* (5.14) *if we let* $m \longrightarrow \infty$? I can not give a general answer; however, what happens in the two particular cases explicitly solved by Corollaries 4 and 5 is easily seen. Indeed, from the Fourier series expansion of $\overline{E}_0(x)$ and by successive integrations according to (8.5) we easily obtain the Fourier expansion of the function (8.6). This expansion shows that

$$\lim_{m\to\infty} X_m(x) = \cos 2\pi x. \tag{8.8}$$

Similarly, the Fourier series of (8.7) shows that

$$\lim_{m\to\infty} Y_m(x) = \frac{1}{2\pi}\sin 2\pi x. \tag{8.9}$$

It seems likely that the $S_{2m}(x)$ of Corollary 2, always converges (as $m \to \infty$) to the interpolating trigonometric polynomial of order $(n-1)/2$ if n is odd, and if n is even, $n = 2k$ say, to the interpolating polynomial having highest order terms of least amplitude, i.e. $a_k^2 + b_k^2 = min$. This, however, is a pure conjecture.

9. A KYNEMATIC INTERPRETATION

Let

$$z = z(t) = x(t) + i\,y(t)\,, \tag{9.1}$$

for $a \leq t \leq b$, denote the motion of a point in the complex plane and let us propose the following problem: We are given $n+1$ values of $t : t_o < t_1 < \ldots < t_n$ in the range $[a, b]$ and we require that

$$z(t_\nu) = z_\nu = x_\nu + i\,y_\nu \qquad (\nu = o, \ldots, n)\,, \tag{9.2}$$

where the points z_ν are specified. Choosing $m \leq n$ and assuming that $z(t) \in F_m[a, b]$, we wish to find among the motions (9.1), subject to (9.2), the one which will solve the problem

$$\int_a^b |z^{(m)}|^2\,dt = \int_a^b (x^{(m)})^2\,dt + \int_a^b (y^{(m)})^2\,dt = minimum\,.$$

For us the solution is obvious, for all we have to do is to solve the two separate interpolation problems

$$x(t_\nu) = x_\nu \quad \text{and} \quad y(t_\nu) = y_\nu$$

by $2m$-splines according to Corollary 1 and Theorem 2. Observe that $z^{(2m-1)}(t)$ is a constant between two consecutive stations while during the "run-in" (a, t_o) and also during the "run-out" (t_n, b), $z^{(m-1)}(t)$ is a constant.

A similar application of Corollary 2 and Theorem 4 can be made for periodic motions passing n given stations at specified times within a period. Corollaries 4 and 5 allow to solve explicitely the following case: Let us find the motion (9.1) of period 1 such that

$$z(o) = 1\,, \quad z'(o) = i\,, \quad z(\tfrac{1}{2}) = -1\,, \tag{9.3}$$

which will solve the problem

$$\int_o^1 |z^{(m)}|^2\,dt = minimum\,, \quad \text{for a given } m \geq 2\,. \tag{9.4}$$

This is a case with three knots per period. Explicitly, in terms of (4.1), we have $n = 3$ and $x_1 = \frac{1}{2}$, $x_2 = x_3 = 1$. To solve (9.4) with the conditions (9.3), we need for this set of knots two periodic $2m$-splines $X_m(t)$ and $Y_m(t)$ satisfying the conditions

$$X_m(o) = 1\ ,\quad X'_m(o) = o\ ,\quad X_m(\tfrac{1}{2}) = -1$$

and

$$Y_m(o) = o\ ,\quad Y'_m(o) = 1\ ,\quad Y_m(\tfrac{1}{2}) = o\ .$$

Since these conditions are satisfied by the functions (8.6) and (8.7), respectively, we obtain our last

COROLLARY 6. *Let $z = z(t)$ be a periodic motion of period 1 such that the conditions* (9.3) *hold. Among such motions the minimum problem* (9.4) *is solved by and only by the motion*

$$Z_m(t) = X_m(t) + i\,Y_m(t)\ , \tag{9.5}$$

where $X_m(t)$ and $Y_m(t)$ are the Euler and Bernoulli functions, respectively, defined by the relations (8.6) *and* (8.7).

In view of the relations (8.8) and (8.9) we see that as $m \longrightarrow \infty$, the motion (9.5) tends to the harmonic motion

$$z(t) = \cos 2\pi t + i\,\frac{\sin 2\pi t}{2\pi}$$

of elliptic path.

For these kynematic minimum problems Holladay's case of $m = 2$, solved by cubic splines, would seem the most interesting: It furnishes the motions which are ***most nearly rectilinear and uniform*** compatible with the passage of assigned stations at specified times.

REFERENCES

1. Ahlberg J.H. and E.N. Nilson: Convergence properties of the spline fit. J.Soc.Indust. Appl. Math. 11 (1963), 95 - 104.
2. Curry H.B. and I.J. Schoenberg: On Pólya frequency functions IV: The spline functions and their limits. Bull. Amer. Math. Soc. 53 (1947), Abstract 380t, 1114.
3. Holladay J.C.: Smoothest curve approximation. Math. Tables Aids Comput. 11 (1957), 233 - 243.
4. Johnson R.S.: On monosplines of least deviation. Trans. Amer. Math. Soc. 96 (1960), 458 - 477.
5. Kuntzmann J.: Méthodes numériques: Interpolation, dérivées (Dunod, Paris 1959).

6. Landis F. and E.N. Nilson: The determination of Thermodynamic properties by direct differentiation techniques. Progress in Internat. Research on Thermodynamic and Transport properties, Amer. Soc. Mech. Engineers, New York, 1962.

7. Nörlund, N.E.: Vorlesungen über Differenzenrechnung. (Berlin 1924).

8. Schoenberg, I.J.: Contributions to the problem of approximation of equidistant data by analytic functions. Quart. Appl. Math., 4 (1946), Part A 45 - 99, Part B 112 - 141.

9. Schoenberg, I.J. and Anne Whitney: On Pólya frequency functions III: The positivity of translation determinants with an application to the interpolation problem by spline curves. Trans. Amer. Math. Soc. 74 (1953), 246 - 259.

10. Schoenberg, I.J.: Spline functions, convex curves and mechanical quadrature. Bull. Amer. Math. Soc. 64 (1958), 352 - 357.

11. Steffensen, J.F.: Interpolation. Baltimore, 1927.

12. Walsh, J.L.; Ahlberg J.H. and E.N. Nilson: Best approximation properties of the spline fit. J.Math. Mech. (Indiana Univ.), 11 (1962), 225 - 234.

The University of Pennsylvania
and
The Institute for Advanced Study

FOOTNOTES

1) Research supported by the Contract NSF G 19839. The author gratefully acknowledges financial support towards the trip to Oberwolfach, Germany, from the National Science Foundation and Deutscher Akademischer Austauschdienst.

2) Professor Walsh informs me in a letter (Sept. 21, 1963) of the following Abstract: Walsh, J.L., Ahlberg J.H. and E.N. Nilson: Best approximation and convergence properties of higher-order spline fits. Amer. Math. Soc. Notices, 10 (February 1963), Abstract 63t - 103, p.202. In this abstract they announce the extension of the results of [12] to higher degree periodic spline functions. Therefore our Theorems 3 and 4 and Corollary 2 (Sections 5 and 6) are to be attributed to these authors who have the double priority of idea and time.

Added in Proof, July 7, 1964: Since this paper was sent to the Editors, the following papers and books concerning spline interpolation have appeared in print:

13. C. de Boor: Best approximation properties of spline functions of odd degree. J. of Math. and Mech., 12 (1963), 747-749.

14. A. Sard: Linear Approximation, Math. Surveys No. 9, Providence, R.I., 1963.

15. I.J. Schoenberg: Spline interpolation and the higher derivatives. Proc. Nat. Acad. of Sciences, 51 (1964), 24-28.

16. I.J. Schoenberg: Spline interpolation and best quadrature formulae. Bull. A.M.S. 70 (1964), 143-148.

17. I.J. Schoenberg: Konikl. Nederl. Akad. van Wet., Proc. Series A, 67 (1964), 155-163.

In [13] C. de Boor has anticipated the results of our Part I with the slight restriction that he assumes that $[a, b] = [x_o, x_n]$. He derives the main results from a general theorem of Schoenberg and Whitney on interpolation by spline functions (see [9]). As this theorem is based on a good deal of Laplace transform theory, our present derivation has the advantage of being direct and elementary. The results of [17] contain those of [16] and uncover the connection with A. Sard's best interpolation formulae (see [14]). Finally, I wish to add to the list of references

18. J. Favard: Sur l'interpolation. J. de Math. (9^e Serie), 19 (1940), 281-306.

In this paper Favard studies the interpolation problem solved by our Theorem 2 (§ 3) with the difference that instead of wanting to minimize

$$\int_a^b (f^{(m)}(x))^2 \, dx ,$$

the object is to minimize the quantity

$$\max_{a \leq x \leq b} |f^{(m)}(x)| .$$

This difference is, of course, profound, but our methods of attack are similar, including the use of divided differences (see [18, page 292]). The Bernoulli and Euler functions of our § 8 also appear in Favard's mémoire: Favard shows that the function $X_m(x)$, defined by (8.6), may also be characterized among the functions $f(x)$, of period 1, with $f(0) = 1$, $f(1/2) = -1$, and having a bounded $(2m-1)$st derivative, as the function which minimizes the quantity $\max |f^{(2m-1)}(x)|$. It would be interesting to see the relevance, if any, of spline interpolation for the problems of approximation which are the primary aim of Favard's investigation.

10

The variational approach to splines

The variational approach to splines plays no role in Schoenberg's early spline papers, except for the fact that the name "spline" is motivated by the observation (credited in [**31***] to L. H. Thomas of Ohio State University) that a draftman's spline, when constrained "to take the shape of a curve which is nearly parallel to the x-axis, ... is a polygonal line composed of cubic arcs which join continuously, with a continuous first and second derivative." Yet it was the variational approach, as commenced by Holladay [**H**] in 1957, that persuaded the computing public, newly equipped with electronic computing machines, that splines rather than polynomials should be used for curve fitting. As Holladay points out, the natural cubic spline s is the smoothest function agreeing with it at its knot sequence $\mathbf{t} = (t_1, \ldots, t_N)$ in the interval $[a, b]$, in the sense that

$$\int_a^b [s''(t)]^2 dt \leq \int_a^b [f''(t)]^2 dt \quad \text{for all } f \text{ with } f = s \text{ on } \mathbf{t}.$$

In the early sixties, several authors, among them Ahlberg & Nilson & Walsh, de Boor, and Schoenberg, wrote papers which generalized this optimality property (and related properties) to odd-degree spline interpolation. Among these papers, Schoenberg's paper [**80***] stands out since it makes essential use of the Curry-Schoenberg B-spline (here still called the "fundamental k-spline"). It also marks the first appearance in Schoenberg's papers of that most beautiful of cardinal splines, the **Euler**-spline.

By this time, the basic paper [**GW**] on best approximation of linear functionals had appeared, in which the optimality of spline interpolation is discussed (unfortunately without any mention of the word "spline") as the most important special case of the following abstract setup. The value μf of the linear functional μ at the function f is to be estimated as closely as possible in terms of the values $\mu_1 f, \ldots, \mu_N f$ at f of certain linear functionals $\mu_1, \ldots, \mu_N$ and the bound r on a certain seminorm $\nu(f)$ of f. If the given information implies any bound on μf at all, then it is always possible to make ν into a norm by the addition of terms involving the μ_i. When the norm derives from an inner product, the best estimate for μf consists of an interval whose length is a linear function of r and whose center is the number $\mu P f$, with Pf the best approximation to f from the span S of representers of the μ_i, i.e., the unique element of S which agrees with f at the μ_i. Consequently, $\mu P f$ can be written as a **rule**, i.e.,

$$\mu P f = \sum_i w_i \mu_i f,$$

and the error in this rule satisfies

$$|\mu f - \mu P f| \leq \|\mu - \sum_i w_i \mu_i\| r = \operatorname{dist}(\mu, \operatorname{span}(\mu_i)_{i=1}^N) r,$$

thus identifying the rule as a **best rule in the sense of Sard**.

From a practical point of view, the most interesting situation is that of point functionals, i.e., $\mu_i f := f(t_i)$. When the seminorm is given by $\nu(f)^2 := \int_a^b (D^m f(t))^2 dt$, then the representer for such a μ_i is the section $G(t_i, \cdot)$ of Green's function for the differential operator D^{2m} with natural boundary conditions. This means that

$$G(t_i, \cdot) = (t_i - \cdot)_+^{2m-1}/(2m-1)! + p$$

for some polynomial p of degree $< 2m$, and the resulting space S is that of natural splines of order $2m$ with knot sequence $t_1, \ldots, t_N$.

Schoenberg was particularly intrigued by the observation that "Newton's idea of the fundamental role of interpolation" [**75***] still applies here, e.g., by the fact that Sard's [S] best quadrature formula in terms of a bound on f'' is identical with the quadrature formula obtained by Holladay [**H**] by integrating the interpolating natural cubic spline. You can see his smile when, on pointing out this coincidence in [**72***], he writes "The author is here performing the meritorious service of decreasing the number of known quadrature formulae by showing that two known sets are identical." Schoenberg had already brought together splines and quadrature rules in [**56**] where he exploits the fact that the Peano kernel for the quadrature error

$$\int_a^b f(t)dt - \sum_{i=1}^N w_i f(t_i)$$

is a **monospline**, i.e., the difference between the monomial t^k and a spline of order k (with knots $t_1, \ldots, t_N$). Subsequent work (in [**83**], [**87**], [**95**], [**101**]) concentrates on mimimizing such monosplines in the 2-norm in order to obtain "best" quadrature rules in the sense of Sard. This shows several time-honored quadrature rules to be "best" rules in this sense, causing Schoenberg to write (in [**83**]): "In the writer's opinion, the 'best' quality of these old formulae speaks for the aptness of Sard's definition of 'best' formulae; for, if what is good is 'good', then 'good' is good, whatever that means."

Other avenues of the variational approach to splines pursued by Schoenberg include the use of derivative functionals $\mu_i : f \mapsto D^j f(t_i)$ (in [**93**]) and of more complex seminorms (in [**77**]). The most interesting development in the area is the topic of [**105***], viz., the use of seminorms which are not (semi-)inner products and, specifically, the use of the max-norm of the m-th derivative. Already in his first paper ([**80***]) on the optimal properties of spline interpolation, Schoenberg points out the fundamental paper [**F**] in which Favard solves the problem of determining the interpolant to given data which minimizes $\max_{a \le t \le b} |D^m f(t)|$ (and uses B-splines in the process much as Schoenberg does in [**80***]). Favard's interpolant turns out to be a spline, but of degree m and with a possibly complicated knot sequence. The existence of such an optimal interpolant s with the additional requirement that it be a **perfect** spline, i.e., $D^m s$ be absolutely constant, (and with not too many knots) was shown eventually by Karlin [K].

Today, the variational approach to splines is subsumed under the heading **Optimal Recovery** of which [**MR**] provides a very good summary. Splines play a major role in the related area of n-width, and, for this, [**P**] provides a very good summary. The original idea of optimal interpolation has brought about a very useful concept of multivariate spline, the **thin-plate** spline of Duchon (see [**M**]), which however lacks a B-spline construct.

Schoenberg's most-quoted paper in the variational approach to splines, [**76***], deals with the problem of smoothing or graduation in which he follows Whittaker's idea of obtaining a smoothed data sequence $(s(t_i))_1^N$ from the given data sequence $(y(t_i))_1^N$ as the minimizer of

$$\sum_{i=1}^N (y(t_i) - s(t_i))^2 + \epsilon \sum_{i=1}^{N-m} ([t_i, \ldots, t_{i+m}]s)^2,$$

but replaces the "smoothness measure" involving the divided differences $[t_i, \ldots, t_{i+m}]s$ by the much happier choice

$$\int_a^b [D^m s(t)]^2 dt$$

which allows him to identify the unique solution as a natural spline of order $2m$ and so facilitates its computation in dependence on the "smoothing parameter" ϵ. The resulting smoothing method, particularly in the form of Reinsch's independently developed algorithm [**R**] and subsequent extensions and refinements, is probably the most widely used spline approximation method today.

References

[**F**] J. Favard (1940), Sur l'interpolation, *J. de Math. (9^e Serie)* **19**, 281-306.

[**GW**] M. Golomb and H. Weinberger (1959), Optimal approximation and error bounds, in *On numerical approximation*, R. E. Langer ed., The University of Wisconsin Press, Madison WI, 117–190.

[**H**] J. C. Holladay (1957), Smoothest curve approximation, *Math. Tables Aids Comput.* **11**, 233-243.

[**K**] S. Karlin (1975), Interpolation properties of generalized perfect splines and the solution of certain extremal problems, *Trans. Amer. Math. Soc.***206**, 25-66.

[**M**] J. Meinguet (1979), Multivariate interpolation at arbitrary points made simple, *J. Appl. Math. Phys. (ZAMP)* **30**, 292-304.

[**MR**] C. A. Micchelli and T. J. Rivlin eds. (1977), *Optimal Estimation in Approximation Theory*, Plenum Press, New York.

[**R**] C. Reinsch, Smoothing by spline functions, *Numer. Math.* **10** (1967), 177-183.

[**S**] A. Sard, (1963), *Linear Approximations*, Math.Surveys No.9, Amer. Math. Soc., Providence RI.

Carl de Boor
University of Wisconsin, Madison

HERMITE-BIRKHOFF INTERPOLATION

Reprinted from JOURNAL OF MATHEMATICAL ANALYSIS AND APPLICATIONS Vol. 16, No 3, December 1966
 Printed in Belgium

On Hermite-Birkhoff Interpolation[1]

I. J. SCHOENBERG

Mathematics Research Center, U.S. Army, University of Wisconsin, Madison, Wisconsin

Dedicated to George Polya for his Eightieth birthday on December 13, 1967

Submitted by Richard Bellman

1. INTRODUCTION AND STATEMENT OF RESULTS

Let k and n be natural numbers and let

$$E = \| \epsilon_{ij} \|, \qquad (i = 1,..., k;\ j = 0, 1,..., n - 1),$$

be a matrix with k rows and n columns having elements

$$\epsilon_{ij} = 0 \qquad \text{or} \qquad 1,$$

which are such that

$$\sum_{i,j} \epsilon_{ij} = n.$$

We shall also assume that E has no row entirely composed of zeros. Let also

$$x_1 < x_2 < \cdots < x_k$$

be increasing reals. We also need the set of ordered pairs

$$e = \{(i, j) \mid \epsilon_{ij} = 1\}.$$

The reals x_i and the "incidence matrix" E describe the interpolation problem

$$f^{(j)}(x_i) = y_i^{(j)} \qquad \text{for} \qquad (i, j) \in e. \tag{1}$$

Here the $y_i^{(j)}$ are prescribed and the problem is to find function $f(x)$ which satisfy the n interpolatory conditions (1). Such interpolation problems were first studied by G. D. Birkhoff [1].

---

[1] Sponsored by the Mathematics Research Center, United States Army, Madison, Wisconsin, under Contract No.: DA-11-022-ORD-2059.

A noteworthy special case is obtained if we assume that E has the following property:

$$\text{If } 0 \leqslant j' < j \quad \text{and} \quad \epsilon_{ij} = 1 \quad \text{then } \epsilon_{i,j'} = 1. \tag{2}$$

It is then seen that at each node x_i the system (1) prescribes the value $f(x_i)$ and perhaps also a certain number of consecutive derivatives $f^{(j)}(x_i)$, for $j = 1,\dots, \alpha_i - 1$, say. (1) is then what we may call an *Hermite interpolation problem* [2]. It is therefore appropriate to refer to (1) as an *Hermite-Birkhoff interpolation problem*, which we shall abbreviate to HB-problem.

A few further special cases are the following: (i) If $\epsilon_{i0} = 1$ for all i, while $\epsilon_{ij} = 0$ if $j > 0$, then $k = n$ and we obtain the *Lagrange interpolation problem* which is a special case of Hermite's problem. (ii) If each column of E has exactly one nonvanishing element, we obtain an *Abel interpolation problem* because Abel studied the case when E is the unit matrix I while the x_i are in arithmetic progression.

We are here concerned with the question whether the HB-problem (1) has a *unique* solution which is a polynomial of degree $n - 1$ for arbitrary values of the $y_i^{(j)}$. Evidently the following definition is relevant.

DEFINITION 1. *We shall say that the HB-problem* (1) *is poised, provided that if*

$$P(x) \in \pi_{n-1} \quad \textit{and} \quad P^{(j)}(x_i) = 0 \quad \text{for} \quad (i, j) \in e, \tag{3}$$

then $P(x) = 0$ *for all* x.

An important example: Every Hermite problem (1) is poised (see [2]).

As we have n linear equations in n unknowns, it is clear that (1) has a unique solution in π_{n-1} for any values of the $y_i^{(j)}$, if and only if (1) is poised.

The condition that (1) be poised can be easily expressed as follows: If

$$P(x) = \sum_0^{n-1} a_\nu \frac{x^\nu}{\nu!}$$

then the equations (3) are

$$\sum_{\nu=0}^{n-1} a_\nu \frac{x_i^{\nu-j}}{(\nu - j)!} = 0 \quad \text{for} \quad (i, j) \in e,$$

where we write $x^r/r! = 1$ if $r = 0$ and $=0$ if $r < 0$.

Therefore (1) is poised if and only if

$$\Delta = \det \left\| \frac{x_i^{\nu-j}}{(\nu - j)!} \right\| \neq 0, \tag{4}$$

where $\nu = 0,\ldots, n-1$ indicates a column, while to each $(i,j) \in e$ corresponds a row of the determinant.

A glance at (4) shows that whether (1) is poised or not depends both on the matrix E and on the location of the nodes x_i. Thus the problem (1) corresponding to

$$E = \begin{Vmatrix} 1 & 0 & 0 \\ 0 & 1 & 0 \\ 1 & 0 & 0 \end{Vmatrix}, \tag{5}$$

where $k = 3$ and $n = 3$, is poised if and only if $x_2 \neq \frac{1}{2}(x_1 + x_3)$. For if $x_2 = \frac{1}{2}(x_1 + x_3)$ then the quadratic polynomial $P(x) = (x - x_1)(x - x_2)$ satisfies all conditions (3) without vanishing identically.

The situation is different if $k = 1$ or $k = 2$. If $k = 1$ then all $\epsilon_{1j} = 1$ and we obtain the *Taylor problem* which is always poised. This is no longer the case if $k = 2$. However, if $k = 2$ it is clear that only the matrix E will matter while the location of the nodes x_1 and x_2 must be irrelevant (change of origin and scale!). For the case when $k = 2$ the necessary and sufficient conditions on E in order that the HB-problem (1) be poised, are due to Polya and are described by him as follows: For our general matrix E we introduce the column-sums

$$m_j = \epsilon_{1j} + \epsilon_{2j} + \cdots + \epsilon_{kj}, \qquad (j = 0,\ldots, n-1),$$

and their partial sums

$$M_j = m_0 + m_1 + \cdots + m_j, \qquad (j = 0,\ldots, n-1).$$

Polya's result is as follows.

POLYA'S THEOREM. *The Hermite-Birkhoff problem* (1), *for* $k = 2$, *is poised if and only if the following inequalities hold:*

$$M_0 \geqslant 1,\ M_1 \geqslant 2,\ldots, M_{n-2} \geqslant n-1. \tag{6}$$

Our first remark is

THEOREM 1. *Polya's conditions* (6) *are necessary for the Hermite-Birkhoff problem* (1) *to be poised.*

Indeed, Polya's proof of this statement for $k = 2$ applies equally well in the general case and we reproduce it here: If (6) do *not* hold then there is a number p, $0 \leqslant p \leqslant n-2$ such that $M_p < p + 1$. Let us try to find a *non-trivial* $P(x) \in \pi_p$ satisfying the conditions (3). The equations (3) with

$j > p$ are automatically satisfied, $P(x)$ being of degree p. The number of the equations (3) with $j \leqslant p$ is

$$m_0 + m_1 + \cdots + m_p = M_p < p + 1.$$

As we have fewer equations than unknowns, the existence of a nontrivial $P(x)$ is assured.

There is another case, besides Polya's case when $k = 2$, in which the poised quality of (1) depends only on E and not on the nodes x_i.

Definition 2. *We say that the problem* (1) *is quasi-Hermite, provided that the property* (2) *is required only if* $i = 2, 3, \ldots, k - 1$.

In other words, the interior nodes $x_2, \ldots, x_{k-1}$ are each provided with data of the Hermite type, with no restriction on the data carried by the extreme nodes x_1 and x_k.

Theorem 2. *A quasi-Hermite problem* (1) *is poised if and only if Polya's conditions* (6) *hold.*

This is a generalization of Polya's theorem which is obtained if we assume that $k = 2$ in Theorem 2.

The connections between the subject of this note and spline functions, in particular with the recent remarkable generalization of spline functions due to Ahlberg and Nilson [3], are very close and will be discussed elsewhere. Here we have studied only polynomial interpolation.

2. A Proof of Theorem 2

1. Auxiliary Theorems. Our proof of Theorem 2 will be a slight elaboration of Polya's beautiful proof of his theorem. The necessity of the condition is already implied by Theorem 1. Let us therefore assume that (1) is a quasi-Hermite problem (Def. 2) whose matrix E satisfied (6) and we are to show that (1) is poised (Def. 1).

We establish two lemmas:

Lemma 1. *If the conditions* (6) *hold and the function* $\varphi(x)$ *satisfies the homogeneous quasi-Hermite system*

$$\varphi^{(j)}(x_i) = 0 \quad \text{if} \quad (i, j) \in e, \tag{7}$$

then each of the functions

$$\varphi(x), \varphi'(x), \ldots, \varphi^{(n-1)}(x)$$

vanishes in at least one point of the closed interval $I = [x_1, x_k]$.

It will be found easier to establish a sharper result which will imply Lemma 1. We first define for each $p = 0, 1, \ldots, n-1$ the following two sets:

$$Z_p = \{x_i \mid 1 < i < k, \epsilon_{ip} = 1\}$$

and

$$\Omega_p = (x_1, x_k) - Z_p .$$

In other words, Z_p *is the set of zeros of* $\varphi^{(p)}(x)$ *which are in* (x_1, x_k) *and are exhibited by the system* (7).

The quasi-Hermite property of (1) may be also described by the equivalent condition

$$Z_0 \supset Z_1 \supset Z_2 \supset \cdots \supset Z_{n-1} .$$

LEMMA 2. *The assumptions of Lemma* 1 *imply that each*

$$\varphi^{(p)}(x) \qquad (p = 0, \ldots, n-1),$$

vanishes in at least $M_{p-1} - p$ *distinct points of the open set* Ω_p, *where by definition* $M_{-1} = 0$.

2. PROOF OF LEMMA 2. We use induction by p. If $p = 0$ then the statement of Lemma 2 is trivially true because then $M_{p-1} - p = 0$. Assume that the statement is true for a certain value of p, $0 \leqslant p \leqslant n-2$, and let us prove it for $p + 1$.

By assumption $\varphi^{(p)}(x)$ vanishes in at least $M_{p-1} - p$ distinct points of Ω_p. If we add its m_p zeros appearing in (7) (for $j = p$) we conclude that $\varphi^{(p)}(x)$ vanishes in at least

$$M_{p-1} - p + m_p = M_p - p \tag{9}$$

distinct points of $[x_1, x_k]$.

We notice that $M_p - p \geqslant 1$ by (6). If $M_p - p = 1$ then $M_p - (p+1) = 0$ and the statement of Lemma 2 concerning the zeros of $\varphi^{(p+1)}(x)$ is trivially true. If $M_p - p \geqslant 2$, we may apply Rolle's theorem to conclude that $\varphi^{(p+1)}(x)$ vanishes in at least $M_p - p - 1$ distinct points and let these be

$$r_\nu \qquad (\nu = 1, \ldots, M_p - p - 1). \tag{10}$$

However, the r_ν are *intermediate* between the consecutive zeros of $\varphi^{(p)}(x)$ among which are *all* points of Z_p. Therefore no r_ν may lie in Z_p. By (8) we know that $Z_{p+1} \subset Z_p$ and conclude that none of the r_ν are in Z_{p+1}. Therefore *the* $M_p - (p+1)$ *distinct zeros* (10) *of* $\varphi^{(p+1)}(x)$ *are contained in* Ω_{p+1} and the proof by induction is complete.

REMARK. Observe that the matrix (5) satisfies the conditions (6) *without* the corresponding HB problem being always poised. Notice, however, that E is not of the quasi-Hermite type. If $x_2 = \frac{1}{2}(x_1 + x_3)$ and $\varphi(x) = (x - x_1)(x - x_2)$ then the last step of our induction proof breaks down: For $p = 0$ the zero of $\varphi^{(p+1)}(x) = \varphi'(x)$ is $x = x_2$ and actually belongs to $Z_{p+1} = Z_1$.

3. LEMMA 2 IMPLIES LEMMA 1. Indeed, by (9) $\varphi^{(p)}(x)$ vanishes in I in at least $M_p - p$ distinct points, a number which was seen to exceed unity for $p = 0,..., n - 1$.

4. PROOF OF THEOREM 2. If the polynomial $P(x) \in \pi_{n-1}$ satisfies the quasi-Hermite system (7) then by Lemma 1 each of the functions $P(x)$, $P'(x),..., P^{(n-1)}(x)$ vanishes at some point of $[x_1, x_k]$. This, however, implies that $P(x)$ vanishes identically (Abel systems are poised!).

REFERENCES

1. G. D. BIRKHOFF. General mean value and remainder theorems with applications to mechanical differentiation and integration. *Trans. Amer. Math. Soc.* **7** (1906), 107-136.
2. CH. HERMITE. Sur la formule d'interpolation de Lagrange. *J. Reine Angew. Math.* **84** (1878), 70-79; Œuvres, Vol. 3, 432-443.
3. J. H. AHLBERG AND E. N. NILSON. The approximation of linear functionals. *SIAM J. Numerical Anal.*, **3** (1966), 173-182.
4. G. POLYA. Bemerkungen zur Interpolation und zur Näherungstheorie der Balkenbiegung. *Z. Angew. Math. Mech.* **11** (1931), 445-449.

Schoenberg's work on Hermite–Birkhoff interpolation

We will discuss the influence of Schoenberg's paper [89*] on this subject. For detailed references, see the book [**LJR**].

1. Schoenberg's paper [**89***] of 1966 had an enormous influence on interpolation, much beyond its size. It revived the problem dormant since the paper [**B**] of G. D. Birkhoff of 1906 and the work of Túran and his school in 1955-58. (Birkhoff's paper, which started his scientific career, had been submitted by him to the American Mathematical Society when he was not yet 20 years old.)

What did Schoenberg say in that paper? He formalized the problem. We are facing a system of $n+1$ interpolation equations for a polynomial $P_n(x) = \sum_{k=0}^{n} a_k x^k$ of degree $\leq n$,

$$P_n^{(k)}(x_i) = c_{i,k} \text{ for some pairs } i,k, \tag{1}$$

where x_i are the nodes, $c_{i,k}$ are the given data. There is a profound difference between the *Hermite case*, when for each node x_i, the orders of the derivatives form a complete sequence $k = 0, 1, \ldots, k_0$, (with $k_0 \leq n$), and the *Birkhoff case*, when they are an arbitrary subset of $0, 1, \ldots, n$. In the first case, a solution always exists, in the second case, in general, only for certain selections of nodes.

Schoenberg introduced an $m \times (n+1)$ *interpolation matrix*

$$E = (e_{i,k})_{i=1,k=0}^{m\ \ \ n}, \quad e_{i,k} = 0 \text{ or } 1 \tag{2}$$

to describe (1). To a pair (i,k) there corresponds an equation (1) if and only if $e_{i,k} = 1$. He wanted to know when a matrix E is *regular*, that is, when the system (1) is solvable for all selections of nodes $x_1 < \ldots < x_m$ and data $c_{i,k}$, and also when it is *almost regular*, that is, solvable for at least one selection of nodes and all data. In the latter case, in fact, one has solvability for x_i "in general position", that is, for almost all $x_1, \ldots, x_m$ in $\mathbb{R}^m$ in the sense of the Lebesgue measure.

Schoenberg actually found a necessary condition for almost regularity, which he called the *Pólya condition*. It requires that the first j columns of E, for $j = 1, \ldots, n+1$, should contain at least j ones.

2. Schoenberg's problems influenced his friends and associates, among them his Ph.D. student, D. R. Ferguson, and A. Sharma.

Before we explain what came out of Schoenberg's ideas, we should say what were the achievements of Birkhoff in [**B**]. Assuming that the problem (1) is solvable for the data $c_{i,k} = f^{(k)}(x_i)$ provided by a function f, he wanted an expression for the remainder of interpolation, $f(x) - P_n(x)$. He obtained

$$f(x) - P_n(x) = \int_a^b f^{(n+1)}(t) K_n(t)\, dt, \tag{3}$$

and a formula for the kernel K_n. (He actually found Peano kernels several years before Peano!) This kernel K_n is a spline, not of the smooth Schoenberg kind, but what we call now a "Birkhoff spline".

The main result of Birkhoff is: (I) K_n has $\leq N$ charges of sign, where N is the number of odd supported sequences (see **3.**) of the matrix E. He also tried to prove: (II) One has almost regularity of the problem if and only if the Pólya condition is satisfied, but his proof contained an unbridgeable gap.

3. What happened after Schoenberg's paper? In his dissertation, Ferguson proved (II). He, Atkinson and Sharma proved: (III) The problem is regular, if the matrix E satisfies the Pólya condition and *has no odd supported sequences.*

Let us explain this. Let the i-th row of E begin with 0110001110.... The second sequence of ones, which begins at position $k = 6$, is odd. The sequence is supported if both above and below the row i there are ones in positions $k < 6$.

Some achievements of the latter authors: A striking singularity theorem was proved. The matrix E is singular (that is, not almost regular) if it has a row with exactly one odd supported sequence. Remarkably enough, the behavior of the other rows of E does not matter. The determinant D of the system (1) is a function of the variables $x_1, \ldots, x_n$. Karlin and Karon stated, and Lorentz and Zeller proved (by means of shifts of matrices) a formula giving the degree of the first non-zero term in the Taylor expansion of D for $x_1 \to x_2$. There is a stronger version of Birkhoff's Theorem I: the *number of zeros* (usually larger than the number of changes of sign) of K_n does not exceed N. Finally, there were many publications (by Turán, Sharma, Riemenschneider and others) about Birkhoff interpolation at special sets of nodes.

Birkhoff interpolation has also been applied to splines, first by Melkman, and later in a powerful theorem by Goodman.

What methods have been used in these achievements? It is very remarkable that prominent among them was the seemingly weak Rolle's lemma, first for polynomials, and later (in a special form that has been developed for this purpose) for splines. One needs the latter even when dealing with interpolation by ordinary polynomials.

It is, of course, possible to study interpolation at a fixed special set of nodes if the interpolation problem provided by the matrix E is regular at these nodes.

Investigations of Túran and his school [**ST**] [**BT**] belong here. Very cleverly they select very special nodes, for example the zeros of some orthogonal polynomials. For this special choice, they are able to go much farther than is possible in general. Also this part of our theory enjoyed excellent development; see [**LJR**].

4. What are the applications of Birkhoff interpolation? I will mention two examples.

The first example are theorems of existence, of uniqueness and characterization of the polynomial of best approximation with constraints, which restrict the values of the derivatives of the polynomial (R. A. Lorentz has such theorems). Another example is ordinary bivariate interpolation, when one tries to match given data with the values of the polynomial (of two variables) and of its partial derivatives at the nodes (see [**LL**]).

In two variables, no matter what is the corresponding matrix, the direct methods of Hermite interpolation fail. The only general methods that we have are those of Birkhoff interpolation, in particular the shift method (mentioned earlier). It is remarkable that the problem of regularity becomes trivial in two dimensions (there are only very few regular methods), but that there are many *sufficient* conditions for almost regularity. In more complicated cases, a theoretical mathematician and a numerical analyst should combine their efforts: the first shows that the given interpolation method is almost regular (solvable in general); the second finds interpolation nodes (perhaps experimentally) which realize this solvability.

5. What would Schoenberg himself say about these developments? It seems to me that he would like to see some necessary and some sufficient conditions for the regularity in the unvariate case, that are not far from each other. We have not quite achieved this, but what we have, would be impossible without Schoenberg's seminal paper [**89***].

In 1984, I talked with Schoenberg about the name: Hermite–Birkhoff interpolation versus Birkhoff interpolation. He agreed that it would be desirable to have distinct names for two different problems, Hermite and Birkhoff, but suggested also a third name. He calls (1) a ***quasi-Hermite interpolation problem*** if all interior rows of E are Hermite, but the two extremal rows (which correspond to the boundary of the interval of interpolation) are Birkhoffian. This case is particularly dear to his heart, and with it he has achieved pretty results, for example for quadrature formulas [**101**] and for spline interpolation [**93**].

References

[**BT**] J. Balázs and P. Túran, Notes on interpolation II: Explicit formulae; III: Convergence, *Acta Math. Acad. Sci. Hungar.* **8** (1957), 201-215; **9** (1958), 195-214.

[**B**] G. D. Birkhoff, General mean value and remainder theorems with applications to mechanical differentiation and quadrature, *Trans. Amer. Math. Soc.* **7** (1906), 107–136.

[**LJR**] G. G. Lorentz, K. Jetter and S. D. Riemenschneider, *Birkhoff interpolation*, Encyclopedia of Mathematics and Its Applications, vol. **19**, Addison Wesley, 1983, Reading.

[**LL**] G. G. Lorentz and R. A. Lorentz, Solvability problems of bivariate interpolation. I, *Constr. Approx.* **2** (1986), 153–169.

[**ST**] J. Surányi and P. Túran, Notes on interpolation I: On some interpolatorial properties of ultraspherical polynomials, *Acta Math. Acad. Sci. Hungar.* **6** (1955), 67-80.

G. G. Lorentz
University of Texas at Austin

Permissions

Birkhäuser Boston thanks the original publishers of the papers of I.J. Schoenberg for granting permission to reprint specific papers in this collection

Cardinal Splines

[31–32] Reprinted from *Quarterly App. Math.* **4,** © 1946 by Brown University

[110] Reprinted from *Linear Operators and Approximation,* © 1972 by Birkhäuser Verlag

[113] Reprinted from *J. d'Analyse Math.* **27,** © 1974 by the Weizmann Science Press

[129] Reprinted from *Studies in Spline Functions and Approximation Theory,* © 1976 by Academic Press

Smoothing

[48] Reprinted from *Bull. AMS* **59,** © 1953 by the American Mathematical Society

Total Positivity

[4] Reprinted from *Math. Zeit.* **32,** © 1930 by Springer-Verlag

[15] Reprinted from *Math. Zeit.* **38,** © 1934 by Springer-Verlag

[45] Reprinted from *J. d'Analyse Math.* **2,** © 1952 by the Weizmann Science Press

[90] Reprinted from *Mathematica* **8,** Romanian Academy of Sciences, © 1966

Polya Frequency Functions

[43] Reprinted from *J. d'Analyse Math.* **1,** © 1951 by the Weizmann Science Press

[47] Reprinted from *Trans. AMS* **74,** © 1953 by the American Mathematical Society

[86] Reprinted from *J. d'Analyse Math.* **17,** © 1966 by the Weizmann Science Press

The Variational Approach

[72] Reprinted from *Bull AMS* **70,** © 1964 by the American Mathematical Society

[75] Reprinted from *Indagationes Math.* **26,** © 1964 by Springer-Verlag

[80] Reprinted from *On Approximation Theory,* © 1964 by Birkhäuser Verlag

Hermite-Berkhoff Interpolation

[89] Reprinted from *J. Math. Anal. Appl.* **16,** © 1966 by Academic Press

GPSR Compliance

The European Union's (EU) General Product Safety Regulation (GPSR) is a set of rules that requires consumer products to be safe and our obligations to ensure this.

If you have any concerns about our products, you can contact us on ProductSafety@springernature.com

In case Publisher is established outside the EU, the EU authorized representative is:

Springer Nature Customer Service Center GmbH
Europaplatz 3
69115 Heidelberg, Germany

Batch number: 10578397

Printed by Printforce, the Netherlands